Daniela Doina Cioboată
Editor

International Conference on Reliable Systems Engineering (ICoRSE) - 2022

Springer

Editor
Daniela Doina Cioboată
INCDMTM (National Institute of Research
and Development in Mechatronics
and Measurement Technique)
Bucharest, Romania

ISSN 2367-3370 ISSN 2367-3389 (electronic)
Lecture Notes in Networks and Systems
ISBN 978-3-031-15943-5 ISBN 978-3-031-15944-2 (eBook)
https://doi.org/10.1007/978-3-031-15944-2

This Springer imprint is published by the registered company Springer Nature Switzerland AG
The registered company address is: Gewerbestrasse 11, 6330 Cham, Switzerland

Contents

Determination of Additional Braking Force for Hydraulic Cylinder Piston

Volodymyr Sokolov(✉)

Volodymyr Dahl East Ukrainian National University, 59-a, pr. Tsentralnyi, Severodonetsk 93400, Ukraine
sokolov.snu.edu@gmail.com

Abstract. The braking process of the hydraulic cylinder piston due to the displacement of the working fluid from the end gap between the side surfaces of the piston and the cover is considered. To describe the fluid flow, the cylindrical coordinate system is used; the flow is considered radial and laminar. To study the flow, the inertia of the flow, the compressibility of the fluid and the action of mass forces are neglected. As the mathematical model for the process of displacing the working fluid from the end gap between the side surfaces of the piston and the cover, the differential equation of the laminar axisymmetric motion of the incompressible viscous fluid is considered. By integrating the obtained equation of motion using Newton's law for fluid friction, the distribution for the velocity of the working fluid in the gap between the side surfaces of the piston and the cover is established. Based on the distribution of the fluid velocity in the gap, ordinary differential equation was obtained for the pressure distribution along the radius of the hydraulic cylinder piston. According to the established pressure distribution, the dependence was determined for calculating the additional braking force of the hydraulic cylinder piston. The analysis of the obtained dependence is carried out, the main parameters are established that determine the braking force due to the displacement of the working fluid from the end gap between the side surfaces of the piston and the cover of the hydraulic cylinder. The example of calculation is presented.

Keywords: Hydraulic cylinder · Braking force · Equation of motion · Flow velocity · Viscosity

1 Introduction

Hydraulic drives are widely used to move the working bodies of various machines. Hydraulic drives are especially widely used in automatic control systems of the working bodies of machines included in a closed technological cycle. There are automatic control systems for metal-cutting machines and automatic lines, robotic manipulators and presses, technological machines for metallurgical, food, light industries, etc. [1–5].

The widespread use of hydraulic drives in the considered areas is determined by their important advantages, which primarily include the ability to obtain large forces and torques with relatively small sizes of hydraulic motors, the smooth movement and

D. D. Cioboată (Ed.): ICoRSE 2022, LNNS 534, pp. 1–9, 2023.
https://doi.org/10.1007/978-3-031-15944-2_1

stepless speed control in the large range, the low inertia, the ability to control modes processing during the movement of the working bodies, the simplicity of the implementation of rectilinear reciprocating movements and automatic control of the working bodies, the ease of protection against overloads and high operational reliability [6–10].

High layout properties of hydraulic systems, based on the constructive independence of the location of individual units, make it possible to create machines that are distinguished by high productivity, reliability and low material consumption. Machine tool construction belongs to those industries where hydraulic drives are traditionally used. Now in metal-cutting tools and forging equipment, the hydraulic drive is used to carry out both main and auxiliary movements, including automatic tracking movements of actuators, drive of working bodies, robotic manipulators, clamping, fixing and transport devices [11–15].

2 Literature Review

When designing hydraulic drives of systems, it is very important to assess the reliability, safety and quality of the system created on their basis. Such assessment can be given by studding of the dynamics of the hydraulic drive and technological equipment as whole, which is the final computational and design stage of creating equipment, automatic control systems with hydraulic drives. The main purpose of the dynamic research for hydraulic systems is to test the operability of the drive or control system based on one or another drive under typical external disturbing factors, as well as under given input (control) influences [16–20].

It is advisable to study the dynamics of hydraulic drives by means of mathematical modeling, which is based on the creation of a mathematical model, taking into account all the features of the drive and adequately reflecting its behavior in the dynamics of a system designed on its basis. The creation of a mathematical model and its research is based on a systematic approach to the description of all elements of a hydraulic scheme or drive, taking into account their dynamic characteristics, based on the methods of decomposition for complex interconnected elements of hydraulics, hydraulic devices and electrical equipment. The mathematical description of all elements is performed in the form of algebraic, logical, differential-integral equations and the representation of the latter in a form convenient for further research using software [21–25].

Despite the fact that the literature contains extensive material [26–30] on the mathematical description of working processes in hydraulic cylinders, there are a number of factors that affect their dynamic properties and, at the same time, are not fully understood. Among these factors is the additional braking force of the hydraulic cylinder piston, which arises due to the displacement of the working fluid from the end gap between the side surfaces of the piston and the cover.

The purpose of this paper is to study the braking process of the hydraulic cylinder piston by displacing the working fluid from the end gap between the side surfaces of the piston and the cover, determining the dependence for calculating the additional braking force of the hydraulic cylinder piston.

3 Study the Braking Process of the Hydraulic Cylinder Piston

In Figs. 1, 2 show typical designs of the single-rod and double-rod hydraulic cylinders, respectively. In these designs, the positioning of the hydraulic cylinder piston is carried out on rigid stops, which are the inner surfaces of the side covers.

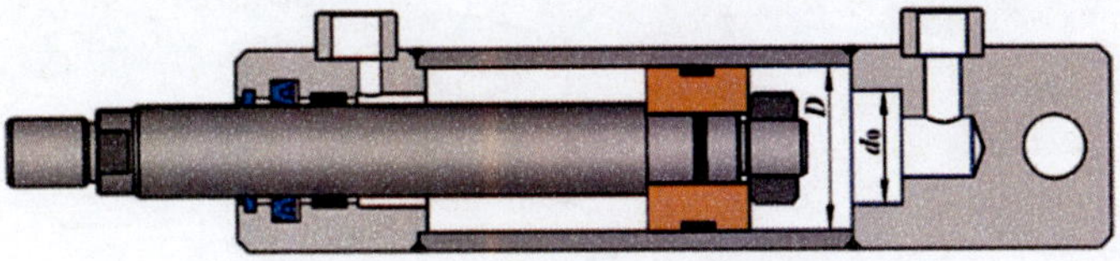

Fig. 1. The single-rod hydraulic cylinder.

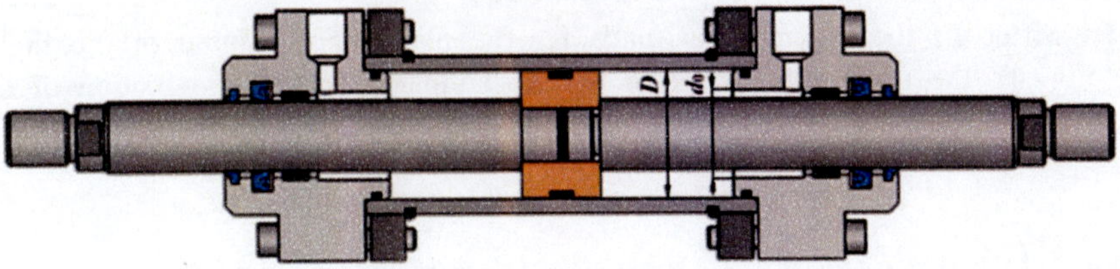

Fig. 2. The double-rod hydraulic cylinder.

Before the piston reaches the extreme position (for example, the extreme right one in Figs. 1, 2), the process of its braking occurs due to the displacement of the working fluid from the end gap between the side surfaces of the piston and the cover. The design scheme of the fluid movement in the variable end gap is shown in Fig. 3.

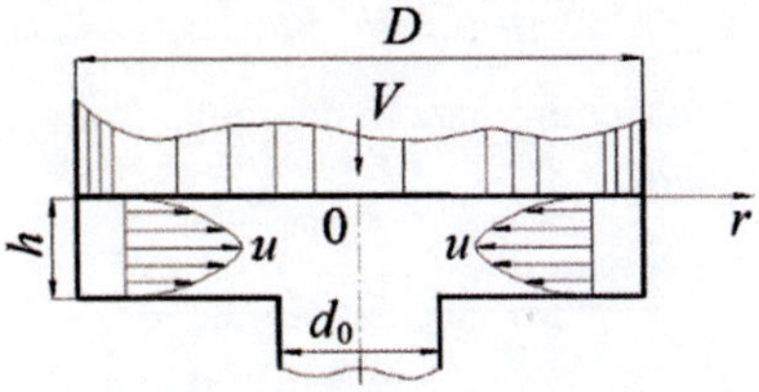

Fig. 3. The design scheme of the fluid movement in the variable end gap.

The diagram shows: V – velocity of the hydraulic cylinder piston; u – velocity of the working fluid in the end gap between the side surfaces of the piston and the cover; D – piston diameter; d_0 – diameter of the channel in the hydraulic cylinder cover; h – height of the variable end gap; r – radial coordinate.

It is quite appropriate to assume that the flow is axisymmetric and, therefore, to use the cylindrical coordinate system, placing the z axis along the piston axis (Fig. 4). We consider the flow to be laminar. We assume that the flow is quasi-radial, i.e., the tangential component of the velocity u_φ is equal to 0, and the axial component of the v locity u_z is much less than the radial u_r. We also neglect the flow inertia, fluid compressibility and the action of mass forces. Thus, the velocity of the working fluid $u = u_r$ in the end gap u is the function of the coordinates z and r.

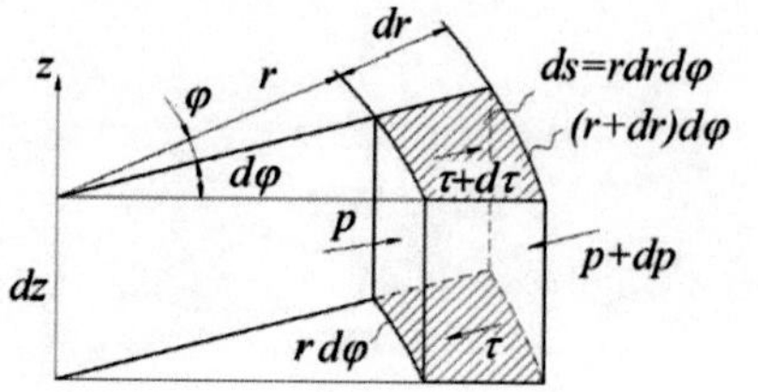

Fig. 4. The elementary volume of the fluid in the end gap.

By virtue of the assumptions made for the elementary volume of the liquid (see Fig. 4), the differential equation of the laminar axisymmetric motion of the incompressible viscous fluid is valid

$$(\tau + d\tau)ds = (p + dp)(r + dr)d\varphi dz + prd\varphi dz, \tag{1}$$

where p – hydrostatic pressure; τ – shear stress; $ds = rd\varphi dr$ – elementary area on which the shear stress acts; dp, $d\tau$, dz, dr, $d\varphi$ – increments of variables and coordinates.

Neglecting the terms of the equation of a grater order of smallness, we further simplify expression (1)

$$rd\tau drd\varphi = pdzdrd\varphi + rdpdzd\varphi. \tag{2}$$

We transform (2) and obtain the following differential equation

$$\frac{d\tau}{dz} = \frac{p}{r} + \frac{dp}{dr}. \tag{3}$$

We integrate (3) along the z coordinate in the range from zero to the size of the gap h, taking into account that $dp/dz = 0$,

$$\frac{\tau|_{z=h} - \tau|_{z=0}}{h} = \frac{p}{r} + \frac{dp}{dr}. \tag{4}$$

To calculate the shear stress on the piston and the cover, we obtain the velocity distribution over the gap. For this, we transform expression (2) taking into account the Newton's law of fluid friction

$$\tau = \rho\nu\frac{du}{dz}, \tag{5}$$

where ρ, ν – density and kinematic viscosity of the working fluid.

Then from (3)

$$\rho\nu\frac{d^2u}{dz^2} = \left(\frac{p}{r} + \frac{dp}{dr}\right). \tag{6}$$

Since $dp/dz = 0$, we have

$$\frac{d^2u}{dz^2} = A, \quad (7)$$

where A – parameter that does not depend on the z coordinate.

Therefore, the velocity distribution is a parabola, the equation of which can be represented as

$$u = -6u_0\left(1 - \frac{z}{h}\right)\frac{z}{h}, \quad (8)$$

where u_0 – average velocity over the height of the gap, and the sign "-" indicates the direction of the velocity to the coordinate center.

Without taking into account the compressibility, the flow rate of the working fluid through the annular slot $2\pi rh$ with the area is equal to the flow rate displaced by the piston in the area $\pi\left(D^2/4 - r^2\right)$, therefore

$$u_0 = \frac{V\pi\left(\frac{D^2}{4} - r^2\right)}{2\pi rh} = \frac{V}{2h}\left(\frac{D^2}{4r} - r\right). \quad (9)$$

Then

$$u = -\frac{3V}{h}\left(\frac{D^2}{4r} - r\right)\left(1 - \frac{z}{h}\right)\frac{z}{h}. \quad (10)$$

According to (5), we define

$$\tau|_{z=0} = \rho\nu\frac{du}{dz}|_{z=0} = -\frac{3V\rho\nu}{h^2}\left(\frac{D^2}{4r} - r\right), \quad (11)$$

$$\tau|_{z=h} = \rho\nu\frac{du}{dz}|_{z=h} = \frac{3V\rho\nu}{h^2}\left(\frac{D^2}{4r} - r\right). \quad (12)$$

Substituting (10), (11) into (4), we obtain the ordinary differential equation of the first order for the pressure distribution along the piston radius

$$\frac{dp}{dr} = \frac{6V\rho\nu}{h^2}\left(\frac{D^2}{4r} - r\right) - \frac{p}{r}, \quad (13)$$

which we integrate with the initial condition $p = 0$ at $r = d_0/2$.

We get the following distribution

$$p = \frac{6V\rho\nu}{h^2}\left(\frac{D^2}{4}\left(1 - \frac{d_0}{2r}\right) - \frac{r^2}{3}\left(1 - \frac{d_0^3}{8r^3}\right)\right). \quad (14)$$

For the additional braking force for the hydraulic cylinder piston, the integral expression is valid

$$F_b = \int_0^{2\pi} d\varphi \int_{d_0/2}^{D/2} prdr. \quad (15)$$

We substitute (13) into (14) and integrate, after which we obtain

$$F_b = k_d \frac{\rho \nu V D^4}{h^3}, \tag{16}$$

where k_d – dimensionless coefficient determined by the ratio of diameters d_0/D

$$k_d = \frac{3\pi}{4}\left(\frac{1}{2}\left(1 - \frac{d_0}{D}\right)^2 - \frac{1}{3}\left(\frac{1}{4}\left(1 - \frac{d_0^4}{D^4}\right) - \frac{d_0^3}{D^3}\left(1 - \frac{d_0}{D}\right)\right)\right). \tag{17}$$

The dependence $k_d(d_0/D)$ is shown in Fig. 5. In Fig. 6 shows the example of calculating the additional braking force for the hydraulic cylinder piston due to the displacement of the working fluid from the end gap between the side surfaces of the piston and the cover for the following parameters: $D = 100$ mm; $d_0 = 50$ mm; $\rho = 900$ kg/m^3; $\nu =$ 30 cSt.

It should be noted that in dynamics the velocity of the piston movement does not remain constant, but decreases, therefore, the value of the braking force also decreases.

The obtained dependence for the additional braking force (15) can be used to study the dynamics of the hydraulic cylinder piston, the dynamic characteristics of the hydraulic drive as a whole, as well as to make a decision on the need to install additional braking devices for the output link of the hydraulic drive.

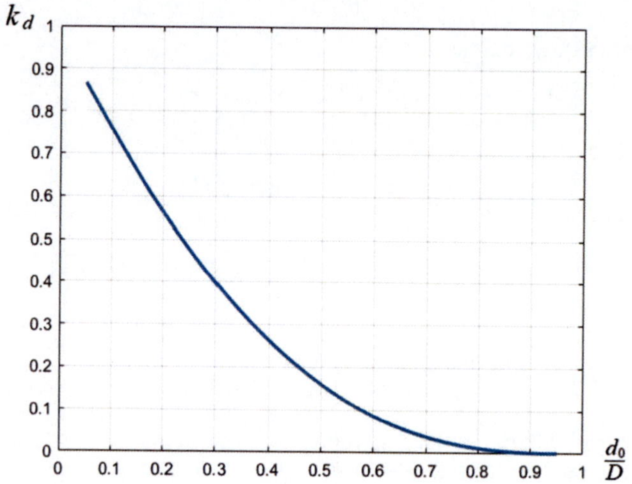

Fig. 5. The dependence for the coefficient k_d.

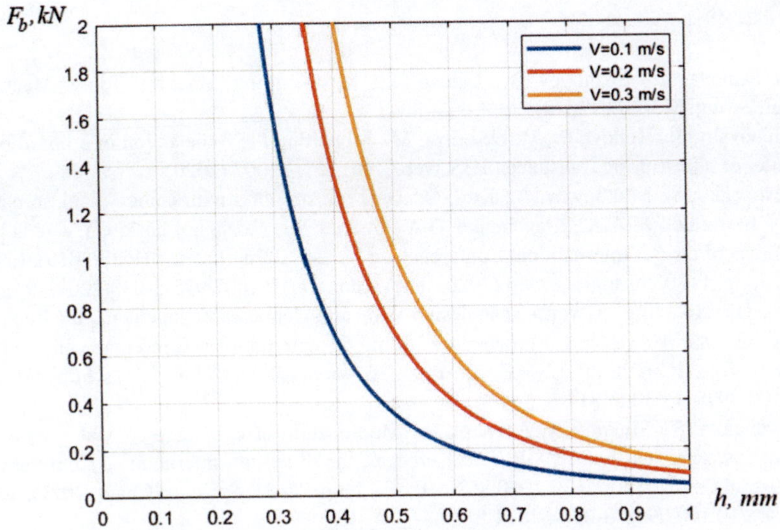

Fig. 6. To determine the velocity distribution in a flat slot.

4 Conclusions

Thus, the braking process of the hydraulic cylinder piston due to the displacement of the working fluid from the end gap between the side surfaces of the piston and the cover is considered. To describe the fluid flow, the cylindrical coordinate system was used; the flow is considered radial and laminar. To study the flow, the inertia of the flow, the compressibility of the fluid and the action of mass forces are neglected.

As the mathematical model for the process of displacing the working fluid from the end gap between the side surfaces of the piston and the cover, the differential equation of the laminar axisymmetric motion of the incompressible viscous fluid is considered. By integrating the obtained equation of motion using Newton's law for fluid friction, the distribution for the velocity of the working fluid in the gap between the side surfaces of the piston and the cover is established. Based on the distribution of the fluid velocity in the gap, ordinary differential equation was obtained for the pressure distribution along the radius of the hydraulic cylinder piston.

According to the established pressure distribution, the dependence was determined for calculating the additional braking force of the hydraulic cylinder piston. The analysis of the obtained dependence is carried out, the main parameters are established that determine the braking force due to the displacement of the working fluid from the end gap between the side surfaces of the piston and the cover of the hydraulic cylinder. The example of calculation is presented.

The obtained dependence for the additional braking force can be used to study the dynamics of the hydraulic cylinder piston, the dynamic characteristics of the hydraulic drive as a whole, as well as to make a decision on the need to install additional braking devices for the output link of the hydraulic drive.

References

1. Abrahamova, T., Bushuyev, V., Gilova, L.: Metal-cutting Machine Tools. Machinery Engineering, Moscow (2018)
2. Golubovsky, V., Konovalov, V., Doncova, M.: Modelling the force action of a liquid on the shutter of a measuring transducer. E3S Web Conf. **157**, 02007 (2020)
3. Shevchenko, S., Mukhovaty, A., Krol, O.: Gear transmission with conic Axoid on parallel axes. In: Radionov, A.A., Kravchenko, O.A., Guzeev, V.I., Rozhdestvenskiy, Y.V. (eds.) Proceedings of the 5th International Conference on Industrial Engineering (ICIE 2019). LNME, vol. 1, pp. 1–10. Springer, Cham (2020). https://doi.org/10.1007/978-3-030-22041-9_1
4. Krol, O.: Modeling of worm gear design with non-clearance engagement. In: Radionov, A.A., Gasiyarov, V.R. (eds.) Proceedings of the 6th International Conference on Industrial Engineering (ICIE 2020). LNME, vol. 1, pp. 36–46. Springer, Cham (2021). https://doi.org/10.1007/978-3-030-54814-8_5
5. Shevchenko, S., Mukhovaty, A., Krol, O.: Modification of two-stage coaxial gearbox. In: Radionov, A.A., Gasiyarov, V.R. (eds.) Proceedings of the 6th International Conference on Industrial Engineering (ICIE 2020). LNME, vol. 1, pp. 28–35. Springer, Cham (2021). https://doi.org/10.1007/978-3-030-54814-8_4
6. Sveshnikov, V.: Hydrodrives of Tools. Machinery Engineering, Moscow (2008)
7. Sokolov, V., Porkuian, O., Krol, O., Baturin, Y.: Design calculation of electrohydraulic servo drive for technological equipment. In: Ivanov, V., Trojanowska, J., Pavlenko, I., Zajac, J., Peraković, D. (eds.) Advances in Design, Simulation and Manufacturing III. LNME, vol. 1, pp. 75–84. Springer, Cham (2020). https://doi.org/10.1007/978-3-030-50794-7_8
8. Sokolov, V., Porkuian, O., Krol, O., Stepanova, O.: Design calculation of automatic rotary motion electrohydraulic drive for technological equipment. In: Ivanov, V., Trojanowska, J., Pavlenko, I., Zajac, J., Peraković, D. (eds.) Advances in Design, Simulation and Manufacturing IV. LNME, vol. 1, pp. 133–142. Springer, Cham (2021). https://doi.org/10.1007/978-3-030-77719-7_14
9. Shevchenko, S., Mukhovaty, A., Krol, O.: Geometric aspects of modifications of tapered roller bearings. Procedia Eng. **150**, 1107–1112 (2016)
10. Shevchenko, S., Mukhovaty, A., Krol, O.: Gear clutch with modified tooth profiles. Procedia Eng. **206**, 979–984 (2017)
11. Popov, D.: Mechanics of Hydro- and Pneumodrives. MGTU, Moscow (205)
12. Kundrák, J., Mitsyk, A., Fedorovich, V., Markopoulos, A., Grabchenko, A.: Simulation of the circulating motion of the working medium and metal removal during multi-energy processing under the action of vibration and centrifugal forces. Machines **9**(6), 1–22 (2021)
13. Kundrák, J., Mitsyk, A., Fedorovich, V., Markopoulos, A., Grabchenko, A.: Modeling the energy action of vibration and centrifugal forces on the working medium and parts in a vibration machine oscillating reservoir with an impeller. Manuf. Technol. **21**(3), 364–372 (2021)
14. Shevchenko, S., Mukhovaty, A., Krol, O.: Modified belt transmission with enhanced technical characteristics. In: Radionov, A.A., Gasiyarov, V.R. (eds.) Proceedings of the 7th International Conference on Industrial Engineering (ICIE 2021). LNME, vol. 1, pp. 42–51. Springer, Cham (2022). https://doi.org/10.1007/978-3-030-85233-7_5
15. Shevchenko, S., Mukhovaty, A., Krol, O.: Research of the modified tooth-belt drive for the machining center. In: Radionov, A.A., Gasiyarov, V.R. (eds.) Proceedings of the 7th International Conference on Industrial Engineering (ICIE 2021). LNME, vol. 1, pp. 32–41. Springer, Cham (2022). https://doi.org/10.1007/978-3-030-85233-7_4
16. Sokolov, V., Krol, O., Stepanova, O.: Nonlinear simulation of electrohydraulic drive for technological equipment. J. Phys. Conf. Ser. **1278**, 1012003 (2019)

17. Krol, O., Porkuian, O., Sokolov, V., Tsankov, P.: Vibration stability of spindle nodes in the zone of tool equipment optimal parameters. Comptes rendus de l'Acade'mie bulgare des Sci. **72**(11), 1546–1556 (2019)
18. Sokolov, V., Krol, O.: Installations criterion of deceleration device in volumetric hydraulic drive. Procedia Eng. **206**, 936–943 (2017)
19. Fomin, O., Lovska, A.: Establishing patterns in determining the dynamics and strength of a covered freight car, which exhausted its resource. Eastern-Eur. J. Enterp. Technol. **6**(7), 21–29 (2020)
20. Lovska, A., Fomin, O.: A new fastener to ensure the reliability of a passenger car body on a train ferry. Acta Polytechn. **60**(6), 478–485 (2020)
21. Popov, D., Panaiotti, S., Ryabinin, M.: Hydromechanics. MSTU, Moscow (2014)
22. Krol, O., Sokolov, V.: Research of modified gear drive for multioperational machine with increased load capacity. Diagnostyka **21**(3), 87–93 (2020)
23. Kovalevskyy, S., Kovalevska, O., Radosavljević, M., Anđelković, M.: Ensuring the life cycle of objects on the basis of a signature approach. In: Karabegović, I. (ed.) New Technologies, Development and Application IV. LNNS, vol. 233, pp. 458–468. Springer, Cham (2021). https://doi.org/10.1007/978-3-030-75275-0_51
24. Ivanov, V., Pavlenko, I., Liaposhchenko, O., Gusak, O., Pavlenko, V.: (2021) Determination of contact points between workpiece and fixture elements as a tool for augmented reality in fixture design. Wirel. Netw. **27**(3), 1657–1664 (2021)
25. Liaposhchenko, O., Pavlenko, I., Monkova, K., Demianenko, M, Starynskyi, O.: Numerical simulation of aeroelastic interaction between gas-liquid flow and deformable elements in modular separation devices. In: Ivanov, V., et al. (eds.) Advances in Design, Simulation and Manufacturing II. LNME, pp. 765–774. Springer, Cham (2020). https://doi.org/10.1007/978-3-030-22365-6_76
26. Rogovyi, A., Korohodskyi, V., Medvediev, Y.: Influence of Bingham fluid viscosity on energy performances of a vortex chamber pump. Energy **218**, 119432 (2021)
27. Rogovyi, A., Korohodskyi, V., Khovanskyi, S., Hrechka, I., Medvediev, Y.: Optimal design of vortex chamber pump. J. Phys: Conf. Ser. **1741**, 012018 (2021)
28. Tsankov, P.: Modeling of vertical spindle head for machining center. J. Phys: Conf. Ser. **1553**, 012012 (2020)
29. Kotliar, A., et al.: Ensuring the economic efficiency of enterprises by multi-criteria selection of the optimal manufacturing process. Manag. Prod. Eng. Rev. **11**(1), 52–61 (2020)
30. Andrenko, P., Rogovyi, A., Hrechka, I., Khovanskyi, S., Svynarenko, M.: The influence of the gas content in the working fluid on parameters of the hydraulic motor's axial piston. In: Ivanov, V., Pavlenko, I., Liaposhchenko, O., Machado, J., Edl, M. (eds.) Advances in Design, Simulation and Manufacturing IV. LNME, pp. 97–106. Springer, Cham (2021). https://doi.org/10.1007/978-3-030-77823-1_10

Modular Spindle Tooling of the Machining Center with Increased Resource

Oleg Krol(✉) and Volodymyr Sokolov

Volodymyr Dahl East Ukrainian National University, 59-a Tsentralnyi Prospect, Severodonetsk 93400, Ukraine
krolos@snu.edu.ua

Abstract. The design of spindle heads of multi-axis metal-cutting machines with driven bevel gears is considered. The kinematic diagrams of the main movement drive for multioperational machines equipped with spindle heads of different technological capabilities have been constructed. Three-dimensional modeling of horizontal, vertical and angular spindle heads in the integrated computer-aided design system KOMPAS-3D is carried out. 3D models of complex housing parts of heads using an extended range of graphic primitives and new versions of the geometric and parametric cores for the system have been built. The new functional of the specialized applied application of the KOMPAS system is used in the procedure for express-building of high-precision models of bevel gear rims with geometrically correct tooth surfaces. The rendering of the spindle head structures in the Artisan rendering module is performed. A variant of solving the problem of increasing the two-stage drive mechanisms service life of spindle heads according to the criterion of effective bevel gears lubrication is proposed. An analytical apparatus for partitioning the total gear ratio of a two-stage drive for cases of using gears with straight and circular teeth is presented. A derivative analytical dependence, which approximates the basic expression reflecting the ratio of the gear ratios of the two-stage drive mechanism is introduced. Variants of a possible combination of hardness of bevel gear teeth on the first and second stages of the spindle head drive mechanism are considered.

Keywords: Spindle head · 3D modeling · Bevel gear · Transmission resource · Gear ratios

1 Introduction

In the process of creating new designs of multioperational machine tools (MMT), special influence is paid to the design of spindle heads and spindle assemblies, which are the main structural elements of any milling, drilling and boring machine. These particular characteristics that mainly affect the choice of a machine tool and are always the main ones, since the quality of the finished product depends on the reliability and accuracy of these elements.

One of the main trends in the growth of machining efficiency is the increase in the technological capabilities of machine tools. In this regard, various spindle heads for

D. D. Cioboată (Ed.): ICoRSE 2022, LNNS 534, pp. 10–21, 2023.
https://doi.org/10.1007/978-3-031-15944-2_2

MMT of the drilling, milling and boring group such as machining centers are used [1–3]. It is here that the principle of modular construction of machine tools based on unified and specialized designs of spindle heads is most fully implemented. As a result, there is an effect of increasing the technological possibilities of using the machine tool [4, 5] and increasing the variety of part machining, without changing the basic (bearing system, drives) of the machine structure.

A lot of research and development is currently underway on additional modular tooling for modern horizontal machining centers. The design of such machines is based on the idea associated with the possibility of changing the working position of the tool blocks in space. As an example, we consider removable angular spindle heads (ASH) with a variable angle of the spindle part with a working tool, which are attached to the spindle unit of the machine [6]. The consequence of this is an increase in the number of differently located surfaces of the part that can be processed from one setup. Taking into account the frequent change in the range of processed parts, the variety of their geometric shapes, the development of new spindle heads designs is an urgent task.

2 Literature View

The issues of shaping spindle unit's development for MMT are presented in works [7–9].

The work [7, 8] considers the analysis of the functioning of a multi-axis milling machine with various technological equipment. The authors' approach to the procedure for revealing the influence of the machine tool layout with various configurations of the rotation axes of the spindle assemblies and rotary tables with workpieces mounted on them is interesting. In the developed databases, various alternative options for the layouts and configurations of the coordinate axes are presented. Databases were structured in the form of separate sections: a) spindle heads; b) instrumental blocks; c) rotary tables, etc. Along with the design component, a complex of analytical models and experimental data arrays was formed. At the same time, the issues of complex 3D modeling of spindle heads using unified components and specialized applications are not touched upon in this work.

In [9], 2D and 3D modeling of a horizontal spindle assembly mounted on two duplex bearings on the front support and a single bearing on the rear support was carried out. An iterative procedure for calculating the stiffness of a spindle presented in the form of a beam element using the FEM toolkit has been implemented [10–12]. At the same time, the issues of modeling the vertical and angular heads of machining centers within the framework of creating a three-dimensional models section of spindle units were not touched upon in the work.

High efficiency and reliability of spindle heads depends on the resource and uninterrupted operation of drive gears [13, 14]. Among the various types of gears, bevel gears intended for transmit motion with a change in direction should be distinguished. This is especially important for machining centers with multi-axis machining capabilities. Therefore, the issues of increasing the durability, load capacity of the gear wheels of the main motion drive by improving design solutions are one of the dominant areas of research and development.

When designing gears at the initial stages of research [13], the search for an analytical apparatus for describing the profile of the lateral surface of the bevel gears teeth is carried out. This description will be the basis for obtaining optimized contact conditions over a wide range of gear transmission operation.

In the fundamental work [14], a new design of gears with increased load capacity for contact stresses is proposed. The design component is based on material-physical ratios (in accordance with the accepted normative calculation ANSI/AGMA 2003-B97) and allows you to assess the effect of the material on the subsurface layer. The proposed computational model makes it possible to assess the level of risk of side surface breakage and pitting corrosion based on load Tooth Contact Analysis. As for the geometric relationships, it should be noted that the new design of gear wheels is characterized by a slight decrease in the normal modulus and the total overlap ratio with an increased average pitch diameter.

Indicators of the load capacity and resource of gear transmissions depend not only on the improvement of their design, but also on the lubrication system. Under the conditions of the most common oil-bath lubrication system at speeds up to 12.5 m/s, a suspension of oil particles that uniformly cover the surfaces of the gear wheels ensure their continuous lubrication. The level of gear wheel immersion h_o an oil reservoir $3m \leq h_0 \leq 0.25d_{2l}$ is also regulated: depending on the gearing module m and the diameter of the low-speed stage d_{2l} [15]. For bevel gears, a significant difference in the diameters of the driven wheels of the two-stage transmission is possible. Moreover, the smaller of these wheels is not directly immersed in the oil reservoir, which worsens the contact conditions of the teeth.

The procedure of three-dimensional representation of the designed mechanism as an integrating link in the scale of the life cycle of creating a new gear mechanism is considered. According to [16], one of the main goals is to create a working 3D model for the analysis of the contact geometry, visual identification and assessment of the relationship of the set of the tooth geometric parameters. Important parameters include the parameters of the groove profile, which are largely determined by the shape of the longitudinal generatrix of the gear and wheel axoids. The design toolkit should include a method for creating a three-dimensional model of a gear in the corresponding integrated CAD systems, such as KOMPAS-3D [17–19].

Based on the analysis of the considered problems of creating an extended section of shaping spindle heads and increasing the resource of their 2-stage gear drive, we form the purpose of the research:

To carry out 3D modeling of the spindle head structures of metal-cutting machines with an increased resource of gear transmissions.

To achieve this purpose, it is necessary to solve the following tasks:

1. To develop a complex of three-dimensional models of MMT spindle heads in the CAD KOMPAS-3D environment using a specialized application program.

2. Research and create a modernized design of a gear drive with an increased resource.

3 Research Methodology. 3D Modeling of Spindle Heads

To increase the technological capabilities of MMT drilling, milling and boring type, additional technological fixtures is introduced into their composition [20, 21]. Such quick-change modular units as horizontal (HSH), vertical (VSH) and angular (ASH) spindle heads provide high-performance machining of any complexity parts and are effective in various types of production from single to mass production.

Consider a universal multioperational machine model SF68VF4, the kinematics of which main motion drive is shown in Fig. 1.

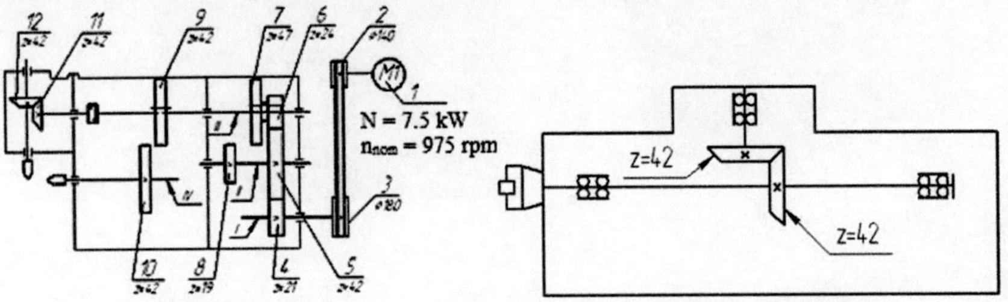

a b

Fig. 1. Kinematics of the main movement drive: a – gearboxes with vertical spindle head; b – kinematics of the angular spindle head

According to the kinematic diagram, the rotation from the electric motor through the poly V-belt is transmitted to the input shaft and then through a two-stage gearbox to the output shaft from which rotation is transmitted to the vertical head- and angular head clutch, or to the horizontal spindle pinion.

In order to carry out a comprehensive procedure for researching the modular tooling of a machine tool and obtain numerical estimates of the stress-strain state, it is necessary at the initial stages to build computational solid models using CAD KOMPAC-3D [18, 22].

Let's use the "bottom-up" principle, i.e. first, we will build 3D-models (Fig. 2) of the constituent parts (housing, shafts, gears) with their subsequent integration into the assembly structure of the angular and vertical spindle heads (Fig. 2). Three-dimensional modeling of the considered parts was carried out in the environment of the integrated CAD using an application of the system "Shafts and mechanical transmissions 3D".

The system software includes: geometric and parametric cores of C3D Modeler and C3D Solver, respectively. The geometric core of C3D Modeler allows you to effectively implement the construction of models of such complex housing parts (Fig. 2, a; b) and calculate the geometric characteristics of the objects being modeled. New functionality introduced in C3D Modeler designed to remove holes and fillets from the model, which are typical for cast iron housing workpieces (Fig. 2, a; b). This leads to a simplification of the 3D model, which is prepared for further calculation in the CAE system [23–25].

It is important to be able to operate not with geometric primitives of a plane or elements of space, but with combined objects and parts in general in the application environment "Shafts and Mechanical Transmissions 3D". The procedure for express-building of high-precision models of bevel gear- and pinions rims with geometrically correct tooth surfaces has been improved [26–28]. This is achieved by using the bevel gear hobbing simulation method and a new type of "Conical surface section". At the same time, in the sense of kinematics, this is realized by moving the curve of the conical

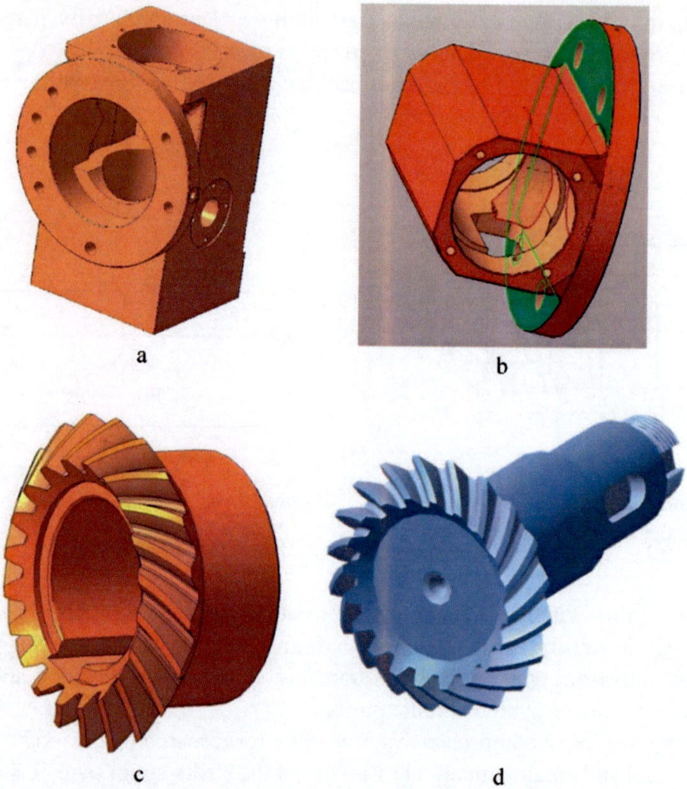

Fig. 2. Solid models of machine drive parts: a – VSH housing; b – ASH housing; c – bevel gear; d – bevel pinion shaft

section along two guides with the possibility of changing the parameters of this section. As a result, a very smooth surface is formed along its entire length. In Fig. 2, c; d shows 3D models of a bevel gear and pinion shaft.

The three-dimensional models of the spindle assemblies for the HSH, VSH and ASH developed in the CAD KOMPAS environment are shown in Fig. 3.

Based on the constructed 3D models of the spindle heads (Fig. 3), rendering was performed in the Artisan Rendering module [29, 30], which is integrated into the KOMPAS-3D system.

This creation of a photorealistic image and analysis of the appearance of the spindle heads design form an integral idea of the design, while it becomes possible to select materials taking into account the colors and textures. It is important to create subsequent feedback in the course of adjusting the geometry of the product for its improvement. In Fig. 4 shows a rendering of the HSH, VSH and ASH assemblies.

In the considered designs of the spindle heads for a multioperational machine, which is part of the horizontal machining centers group, an important role is played by a gear drive, considered as a gearbox with two bevel gears BG2. At the first and second stages of converting motion from horizontal HSH to vertical – VSH and from vertical

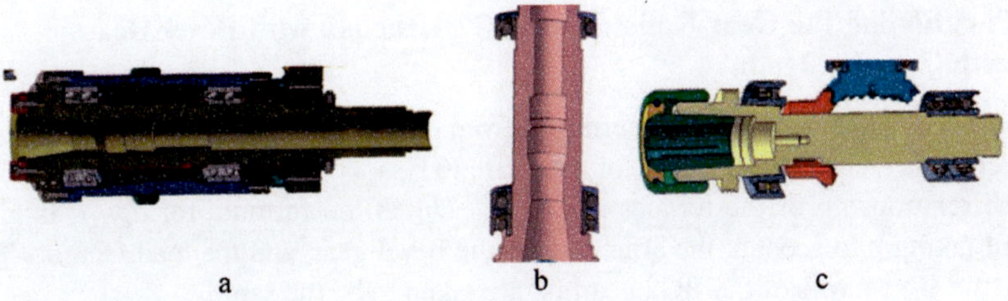

Fig. 3. Three-dimensional models of spindle assemblies: a – horizontal; b – vertical; c – angular

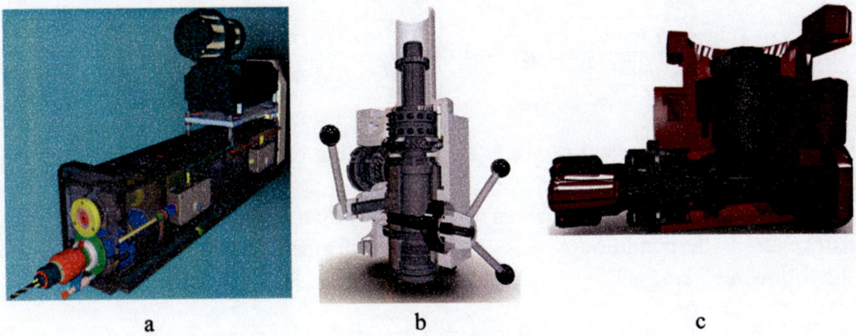

Fig. 4. Rendering of spindle heads: a – horizontal; b – vertical; c – angular

to angular ASH) that transform the movement of the machine forming units in the machine coordinate system. The improvement of their designs by technological and design methods is carried out. Equally important are the issues of effective lubrication of bevel gears.

4 Results. Increase in the Resource of the Gear Drive

As you know, the main criterion for evaluating the performance of bevel gears with straight and circular teeth is the characteristic of contact endurance, which is used to calculate the basic parameters – the outer pitch diameters in the first $d_{e2(I)}$ and second $d_{e2(II)}$ stages. Calculations show that the diameters of the driven wheels of BG2 gearbox with both stages can differ significantly in this case. This results in the smaller of these wheels not being directly immersed in the oil reservoir. Accordingly, in the engagement of this stage, the contact conditions of the teeth deteriorate, which ultimately reduces both its resource and the resource of the gearbox as a whole. This problem can be solved by such partitioning of the BG2 gearbox gear ratio U according to its stages I and II, in which the outer pitch diameters of the wheels of both stages – $d_{e2(I)}$ and $d_{e2(II)}$, will be the same, that is, to implement the condition $d_{e2(I)} = d_{e2(II)}$. Consider the problem of partitioning the gear ratio in two versions – wheels with circular and straight teeth.

4.1 Partitioning the Gear Ratio of the BG2 Gearbox with Bevel Gears with Circular Teeth

Parameters $d_{e2(I)}$ and $d_{e2(II)}$ are determined from the main criterion for the performance of bevel gears – contact endurance of the teeth, [31]:

The relationship of the torques $T_3 \approx T_2 \cdot U_{II}$ in the formula for $d_{e2(II)}$ is given without taking into account the efficiency of the bevel gear, and the load factors K_H in the Ist and IInd stages of the BG2 gearbox are taken to be the same.

Coefficients of the influence of the longitudinal shape of the teeth on the contact stresses of the bevel gears circular teeth, [15, 31]:

$$d_{e2(I)} = 1650 \cdot \sqrt[3]{\frac{K_H \cdot T_2 \cdot U_I}{\theta_{H(I)} \cdot [\sigma_H]_I^2}};\ d_{e2(II)} = 1650 \cdot \sqrt[3]{\frac{K_H \cdot T_3 \cdot U_{II}}{\theta_{H(II)} \cdot [\sigma_H]_{II}^2}} = 1650 \cdot \sqrt[3]{\frac{K_H \cdot T_2 \cdot U_{II}^2}{\theta_{H(II)} \cdot [\sigma_H]_{II}^2}}. \tag{1}$$

where the constants $c_{1(i)}$ and $c_{2(i)}$ are taken depending on the hardness of the pinion- and wheel teeth of i-stage.

After substitution $\theta_{H(i)}$ in the given dependencies for $d_{e2(I)}$ and $d_{e2(II)}$ corresponding transformations, the condition $d_{e2(I)} = d_{e2(II)}$ is transformed into a function $U_I = U_I(U)$ of the following form:

$$U_I^3 + a \cdot U_I^2 + b \cdot U_I + c = 0 \tag{2}$$

where $a = \frac{c_{2(II)} \cdot U}{c_{1(II)}}$; $b = -\frac{U^2 \cdot c_{2(I)} \cdot [\sigma_H]_I^2}{c_{1(II)} \cdot [\sigma_H]_{II}^2}$; $c = -\frac{U^2 \cdot c_{1(I)} \cdot [\sigma_H]_I^2}{c_{1(II)} \cdot [\sigma_H]_{II}^2}$.

It should be noted that during operation, the forces and contact stresses in the engagement are changed, which leads to a certain level of error in the center distance. This phenomenon is associated with changes in the distance between gearbox shafts. An interesting approach was proposed in [32] to estimate such an error. Based on the analysis of the change in the distance between the shafts of two gears, a procedure for determining the influence of various geometric errors of the gears. This task is accomplished by evaluating the sinusoidal component of the variation curve, obtained by means of the Fourier transform was proposed.

The solution of the cubic Eq. (2) is performed for 3 variants of the combination of the teeth hardness of the Ist and IInd stages for the BG2 gearbox:

a) both stages are "soft", $\{\overline{H}_I < 350HB,\ \overline{H}_{II} < 350HB\}$: $c_{1(I)} = c_{1(II)} = 1.22$; $c_{2(I)} = c_{2(II)} = 0.21$; $[\sigma_H]_I = [\sigma_H]_{II} = 507\,MPa$;
b) -Ist stage "soft", $\{\overline{H}_I < 350HB\}$; IInd stage "hard", $\{\overline{H}_{II} > 350HB\}$: $c_{1(I)} = 1.22$; $c_{2(I)} = 0.21$; $c_{1(II)} = 0.81$; $c_{2(II)} = 0.15$; $[\sigma_H]_I = 507\,MPa$; $[\sigma_H]_{II} = 814\,MPa$;
c) both stages are "solid", $\{\overline{H}_I > 350HB,\ \overline{H}_{II} > 350HB\}$; $c_{1(I)} = c_{1(II)} = 0.81$; $c_{2(I)} = c_{2(II)} = 0.15$; $[\sigma_H]_I = [\sigma_H]_{II} = 814\,MPa$,

here $\overline{H}_I$ and $\overline{H}_{II}$ are the average hardness of the teeth at the Ist and IInd stages.

For the convenience of using dependence (1) in computational practice, we approximate its power function $U_I = k \cdot U^x$: option a): $U_I \approx 0.87 \cdot U^{0.78}$; option b): $U_I \approx 0.7 \cdot U^{0.76}$; option c): $U_I \approx 0.86 \cdot U^{0.79}$ (approximation error less than 0.5%).

For the found value U_I the amount U_{II} is found by the usual method: $U_{II} = U/U_I$.

For clarity of the analysis of functions $U_I = U_I(U)$, the calculations according to (1) and (2) were performed with some excess of the maximum recommended value of the gear ratio U_I in standard bevel gears [5] (in Fig. 5 the limiting value U_I is marked by the size $U_{i(\max)} = 6.3$).

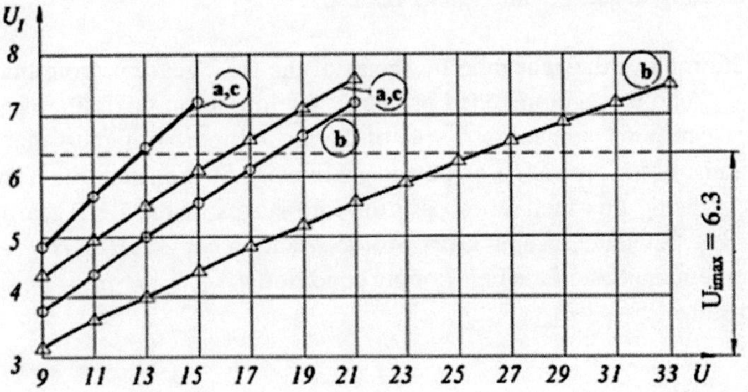

Fig. 5. Graphs of the function $U_I = U_I(U)$: lines with symbols ○ – for circular teeth, formula (2); with symbols Δ – for straight teeth, formula (3)

4.2 Partitioning the Gear Ratio of the BG2 Gearbox with Bevel Gears with Straight Teeth

In spur bevel gears, the coefficients of influence $\theta_{H(I)} = \theta_{H(II)} = 0.85$, [15, 31]. As a result, the function $U_I = U_I(U)$ for partitioning U into stages of the BG2 gearbox from the condition $d_{e2(I)} = d_{e2(II)}$ takes on a simpler form:

$$U_I = k \cdot \sqrt[3]{U^2} \tag{3}$$

where $k = \sqrt[3]{[\sigma_H]_I^2/[\sigma_H]_{II}^2}$.

Graphic functions (1) and (2) are shown in Fig. 5 for options a), b), c) in relation to bevel gears with circular and straight teeth.

Based on the calculations performed, we will analyze the features of the procedure for partitioning the overall gear ratio.

1) For the combination $\{\overline{H}_I < 350HB; \overline{H}_{II} > 350HB\}$ – these are lines b) for circular and straight teeth in Fig. 5. Values U_I and U_{II}, which do not exceed the maximum recommended value $U_{i(\max)} = 6.3$, are obtained for spur gears at gear ratios of the gearbox $U \leq (25 \div 26)$, and for circular teeth at $U \leq 18$. Hence, the conclusion is that a partitioning U by stages of the BG2 gearbox from the condition $d_{e2(I)} = d_{e2(II)}$ for this combination of hardness of the teeth of stages I and II is possible, but for straight teeth it is feasible over a larger range of values U.
2) The smallest range $U : U \leq 13$, at which it is possible to partition it into steps from the condition $d_{e2(I)} = d_{e2(II)}$ with observance of the limitation $U_i \leq U_{i(\max)} =$

6.3, is given by gears with circular teeth at $\{\overline{H}_I < 350HB;\ \overline{H}_{II} < 350HB\}$ and $\{\overline{H}_I > 350HB;\ \overline{H}_{II} > 350HB\}$ – the line a) and c) with symbols "o" in Fig. 5. The values U_I and U_{II} for these combinations of hardness practically coincide, therefore they are shown by one line.

3) A conclusion similar to 2) can be made in relation to the same combination of hardness for straight teeth, line a), c) with the symbols "Δ" in Fig. 5. The difference is only in a slightly larger maximum value $U:\ U \leq 15$.

The partitioning of the gear ratio by stages of the BG2 gearbox from the condition $d_{e2(I)} = d_{e2(II)}$ can be recommended primarily for long-term operating modes, when abundant lubrication of the gearing is one of the most important factors that positively affects the transmission resource of power gearboxes. At the same time, the greatest possibilities of using this method of partition into stages of the BG2 gearbox, which follows from Fig. 5, will be for spur gears with a variant $\{\overline{H}_I < 350HB;\ \overline{H}_{II} > 350HB\}$, where the implementation of the partitioning condition $d_{e2(I)} = d_{e2(II)}$ is possible for U reaching the value $25 \div 26$.

5 Conclusions

In this work, the following results are obtained:

1. The complex of 3D models of horizontal, vertical and angular spindle heads for multioperational drilling, milling and boring machine with a two-stage gear drive in the CAD KOMPAS-3D environment has been developed. This 3D project became the winner of the International Competition "Future Aces of Computer 3D Modeling.
2. Created three-dimensional models of housing parts of machine tool spindle heads with complex spatial shape. The strategy of creating solid models based on an extended range of geometric primitives of the KOMPAS system in the form of combined functional elements and parts as a whole has been effectively used. This makes it possible to increase the productivity of the designer at the stage of creating a 3D project for the designed product.
3. In the specialized application "Shafts and mechanical transmissions 3D", three-dimensional models of bevel gears with circular teeth are created. In the process of creation, a unique procedure for constructing gear rims was used by the imitation method of gear milling for bevel gears and a new functionality of the geometric core module "Conical section surface".
4. In the Artisan Rendering module, the machine tool heads are rendering using high-quality hardware OpenGL, the advantages of which are the simplicity and speed of installing a complete scene (Snapshot) and the ability to view and generate several Snapshots for software rendering. This realistic image of the projected spindle heads allows you to get a more complete view of the designs and outline ways to improve them.
5. A search procedure for the design of a two-stage gear drive with an increased parameter of teeth contact endurance due to improved lubrication of the main drive elements has been developed. For this purpose, analytical dependences are proposed for partitioning the gear ratio of a two-stage drive according to its stages I and II, at which

the outer pitch diameters of the wheels of both stages will be the same. A derivative analytical dependence $U_I = k \cdot U^x$ of the first stage gear ratio is introduced for three variants of the gear materials hardness, which makes it possible to effectively solve the basic cubic equation of the first stage gear ratio. The approximation error was less than 0.5%. The results obtained are recommended primarily for long-term operating conditions, when abundant lubrication of the gearing is one of the most important factors that positively affect the gear resource of the drive mechanisms for spindle heads of various designs.

References

1. Push, A.V., Zverev, I.A.: Spindle Nodes. Designing and Research. Stankin, Moscow (2000)
2. Balmont, V.B.: Calculations of High-Speed Spindle Units. VNIITEMR, Moscow (1987)
3. Kong, J., Cheng, X.: Modal analysis of CNC lathe's spindle based on finite element. Adv. Eng. Res. (AER) **148**, 318–321 (2017)
4. Sokolov, V., Rasskazova, Y.: Automation of control processes of technological equipment with rotary hydraulic drive. Eastern-Eur. J. Enterp. Technol. **2/2**(80), 44–50 (2016). https://doi.org/10.15587/1729-4061.2016.637
5. Sokolov, V.: Increased measurement accuracy of average velocity for turbulent flows in channels of ventilation systems. In: Radionov, A.A., Gasiyarov, V.R. (eds.) Proceedings of the 6th International Conference on Industrial Engineering (ICIE 2020). LNME, vol. 2, pp. 1182–1190. Springer, Cham (2021). https://doi.org/10.1007/978-3-030-54817-9_137
6. Gao, X., Li, B., Hong, J., Guo, J.: Stiffness modeling of machine tools based on machining space analysis. Int. J. Adv. Manuf. Technol. **86**(5–8), 2093–2106 (2016). https://doi.org/10.1007/s00170-015-8336-z
7. Brecher, C., Fey, M.: Daniels modeling of position-tool- and workpiece-dependent milling machine dynamics. High Sped Mach. **2**, 15–25 (2016). https://doi.org/10.1515/hsm-2016-0003
8. Brecher, C., Bäumler, S., Daniels, M.: Prediction of dynamics of modified machine tool by experimental substructuring. In: Allen, M., Mayes, R., Rixen, D. (eds.) Dynamics of Coupled Structures, Volume 1. CPSEMS, pp. 297–305. Springer, Cham (2014). https://doi.org/10.1007/978-3-319-04501-6_28
9. Kumbar, S., Birangane, V.: Design and analysis of machine tool spindle. Int. J. Eng. Trends Technol. **48**(7), 387–392 (2017)
10. Sokolov, V.: Transfer functions for shearing stress in nonstationary fluid friction. In: Radionov, A.A., Kravchenko, O.A., Guzeev, V.I., Rozhdestvenskiy, Y.V. (eds.) Proceedings of the 5th International Conference on Industrial Engineering (ICIE 2019). LNME, vol. 1, pp. 707–715. Springer, Cham (2020). https://doi.org/10.1007/978-3-030-22041-9_76
11. Krol, O.: Modeling of worm gear design with non-clearance engagement. In: Radionov, A.A., Gasiyarov, V.R. (eds.) Proceedings of the 6th International Conference on Industrial Engineering (ICIE 2020). LNME, vol. 2, pp. 36–46. Springer, Cham (2021). https://doi.org/10.1007/978-3-030-54814-8_5
12. Sokolov, V., Krol, O., Stepanova, O.: Nonlinear simulation of electrohydraulic drive for technological equipment. J. Phys: Conf. Ser. **1278**, 012003 (2019). https://doi.org/10.1088/1742-6596/1278/1/012003
13. Kushnir, E., Portman, V.T., Aguilar, A., Clark, W.: Layout evaluation at earlier stages of machine tool design: form-shaping function-based approach. Int. J. Adv. Manuf. Technol. **90**(9–12), 3333–3346 (2016). https://doi.org/10.1007/s00170-016-9667-0

14. Wirth, C., Höhn, B-R., Braykoff, C.: New methods for the calculation of the load capacity of bevel and hypoid gears. Gear Technology, June/July, pp. 44–54 (2013)
15. Chernavsky, S.A., Itskovich, G.M., Kiselev, V.A., et al.: Design of Mechanical Transmissions: Textbook. Mashinostroenie, Moscow (1976)
16. Amendola, J.B., Amendola, J.B., III., Yatzook, D.: Longitudinal tooth contact pattern shift. Geartechnology **2012**(May), 63–67 (2012)
17. Krol, O., Porkuian, O., Sokolov, V., Tsankov, P.: Vibration stability of spindle nodes in the zone of tool equipment optimal parameters. Comptes rendus de l academie bulgare des sciences **72**(11), 1546–1556 (2019). https://doi.org/10.7546/CRABS.2019.11.12
18. Nikonov, V.V.: KOMPAS 3D. Creation of Models. Peter, St. Petersburg (2020)
19. Krol, O., Sokolov, V.: Modelling of spindle nodes for machining centers. J. Phys. Conf. Ser. **1084**, 012007 (2018). https://doi.org/10.1088/1742-6596/1084/1/012007
20. Sokolov, V.: Diffusion of circular source in the channels of ventilation systems. In: Fujita, H., Nguyen, D.C., Vu, N.P., Banh, T.L., Puta, H.H. (eds.) Advances in Engineering Research and Application. LNNS, vol. 63, pp. 278–283. Springer, Cham (2019). https://doi.org/10.1007/978-3-030-04792-4_37
21. Sokolov, V.: Hydrodynamics of flow in a flat slot with boundary change of viscosity. In: Radionov, A.A., Gasiyarov, V.R. (eds.) Proceedings of the 6th International Conference on Industrial Engineering (ICIE 2020). LNME, vol. 2, pp. 1172–1181. Springer, Cham (2021). https://doi.org/10.1007/978-3-030-54817-9_136
22. Chagina, A.V., Bolshakov, V.P.: 3D modeling in KOMPAS-3D v17 and higher. Textbook for universities. Peter, St. Petersburg (2021)
23. Sokolov, V.: Dynamics of positioning process for hydraulic drive output link by distributor with closed center. In: Radionov, A.A., Gasiyarov, V.R. (eds.) Proceedings of the 7th International Conference on Industrial Engineering (ICIE 2021). LNME, vol. 2, pp. 715–723. Springer, Cham (2022). https://doi.org/10.1007/978-3-030-85230-6_84
24. Krol, O., Sokolov, V.: Research of toothed belt transmission with arched teeth. Diagnostyka. **21**(4), 15–22 (2020). https://doi.org/10.29354/diag/127193
25. Ivanov, V., Dehtiarov, I., Pavlenko, I., Liaposhchenko, O., Zaloga, V.: Parametric Optimization of Fixtures for Multiaxis Machining of Parts. In: Hamrol, A., Kujawińska, A., Barraza, M.F.S. (eds.) Advances in Manufacturing II. LNME, vol. 2, pp. 335–347. Springer, Cham (2019). https://doi.org/10.1007/978-3-030-18789-7_28
26. Shevchenko, S., Mukhovaty, A., Krol, O.: Gear Transmission with conic Axoid on parallel axes. In: Radionov, A.A., Kravchenko, O.A., Guzeev, V.I., Rozhdestvenskiy, Y.V. (eds.) Proceedings of the 5th International Conference on Industrial Engineering (ICIE 2019). LNME, vol. 2, pp. 1–10. Springer, Cham (2020). https://doi.org/10.1007/978-3-030-22041-9_1
27. Krol, O., Sokolov, V.: Research of modified gear drive for multioperational machine with increased load capacity. Diagnostyka. **21**(3), 87–93 (2020). https://doi.org/10.29354/diag/126026
28. Pastukhov, A.G., Timashov, E.P., Kravchenko, I.N.: Thermal diagnostics of mechanical gear elements of combine harvester. IOP Conf. Ser. Earth. Environ. Sci. **845**, 012137 (2021). https://doi.org/10.1088/1755-1315/845/1/012137
29. Tsankov, P.: Modeling of vertical spindle head for machining center. J. Phys. Conf. Ser. **1553**, 012012 (2020). https://doi.org/10.1088/1742-6596/1553/1/012012
30. Sokolov, V.: Criteria analysis of diffusion processes in channels of industrial ventilation systems. In: Radionov, A.A., Gasiyarov, V.R. (eds.) Proceedings of the 7th International Conference on Industrial Engineering (ICIE 2021). LNME, vol. 2, pp. 725–731. Springer, Cham (2022). https://doi.org/10.1007/978-3-030-85230-6_85

31. Litvin, F.L.: The Theory of Gearing. Nauka, Moscow (1968)
32. Stanciu, D.-I., Cheorghe, G.I., Ciobotă, D.-D.: Master error elimination in forced gear engagement testing machine using harmonic analysis through fast Fourier transform. Int. J. Mechatron. Appl. Mech. **2**(10), 238–247 (2021)

Check for updates

Optical Fiber Behavior Under Inert Atmosphere

R. El Abdi[1(✉)], R. Leite Pinto[1], G. Guérard[2], and C. Capena[2]

[1] Univ. Rennes-CNRS, Institut de Physique de Rennes, UMR 6251, 35000 Rennes, France
relabdi@univ-rennes1.fr

[2] Entreprise Acome-Usines de Mortain, 50140 Mortain, France

Abstract. Optical fibers are made from fused silica glass surrounded by a protective coating and the optical strength is almost entirely determined by the strength of the glass. In telecommunications applications the single biggest cause of system failure is failure of the cable and it is not easy to estimate the theoretical maximum strength of fibers which depend on applied stresses and used environment. On the other hand, glass optical fibers were almost always coated with a polymer immediately after drawing to protect them from subsequent handling damage and from chemical environment. For the study of fiber strength, it was useful to be able to remove this coating in order to directly observe the fatigue properties of the glass in immediate contact with the environment.

The mechanical strength of coated and stripped silica optical fibers has been investigated through aging in the air and in an inert atmosphere. Dynamic tests for different faceplate velocities were implemented using a two-point bending testing device in order to determine the median strength for coated and stripped fibers from Weibull statistics.

The analysis of the Weibull curves demonstrates the influence of the used atmosphere on the fiber strength as well as the role of the polymeric coatings.

Keywords: Optical fibers · Weibull distributions · Dynamic tests · Inert atmosphere

1 Introduction

A lot of progress has been made on the ability to produce long lengths of strong optical fibers for telecommunication networks. Thereby, some important advances have also been made in the protection of the external surface of the fiber from environment damage through the development of new coatings. Among these coatings, epoxy acrylate coatings were used in telecommunication networks; during the drawing, the application of the polymer on the fiber was easy and its polymerization was very fast.

Polymer coatings are currently applied to optical fibers to prevent the formation of surface defects through scratches and abrasion and to minimize the influence of the pre-existing defects. They also act as a diffusion barrier against the surrounding humidity reaching the glass surface. Water is known to be one major factor of the propagation of cracks at fiber glass surface because it makes much easier the breaking of the Si-O bonds which build the vitreous network [1]. Accordingly, fiber strength is closely related to the

D. D. Cioboată (Ed.): ICoRSE 2022, LNNS 534, pp. 22–29, 2023.
https://doi.org/10.1007/978-3-031-15944-2_3

water concentration at the glass surface [2]. It is well known that flaws in glass subject to stress in humid conditions grow subcritically. Crack velocity is related to applied stress and also to relative humidity [3]. It has also been reported that the kinetics of the reaction between silica and water changes at very low water concentration [4–6].

However, there is not a sufficiently detailed understanding of the influence of aging conditions on the fiber strength when these types of fibers were aged in an aqueous environment and under mechanical stress. It is important to estimate the life duration of these fibers which were degraded under moisture, inert environment and mechanical stresses.

In this paper, the influence of chemical environment and that of the polymer coating are analysed. Silica optical fibers were stripped using a new stripping gel to remove the optical fiber coating materials that leaves the surface of the glass cladding intact. Optical fibers with and without polymer coatings were submitted to bending tests in two atmospheres: air atmosphere and inert atmosphere to measure the humidity influence on fiber strength.

2 Fiber and Test Bench Used

The used monomode silica fiber has an acrylate coating. This fiber was manufactured using the Plasma activated Chemical Vapor Deposition (PCVD) process which produces a totally synthetic, ultra-pure fiber. The combined coating diameter is $242 \pm 5\ \mu m$, the clad diameter is $125 \pm 0.7\ \mu m$ and the coating thickness is $58.5 \pm 0.5\ \mu m$.

A two points bending bench made up of a displacement plate which is mounted on an aluminium plate. The first thrust block is movable and mounted on the displacement plate, while the second thrust block is fixed on a force sensor. The optical fiber is positioned between the two thrust blocks in such a way that it forms a "U". To avoid slipping, the fiber is positioned in the grooves of the thrust blocks. During the test, load and displacement are recorded, allowing the load/displacement curve to be obtained (Fig. 1).

The test bench was introduced inside a compact glove box with purification system ($H_2O < 1$ ppm, $O_2 < 1$ ppm, 95% N_2 or Ar + 5% H_2) (Fig. 2). External gloves allow performing bending tests inside the box.

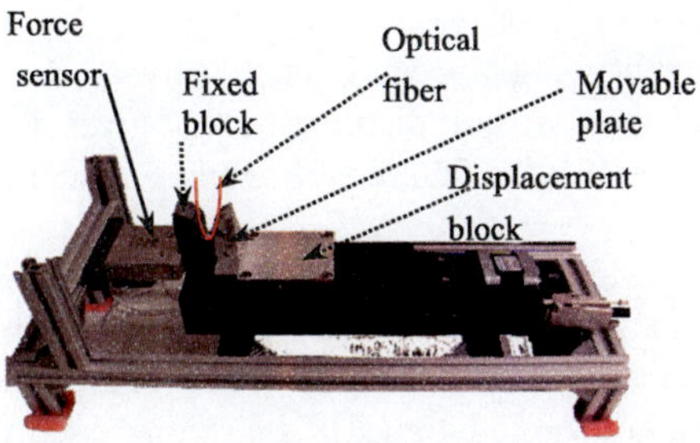

Fig. 1. Bending bench used

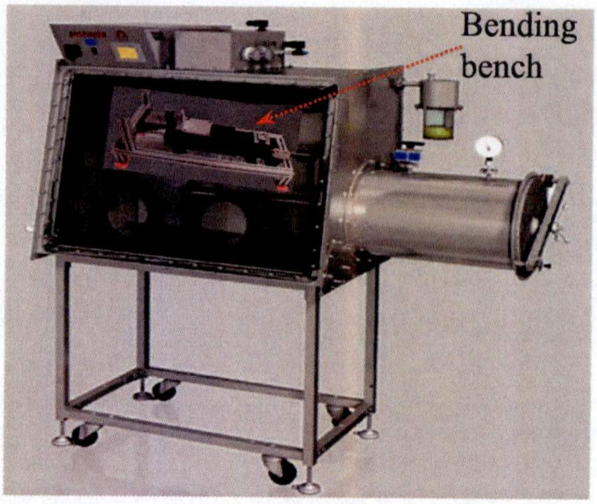

Fig. 2. Compact glove box with purification system (Jacomex GP campus T2)

The stripping gel used contains methylene chloride (CH_2Cl_2: > 60%) (used as a solvent in manufacturing) and other elements such as trichloroethylene ($CHCl_3$) and heavy naphtha.

Figure 3 shows that the coating has been removed and the glass cladding surface has remained intact.

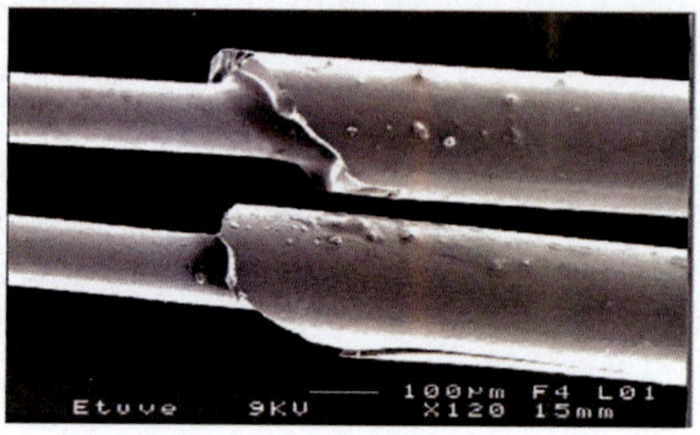

Fig. 3. Stripped fibers

3 Flaws Distribution

The influence of the different aging procedures on the fiber mechanical strength has been evidenced using the statistical Weibull distribution that gives the relation between the logarithm function of the cumulative failure probability F related to the logarithm of the fracture stress [7].

The relations in the previous section have been determined for single flaws.

Because many flaws may be present on an optical fiber, a statistical distribution must be applied. In this section a discription using a Weibull distribution will be given. This is illustrated for static and dynamic fatigue. It is also discussed that two distributions, the intrinsic and extrinsic (Fig. 4), exist for optical fibers for short and long lengths, respectively.

According to the weakest-link principle the probability of failure F on length L can be found from the average number of flaws N (per unit of length) with strength below the value σ [7]:

$$F = 1 - \exp[-L.N(\sigma)/L_0] \quad (1)$$

with L_0 a scale factor. The initial strength distribution is assumed to be a Weibull distribution, which is commonly used in the fracture mechanics of glass and ceramic materials:

$$N(\sigma_1) = (\sigma_1/\sigma_0)^m \quad (2)$$

With σ_0 a scale factor and m the Weibull parameter. The m-value represents the sharpness of the distribution [8]. This distribution is nothing more than a straight-line fit function for a double-logarithmic plot of ln (1-F) against σ_1 and can be plotted on a Weibull scale. The F parameter is shown how the failure probability F on the Weibull scale is obtained from measurements [7].

When longer samples are tested these lines shift upwards, as follows with (1) and (2). Hence the probability of failure increases or, for constant failure, the strength decreases for longer test lengths.

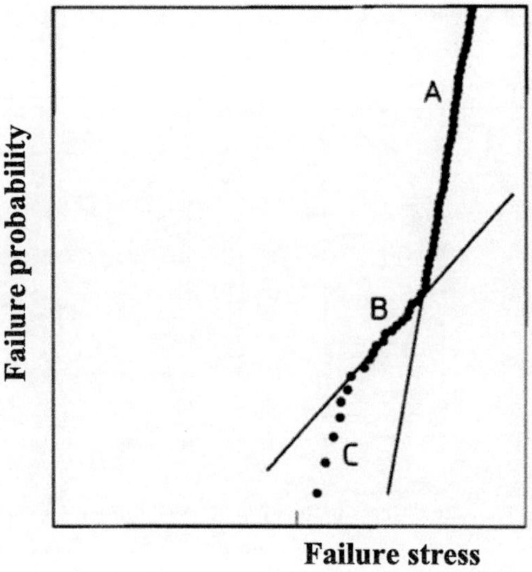

Fig. 4. Weibull plot of dynamic failure stress for tested fibers. Region A is the intrinsic (high strength mode) and region B is the extrinsic (weak spot). Region C is the effect of proof test truncation [7].

4 Results and Discussion

4.1 Tests Under Air Atmosphere

Figures 5 and 6 give Weibull curves for different faceplate velocities when the tests were undertaken in the air for coated and stripped fibers respectively. Overall, coated and uncoated fibers exhibit the same mechanical behaviour. The failure stress increases with the faceplate velocity. The failure stresses of coated and stripped fibers were similar; the coated fibers have slightly higher values than the uncoated fibers which don't benefit from the coating rigidity (Fig. 7).

4.2 Tests Under Inert Atmosphere

Figures 8 and 9 give Weibull curves for different faceplate velocities when the tests were undertaken in the inert atmosphere for coated and stripped fibers respectively.

Here again, the failure stresses of coated fibers are slightly higher than those of stripped fibers (Fig. 10).

Table 1 summarizes the results for both atmospheres for coated and stripped fibers. It should be noted that the median failure strength doesn't varies significantly when the coating was removed. On the other hand, the test environment had an important influence. Indeed, the failure fiber strength can drop by 50% due to the air humidity which weakens the silica bonds [3, 6].

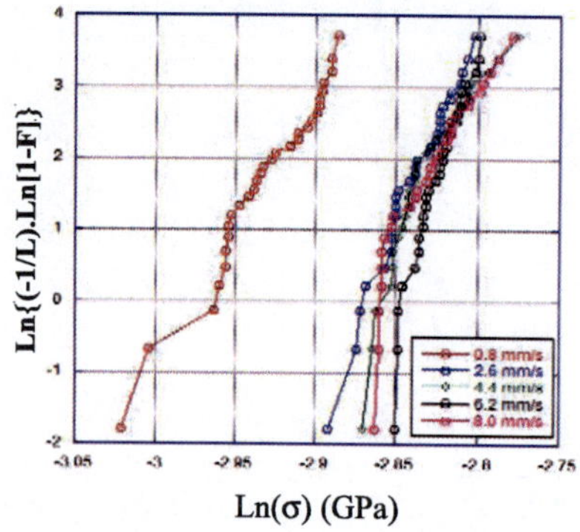

Fig. 5. Weibull curves for coated fibers in the air

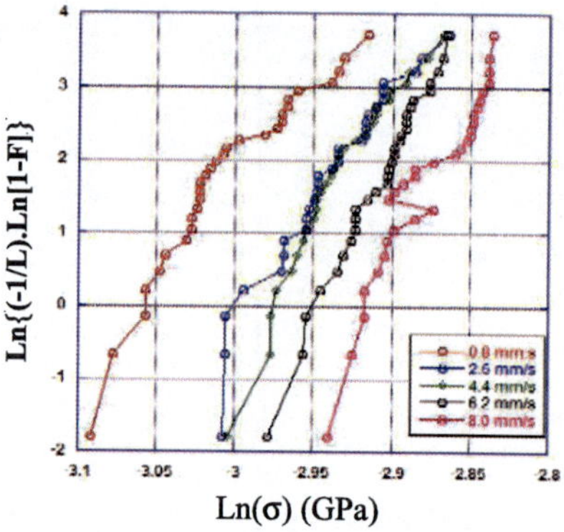

Fig. 6. Weibull curves for stripped fibers in the air

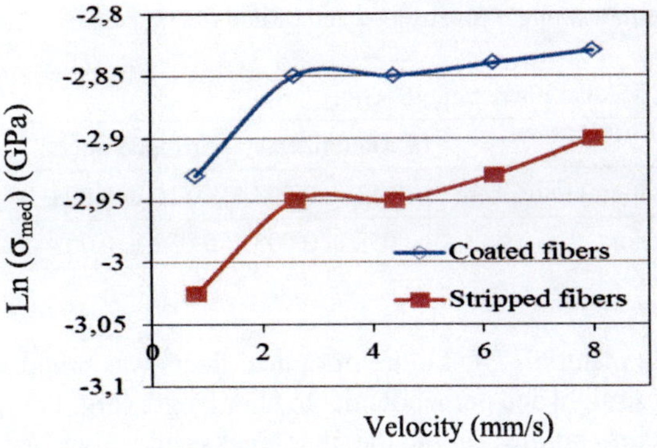

Fig. 7. Median strength for coated and stripped fibers in the air versus faceplate velocity

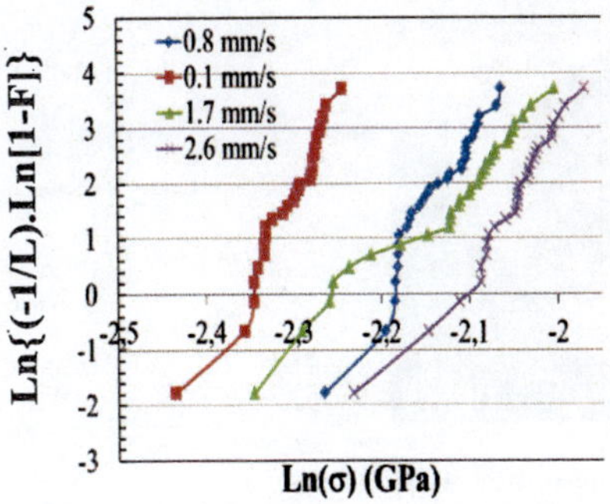

Fig. 8. Weibull curves for coated fibers in inert atmosphere

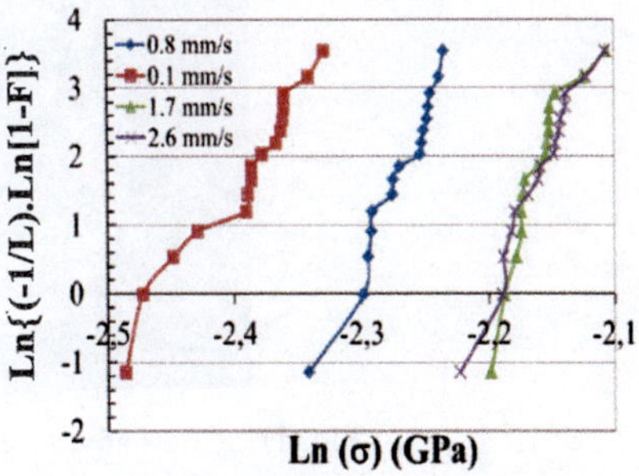

Fig. 9. Weibull curves for stripped fibers in inert atmosphere

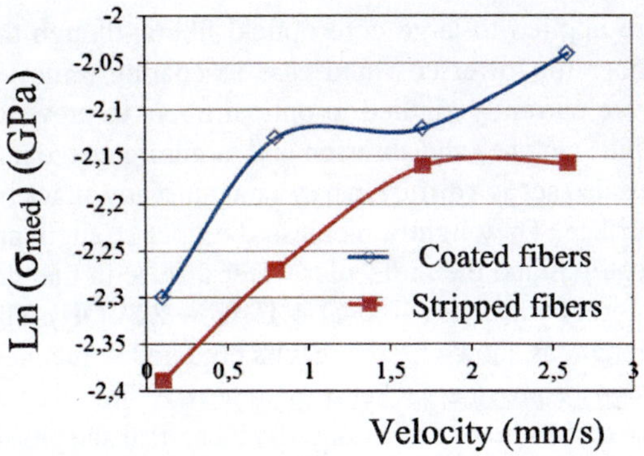

Fig. 10. Median strength for coated and stripped fibers under inert atmosphere versus faceplate velocity

Table 1. Median strength for stripped and coated fibers under two atmospheres

Median fiber strength (GPa)		
	Coated fibers	Stripped fibers
Inert atmosphere	0.11 ± 0.003	0.10 ± 0.003
Air	0.058 ± 0.003	0.052 ± 0.003

Silica is a brittle material. The failure of optical fibers was brutal. The failure cross section surface was straight and perpendicular to fiber length (Fig. 11). Some glass flakes were found on the cross failure section; the fiber breaks suddenly when the failure stress was reached.

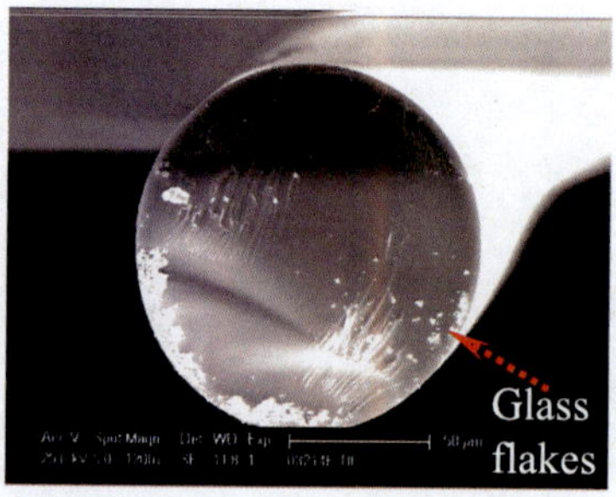

Fig. 11. Cross area of the stripped fiber after failure.

5 Conclusion

Acrylate coatings are applied to large core optical fibers, though they are applied to nearly all telecom fiber for lower cost and ease to coating removal for termination. Acrylate coatings were currently applied to optical fibers to prevent the formation of surface defects through scratches and abrasion and to minimize the influence of the pre-existing defects. They also act as a diffusion barrier against any attack by chemical agents reaching the glass surface. They lightly increase the fiber strength and limit the water attack but they can't stop it and the molecular water diffuse in the glass and react with the glass network to form hydroxyls (Si-O-Si + $H_2O \leftrightarrow$ 2 Si-OH). This water diffusion weakens the fiber strength as shown by the results obtained in the inert atmosphere.

In some fiber designs, a polymer jacket is extruded over initial coating for extra protection. But the choice of the fiber is conditioned by its optical and physical performances but also by its cost.

References

1. Charles, R.J., Hillig, W.B.: Union Scientifique Continentale du Verre, Charleroi, Belgium (1962)

2. Gurney, C.: Delayed fracture in glass. Proc. Phys. Soc. London. **59**, 169–185 (1947)
3. Wiederhorn, S.M.: Influence of water vapor on crack propagation in soda-lime glass. J. Am. Ceram. Soc. **50**, 407–414 (1967)
4. Duncan, W.J., France, P.W., Craig, S.P.: Strength of inorganic glass. In: Kurkjian, C.R., (ed.) The Effect of Environment on the Strength of Optical Fiber, 351p. Plenum Press, New-York (1985)
5. Evano, N., El Abdi, R., Poulain, M.: Aging of silica optical fiber in static fatigue tests. Appl. Mech. Mater. **859**, 3–8 (2016)
6. Mrotek, J.L., Matthewson, M.J., Kurkjian, C.R.: Diffusion of moisture through optical fiber coatings. J. Lightwave Tech. **19**(7), 988–993 (2001)
7. Griffieon, W.: Optical fiber mechanical reliability. Ph.D. thesis. Royal PTT Netherland NV, PTT Research, Leidschendam (1994)
8. Zhao, F.M., Okabe, T., Takeda, N.: The estimation of the statistical fiber strength by fragmentation tests of single-fiber composites. Comp. Sci. Tech. **60**, 1965–1974 (2000)

A Computer System for Reliable Operation of a Diesel Generator on the Basis of Indirect Measurement Data Processing

Oleksandr Yenikieiev[1], Dmytro Zakharenkov[1], Yevhen Korotenko[1], Olexii Razzhyvin[1], Ihor Yakovenko[2], Fatima Yevsyukova[2](✉), and Olena Naboka[2]

[1] Donbas State Engineering Academy, 72, Akademicheskaia str., Kramatorsk 84313, Ukraine
[2] National Technical University "Kharkiv Polytechnic Institute", 2, Kyrpychova str., Kharkiv 61002, Ukraine
fatimaevsukova@gmail.com

Abstract. Algorithmic software for monitoring the cylinder power of a diesel generator based on the processing of indirect measurement data has been developed. The instantaneous speed of the crankshaft is used as input information. The kinematic scheme of the engine is given by a mechanical system with four degrees of freedom, subject to friction. The motions of the model masses are described by a system of differential equations, the parameters of which are normalized on the basis of the methods of similarity theory. The Laplace transform under zero initial conditions as a mathematical apparatus is applied. The method of determinants was used to establish information connections between the torques of the cylinders and the signal of the measuring information. In the Matlab software environment with the Simulink extension, a scheme of computer simulation of the signal of fluctuations of the speed of rotation of the first mass is built. The method of adjusting the length of information links between its components is used to adjust the parameters of the scheme. The information technology of monitoring of cylinder powers is developed for frequency representation of a signal of fluctuations, transfer functions and the approximated torques of cylinders. The least squares method is used to minimize the incoherence of the system of algebraic equations. Additive random interference is used in computer simulation of the procedure for solving the system of equations. Requirements for metrological characteristics of the measuring transducer of the frequency-modulated signal of the crankshaft speed are formed.

Keywords: Computer system · Information technology · Software · Mathematical and computer modeling · Metrological characteristics · Fluctuation signal

1 Introduction

Technical and economic indicators of diesel generators (DG) determine the settings of fuel and air supply to the cylinders [1, 2]. Known mechanical systems for these settings employ crankshafts and camshafts connected to each other by gears. The uncertainty

D. D. Cioboată (Ed.): ICoRSE 2022, LNNS 534, pp. 30–44, 2023.
https://doi.org/10.1007/978-3-031-15944-2_4

interval around the optimal angles of fuel and air supply is formed as a result of the kinematic uncertainty of the gears and regulatory tolerances on the geometry of the shafts. To reduce it, hardware means are employed to set individual angles of fuel and air supply. A well-known method of establishing the cylinder power monitoring involves measuring the pressure of the cylinders and construction of indicator diagrams. Comparison of these diagrams allows to set deviations from the medium value and make appropriate adjustments. Application of manual labor, a large number of cylinders and the lack of electrical output signals from the primary pressure transducers significantly limit the performance of this evaluation method. There is an idea to employ the crankshaft instantaneous speed signal for indirect measurements of cylinder power. The computer system (CS) determines their distribution and generates, if necessary, appropriate changes in the settings of fuel supply phases. Solution of this problem will save fuel up to 5% [3], reduce the likelihood of overloading individual cylinders, significantly increase the life of the unit and reduce the cost of maintenance, service and repair. Therefore, creation of algorithmic and applicable software, that allows to build hardware with a given uncertainty and productivity of input information, determines the relevance of this scientific and applied problem.

2 Literature Review

Technical literature pays sufficient attention to the issue of ensuring DG reliable operation based on the processing of crankshaft non-uniformity signal. And, as noted at the work [4, 5] the intensifying of the manufacturing process and increasing the efficiency of production planning of precise and non-rigid parts, mainly crankshafts, are the first-priority task in modern manufacturing. In that regard in paper [6], a method for diagnosing multi-cylinder engines using the crankshaft non-uniformity signal obtained as a result of computer simulation is proposed. The model parameters were identified on the basis of experimental data processing. The technique takes into account the peculiarities of the engines in the case of adjacent torques imposition. In order to reduce the impact of random interference on the information signal of uneven rotation of the power unit crankshaft in [7] it is proposed to use a high-pass filter with finite pulse response. A method for the processing of measuring signal using the capabilities of the Matlab software environment has been developed. In [8] the influence of torque non-uniformity on dynamic characteristics and engine power indicators is given research. The connection between indicators of non-uniformity of indicator torque and non-uniformity of crankshaft rotation is established and their influence on the change of indicator power of four-stroke auto tractor internal combustion engines is determined.

In [9], an expert system is proposed for identifying the state of a tractor engine based on the use of a self-learning computer model. Expressions of the amplitude-frequency and energy spectra and the auto-correlation function for the angular acceleration in the free acceleration mode, which is averaged over all cylinders, are obtained. Criteria are defined for assessing the unevenness and tightness of the cylinders, as well as the state of the angles of advance of the fuel supply. In [10], sensitivity of several criteria for estimating the non-uniformity of the angular velocity of the crankshaft by maximum pressure of the indicator diagrams of successive operating cycles is considered. Research is given to the coefficient of variation of angular velocity increments over several operating cycles, to the coefficient of variation of its amplitude values, to the coefficient of variation of the set of the largest maximum and the smallest minimum values, as well as to the ratio criterion of real and kinematic non-uniformity of angular velocity.

In [11], a method is proposed to reduce the uncertainty of fuel supply control based on measurements of the amplitude of oscillations of angular velocity and displacements in the phase of their extremes relative to the top dead center of the corresponding cylinder. In [12], a method of adjusting the parameters of cyclic fuel supply is proposed, the identity of which is established on the basis of processing the non-uniformity signal of crankshaft rotation. In [13], a method for diagnosing the state of an internal combustion engine by establishing indicators of non-uniformity of its operation according to the parameters of vibroacoustic emission signals is proposed. Appropriate hardware has been developed. It is proved that the use of digital measuring devices for processing input signals provides the possibility of real-time control. In [14], the expressions of forced oscillations of a discrete multi-mass model are given, taking into account the torque matrix. Measuring the amplitude of oscillations provides an opportunity to set the parameters of torque, as well as to assess the possibility of the resonant phenomena. The line of trend of linear regression of degradation of technical condition of a cranked shaft is given, which allows to establish zones of satisfactory and unsatisfactory conditions.

In [15], the use of the method of analytical dependences to establish the dynamic characteristics of a machine unit is proposed. Consideration of the factors influencing the movement of the chains reduces the uneven rotation of the main motion drive. The disadvantage of the known methods of controlling the fuel supply to the DG cylinders is unsatisfactory variance of measurements and performance of hardware for determination of cylinder power, lack of algorithmic and applied software for frequency-modulated signal processing.

3 Research Methodology

The purpose of the research is to reduce the uncertainty of the crankshaft speed fluctuation signal processing and to improve the hardware productivity so that to ensure the DG reliable operation.

To achieve this goal the following tasks, need to be solved:

- develop the architecture of the CS hardware;
- build the deterministic mathematical model of the torsional circuit of the power unit and perform the procedure of identification of its parameters;
- establish information connections between the torques of the cylinders and the signal of the measuring information;
- study the transfer functions and adjust the frequency characteristics of hardware using the methods of neural network technology;
- build a scheme of computer simulation of the signal of non-uniformity of rotation of the first mass of the DG crankshaft;
- develop an information technology for establishing the distribution of cylinder power based on the processing of the signal of the crankshaft rotation speed fluctuations;
- create requirements for metrological characteristics of the measuring transducer of frequency-modulated signal.

3.1 Development of Hardware Architecture

The idea is proposed for controlling the fuel supply to the DG 3TD-1 cylinders based on processing of the signal of the crankshaft instantaneous speed. Synchronization of hardware with the phase of the crankshaft rotation provides a signal of the top dead center of the first cylinder. Electro-hydraulic injectors are used as actuating mechanisms for setting the fuel supply to the cylinders of the power unit. The architecture of the CS hardware equipment (Fig. 1) consists of the following blocks: sensor of the instantaneous speed of crankshaft rotation (S1), sensor of the top dead center of the first cylinder (S2), measuring transducer (MT), microcomputer and three actuators (A1, A2 and A3).

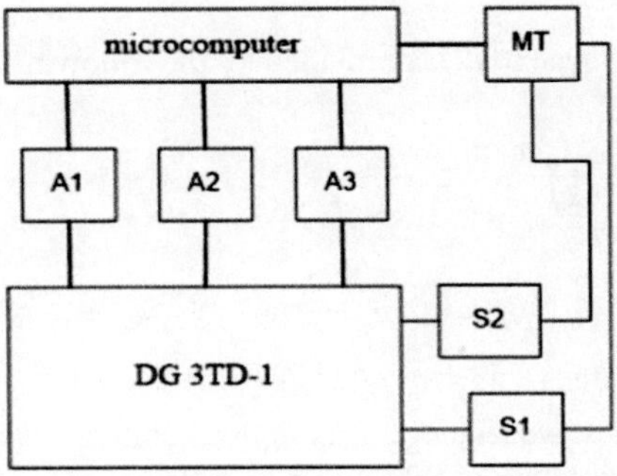

Fig. 1. CS of software control of fuel and air supply

3.2 Mathematical Modeling of the Torsional Circuit of the Power Unit

The DG kinematic scheme with three-cylinder two-stroke 3TD-1 diesel is developed in [16] and its main mechanical parameters are established. As a result of the DG model analysis, the authors propose: to obtain input information the intake crankshaft should be applied, which significantly limits the number of masses of the mechanical model; the primary converter should be installed near the first cylinder; subject to friction and oscillations between the masses, the torsional circuit should be presented in the form of a mechanical system with four degrees of freedom. Dynamics of rotation of cylindrical masses of a mechanical system is described by the following system of differential equations [3, 16].

$$J_i\phi_i^{//}(t) + \beta\phi_i^{/}(t) - e^{-1}\left[\phi_{i+1}(t) - \phi_i(t)\right] + e^{-1}\left[\phi_i(t) - \phi_{i-1}(t)\right] = M_i(t), \qquad (1)$$

where $i = 1,\ 2,\ \ldots\ 4$; $\phi_i(t)$ is mass rotation angle; $M_i(t)$ is torque; $\beta = 4.2$ Nms is friction; $e = 3.84 \cdot 10^{-7}$ $(\mathrm{Nm})^{-1}$ is flexibility of connections; $J_1 = 0.715$ Nm ^{2}, $J_4 = 0.225$ Nm ^{2} are moments of inertia of the first and fourth masses. The rotation of the shaft of the two-stroke 3TD-1 engine is in the frequency range $80 \div 280$ s^{-1}. Maximum power 220 kW (300 HP) at nominal torque 828 Nm is achieved at nominal speed 260 s^{-1}; maximum torque 912 Nm occurs at speed 195 s^{-1}.

To generalize the research results, we reduce the system of differential Eqs. (1) to dimensionless form using the theorems of similarity theory. Let's choose the following basic values $J_б, \beta_б, e_б, M_б, \varphi_б, t_б$. We introduce the following notation [3].

$$J^* = J/J_б, \beta^* = \beta/\beta_б, e^* = e/e_б, M^* = M/M_б, \varphi^* = \varphi/\varphi_б, t^* = t/t_б .$$

In this case, all values with superscript will be dimensionless. Then multiply each parameter of the system of Eqs. (1) by the ratio of the basic values. This procedure does not change the system of differential equations. After multiplication and taking into account the entered notations we have the following expression

$$\frac{J_i^* J_б \varphi_б}{M_б t_б^2}\varphi_i^{//} + \frac{\beta^* \beta_б \varphi_б}{M_б t_б}\varphi_i^{/} - \frac{\varphi_б}{e^* e_б M_б}(\varphi_{i+1} - \varphi_i) + \frac{\varphi_б}{e^* e_б M_б}(\varphi_i - \varphi_{i-1}) = M_i^* . \quad (2)$$

Simplification of the system (2) is provided by the following conditions

$$\frac{J_б \varphi_б}{M_б t_б^2} = 1, \quad \frac{\beta_б \varphi_б}{M_б t_б} = 1, \quad \frac{\varphi_б}{e_б M_б} = 1. \quad (3)$$

By choosing $J_б = 0.715, \beta_б = 4.2, e_б = 3.84 \cdot 10^{-7}, \varphi_б = 1 rad$,, based on conditions (3) we set numerical values $M_б = 2.604 \cdot 10^6, t_б = 5.24 \cdot 10^{-4}$.. Thus, by appropriate selection of basic values we achieve a complete coincidence of dimensional and dimensionless forms of notation of the system of differential Eqs. (1).

As a mathematical apparatus of analysis we use the Laplace transform under zero initial conditions [17, 18]. The system of differential Eqs. (2) takes the following form

$$\left(J_i p^2 + p + 2\right)\varphi_i(p) - \varphi_{i+1}(p) - \varphi_{i-1}(p) = M_i(p). \quad (4)$$

In order to simplify further transformations, we introduce the following notation

$$a = \frac{1}{p^2 + p + 1}, \quad b = \frac{1}{p^2 + p + 1}, \quad c = \frac{1}{p^2 + p + 2},$$
$$d = \frac{1}{p^2 + p + 2}, \quad f = \frac{1}{0.315p^2 + p + 1}.$$

Taking into account these notations, the system of Eqs. (4) takes the following form

$$\begin{cases} \varphi_1 - a\varphi_2 = bM_1 \\ \varphi_2 - d\phi_3 - d\varphi_1 = cM_2 \\ \varphi_3 - d\varphi_4 - d\varphi_2 = cM_3 \\ \varphi_4 - f\varphi_3 = 0 \end{cases} \quad (5)$$

In matrix form it is written as follows

$$\varphi = A\varphi + M, \quad (6)$$

where φ is vector-column of angles of oscillation of masses, A is matrix of angles, M is torque vector column.

After simple mathematical transformations, the system of Eqs. (4) takes the following form

$$\varphi = (I - A)^{-1} M, \tag{7}$$

where I is single matrix.

3.3 Investigation of Torque Transfer Functions

Determinants of the system of Eqs. (5) are set using the Mathcad software environment as follows

$$\begin{aligned} &\Delta = ab^2 f - ad - df - d^2 + 1, \quad \Delta_1 = b - bd^2 + ac + acd - bdf - acdf; \\ &\Delta_2 = c + bd + cd - cdf - bd^2 f, \quad \Delta_3 = c + bd^2 + cd - acd. \end{aligned} \tag{8}$$

Under this condition, the system of algebraic Eqs. (7) takes the following form

$$\varphi_1(p) = \sum_{j=1}^{3} \frac{\Delta_j}{\Delta} M_j(p) = \sum_{j=1}^{3} W_j M_j(p). \tag{9}$$

where $\varphi_1(p)$ is Laplace transform of the signal of fluctuations of the rotation speed of the first mass, W_j are transmission functions that establish information links between cylinder torques and fluctuations in the speed of rotation of the first mass.

The procedure of finding the zeros and poles of the transfer functions of the channels "cylinders-first mass of the crankshaft" based on the use of Matlab software environment [19, 20] allows to present them in the form of a series connection of elementary circuits. This simplifies the expressions for transfer functions by cancellation the roots of the numerator and denominator, as well as the rejection of unstable and the roots of the second order of smallness [21]. As a result of the analysis of zeros and poles of transfer functions the values of roots are established (Table 1) which correspond these conditions.

Table 1. Roots of transfer functions

W_1	W_2	W_3
Zero		
–0.4331	–0.5871	–0.4581 + 0.7852i
		–0.4581 – 0.7852i
Pole		
0	0	0
–0.5545 + 0.7589i	–0.7684 + 0.5430i	–2.9807
–0.5545 – 0.7589i	–0.7684 – 0.5430i	–0.7684 + 0.5430i
		–0.7684 – 0.5430i

Based on these roots, the transfer functions take on this form

$$W_4(p) = \frac{p + a_1}{p\left[(p + a_4)^2 + b_4^2\right]}, \quad W_6(p) = \frac{(p + a_3)^2 + b_3^2}{p\left[(p + a_5)^2 + b_5^2\right](p + a_6)}. \tag{10}$$

After simple mathematical transformations, expressions for the transfer functions of the first and the third cylinders are obtained

$$W_4(p) = \frac{k_1\left(a_1^{-1}p + 1\right)}{\frac{1}{a_4^2 + b_4^2}p^3 + \frac{2a_4}{a_4^2 + b_4^2}p^2 + p}, \tag{11}$$

$$W_6(p) = \frac{k_3\left(\frac{1}{a_3^2 + b_3^2}p^2 + \frac{2a_3}{a_3^2 + b_3^2}p + 1\right)}{\frac{1}{a_6\left(a_5^2 + b_5^2\right)}p^4 + \frac{a_6 + 2a_5}{a_6\left(a_5^2 + b_5^2\right)}p^3 + \frac{2a_5a_6 + a_5^2 + b_5^2}{a_6\left(a_5^2 + b_5^2\right)}p^2 + p}, \tag{12}$$

where $k_1 = \frac{a_1}{a_4^2 + b_4^2}$, $k_3 = \frac{a_3^2 + b_3^2}{a_6\left(a_5^2 + b_5^2\right)}$ are conversion factors.

The command line for setting the expressions of the transfer functions of the channels "cylinders-first mass of the crankshaft" in the Matlab software environment is as follows.

W4 = tf([1.132 0.49], [1.132 1.255 1 0]); W5 = tf([1.13 0.663], [1.13 1.736 1 0]); W6 = tf([0.379 0.347 0.313], [0.379 1.712 2.071 1 0]).

Calculation of logarithmic amplitude-frequency characteristics (LAFC) of the "cylinders-first mass of the crankshaft" channels has been performed. As a result of the LAFC analysis it is established that within the frequency band 9–50 Hz slight oscillations are observed. Therefore, transmission of individual cylinder torque through the communication channel is determined by some instability of behavior. To correct the expressions of transfer functions, the authors propose a developed technique using the method of regulation of the length of information links of the neural network technology:

- decompose the expressions of transfer functions into simple fractions;
- summation of simple fractions to determine the transfer function is performed taking into account the coefficients μ_k, where k is the number of components in its expression;
- summation of fluctuation signals of individual masses is performed taking into account the coefficients ν_n, where n is the number of masses of the 3TD-1 model.

Decomposition of the transfer function W6 into prime factors gives the following expressions

$$W_7(p) = \frac{A_1}{p} + \frac{A_2}{p + a_6} + \frac{A_3}{p^2 + 2a_5p + a_5^2 + b_5^2}. \tag{13}$$

Coefficients A_i are defined by the following expressions.

$$A_1 = \frac{0.313}{a_6\left(a_5^2 + b_5^2\right)}, \quad A_2 = -A_1, \quad A_3 = \frac{0.347 - a_5a_6A_1}{a_6}.$$

As the length of the relationships between prime factors changes, expression (13) takes the following form

$$W_7(p) = \frac{0.119\mu_1}{p} - \frac{0.119\mu_2}{p + a_6} + \frac{0.025\mu_3}{p^2 + 2a_5 p + a_5^2 + b_5^2}. \tag{14}$$

3.4 Approximation of Cylinder Torques

The torque created by a separate cylinder on the crankshaft is obtained as a result of processing the experimental pressure data of the DG 3TD-1 first cylinder. The authors employed the indicator and compression diagram as the initial information when setting the torque graphs. The differential torque that provides rotation of the DG 3TD-1 crankshaft with a given angular velocity is determined using the expression

$$M_{i,p}(t) = M_{i,1}(t) - M_{i,2}(t), \tag{15}$$

where $M_{i,1}(t)$, $M_{i,2}(t)$ are torques according to the indicator and compression diagram.

In this case, it is fundamentally possible to organize changes in the settings of fuel and air supply to the power unit cylinders. It is proposed to present the distribution of cylinder power in the form of amplitude coefficients $D_i = 0 \ldots 1$. Based on this, expression (15) takes the following form

$$D_i M_{i,p}(t) = M_{i,1}(t) - M_{i,2}(t). \tag{16}$$

Phase delay of the cylinders relative to the first one is a multiple of 120° and is calculated taking into account the following sequence of their operation: 1–2–3. For further research, we use the Fourier transform as a mathematical apparatus. Under such conditions, expression (10) takes the following form

$$\varphi_1(j\omega) = \sum_{i=1}^{3} W_i D_i M_{i,p}(j\omega) + \sum_{i=1}^{3} W_i M_{i,2}(j\omega). \tag{17}$$

3.5 Development of Computer Simulation Scheme

The scheme of computer modeling of the movements of the first mass of DG 3TD-1 is made taking into account the capabilities of the Matlab software environment with Simulink extension (Fig. 2) based on expressions 15 and 17, as well as on transfer functions W4, W5 and W6.

The output signal is the fluctuation of its rotation speed. Mathematical models of torques of the second and the third cylinders (Model M2 and Model M3) are similar to the first one, taking into account the corresponding phase delay. The torque models take into account the amplitude normalization by multiplication $M_6^{-1} = 3.84 \cdot 10^{-7}$ .. When compiling the computer simulation scheme, the possibility of adjusting the contributions of individual cylinders to the signal of fluctuations of the first mass using weights ν_n [19] is also taken into account. Numerical values of these coefficients are established when identifying the parameters of the mathematical model of the mechanical system with four degrees of freedom. To perform this procedure, the method of self-learning model and the experimental data of the first mass instantaneous speed of rotation are employed.

The information and measuring device is developed in [20] and its metrological characteristics investigated. Suitability of the device for measuring the signal of the instantaneous rotation speed of the DG crankshaft type 10D100 is determined. The information technology of processing the array of experimental data of instantaneous speed involves the following computational actions:

- calculate the average value of the period of the input signal of the measuring transducer;
- fluctuation signal is the difference between the mean and instantaneous period;
- determine the array of fluctuations within the total amount of research data;
- perform averaging and form an array of fluctuations within one revolution of the crankshaft;
- this signal is represented in the form of a limited Fourier series.

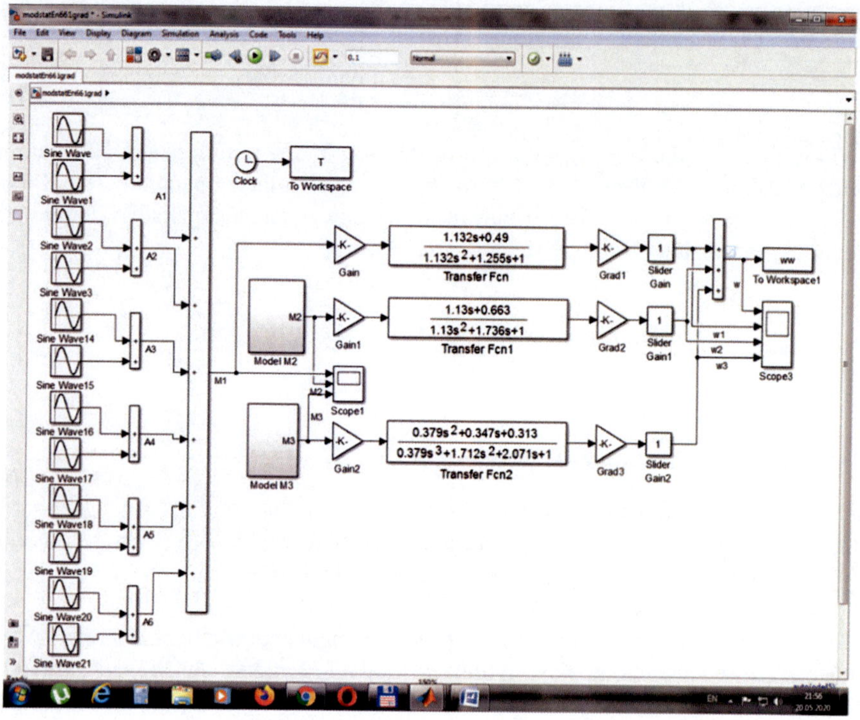

Fig. 2. Computer simulation scheme

3.6 Algorithmic Hardware Support

Algorithmic support for estimating the distribution of DG 3TD-1 cylinder power is based on establishing the values of amplitude coefficients D_i cylinders. To do this, we solve the following system of algebraic equations

$$BD = \varphi_1 - \varphi_{1,0}, \tag{18}$$

where B is matrix, the coefficients of which are determined on the basis of LAFC transfer functions of the paths "cylinders-first mass of the crankshaft" and the torques of the cylinders, depending on the chosen method of calculation; D is vector-column of amplitude coefficients; ϕ_1 is vector column of the frequency representation of the signal of fluctuations of the first mass; $\phi_{1,0}$ is vector-column of the fluctuation signal in the absence of fuel supply to the cylinders, which is obtained in advance on the basis of the frequency representation of torques in the absence of fuel supply to the cylinders and the obtained transfer functions. At frequency representation of the information signal, matrix coefficients are defined as follows

$$B_{i,j} = \sum_{i=1}^{3} \nu_i W_i(j\Omega) M_{i,p}(j\Omega). \tag{19}$$

If the frequency spectrum of the signal of fluctuations of the rotation speed of the first mass exceeds 3 harmonic components, then the system of algebraic Eqs. (18) is over determined. To calculate the optimal values of the coefficients D_i a standard subroutine of "LLSQ" library of mathematical software is used, which implements the algorithm to minimize decoupling based on the method of least squares. According to the calculation results, the CS forms software changes in the settings of fuel and air supply to the DG 3TD-1 cylinders.

4 Results

The method for adjusting the frequency characteristics of the channels "first cylinder-first mass of the crankshaft" is considered on the example of LAFC at the following values of the coefficients $\mu_1 = 1.5,\ \mu_2 = 1,\ \mu_3 = 1,\ \mu_1 = 1,\ \mu_2 = 1.5,\ \mu_3 = 1$ and $\mu_1 = 1,\ \mu_2 = 1,\ \mu_3 = 1.5$. The command line for setting in the Matlab transfer function environment is as follows.

W8 = tf([0.1785], [1 0]); W9 = tf([0.354], [0.335 1]); W10 = tf([0.022], [1.13 1.736 1]);

W11 = W8 + W9 + W10; W12 = tf([0.119], [1 0]); W13 = tf([0.531], [0.335 1]);

W14 = tf([0.022], [1.13 1.736 1]); W15 = W12 + W13 + W14; W16 = tf([0.119], [1 0]);

W17 = tf([0.354], [0.335 1]); W18 = tf([0.033], [1.13 1.736 1]); W19 = W16+W17 + W18.

The comparison is performed using the following command: bode (W6, W11, W15, W19); grid. The results of the comparison are shown in Fig. 3 as graphs, the analysis of which allows to draw the following conclusions. Growth of coefficients A1, A2 and A3 reduces the slope of the LAFC torque transmission channels. Also, increase of the coefficients A1, A2 and A3 leads to the adjustment of the LAFC of the torque transmission channels by significantly reducing the phase delay (from 180° to 90°). Increase of the coefficient A3 does not affect the LAFC torque transmission channels, so adjustment of transfer functions by means of it is inexpedient. Increase of the coefficients A1 and A2 leads to the decrease in the gain of the torque transmission channels in the low frequency band and increases it in the medium and high frequency band. Reduction of the coefficient A1 does not affect the LAFC channel transmission torque of the first cylinder. Also, reduction of the coefficients A1 and A2 leads to the adjustment of the LAFC of the torque transmission channels of the second and the third cylinders by significantly reducing the phase delay (from 180° to 90°). Reduction of the coefficients A1 and A2 reduces the gain of the torque transmission channels in the low frequency band and increases it in the medium and high frequency band.

Some results of computer simulation are shown in Fig. 4. Their analysis allows to establish the following:

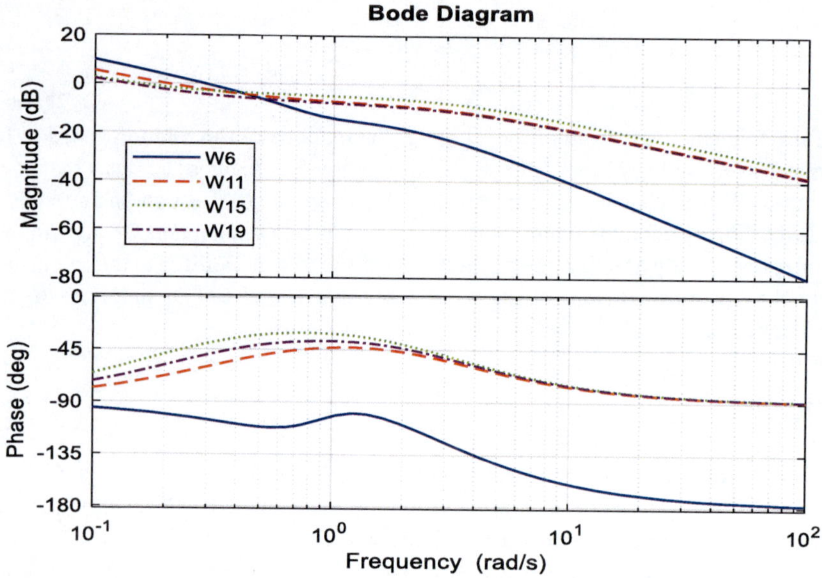

Fig. 3. Comparison of LAFC

- the transition process is approximately 3.668 ms. Upon its completion, a steady signal of the first mass rotational velocity fluctuations is observed;
- in Fig. 4 graphs of fluctuations are presented as the following values of weight coefficients $\nu_1 = 1, \nu_2 = 1, \nu_3 = 1, \nu_1 = 1, \nu_2 = 0.75, \nu_3 = 1$ and $\nu_1 = 1, \nu_2 = 1, \nu_3 = 0.75$. Alteration of the values of the weights allows adjusting the output signal of the computer simulation scheme. Identification procedure of the

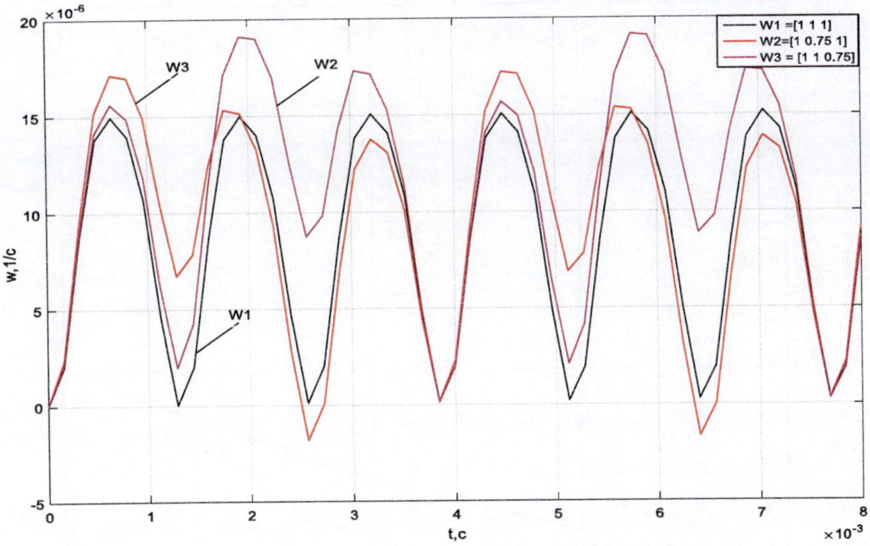

Fig. 4. Computer simulation results

mathematical model parameters of the mechanical system with four degrees of freedom is carried out by comparing its output signal with the signal obtained as a result of the experimental data processing;

- by choosing weight factors ν_n, the values of which are found using the computer simulation scheme, the signal of fluctuations in the rotation speed of the first mass is determined from the action of the torques of the compression diagrams.

Requirements for metrological characteristics of the measuring transducer are formed on the basis of computer modeling. The subject of research is the computational procedure for solving the system of algebraic Eqs. (18) under the conditions of additive random interference on the signal of fluctuations in the rotation speed of the first mass. The information technology for processing the results of calculations consists of the following actions: we determine the average value of the real and imaginary part of the sets of amplitude coefficients of the cylinders D_i, calculate the absolute and relative value of uncertainty. Graphs of coefficient calculation variance D_i at different levels (δ) of additive random interference is presented in Fig. 5. Requirements for the uncertainty of the measuring transducer of the fluctuation signal are set by the variance value.

As a result of theoretical research, the information links between the torques of the cylinders and the signal of fluctuations of rotation speed of the first mass of a cranked shaft are established in the form of transfer functions. The information technology of analysis and correction of frequency characteristics of mechanical channels of transmission of torques of cylinders is constructed taking into account opportunities provided by the Matlab environment. Uncertainties of information technology of monitoring distribution of cylinder capacities are investigated.

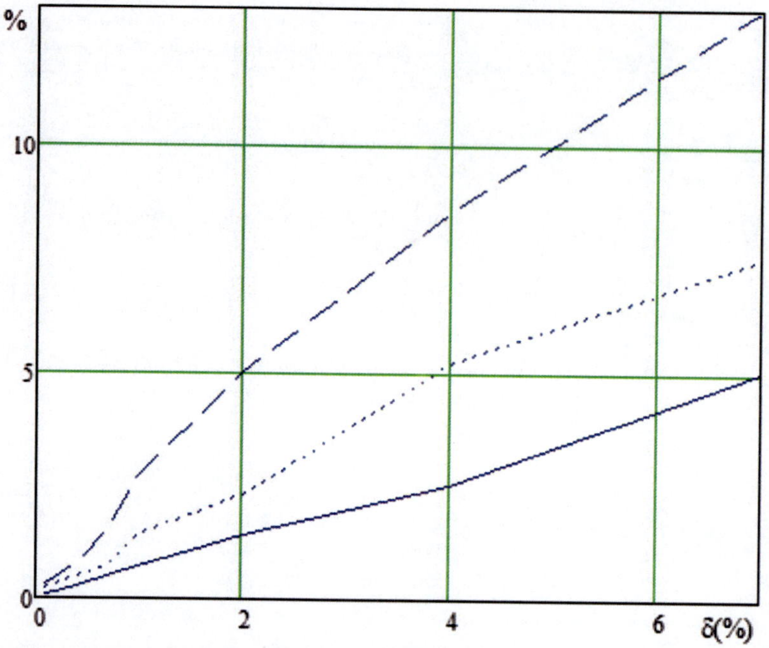

Fig. 5. Graphs of variance of coefficients at different δ (D_1 – solid, D_2 – dash, D_3 – dot)

5 Conclusions

A computer control system for fuel and air supply has been developed to ensure reliable operation of the 3TD-1 diesel generator. The crankshaft instantaneous rotation speed signal is used as the input information. A productive information technology for processing research data in order to obtain a fluctuation signal has been developed.

A mechanical system with four degrees of freedom under conditions of friction is used as a deterministic mathematical model of the 3TD-1 diesel generator. The oscillatory motions of the masses of the mathematical model are described by a linear system of differential equations, the parameters of which are normalized. To simplify further computational procedures, the Laplace transform under zero initial conditions is employed, and as a result of mathematical transformations, a system of algebraic equations has been obtained.

Transfer functions, which establish information links between the torques of the cylinders and the signal of fluctuations in the rotation speed of the first mass, are obtained as the ratio of the determinants of the system of algebraic equations. It is established that increase of the coefficients A1, A2 and A3 reduces the slope of the LAFC channels of torque transmission and significantly reduces the phase delay. Reduction of the coefficient A1 does not affect the LAFC channel transmission torque of the first cylinder. Also, a decrease in the coefficients A1 and A2 leads to a decrease in the gain of the torque transmission channels in the low frequency band and increases it in the mid and high frequency band.

The scheme of computer modeling of movements of the first mass with reception of an output signal in the form of fluctuations of speed of rotation is constructed. The procedure for identifying the parameters of the mathematical model is performed on the basis of comparison with the experimental data. The parameters of the computer simulation scheme are adjusted using the method of adjusting the length of information connections of the neural network technology.

On the basis of frequency representation of a fluctuation signal the information technology of establishment of distribution of cylindrical powers of the diesel generator is developed. By solving a over determined system of algebraic equations using the algorithm to minimize non-binding, we set the amplitude coefficients of the cylinders. By the value of these coefficients, the computer system performs software changes to the settings of fuel and air supply to the cylinders of the power unit.

Computer modeling of the procedure for solving the system of algebraic equations under the conditions of additive random interference on the signal of fluctuations of the rotation speed of the first mass has been carried out. Graphs of variance of calculation of coefficients are obtained, the analysis of which allows to form requirements concerning metrological characteristics of the measuring converter of the information signal.

References

1. Gawande, S.H., Navale, L.G., Nandgaonkar, M.R., Butala, D.S., Kunamalla, S.: cylinder imbalance detection of six cylinder di diesel engine using pressure variation. Int. J. Eng. Sci. Technol. **2**(3), 433–441 (2010)
2. Gawande, S.H., Navale, L.G., Nandgaonkar, M.R., Butala, D.S.: Harmonic frequency analysis of multi-cylinder inline diesel engine genset for detecting imbalance. Int. Rev. Mech. Eng. **3**(6), 782–787 (2009)
3. Enikeev, A.F., Borisenko, A.N., Samsonov, V.P., Kiseleva, G.M.: Diagnosis of a diesel generator by the deviation in shaft speed. Measure. Techn. USSR. **31**(9), 868–871 (1988). https://doi.org/10.1007/BF00863884
4. Pavlenko, I., et al.: Parameter Identification of Cutting Forces in Crankshaft Grinding Using Artificial Neural Networks. Materials (Basel). **13**(23), 5357 (2020). https://doi.org/10.3390/ma13235357
5. Kotliar, A., Basova, Y., Ivanova, M., Gasanov, M., Sazhniev, I.: Technological assurance of machining accuracy of crankshaft. In: Diering, M., Wieczorowski, M., Brown, C.A. (eds.) MANUFACTURING 2019. LNME, pp. 37–51. Springer, Cham (2019). https://doi.org/10.1007/978-3-030-18682-1_4
6. Sivyakov, B.K., Truber, S.S.: Diagnosis of multi-cylinder engines using non-uniformity of speed. Vestnik SGTU **1**(44), 76–82 (2010)
7. Bodnar, B.E., Ochkasov, O.B., Chernyaev, D.V.: Determination of the method of filtering the signal of non-uniformity of the diesel crankshaft speed. Visnyk DNUZT. **1**(43), 113–118 (2013)
8. Trembling, N.M.: Influence of torque non-uniformity on dynamic and power indicators of internal combustion engines of wheeled machines. Uchenye zapiski KIPU. **38**, 18–24 (2013)
9. Savchenko, O.F., Dobrolyubov, I.P.: Modeling the process of identifying the state of tractor engines. Probl. Comput. Appl. Math. **4**(6), 4–12 (2016)
10. Safronov, P.V., Kudinova, A.V.: Estimation of sensitivity of criteria of non-uniformity of frequency of rotation to non-reproducibility of working cycles in the cylinder of the internal combustion engine in the idling mode. In: Proceedings of the International Scientific and Technical Conference. MADI, pp. 336–347 (2019)

11. Grebennikov, S.A., Grebennikov, A.S., Nikitin, A.V.: Adaptive control of fuel supply by ICE according to indices of uneven rotation of the crankshaft. Vestnik SGTU. **2**(72), 80–83 (2013)
12. Bashirov, R.M., Insafuddinov, S.Z., Safin, F.R.: Uneven fuel supply in diesels: problems and methods of their solution. Izvestiya OGAU. **1**(75), 78–82 (2019)
13. Bykov, V.V., Mokrushin, D.A., Petrov, A.I.: Diagnostics of the internal combustion engine using the method of vibroacoustic emission. In: Proceedings of the International Scientific and Technical Conference on Problems and Prospects of Development of the Motor Transport Complex, pp. 526–537 (2019)
14. Glushkov, S.P., Glushkov, S.S., Kochergin, V.I., Lebedev, B.O.: Analysis of dynamic characteristics of torsional-oscillatory systems of ship power plants. Marine Intell. Technol. **2**(40), 59–66 (2018)
15. Bukina, S.V., Sitnikova, T.A.: Dynamic analysis of the machine unit by means of the automated calculation taking into account non-uniformity of rotation of the main shaft of the machine. Technol. Qual. **39**(1), 28–35 (2018)
16. Zakharenkov, D.Y., Shatokhin, V.M.: Analysis of frequency characteristics of the mathematical model of the diesel generator 3TD-1. Bull. DSEA. **1**(45), 115–120 (2019)
17. Anetor, L., et al.: Combustion dynamics at the top dead center position of a spark ignition engine. FME Trans. **45**(4), 548–558 (2017). https://doi.org/10.5937/fmet1704548A
18. Gritsenko, A., Shepelev, V., Zadorozhnaya, E., Almetova, Z., Burzev, A.: The advancement of methods of vibro-acoustic control of ICE gas distribution mechanism. FME Trans. **48**(1), 127–136 (2020). https://doi.org/10.5937/fmet2001127G
19. Yenikieiev O., Yevsiukova F., Prihodko O., Ivanova M., Basova Ye., Gasanov M. Analysis of the frequensy characteristics of the automatic control system of manufacturing prosess parameters. Acta Techn. Naposensis. **62**(111), 473–482 (2019)
20. Yenikieiev O., Shcherbak L. Information technology for protecting diesel-electric station reliable operation. Tekhnichna Elektrodynamika. **4**, 85–91 (2019). https://doi.org/10.15407/techned2019.04.085
21. Chaparro L.F.: Signals and systems using MATLAB. Elsevier Inc. (2011). ISBN 978-0-12-374716-7

Research on Optimizing the Hardening Process of Lamellar Spring Sheets Using the Factorial Experiment Method

Aurel Mihail Țîțu[1](✉) and Alina Bianca Pop[2]

[1] Faculty of Engineering, Industrial Engineering and Management Department, "Lucian Blaga" University of Sibiu, 10 Victoriei Street, 550024 Sibiu, Romania
mihail.titu@ulbsibiu.ro

[2] Faculty of Engineering, Department of Engineering and Technology Management, Technical University of Cluj-Napoca, Northern University Centre of Baia Mare, 62A, Victor Babes Street, 430083 Baia Mare, Maramures, Romania

Abstract. This scientific paper was carried out within an industrial organization with activity in the production and marketing of lamellar spring. The object of the present study is to highlight the way of obtaining certain information, their level of veracity, the degree of interpretation and the final purpose for which these measurements and interpretations are made. The focus will be on a study related to the heat treatments within the company, namely on the hardness of spring sheets obtained after their hardening process. The purpose of this experimental data processing is to optimize the quality of the hardening process, this means obtaining hardness of the quality sheets as close as possible to the nominal technological parameters, and the destination of this data processing is to obtain a more accurate picture of the quality. This process is beneficial both for the company in which the study was conducted and in the case of external audits which within the analyzed company are quite frequent given the fact that one of its most important clients is Ford.

Keywords: Research · Organization · Spring · Hardness · Hardening process · Optimization · Factorial experiment

1 Introduction

The processing of data and the related conclusions that can be drawn are most often based on experimental results obtained from extensive tracking and measurement processes [1]. As any result obtained from physical measurements is affected by random measurement errors, reading the results, formulating the correct conclusions, and taking the most appropriate measurements must be done in conditions that consider the existence of these possible errors, i.e., never one can speak of the existence of a 100% certainty [2].

The object of the present study is precisely the role of highlighting the way of obtaining certain information, their level of veracity, the degree of interpretation and the final purpose for which these measurements and interpretations are made [3].

D. D. Cioboată (Ed.): ICoRSE 2022, LNNS 534, pp. 45–56, 2023.
https://doi.org/10.1007/978-3-031-15944-2_5

Attention will be focused on a study related to heat treatments within an organization, namely the hardness of spring sheets obtained after their hardening process [4].

An essential condition for the correct processing of a series of experimental data (in this case obtained from a hardness measurement process - HB) [5, 6], is that the one who performs this processing to know thoroughly the process from which the data were extracted, as well as all the details regarding the manner and conditions in which the measurements were carried out, the purpose of the experiment, the physical nature of the measured or neglected parameters, the purpose of processing this data and their destination [7, 8].

The purpose of this experimental data processing is to optimize the quality of the hardening process, which means obtaining hardened sheets as close as possible to the nominal technological parameters [9, 10]. The purpose of this data processing is to obtain a more accurate picture of the quality of the process, which is beneficial both for the company in which the study was conducted and in the case of external audits which within the analyzed company are quite frequent given the fact that one of its most important customers is Ford.

The company in which the research was carried out has eight production points (factories): in Austria, Finland, France, Germany, Portugal, Romania, and Slovenia.

The form of ownership of the company is, as it appears from the statute, a joint stock company, traded on both the domestic and international capital market, it is a Romanian legal entity carrying out its activity in accordance with the Romanian legislation in the field.

The object of activity of the company defined by the statute is the production and sale of lamellar springs (trapezoidal and parabolic), with a maximum section of up to 150 × 50 mm square, the current share having the field of light parabolic spring manufacturing.

This company can manufacture parabolic springs and trapezoids with a maximum cross section of up to 150 × 50 mm, the current share being in the field of light parabolic spring manufacturing. The most important customers of the organization are Ford (transit), Dacia (pick-up), Mercedes, Daimler Chrysler, and Iveco (heavy and medium trucks).

The production of the organization consists of a line for parabolic springs, also has among others two rolling mills for parabolic (HILLE, EUMUCO 2), three rolling machines end of sheets (AFS 1, AFS 2, LEEDS), two lines for heat treatments (HEUSER and REYNA 2), an anti-tension blasting installation as well as two automatic installation and control installations (PROBAT 1 and 2), to these are added an automated painting installation (INDIVIDUAL), and a warehouse with a storage capacity of 10000t finished products.

2 SWOT Analysis of the Company

SWOT stands for "Strengths", "Weaknesses", "Opportunities" and "Threats". The first two look at the organization and reflect its situation, and the next two look at the environment and reflect its impact on the organization's activity [11–13].

Below are the four points characteristic of SWOT analysis, grouped after an internal and an external analysis.

2.1 Internal Analysis

Strengths:

- The organization is privatized through the method of foreign investments, this ensuring the necessary funds for investments especially in modern technology and refurbishment.
- Privatization by buying by the German concern, increased concern for raising the quality of products obtained and increasing production capacity, even though own efforts.
- The organization benefits from a modern management.
- The organizational structure of the organization is modern comprising the sectors, sections, and compartments necessary for a modern enterprise.
- The organization has its own research structure.
- Career advancement opportunities.
- Through the modern technologies and machines, it is equipped with, the organization has the possibility to efficiently monitor the quality of the products, this being a constant concern of the top management.
- The team of specialists at all levels have always shown a very good ability to adapt to technical changes along the way.
- Facilities offered by the organization in terms of transportation, bonuses and bonuses, vacation, and treatment.

Weaknesses:

- 100% of the company's shares are owned by the German concern.
- The organization does not benefit from its own transport network.
- Due to the poor quality of Romanian steels, the organization must import them at high prices.
- Lack of an efficient schooling system.
- Technological substitution, standardization, lack of continuity, lack of a group spirit caused by cultural differences, limited market by shared expansion, major goals pursued by the group, investments necessary for the development of factories that have much too little production capacity in the face of demand market.

2.2 External Analysis

Opportunities:

- The organization is the only one with the capacity to meet the demands of demanding customers, and it must also be borne in mind that it is one of the few factories in SE Europe that has undergone a deep modernization and can expand its production network at any time.
- The organization has the possibility to extend the product nomenclature either by new spring landmarks or by modifying existing ones.
- The organization through its potential and the economic potential of the area in which it is located has the possibility of vertical integration and diversification.
- Excellent reputation for Quality, the goal of satisfying a market with ever-increasing demand, very good cooperative relationships with very strong financial customers

Threats:

- Increasing pressure from competition.
- Increasing the qualitative and especially quantitative demands of large customers.
- Fluctuations of the national currency against the single European currency - euro or dollar.
- The lack of an efficient and permanent schooling.
- The policy of reducing or eliminating staff training investments.

Following the identification of strengths, weaknesses, opportunities and threats, it can be concluded that the organization:

- is a profitable organization;
- has a modern management;
- has an optimal organizational structure;
- possesses a modern technical-material base in a proportion of approximately 80%;
- has a good organizational culture;
- has diversified and good quality products;
- has a competitive cost of products;
- it has a stable market and it is possible to expand it;
- made steadily increasing profits;
- has the possibility of vertical integration of staff;
- has the possibility and conditions for product diversification;
- it has the potential to remain among the national and even world market leaders.

In order to maintain and accentuate the development of the society, it is possible to act through:

- continuing the modernization of the technical material base;
- development of the sales sector and direct sales of products to customers;
- development of the nomenclature of finished products to unexplored and unexploited foreign markets.

- production and marketing of raw materials or semi-finished products for other automotive companies;
- training and education of specialists for interested companies;
- performing "service" to other enterprises with identical installations;
- Following the analysis of the conclusions by the top management of the company, it can be done:
- redefining the mission, strategic directions and development objectives (where it is intended to be achieved);
- prioritization of development objectives;
- elaboration of development measures and establishment of indicators to be achieved.

Therefore, SWOT analysis is an effective method, used in the case of strategic planning to identify potential dangers, priorities and to create a common vision for achieving the development strategy. In fact, the SWOT analysis must answer the question "where are we?" and "where we want to go!", which involves analyzing the internal environment of the enterprise and the general and specific external environment.

The SWOT analysis provides an opportunity to identify appropriate measures to eliminate and reduce weaknesses (prioritize them) and to largely eliminate surprise in the event of threats that may arise along the way.

3 Research Methodology

To study the obtaining of an optimal hardness of the hardened spring sheets (i.e., framing within the limits:

$V_{min} = 454\ HB$
$V_{nom} = 474.5\ HB$
$V_{max} = 495\ HB$

the aim was to obtain an experimental model.

Following the primary analyzes performed on the hardening process, it was decided to model by ordinal factorial experiment EFC 2^3 [14].

The three factors that have a direct influence on the objective function Y (hardness of the sheets after the hardening process) [HB], which are to be analyzed are:

- X_1 - speed of the supply belts of the hardening frames n [rot/min].
- X_2 - oil temperature in tempering basins g [°C].
- X_3 - the number of sheets placed on each frame used in the hardening process [N-no. of sheets].

The equation of the first order model to be explained after processing the measured data has the form:

$$Y = b_0 + \sum_{j=1}^{3} b_j \cdot x \cdot X_j + \sum_{\substack{u,j=1,\\ u \neq 1}}^{3} b_{gu} \cdot x \cdot X_j \cdot x \cdot X_u + \sum_{\substack{u,t,j=1,\\ u \neq t \neq 1}}^{3} b_{jut} \cdot x \cdot X_j \cdot x \cdot X_u \cdot x \cdot X_t \tag{1}$$

By formulating the above equation, we want to estimate all the effects caused by the interactions between the three factors on the objective function (in our case the hardness).

4 Designing the Experiment and Making the Measurements

Based on the information gathered during August 2021, as well as the previous experiments that were performed, the coordinates of the midpoint of the experiment and the intervals of variation of the three influencing factors were chosen, resulting in the values of the upper and lower level. for each influencing factor shown in Table 1.

Table 1. The coordinates of the midpoint and the intervals of variation of the factors

Parameter	Coded value	Physical value		
		$X_1 <=> n$ [rot/min]	$X_2 <=> g$ [°C]	$X_3 <=> c$ [N-no. of sheets]
Central point X_{j0}	0	17.5	32	2
Variation range,D_J	Δ_J	10	27.5	1.5
Upper level, X_{isup}	+1	30	65	4
Lower level,X_{iinf}	−1	10	10	1

Table 2. The program matrix of the factorial experiment EFC 2^3 (without the columns corresponding to the interactions)

Trial number	Average X_0	Levels of influencing factors			Y_1[HB]	Y_2[HB]
		X_1	X_2	X_3		
1	+1	−1	−1	−1	466	474.5
2	+1	+1	−1	−1	469	474.5
3	+1	−1	+1	−1	471	474.5
4	+1	+1	−1	−1	472.5	474.5
5	+1	−1	−1	+1	473	474.5
6	+1	+1	−1	+1	485	474.5
7	+1	−1	+1	+1	473.5	474.5
8	+1	+1	+1	+1	478	474.5

Table 3. The values of the effects produced by the influencing factors and their interaction

Effect	Value	Effect	Value	Effect	Value	Effect	Value
Er	2.2	E2	0.538	E12	−0.567	E23	−2.312
E_i	0.788	E3	0.687	E13	0.313	E123	−0.713

Objective function Y_1 it is the function resulting from the measurements and the function is objective Y_2 s the function that should be obtained after optimizing the hardening process (Tables 2 and 3).

5 Calculation of the Coefficients of the Regression Polynomial

The amplitude of the effects produced by the influencing factors ($X_1; X_2; X_3$) on objective functions ($Y_1; Y_2$) can be easily appreciated based on the graphical representations below. Figures 1, 2 and 3 show the histograms related to the three factors analyzed and the graphs presented in Figs. 4, 5 and 6 show the effects produced by the three factors.

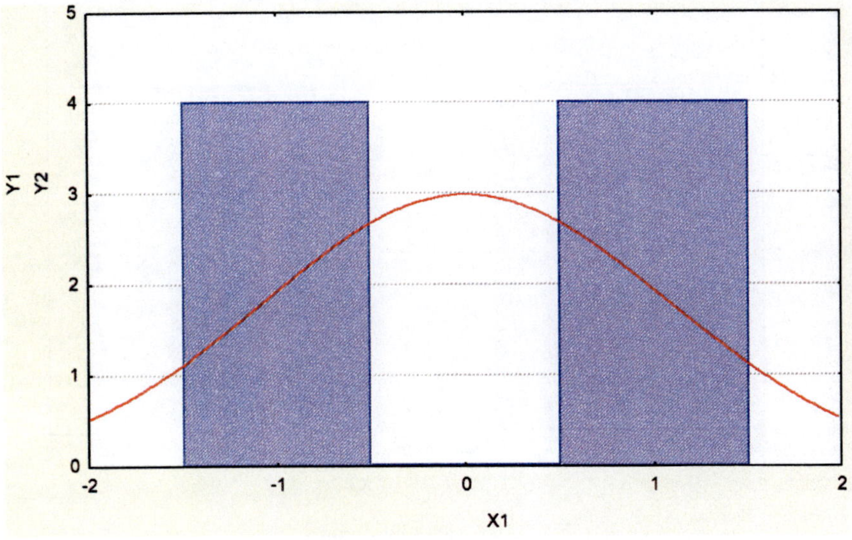

Fig. 1. X_1 factor histogram.

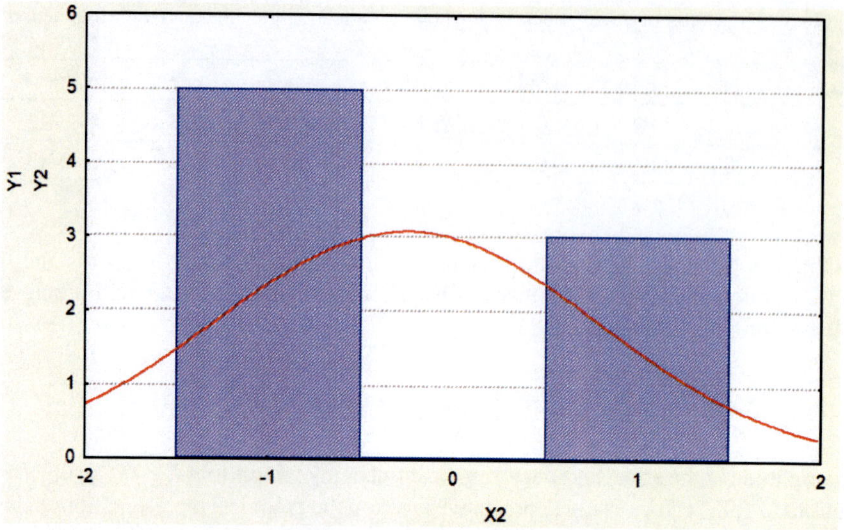

Fig. 2. X_2 factor histogram.

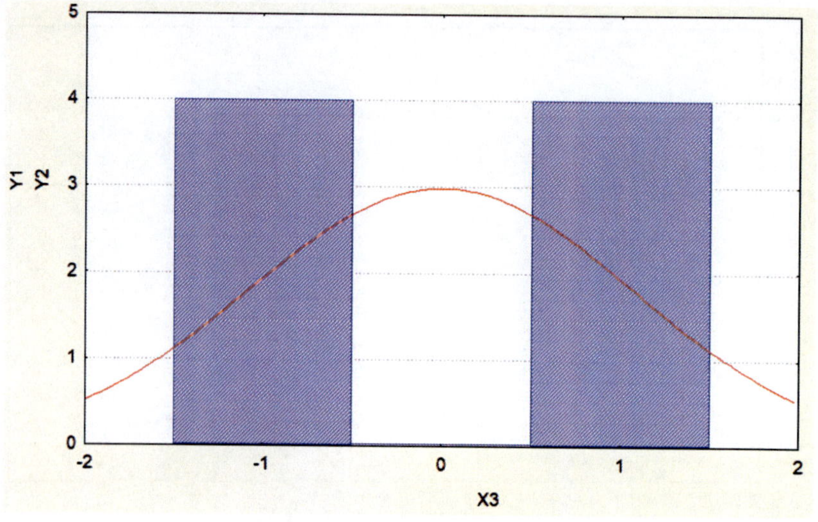

Fig. 3. X_3 factor histogram.

From the graphs above it can be seen that most of the influences on the objective function Y_1 and respectively Y_2 it has the factor X_2 (toil temperature in hardening basins), followed by factor X_1 (feed belt speed), as in the case of the analyzed number of sheets concentrated on each hardening frame remains constant, it can be seen that the graphs referring to the factor X_3 remain constant and in our case the objective function (obtaining an op-time hardness), is influenced only by the factors X_1 and X_2.

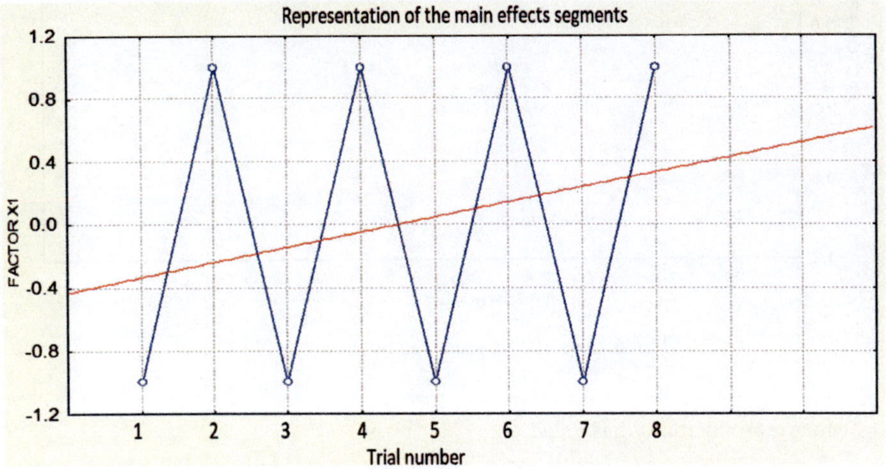

Fig. 4. The effects produced by the factor X_1.

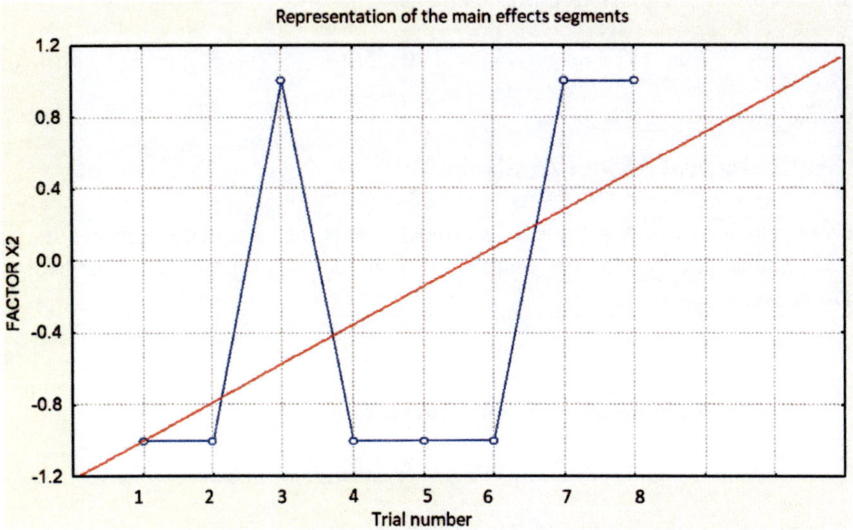

Fig. 5. The effects produced by the factor X_2.

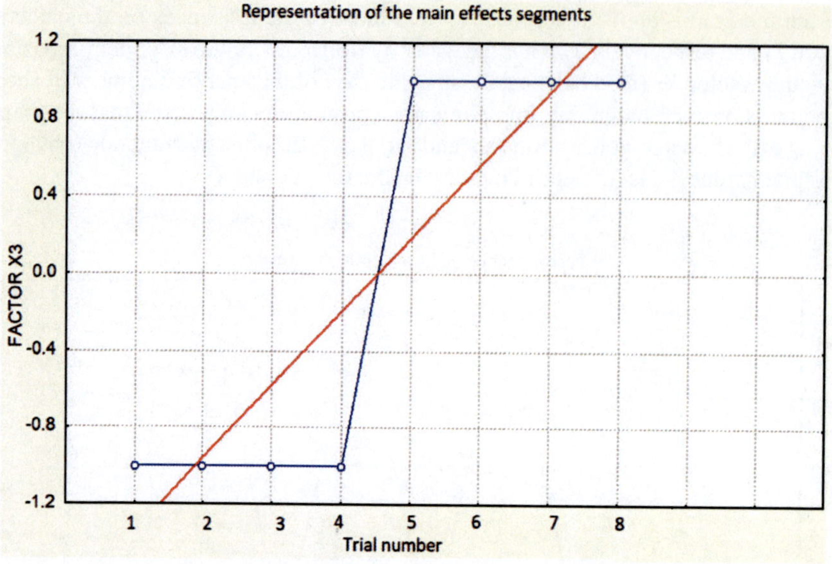

Fig. 6. The effects produced by the factor X_3.

Scatter reproducibility has value:

$$S_0^2 = \frac{\sum_{V=1}^{c} (y_{0v} - y_0)^2}{c} = 0.01 \quad (2)$$

It turns out that the dispersions of the coefficients are:

$$S_{bj} = 0.00125 \quad (3)$$

6 Mathematical Model Analysis

Analyzing the significance of the regression coefficients, the value of the Student criterion for the significance threshold is established $\alpha = 0.05$ and a number of degrees of freedom:

$$v = Q \cdot (c - 1) = 2 \cdot (2 - 1) = 2 \quad (4)$$

So for all the 8 coefficients the comparison is made:

$$|b_j| \geq t_{\alpha,v} \cdot S_{bj} = 8.14 \cdot \sqrt{0.00125} = 0.280 \quad (5)$$

It turns out that the coefficients are statistically significant: b_{12}, b_{13}, b_{123}, therefore the form of the ordinal I model will be the following:

$$\overline{Y} = 2.2 + 0.280 \cdot X_1 + 0.538 \cdot X_2 + 0.685 \cdot X_3 - 2.312 \cdot X_2 \cdot X_3 \quad (6)$$

For the calculation of the dispersion, the values of the objective function were estimated. Because the number of significant coefficients of the regression equation is $k' = 5$, it results:

$$S_{conc}^2 = \frac{\sum_{I=1}^{N} (y' - y_i)^2}{N - k'} = \frac{1.599}{8 - 5} = 0.533 \tag{7}$$

The calculated value of criterion F will be:

$$F_{calc} = \frac{S_{conc}^2}{S_0^2} = \frac{0.533}{0.01} = 53.3 \tag{8}$$

For $\alpha = 0.05$, $v_1 = 3$ and $v_2 = 1$ value $F_{tab} = F_{\alpha,v_1,v_2} = 474.3$, so $F_{calc} < F_{tab}$ resulting in the conclusion that the experimental model describes approximately exactly the results of the measurements.

7 Results Interpretation

When we had constant values of the factors with direct influence on the objective function, the variations of the parameters were very small. It is worth noting that one of the objective functions (obtaining a hardness as close as possible to the nominal prescribed by the technical documentation), could be fulfilled if one of the factors could be kept under control.

The X_2 (temperature of the oil in the hardening basins), is the most important parameter, and in the second place is X_1 (feed belt speed). If these factors are controlled and if the statistical calculation model is followed, then it can be said that the hardening process to obtain an optimal hardness should be unsuccessful.

8 Conclusions

By using this calculation method, it results that approximately 47.43% of the imparting of the objective function can be explained with the help of the 1st order regression model applied above.

Following the analytical processing of the experimental results as well as because of the analysis of the graphical representations, the following conclusions can be formulated regarding the modeling by factorial experiment of ordinal 1 of the hardness of the spring sheets after the hardening process.:

- In the field of the multifactorial space investigated around the central point of the experiment, the real response hardness can be estimated with an accuracy of only 47.43% using the explicit working model.
- The effects generated by all the influencing factors are the interactions of the 2nd order of these found significant, except for the interaction X_2, X_3, they are positive so the increase of the values of the influencing factors determines the increase of the values of the objective functions.

- The most important factors influencing the objective function are the temperature of the oil in the hardening basins and the supply speed of the conveyor belt, as well as the interaction between the two factors.

As a direction of further research, the study could be continued by moving to the optimum value (considered to be the maximum value) of the response surface by the method of ascending in the direction of the maximum slope.

References

1. Hazir, E., Erdinler, E.S., Koc, K.H.: Optimization of CNC cutting parameters using design of experiment (DOE) and desirability function. J. Forest. Res. **29**(5), 1423–1434 (2017). https://doi.org/10.1007/s11676-017-0555-8
2. KhairiFaiz, M., et al.: Effect of sand mold constraint on warpage deformation of lamellar graphite gray cast iron and prediction by elastoplastic-creep finite element analysis. J. Mater. Eng. Perform. **30**(6), 4669–4680 (2021). https://doi.org/10.1007/s11665-021-05716-1
3. Cheerova, M.N., Komarova, T.V., Dubinskii, V.N.: Effect of initial structure on the characteristics of austenite formed under heat treatment of spring steels. Met. Sci. Heat Treat. **63**(1–2), 11–17 (2021). https://doi.org/10.1007/s11041-021-00640-6
4. Derouiche, K., Garois, S., Champaney, V., Daoud, M., Traidi, K., Chinesta, F.: Data-driven modeling for multiphysics parametrized problems-application to induction hardening process. Metals. **11**, 738 (2021)
5. Galetto, M., Genta, G., Maculotti, G., Verna, E.: Defect probability estimation for hardness-optimised parts by selective laser melting. Int. J. Precis. Eng. Manuf. **21**(9), 1739–1753 (2020). https://doi.org/10.1007/s12541-020-00381-1
6. Matyunin, V.M., et al.: Correlation between the ultimate tensile strength and the Brinell hardness of ferrous and nonferrous structural materials. Russ. Metall. **2021**, 1719–1724 (2021)
7. Ghimisi, S.: Study Considerations on the Fretting Phenomenon for Lamellar Springs. Fiability & Durability/Fiabilitate Si Durabilitate, pp. 18–22 (2012)
8. Koike, K., Clarke, K.D., Clarke, A.J.: Microstructural evolution and mechanical properties of heavily cold-rolled and subsequently annealed Cu-3 wt.%Ti alloys with nano-lamellar structure. JOM. **71**, 4789–4798 (2019)
9. Özdemir, M., Dilipak, H., Bostan, B.: Experimental investigation of deformation and spring-back and spring-go amounts of 1.5415 (16MO$_3$) sheet material. Metallogr. Microstruct. Anal. **9**(6), 796–806 (2020). https://doi.org/10.1007/s13632-020-00687-6
10. Jiang, Y., et al.: Influence of multiphase on the strain hardening behavior of 60Si2CrVAT spring steel treated by a Q-P–T process. J. Mater. Sci. **53**, 10396–10410 (2018)
11. Teoli D., Sanvictores. T.: An J. SWOT Analysis. In: StatPearls. StatPearls Publishing, Treasure Island (FL) (2021)
12. Vlados, C.: On a correlative and evolutionary SWOT analysis. J. Strateg. Manag. **12**(3), 347–363 (2019)
13. Benzaghta, M.A., Elwalda, A., Mousa, M.M., Erkan, I., Rahman, M.: SWOT analysis applications: an integrative literature review. J. Global Bus. Insights **6**(1), 55–73 (2021)
14. FeujofackKemda, B.V., Barka, N., Jahazi, M., Osmani, D.: Optimization of resistance spot welding process applied to A36 mild steel and hot dipped galvanized steel based on hardness and nugget geometry. Int. J. Adv. Manuf. Technol. **106**(5–6), 2477–2491 (2019). https://doi.org/10.1007/s00170-019-04707-w

Natural Vibrations of a Turbine Blade During Milling

Sergey Dobrotvorskiy[1], Yevheniia Basova[1](✉), Vitalii Yepifanov[1], Valerii Letiuk[2], Ludmila Dobrovolska[1], and Oleksandr Shelkovyi[1]

[1] National Technical University "Kharkiv Polytechnic Institute", 2 Kyrpychova Street, Kharkiv 61002, Ukraine
Yevheniia.Basova@khpi.edu.ua

[2] JSC "Ukrainian Energy Machines, 199 Moskovsky Avenue, Kharkiv 61037, Ukraine

Abstract. The main problems in processing the surfaces of turbine blades are deformation and oscillatory processes directly during their processing. The same processes further determine both the product life and the noise intensity generated by devices using turbines, including aircraft engines. The aim of the work was to study the regularities of changes in the frequency of natural vibrations of a turbine blade throughout the entire technological cycle from design, manufacture of a workpiece, milling and fixing in the turbine disk. The article developed an approach to the analysis of this problem based on digital models. Computer modeling and determination of natural frequencies of turbine blade vibrations are consistently carried out for a stamped blank, during its milling with fixing in fixtures along the base and apex, as well as in the dovetail of the turbine disk lock. The distribution of stresses and deflections in milling conditions is determined. It is concluded that the natural vibrations of the blade airfoil during milling shift to the region of lower frequencies when the machining allowance is removed. However, for the first two lower vibration modes, technological heredity is preserved. Therefore, at the stage of designing the technological process of processing, it is necessary to determine the natural vibration frequencies and maximum deviations of the blade airfoil in order to avoid the occurrence of resonance vibrations when the processing conditions change and to carry out processing within the tolerance.

Keywords: Turbine blade · Free mode · Mechanical processing · Deviation · Aircraft noise

1 Introduction

The existing programs "Europe-Horizon" in the development of green energy and transport provide, in addition to reducing emissions of carbon dioxide, nitrogen oxides and particulate matter to zero, also the development of programs to reduce aircraft noise [1]. The solution to the problems is planned to be obtained at the expense of the energy of hydrogen, wind, and water. Almost all of the listed energy conversion processes use various turbines and devices containing blades. These blades work in a wide variety of conditions. Aircraft engines that use hydrogen as fuel are of particular interest. In this

D. D. Cioboată (Ed.): ICoRSE 2022, LNNS 534, pp. 57–69, 2023.
https://doi.org/10.1007/978-3-031-15944-2_6

case, the emissions of water and steam create new problems in the atmosphere, and the turbine from a truly gas-dynamic one turns into a hybrid or steam one. Consequently, traditional technologies for the manufacture of steam turbine blades must be considered from the point of view of their use in aircraft engines. Moreover, since the noise of aircraft engines is closely related to the frequency and amplitude of natural oscillations of the turbine blades, scientific interest is formed, in particular, from consideration of the processing conditions of the blades and their influence on the subsequent number of blades. noise. The aforementioned problems give rise to increased interest in the problems of manufacturing turbine blades. The most important step in blade technology is milling, as this removes the maximum layer of stock. However, the determination of the spectrum of natural vibrations of products is carried out only after polishing and often varies greatly, despite the seemingly identical processing conditions. Therefore, it is relevant to consider the peculiarities of the distribution of the natural frequencies of the blade vibrations during the milling operation and the arising deviations of the console from the cutting forces in order to maintain the processing accuracy, as well as a preliminary assessment. the influence of the technology of manufacturing the blade apparatus on the noise level.

An analysis of modern studies of fluctuations shows that the main efforts are directed in two directions. The first is the forced vibrations of the machine part system, which are characteristic of an intermittent milling process. Secondly, there are various types of blade oscillations under conditions of gas-dynamic flow and rotation. Very little is paid to the process of forming and changing the spectrum of natural vibrations of blades at various successive stages of manufacturing technology. Thus, the aim of this study is to fill this research gap.

2 Literature Review

In predicting accuracy and stability during parts' machining, accurate knowledge and information about the dynamics of the machine-tool-part is very important. It was an article [2] performed a Modal Frequency Study of Carbon Steel and Cast-Iron End Mills Using Impact Excitation in order to potentially assess the risk of resonance phenomena during machining since vibrations affect the quality of the workpiece. The experiments were conducted with a carbon steel and cast-iron end mill and workpiece to study modal frequencies using a frequency response function (FRF). Tool, part and cutting tool modeling were investigated using FEA.

As is well known the dynamics of a part is an important factor when planning a machining strategy [3]. On the basis of the analysis which revealed that, typically, the structural dynamic parameters of a part are obtained by experimental modal analysis (EMA) [4]. However, the dynamics of thin-walled workpieces changes due to material removal and tool-to-workpiece adhesion. Online modal analysis (OMA) provides a way to evaluate structural dynamic parameters during operation, but the input excitation of the milling system is mainly periodic milling force, which violates the OMA premise. In article [4] it was requested to the modal identification method only for the derivation of the dynamic parameters of a thin-walled workpiece. The milling force has been analyzed by authors [4] and the analysis results were shown that the milling force contains white noise for OMA. In research [5] it was developed a dynamic model to study the stability of the thin floor milling process by comprehensively considering the flexibility of both static bending defects and dynamic reactions. In paper [6] the authors studied deformations and vibrations arising from side milling of thin-walled parts. The explores deformation and vibration caused by force have been considered.

The examines the effect of radial engagement and milling direction on the stability of thin-walled milling is presented in the paper [7]. The dynamics of the processes of milling thin-walled parts, in particular of face milling operations, in this research is described by a matrix of directivity coefficients, which is compressed to a single coefficient, the average value of which unambiguously depends on the engagement limits and the ratio of the radial-tangential cutting force [7].

The modified Nyquist method to study the effects of spindle speed, depth of cut and structural damping factor on vibration frequencies was used in the works [8, 9]. It was important to note, however, that for more convenient and accurate control of the milling process [10], and adaptable single-point control method based on one acceleration sensor can be used. For increasing resistance to vibration during end milling the development of a holder with damping of a limited layer has been devoted paper [11]. These authors found the dynamic stiffness is proportional to the damping coefficient and static stiffness of the holder, while the natural frequency is proportional to the specific stiffness of the tool materials [11]. On the basis of analysis of the literature researches, it was observed that composite damping materials have been used in the tool-holder clamping industry to suppress vibration [12, 13]. The authors [14] claim that composite materials, widely used in CNC machine beds, have a huge impact on the efficiency and accuracy of a part's processing. In addition, has been noted that the optimization of the parameters of the processing of thin-walled parts is generally carried out on the basis of finite element modeling of oscillations of the workpiece [3, 15]. In paper [15] it was noted that in practice, it is necessary to take into account the change in natural frequencies due to the removal of the allowance and change in the position of the cutter during processing thin-walled parts. Currently, further progress in high-speed machining efficiency is mainly limited by vibrations of the machine and part system, which limits productivity, accuracy, and part quality. The problem becomes critical during finishing. thin-walled parts with tightly limited accuracy. In thin-walled milling, the proposed method of stability petals cannot be applied directly due to dynamic changes during processing [15, 16]. The natural frequencies of the workpiece change in the process of stock removal, which reduces the weight and rigidity of the workpiece. The solution to this problem in research [15] was proposed to study the processing of a part in small zones. In work [16] authors have addressed the prediction of stability boundaries during milling taking into account changes in dynamic parameters and specific coefficients of cutting force. In research [17] the issues address the gap in the variety of the thin-walled parts computer-aided machining parameters calculation solutions that have been addressed. The research [18] touched upon issues related to the pre-scheduling of the feedrate before the start of motion control, which has no constraints on the number of analyzed blocks and the scheduling execution. In this context, it is also important to indicate that the topic of vibration and noise in aviation is constantly in the spotlight [19]. Moreover, since the noise of aircraft engines is closely related to the frequency and amplitude of natural vibrations of turbine blades, scientific interest is formed, among other things, from consideration of the conditions for machining the blades and their influence on the subsequent magnitude of aircraft engine noise. Most significantly, the noise has recently gained relevance due to the new stringent ICAO requirements for noise and pollution at the airport at the level of Chapter 4 requirements according to 7 EPNdB [19, 20]. The most comprehensive overview of reliable aviation research is currently presented in work [21]. In particular, emphasis was placed on aviation applications for noise reduction in aircraft and aircraft propulsion systems. It was noted that the most prominent sources of noise in modern aircraft are associated with aircraft engines.

A generalized European view of aircraft noise is detailed in the source [22]. It should be noted that for a turbofan engine with a by-pass degree of 3 and higher, the turbine noise becomes significant, and research is needed to suppress it.

There are theoretical and empirical models for estimating turbine noise, taking into account various correlation parameters. However, the results of theoretical results of experimental studies show that one of the proposed methods is imperfect and further experimental and analytical studies are needed to study the correlation parameters, which are more related to the physical methods of generation, propagation and emission of noise. Developing a robust model that can predict turbine noise during the design phase is still a valid test.

Analysis of the state of affairs has shown that the role of forced vibrations of the blades in the milling process is being intensively studied. At the same time, attention is paid to reducing the vibrations of the machine bed, holder, tool, forced vibrations in the cutting zone. The relevance of the role of natural and forced oscillations of blades in the composition of turbines is obvious. However, the question of the influence of successive stages of processing, including milling, on the amplitude of natural vibrations of the blades requires further study. The purpose of this article is to fill this gap to some extent.

3 Research Methodology

As well know, the noise range is in the frequency range from 0 to 20,000 Hz. Oscillations in this range can be roughly divided into two groups. Oscillations associated with the rotational frequencies of the rotors are characterized by a frequency range 70… 200 Hz, and blade frequencies 200… 20.0 kHz.

The need to calculate the natural frequencies and the corresponding vibration modes arises when analyzing the dynamic behavior of the blades under the action of variable loads. The most typical situation is when at the design stage it is necessary to check the likelihood of product resonance under operating conditions. Resonance phenomena are observed at frequencies close to the natural frequencies of the blades. Therefore, if during the design of the product it is possible to assess the spectrum of natural frequencies of the blades, then it is possible with a significant degree of probability to predict the risk of resonance phenomena in the known frequency range of external influences. Therefore, the research methodology includes the stages of design, modeling and experimental research. At the first stage, digital 3D models of the blade workpiece and the blade model after the milling process using the SolidWorks computer program were created. Curved blade airfoil surfaces were described using NURBS. At the next stage, the study of the natural oscillation frequencies of the workpiece and blade after milling using the CAE SolidWorks Simulation software and the CAE ANSYS MODAL program with the finite element method was carried out. Duplication of studies, on the one hand, made it possible to exclude random errors in the results, and on the other hand, to expand the possibilities of research. A further step, the study of the natural oscillation frequencies of the processed blade when it is fixed in the turbine disk, a herringbone dovetail lock in order to determine the natural frequencies responsible for creating noise was carried out. The next phase was the modeling of blade deformation during processing, which had been implemented in the CAE ANSYS system using the finite element method. At the final phase, the milling of the blade on a CNC machine, with regard to results-based, was carried out. Thus, the presented study covers the full end-to-end cycle of "design-manufacture" of a blade using its digital model at all phases.

Online calculator Kennametal [23] was used to calculate the tangential cutting force, moment and power.

For research, a blade made of stainless steel AISI 304/X5CrNi18-10 (DIN 1/4301) was chosen. Blade length - 625 mm. Workpiece type - blanking. The allowance for the blade airfoil processing was 3 mm.

Basic material properties: elstic moduls – 2e+11 N/m^2, poisson's ratio – 0.28 N/A, shear modulus – 7.9e+10 N/m^2, mass density – 7900 kg/m^3, tensile strength – 600 000 000 N/m^2, yield strength – 400 000 000 N/m^2, hardness – 170 MPa.

4 Results

The study of the distribution of natural oscillation frequency was carried out in the range of up to 80 modes, which corresponds to the natural oscillation frequency of $2.05 \times 10^{+4}$ Hz. However, the most dangerous frequencies are in the low-frequency region and therefore were considered in more detail. The allocation of natural frequencies for the workpiece is shown in Table 1 and Fig. 1. It was found that the most dangerous are modes 1, 2, 4 and 10. Natural frequencies 585.77, 1202.1 and 1685.2 are characteristic of lateral modes in the directions of the X and Y axes. Based on the research results the mode of frequency 10 is dangerous along the Z axis.

Table 1. Distribution of workpiece's natural frequencies

Mode number	Frequency (Hertz)	X direction	Y direction	Z direction
1	585.77	0.13911	0.19454	4.7461e−06
2	1202.1	0.10385	0.099792	3.4503e−05
4	1685.6	0.10042	0.052854	9.6407e−05
10	4880.2	9.1375e−05	1.8388e−06	0.44696

Beginning with a frequency of 5000 Hz, the vibration amplitudes along all axes become more equiprobable. This was confirmed by the distribution of the value of the cumulative vibration mass of the workpiece (Fig. 2). The construction design of the blade provides for fixing it in the turbine disk with one end, and the other in the shroud ring.

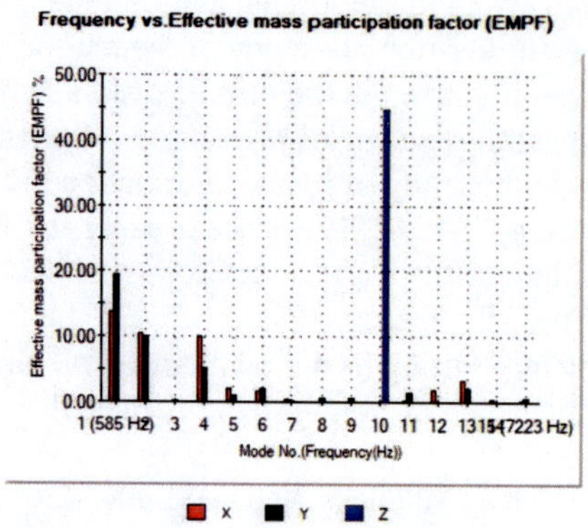

Fig. 1. Distribution of the effective mass depending on the mode of the workpiece

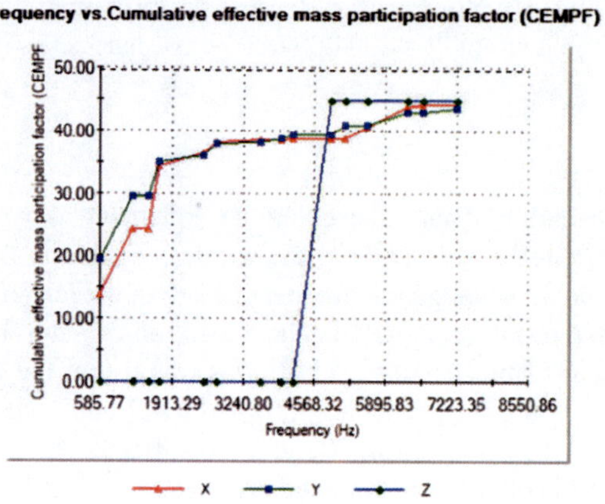

Fig. 2. Distribution of the cumulative effective mass for the workpiece depending on natural frequency of vibration

Therefore, in simulating the processing of a blade, the workpiece was fixed at both ends in accordance with the processing scheme in the metal-cutting machine tool (Fig. 3). Here, the shape and distribution of the amplitude of natural frequencies along the workpiece aerofoil was also presented. The maximum local deviations are 0.72 mm at 585.77 Hz, 1.078 mm at 1202.1 Hz, 1.126 at 1685.6 Hz and 0.54 mm at 4880.2 Hz. After removing the 3 mm allowance from the workpiece, the following results were obtained. The 1, 2 and 4 modes of lateral oscillation have been preserved. However, the natural frequency of the first mode decreased by 23%, the second - by 19%, and the fourth - by 12% (Table 2).

Table 2. The distribution of the detail's natural frequencies after removing the allowance of 3 mm

Mode number	Frequency (Hertz)	X direction	Y direction	Z direction
1	475.27	0.12032	0.20281	1.3239e−06
2	1010.2	0.099691	0.087818	8.4636e−05
4	1493.6	0.10434	0.052689	7.9674e−05
10	3873.5	0.002094	0.012815	8.4446e−05
13	4859.9	4.0549e−05	0.00029477	0.40914

Oscillations in the longitudinal direction Z were replaced from mode 10 to mode 13 (Fig. 4), but the natural frequency of oscillations decreased by only 0.4%.

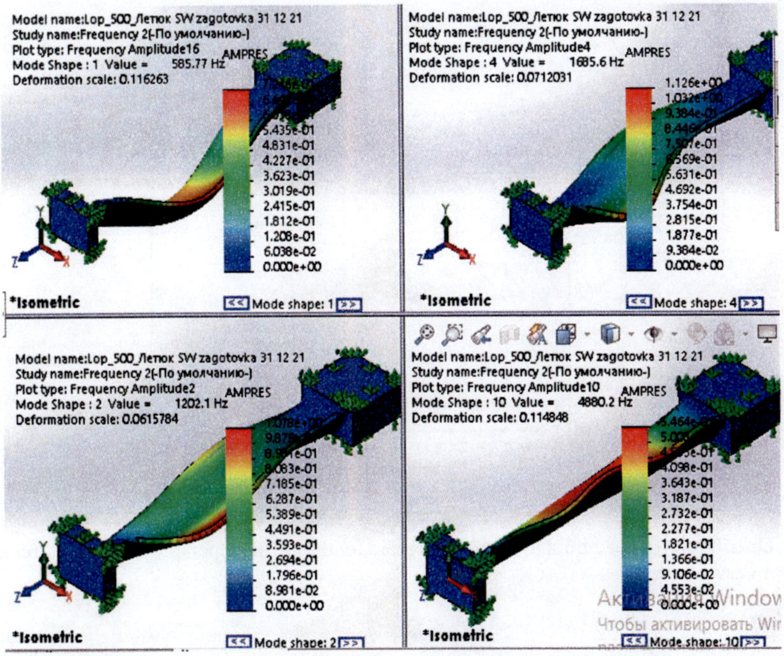

Fig. 3. The shape and distribution of the amplitude of natural frequencies along the workpiece airfoil

This can be explained by the fact that the reduction in the blade stiffness when removing the allowance occurs in the transverse direction, i.e. in the XY plane. Longitudinal stiffness changes insignificantly. This is also confirmed by changes in the distribution of the cumulative effective mass (Fig. 5).

In Fig. 6 was shown a diagram of fastening a blade to a herringbone lock and side flat surfaces in accordance with the processing scheme in the machine-tool fixture. It is important to point out that the shape of the distribution of natural frequencies of the blade airfoil after the end of milling processing is similar to the distribution for the workpiece. However, the amplitude of the natural frequencies increases significantly.

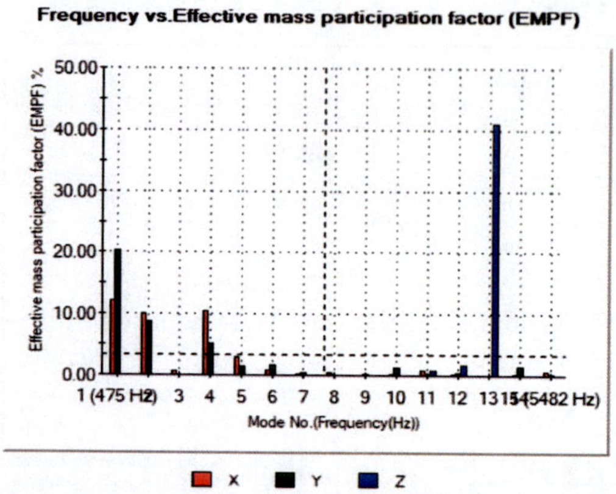

Fig. 4. Distribution of the effective mass depending on the mode of the airfoil

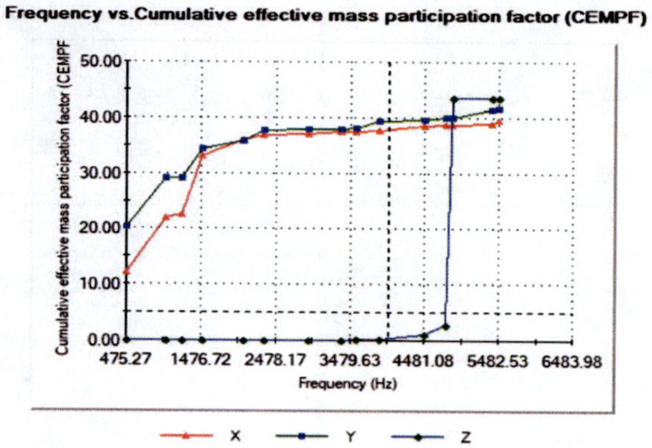

Fig. 5. Distribution of the cumulative effective mass for the airfoil depending on natural frequency of vibration

So, the maximum amplitude of the first mode increases from 0.72 mm to 0.91 mm (26%), and the second mode increases from 1.078 mm to 1.64 mm (52%). This may be an indication that considering only the maximum amplitudes is mandatory in the analysis of machining operation, but not entirely sufficient. It is necessary to coordinate the tool-path in the real form of the distribution of natural frequencies along the blade of the airfoil. For this purpose, an analysis of the stress-strain state of the blade, simulating the loading process during milling, was performed. Since we were interested in precisely the maximum possible deviations, to simplify the calculations, a static load distributed along three axes with a predominance of the tangential load was applied to the blade airfoil (Fig. 7). As a result of modeling, the area of the greatest deformations of the blade airfoil was determined (Fig. 8). From the obtained simulation results, it becomes obvious that the shape of the distribution of the blade natural frequencies oscillations (Fig. 6) is correlated with the shape of the distribution of deformations (Fig. 8). It also became apparent that the processing conditions in this local area needed to be changed. To consider this from the standpoint of natural frequencies, the modal analysis of the workpiece was carried out and the distribution of the isolines of the workpiece deformations along the airfoil blade workpiece, depending on the natural frequency before the start of processing, was obtained (Fig. 9).

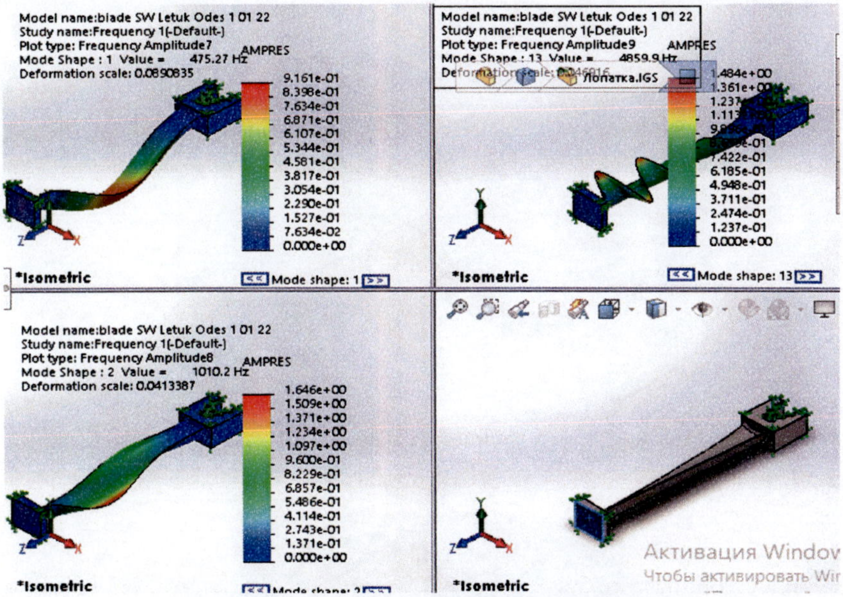

Fig. 6. Distribution of the amplitude and mode of the blade airfoil natural frequency

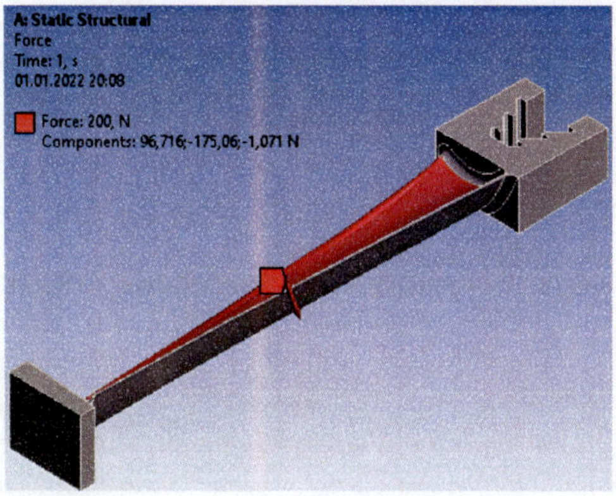

Fig. 7. Local loading of the blade airfoil

Isolines clearly delineate the hazardous and adjacent areas. If we take into account the fact that the change in the natural vibration frequency of the first and second modes during the transition from the workpiece to the blade changes by an average of 20%, then this dangerous local area should also be increased by 20%.

The experimental processing of the blade on a multi-axis CNC machine was performed (Fig. 10).

The following processing parameters were used: Ø 32 mm trident cutter with 12 mm replaceable inserts, spindle speed 1250 rpm, cutting depth 3 mm, cutter attack angle 20°.

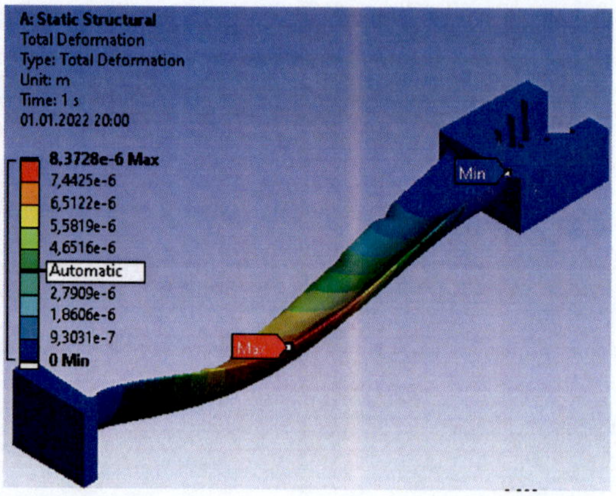

Fig. 8. Type of total blade deformation in the first natural frequency mode

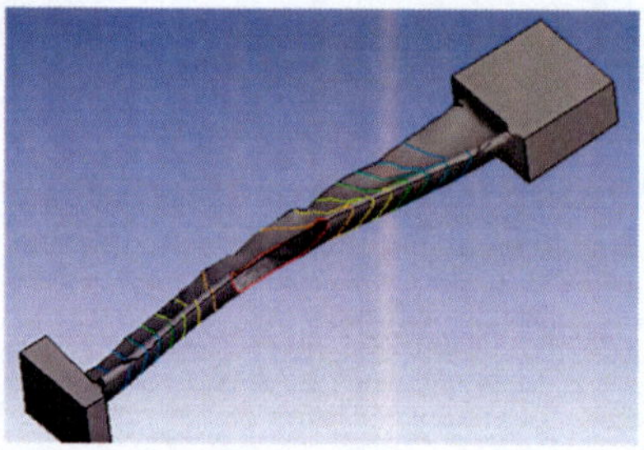

Fig. 9. Distribution of isolines of oscillation natural frequencies of the workpiece (directional deformation - Y Axis, frequency - 590.32 Hz)

In non-hazardous areas, the feed per minute corresponded to 1200 mm/min. In dangerous areas, the feed per minute was changed programmatically. Since the surface was described with Nurbs, justification for retaining the smoothing machining the S-Shape feedrate scheduling method was used [18]. As a result of this approach, the entire blade surface had a uniform roughness of 0.1 mm while maintaining accuracy.

The distribution of natural frequencies of the blade in the turbine disk in working condition is shown in Fig. 11. It should be noted that the error in finding natural frequencies in the SolidWorks and ANSYS programs does not exceed 1%.

Fig. 10. Experimental blade treatment

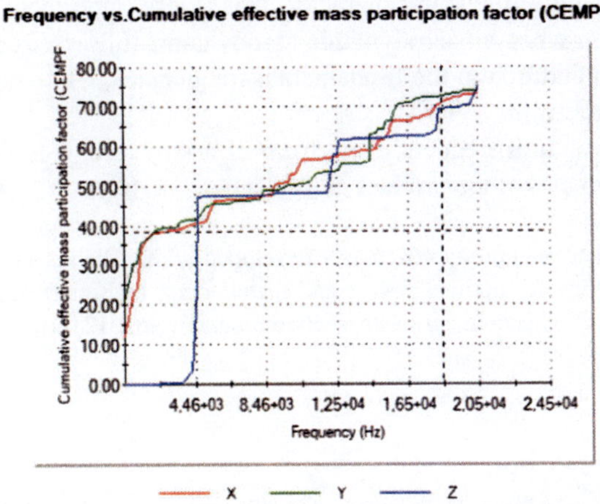

Fig. 11. The distribution of natural frequencies of the blade in the turbine disk in working condition

In the range of more than 5000 Hz, the presence of natural high-frequency oscillations was found. These frequencies will undoubtedly also contribute to turbine noise and are of interest in the high-speed milling field.

5 Conclusions

A detailed study of the change in the distribution of natural frequencies of parts during machining allows to take a new look at the machining process itself, to significantly expand the understanding of the problems that arise and how to eliminate them. The distribution of natural frequencies over the surface of the blade has extrema that correlate with the distribution of deformations. Thus, in contrast to the existing processing strategies, a new strategy based on the processing of local dangerous areas of the blade airfoil using isolines of equal strain and natural frequencies has been proposed.

A technique for computer prediction of the behavior of natural frequencies for both the machining area and the appearance of the noise spectrum is proposed. This approach makes it possible to find dangerous areas without experimental studies with the imposition of forced vibrations and makes it possible to make adjustments to the blade model at the design stage.

It has been established that for the 1st, 2nd and 4th blade modes, the frequency technological heredity will be preserved. These modes carry the most important information in successive stages of processing. To avoid the possibility of resonances, it is necessary that the main part of the low natural frequencies of the structure does not lie in the frequency range of external influences.

However, the traditional method of shifting the frequency of forced oscillations of the tool relative to the natural oscillation frequency by a constant value may be ineffective, since the real distribution of natural frequencies over the surface of the blade airfoil has a complex and non-single-frequency character. This is also important for the formation of the noise spectrum when an aerodynamic aerodynamical flows around it in various areas. As a result, in addition to the fundamental frequencies, white noise components may appear in the spectrum.

To optimize the natural frequency spectrum of a structure, it is first necessary to evaluate these frequencies at the product design stage.

Acknowledgments. The general approach was developed within the research project "Development of a methodology for optimal design and manufacture highly efficient, highly reliable turbomachines, taking into account various operating modes" (No. 0121U107511).

References

1. European Commission. https://ec.europa.eu/info/funding-tenders_en. Accessed 07 Jan 2022
2. Patil, R.A., Amarapureb, S.S.: Investigation of end mill tool modal frequency with carbon steel and cast iron work material using impact excitation. Mater. Today Proc. **52**(3), 1249–1254 (2022). https://doi.org/10.1016/j.matpr.2021.11.047
3. Dobrotvorskiy, S., Basova, Y., Kononenko, S., Dobrovolska, L., Ivanova, M.: Numerical deflections analysis of variable low stiffness of thin-walled parts during milling. In: Ivanov, V., et al. (eds.) Advances in Design, Simulation and Manufacturing II. LNME, pp. 43–53. Springer, Cham (2020). https://doi.org/10.1007/978-3-030-22365-6_5
4. Liu, D., Luo, M., Zhang, Z., Hu, Y., Zhang, D.: Operational modal analysis based dynamic parameters identification in milling of thin-walled workpiece. Mech. Syst. Signal Process. **167**(A), 108469 (2022). https://doi.org/10.1016/j.ymssp.2021.108469
5. Dang, X.-B., Wan, M., Zhang, W.-H., Yang, Y.: Stability analysis of the milling process of the thin floor structures. Mech. Syst. Signal Process. **165**, 108311 (2022). https://doi.org/10.1016/j.ymssp.2021.108311
6. Li, W., Wang, L., Yu, G.: Chatter prediction in flank milling of thin-walled parts considering force-induced deformation. Mech. Syst. Signal Process. **165**, 108314 (2022). https://doi.org/10.1016/j.ymssp.2021.108314
7. Sanz-Calle, M., Munoa, J., Iglesias, A., De Lacalle, L.N.L., Dombovari, Z.: The influence of radial engagement and milling direction for thin wall machining: a semi-analytical study. Procedia CIRP **102**, 180–185 (2021). https://doi.org/10.1016/j.procir.2021.09.031
8. Eynian, M.: Frequency domain study of vibrations above and under stability lobes in machining systems. Procedia CIRP **14**, 164–169 (2014). https://doi.org/10.1016/j.procir.2014.03.068

9. Eynian, M.: Prediction of vibration frequencies in milling using modified Nyquist method. CIRP J. Manuf. Sci. Technol. **11**, 73–81 (2015). https://doi.org/10.1016/j.cirpj.2015.08.006
10. Yao, Z., Luo, M., Mei, J., Zhang, D.: Position dependent vibration evaluation in milling of thin-walled part based on single-point monitoring. Measure. J. Int. Measure. Confeder. **171**, 108810 (2021). https://doi.org/10.1016/j.measurement.2020.108810
11. Liu, Y., Liu, Z., Song, Q., Wang, B.: Development of constrained layer damping toolholder to improve chatter stability in end milling. Int. J. Mech. Sci. **117**, 299–308 (2016). https://doi.org/10.1016/j.ijmecsci.2016.09.003
12. Sokolov, V., Krol, O., Baturin, Y.: Dynamics research and automatic control of technological equipment with electrohydraulic drive. In: 2019 International Russian Automation Conference, RusAutoCon 2019. 152757. IEEE, Sochi, Russia (2019). https://doi.org/10.1109/RUSAUTOCON.2019.8867652
13. Pavlenko, I.V., Simonovskiy, V.I., Demianenko, M.M.: Dynamic analysis of centrifugal machines rotors supported on ball bearings by combined application of 3D and beam finite element models. In: 15th International Scientific and Engineering Conference Hermetic Sealing, Vibration Reliability and Ecological Safety of Pump and Compressor Machinery, HERVICON+PUMPS 2017. Institute of Physics Publishing (2017)
14. Kumar, R., Jain, A., Mishra, S.K., Joshi, K., Singh, K., Jain, R.: Comparative structural analysis of CNC milling machine bed using Al-SIC/graphite, al alloy and Al-SIC composite material. Mater. Today Proc. **51**(1), 735–741 (2022). https://doi.org/10.1016/j.matpr.2021.06.219
15. Bolsunovskiy, S., Vermel, V., Gubanov, G., Kacharava, I., Kudryashov, A.: Thin-walled part machining process parameters optimization based on finite-element modeling of workpiece vibrations. Procedia CIRP **8**, 276–280 (2013)
16. Paliwal, V., Babu, N.R.: Prediction of stability boundaries in milling by considering the variation of dynamic parameters and specific cutting force coefficients. Procedia CIRP **99**, 183–188 (2021). https://doi.org/10.1016/j.procir.2021.03.026
17. Dobrotvorskiy, S., Kononenko, S., Basova, Y., Dobrovolska, L., Edl, M.: Development of optimum thin-walled parts milling parameters calculation technique. In: Ivanov, V., Trojanowska, J., Pavlenko, I., Zajac, J., Peraković, D. (eds.) Advances in Design, Simulation and Manufacturing IV. LNME, pp. 343–352. Springer, Cham (2021). https://doi.org/10.1007/978-3-030-77719-7_34
18. Kombarov, V., Sorokin, V., Tsegelnyk, Y., Plankovskyy, S., Aksonov, Y., Fojtů, O.: S-shape Feedrate scheduling method with smoothly-limited jerk in cyber-physical systems. In: Cioboată, D.D. (ed.) International Conference on Reliable Systems Engineering (ICoRSE) - 2021. LNNS, vol. 305, pp. 54–68. Springer, Cham (2022). https://doi.org/10.1007/978-3-030-83368-8_6
19. Jones, G.M., Sanks, R.L., Tchobanoglous, G., Bosserman, II, B.E.: Pumping Station Design. 3d edn. Elsevier, Butterworth-Heinemann (2008)
20. ICAO ENVIRONMENT. https://www.icao.int/environmental-protection/pages/reduction-of-noise-at-source.aspx. Accessed 11 Jan 2022
21. Sun, X., Wang X.: Fundamentals of Aeroacoustics with Applications to Aeropropulsion Systems: Elsevier and Shanghai Jiao Tong University Press Aerospace Series (Aerospace Engineering). 1st Edition. Academic Press (2020)
22. Camussi, R., Bennett, G.J.: Aeroacoustics research in Europe: the CEAS-ASC report on 2019 highlights. J. Sound Vib. **484**, 115540 (2020). https://doi.org/10.1016/j.jsv.2020.115540
23. Kennametal. https://www.kennametal.com/us/en/resources/engineering-calculators.html. Accessed 03 Jan 2022

Check for updates

Modeling Characteristics of Ventilation Systems with Vortex Regulation Devices

Volodymyr Sokolov(✉)

Volodymyr Dahl East Ukrainian National University, 59-a, pr. Tsentralnyi, Severodonetsk 93400, Ukraine
sokolov.snu.edu@gmail.com

Abstract. Ventilation systems with vortex regulation devices are considered. Three typical schemes for constructing ventilator units are presented: a scheme with the vortex regulation device in the pressure line, a scheme with the vortex regulation device in the parallel line and a scheme with the vortex control device in the suction line. To simulate the characteristics of ventilation systems with vortex regulation devices, the calculation method is presented that allows the analysis of working processes for the presented basic schemes of ventilation units. The ventilator characteristics consider the dependence for the total head on the flow rate and the dependence for the efficiency on the flow rate. When modeling, the ventilator characteristics are approximated by the second-order polynomials. When studying the energy efficiency of various schemes for ventilator units, the dependence for the relative power on the depth of flow rate regulation was taken as the regulation characteristic. The modeling of ventilation systems with the vortex regulation devices was carried out using the example of the ventilator unit with the ventilator VC 4-75 N 2.5. The possibilities of increasing the energy efficiency of the ventilation system by reducing the power consumed by the ventilator unit have been substantiated. The amount of power reduction is determined by the installation scheme of the vortex regulation device and is achieved by reducing the energy losses associated with the regulation process.

Keywords: Ventilation system · Ventilator unit · Vortex regulation device · Throttle · Modeling

1 Introduction

One of the main tasks of regulating mechanical ventilation systems is to ensure the air flow rates provided for by the project in all sections of the ventilation network [1–5]. For regulation and adjustment, a technical test of the ventilator unit is preliminarily performed, in which the actual indicators of its operation are taken.

The most common way to regulate the ventilation system is to change the characteristics of the network, i.e., increase or decrease the total resistance by opening or closing control devices (throttle valves, dampers, air distribution devices, etc.) [6–10]. For the ventilation system, regulation begins with the regulation of the ventilator unit in order to bring the value of its total pressure and capacity into line with the design parameters.

D. D. Cioboată (Ed.): ICoRSE 2022, LNNS 534, pp. 70–79, 2023.
https://doi.org/10.1007/978-3-031-15944-2_7

Network regulation is carried out in the lines located closer to the ventilator. By creating additional resistances in the lines with the help of regulating devices, the productivity of the branches is brought to the design values. Excess air is redistributed over the sections, transferring them to those lines where the values of air flow rates are underestimated.

In the event that the air flow rates through the ventilation system devices, as well as in the head section of the ventilator unit, will be brought to the design parameters or will differ from them by no more than 10%, the regulation is completed. When regulating, the increase in the power consumption of the ventilators is usually taken into account with an increase in their capacity [11–15].

2 Literature Review

Vortex regulation devices (VRD) are increasingly used in ventilation and air heating systems [16–20]. VRD do not contain moving mechanical elements and therefore have higher reliability and durability in comparison with mechanical regulation devices.

The principle of operation of the VRD as a regulating device is as follows [21–25]. The main flow through the VRD (Fig. 1) enters the radial channel and exits through the central hole. In the absence of a control flow signal, which is supplied to the tangential channel, the hydraulic resistance, which creates the VRD to the main flow, is not great. The control flow entering the tangential channel swirls the main flow, while the coefficient of hydraulic resistance of the VRD increases significantly, which leads to change in the flow rate of the ventilator unit. As a control flow, a part of the main flow can be used, which is supplied through a regulating body, which is much smaller than the VRD.

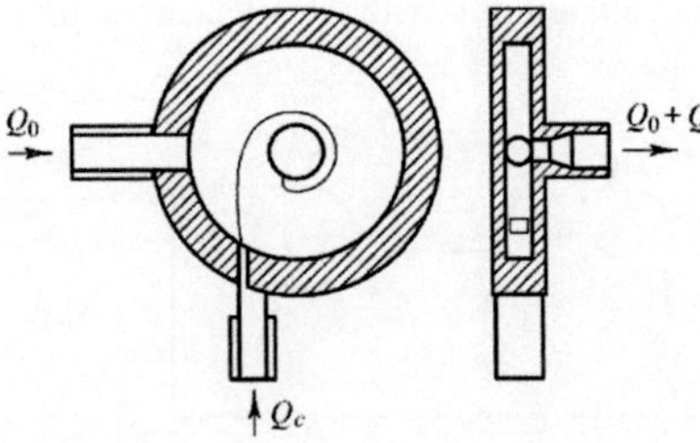

Fig. 1. The vortex regulation device.

The analysis of the efficiency of ventilator units with VRD requires the development of methods for calculating the control characteristics, which make it possible to determine the power costs associated with the regulation process [26–30]. At present, a certain experience has been accumulated in the calculation and design of the VRD, which makes it possible to select the required geometric parameters for the given working aerodynamic characteristics of the VRD. At the same time, there is no information in the literature on the calculation of the characteristics of processes in the joint operation of ventilator units with VRD.

The purpose of this paper is to study the energy efficiency of ventilation systems with VRD, to substantiate the possibility of increasing the energy efficiency of the ventilation system by reducing the power consumption of the ventilator unit.

3 Research Methodology

In general, ventilation units with VRD can be built according to the following three main schemes (Figs. 2, 3 and 4): the ventilation unit with VRD in the pressure line (Fig. 2); the ventilation unit with VRD in the parallel line (Fig. 3); the ventilation unit with VRD in the suction line (Fig. 4). The figures indicate: VT – ventilator; VRD – vortex regulation device; DR – throttle; REC – regulation element in the control channel.

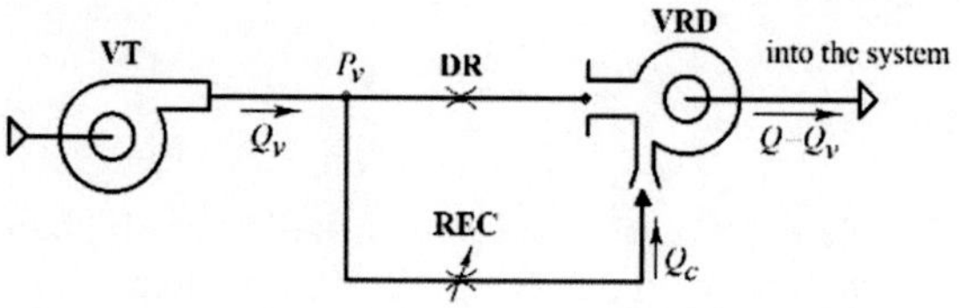

Fig. 2. The ventilator unit with the VRD in the pressure line.

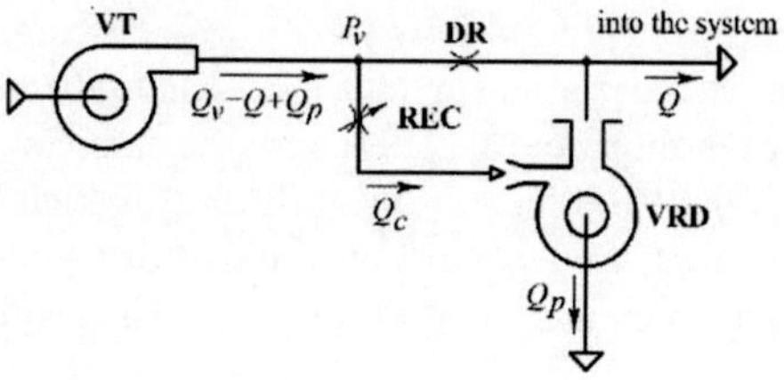

Fig. 3. The ventilator unit with the VRD in the parallel line.

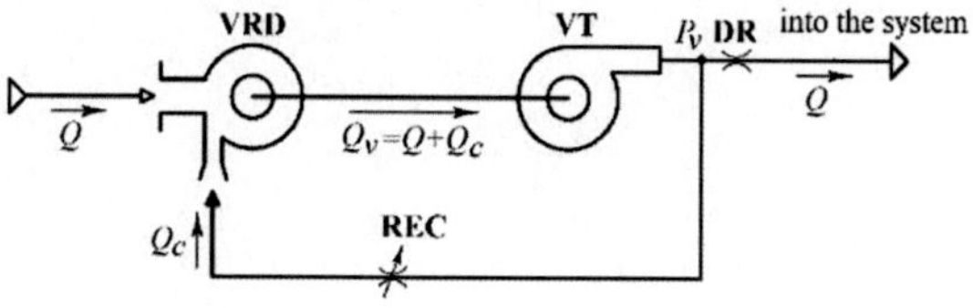

Fig. 4. The ventilator unit with the VRD in the suction line.

The working process of regulating the ventilator unit with the VRD in the pressure line is illustrated in Fig. 5.

For the normal open VRD, the ventilator characteristic $Pv(Q)$ and the initial characteristic of the network, that set by the throttle DR, determine the initial value of the flow rate into the system, which is equal to the flow rate of the ventilator. When the control channel is opened, which is carried out by the REC installed in it, the main flow is vortexed and the aerodynamic resistance of the VRD increases. The characteristic of the network changes due to the additional pressure drop across the VRD ΔP_{VRD}. The value of the flow rate Q entering the system, as well as the flow rate of the ventilator Qv, decreases. For this scheme of ventilator unit $Q = Qv$.

The working process of regulating the ventilator unit with the VRD in the parallel line is illustrated in Fig. 6.

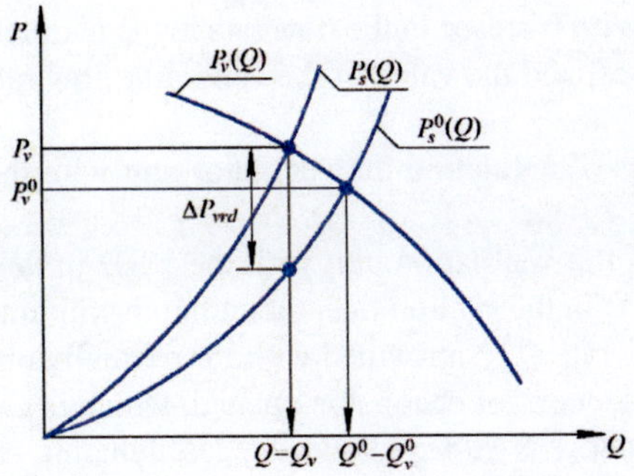

Fig. 5. The regulating the ventilator unit with the VRD in the pressure line.

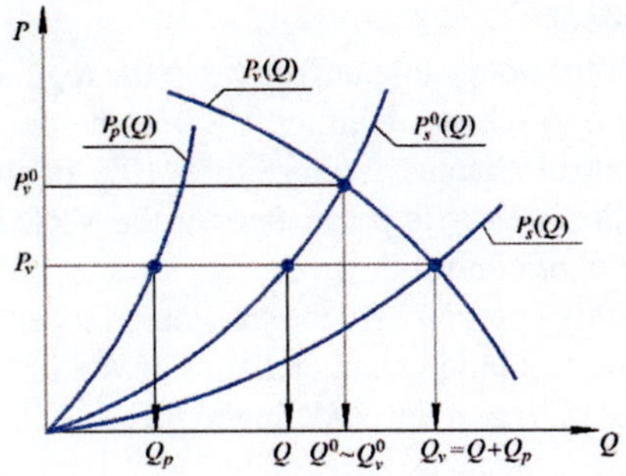

Fig. 6. The regulating the ventilator unit with the VRD in the parallel line.

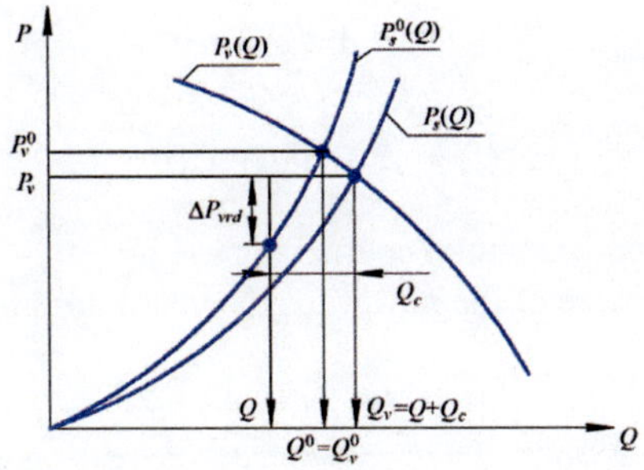

Fig. 7. The regulating the ventilator unit with the VRD in the suction line.

In this scheme, the initial maximum value of the flow rate Q^0 entering the system is provided with the normally closed VRD and is set by the throttle DR. The ventilator flow rate Q_v^0 is greater by the flow rate Q^0 in the control channel Q_c. By choosing the geometric parameters of the VRD, it can be closed by the maximum flow rate in the control channel Q_c max up to 5% of the initial flow rate of the ventilator. Therefore, for the starting operating point can be adopted $Q^0 \approx Q_v^0$.

When the control channel is closed, which is carried out by the REC installed in it, the vortex of the flow in the VRD decreases and, consequently, the aerodynamic resistance of the VRD decreases. With increase in the flow rate in the parallel line Q_p, the flow rate to the system Q decreases, and the value of the ventilator flow rate Q_v increases, while $Q_v = Q + Q_p$.

The working process of regulating the ventilator unit with the VRD in the suction line is illustrated in Fig. 7.

As in the scheme of the ventilation unit with the VRD in the pressure line, and in the scheme with the VRD in the suction line, the initial maximum flow rate Q^0 equal to the initial ventilator flow rate Q_v^0 is provided with the normally open VRD and is set by the throttle DR. When the control channel is opened, which is carried out by the REC installed in it, the main flow is vortexed and the aerodynamic resistance of the VRD increases. The characteristic of the network changes both due to the additional pressure drop across the VRD ΔP_{VRD} and the additional flow rate in the control channel Q_c. The flow rate into the system Q decreases, and the value of the ventilator flow rate Q_v increases, while $Q_v = Q + Q_c$.

It should be noted that the noticeable difference in the regulation characteristics for this scheme, in comparison to scheme with the VRD in the pressure line, will be only if the flow rate in the control channel is more than 10% of the ventilator flow rate. Therefore, the choice of the geometric parameters of the VRD for this scheme should provide an appropriate level of control flow rate.

The analysis of the working processes in the considered schemes allows us to propose methods for calculating the regulation characteristics of ventilator units with VRD.

Consider the ventilator unit with the VRD in the pressure line.

We obtain approximation dependences for the ventilator characteristics: its total pressure $P_v(Q)$ and its efficiency $\eta_v(Q)$. According to the known approaches [1, 6, 16], sufficient accuracy is provided by the approximation by a second-order polynomial:

$$P_v(Q) = a + bQ + cQ^2; \tag{1}$$

$$\eta_v(Q) = A + BQ + CQ^2; \tag{2}$$

where a; b; c; A; B; C – approximation coefficients.

Determine the initial value of the power for the ventilator unit

$$N_0 = \frac{\left(a + bQ + c\left(Q_v^0\right)^2\right)Q_v^0}{a + bQ + c\left(Q_v^0\right)^2}. \tag{3}$$

Setting the depth of regulation k

$$k = Q\big/Q^0 \tag{4}$$

and determine the new value of the flow rate into system

$$Q = kQ^0. \tag{5}$$

Consider for the ventilator the new value of flow rate (capacity) $Q_v = Q$ and determine total pressure P_v and efficiency η_v according to (1), (2). We calculate the value of the power for the ventilator unit and its dimensionless value

$$N = \frac{P_v Q_v}{\eta_v}; \; \overline{N} = \frac{N}{N_0}. \tag{6}$$

We repeat the calculations for new values of the regulation depth k and establish the dependence $\overline{N}(k)$.

Consider the ventilator unit with the VRD in the parallel line.

Setting the depth of regulation k and determine the new value of the flow rate into system Q. For this scheme, for the ventilator the new value of the flow rate (capacity) Q_v, the total pressure P_v and the efficiency η_v are determined as follows. The value of the ventilator total pressure is the required pressure for the system with the flow rate Q. Since the characteristic of the system is the initial characteristic of the network, then $P_v = P_s^0(Q)$.

The initial characteristic of the network can be described by a quadratic dependence

$$P_s^0 = R_0 Q^2; \tag{7}$$

where R_0 – reduced aerodynamic resistance of the system [6, 11].

Therefore

$$R_0 = P_v^0 \Big/ Q_0^2 \tag{8}$$

and the new value of the ventilator total pressure is determined by the expression

$$P_v = R_0 Q^2. \tag{9}$$

Further, according to (1), we determine the new value of the ventilator flow rate

$$Q_v = \frac{-b - \sqrt{b^2 - 4c(a - P_v)}}{2c} \tag{10}$$

and the value of the efficiency η_v according to dependence (2).

Also, we repeat the calculations for new values of the regulation depth k and establish the dependence $\overline{N} = (k)$.

Consider the ventilator unit with the VRD in the suction line.

Setting the depth of regulation k and determine the new value of the flow rate into system Q. We determine for the ventilator the new value of the flow rate (capacity) Q_v, the total pressure P_v and the efficiency η_v, proceeding from the fact that the characteristics of the network will change both due to the additional pressure drop across the VRD and the additional flow in the control channel. Therefore, for the specific VRD, according to the value of Q, we find the value of ΔP_{VRD}, and, taking into account the peculiarities of the operation of the ventilator unit with the VRD in the parallel line, we find the new value of the ventilator total pressure

$$P_v = R_0 Q^2 + \Delta P_{VRD}. \tag{11}$$

Here the value of the reduced aerodynamic resistance of the system R_0 is determined according to (8).

New values of the fan flow rate Q_v and its efficiency η_v are determined, respectively, by expressions (10) and (2). We repeat the calculations for new values of the regulation depth k and establish the dependence $\overline{N}(k)$.

4 Results

The proposed methods were used to calculate the regulation characteristics for ventilation system with the ventilator unit based on the ventilator VC 4-75 N 2.5.

The approximate characteristic of the ventilator and its efficiency are shown in Fig. 8 and have the form (units in the SI system):

$$P_v(Q) = 171.7 + 2.07 \cdot 10^3 Q - 5.82 \cdot 10^3 Q^2; \tag{12}$$

$$\eta_v(Q) = 0.0096 + 6.31Q - 13.3Q^2. \tag{13}$$

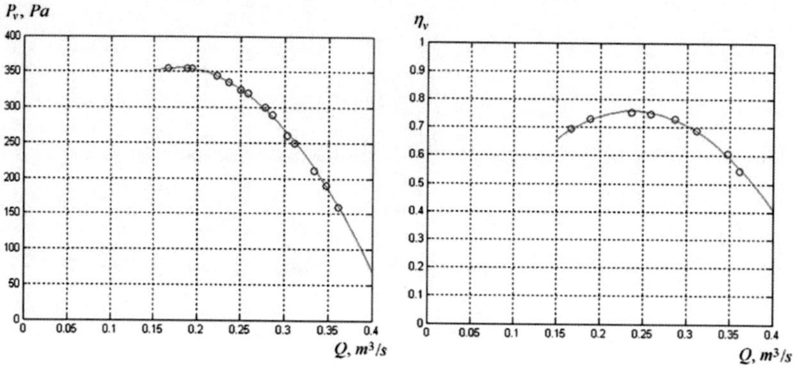

Fig. 8. The characteristics of the ventilator VC 4-75 N 2.5 (o - experiment, — - approximation).

The results of calculating the regulation characteristics for considered schemes of ventilator units with VRD are shown in Fig. 9. The value $Q_0 = 0.32$ m^3/s was taken as the initial value of the flow rate.

The analysis of the results obtained shows that the scheme with the VRD in the pressure line does not increase efficiency due to the reduction in power consumption, since the regulation characteristic corresponds to those known when installing regulation devices in the pressure line. The decrease in power consumption is achieved in the ventilator unit with the VRD in the suction line (curve 3) and, to a greater extent, with the VRD in the parallel line (more than 10% at the regulation depth of up to 0.5 of the initial flow rate, curve 2). When installing the regulation device in the parallel line, the VRD was used with a control flow rate Q_c up to $0.05Q_0$, the VRD was installed in the suction line with the characteristic $\Delta P_{VRD} = 0.6 R_0 Q_c^2$.

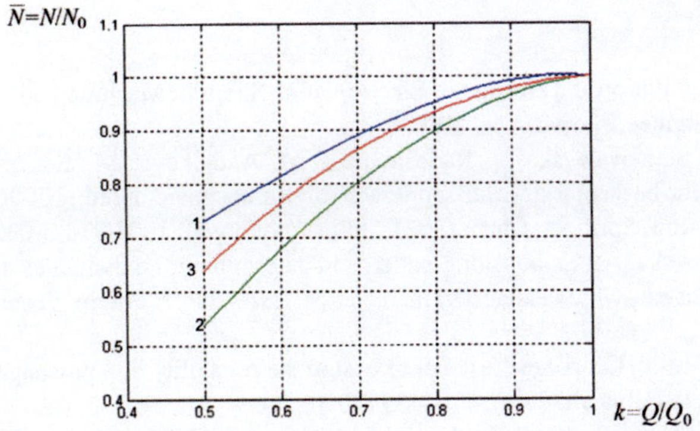

Fig. 9. The regulation characteristics of the ventilator unit with the VRD (1 – the VRD in the pressure line; 2 – the VRD in the parallel line; 3 – the VRD in the suction line).

It should be noted that this behavior of the regulation characteristics takes place for ventilators with backward curved blades. In addition, the amount of power consumption reduction is determined by the characteristics of the specific ventilator, the choice of the initial operating point and other factors.

5 Conclusions

Thus, the modeling and studies of the energy efficiency of ventilation systems with vortex control devices were carried out. Ventilation systems with the vortex control devices are considered. Three typical schemes for constructing ventilator units are presented: a scheme with the vortex regulation device in the pressure line, a scheme with the vortex regulation device in the parallel branch and a scheme with the vortex control device in the suction line.

To model the characteristics of ventilation systems with vortex control devices, the calculation method is presented that allows an analysis of working processes for the presented basic schemes of ventilator units. The ventilator characteristics consider the dependence for the total head on the flow rate and the dependence for the efficiency on the flow rate. In the modeling, the ventilator characteristics are approximated by the second-order polynomials. When studying the energy efficiency of various schemes of ventilator units, the dependence for the relative power on the depth of flow regulation was taken as the regulation characteristic. Under the depth of regulation, the ratio of the set flow rate to its initial (maximum) value was considered, under the relative power, the ratio of the power of the ventilator unit at the given flow rates.

The modeling of ventilation systems with vortex control devices was carried out using the example for the ventilator unit with the ventilator VC 4-75 N 2.5. The possibilities of increasing the energy efficiency for the ventilation system by reducing the power consumed by the ventilation unit are substantiated. The amount of power reduction is determined by the installation scheme for the vortex control device and is achieved by reducing energy losses associated with the control process.

References

1. Elinsky, I.: Ventilation and Heating of Electroplating Shops of Machine-Building Enterprises. Machinery Engineering, Moscow (1989)
2. Kovalevskyy, S., Kovalevska, O., Radosavljević, M., Anđelković, M.: Ensuring the life cycle of objects on the basis of a signature approach. In: Karabegović, I. (ed.) NT 2021. LNNS, vol. 233, pp. 458–468. Springer, Cham (2021). https://doi.org/10.1007/978-3-030-75275-0_51
3. Fomin, O., Lovska, A.: Establishing patterns in determining the dynamics and strength of a covered freight car, which exhausted its resource. East.-Eur. J. Enterp. Technol. **6**(7), 21–29 (2020)
4. Lovska, A., Fomin, O.: A new fastener to ensure the reliability of a passenger car body on a train ferry. Acta Polytech. **60**(6), 478–485 (2020)
5. Popov, D., Panaiotti, S., Ryabinin, M.: Hydromechanics. MSTU, Moscow (2014)
6. Becker, A.: Ventilation Systems. Euroclimate, Moscow (2005)
7. Kotliar, A., et al.: Ensuring the economic efficiency of enterprises by multi-criteria selection of the optimal manufacturing process. Manage. Prod. Eng. Rev. **11**(1), 52–61 (2020)
8. Emtsev, B.: Technical hydromechanics. Machinery Engineering, Moscow (1987)
9. Shevchenko, S., Mukhovaty, A., Krol, O.: Gear transmission with conic axoid on parallel axes. In: Radionov, A.A., Kravchenko, O.A., Guzeev, V.I., Rozhdestvenskiy, Y.V. (eds.) ICIE 2019. LNME, pp. 1–10. Springer, Cham (2020). https://doi.org/10.1007/978-3-030-22041-9_1
10. Kovalenko, A., et al.: Fundamentals of Technical Mechanics of Liquids and Gases. EUSU, Lugansk (1998)
11. Svistunov, V., Pushnyakov, N.: Heating, ventilation and air conditioning facilities of the agro-industrial complex and utilities, St. Petersburg (2007)
12. Liaposhchenko, O., Pavlenko, I., Monkova, K., Demianenko, M, Starynskyi, O.: Numerical simulation of aeroelastic interaction between gas-liquid flow and deformable elements in modular separation devices. In: Ivanov, V., et al. (eds.) DSMIE 2019. LNME, pp. 765–774. Springer, Cham (2020). https://doi.org/10.1007/978-3-030-22365-6_76
13. Popov, D.: Dynamics and Regulation Hydro-and Pneumatic Systems. Machinery Engineering, Moscow (1987)
14. Popov, D.: Mechanics of Hydro- and Pneumodrives. MGTU, Moscow (2005)
15. Romanchenko, O.: Principles of design of specialized technological equipment. Diagnostyka **23**(1), 2022109 (2022)
16. Ananyev, V.: Ventilation and Air Conditioning Systems. Euroclimate, Moscow (2001)
17. Rogovyi, A., Korohodskyi, V., Khovanskyi, S., Hrechka, I., Medvediev, Y.: Optimal design of vortex chamber pump. J. Phys: Conf. Ser. **1741**, 012018 (2021)
18. Andrenko, P., Rogovyi, A., Hrechka, I., Khovanskyi, S., Svynarenko, M.: The influence of the gas content in the working fluid on parameters of the hydraulic motor's axial piston. In: Ivanov, V., Pavlenko, I., Liaposhchenko, O., Machado, J., Edl, M. (eds.) DSMIE 2021. LNME, pp. 97–106. Springer, Cham (2021). https://doi.org/10.1007/978-3-030-77823-1_10
19. Rogovyi, A., Korohodskyi, V., Medvediev, Y.: Influence of Bingham fluid viscosity on energy performances of a vortex chamber pump. Energy **218**, 119432 (2021)
20. Kundrák, J., Mitsyk, A., Fedorovich, V., Markopoulos, A., Grabchenko, A.: Simulation of the circulating motion of the working medium and metal removal during multi-energy processing under the action of vibration and centrifugal forces. Machines **9**(6), 1–22 (2021)
21. Loytsyansky, L.: Mechanics of a Liquid and Gas. Drofa, Moscow (2003)
22. Kundrák, J., Mitsyk, A., Fedorovich, V., Markopoulos, A., Grabchenko, A.: Modeling the energy action of vibration and centrifugal forces on the working medium and parts in a vibration machine oscillating reservoir with an impeller. Manufact. Technol. **21**(3), 364–372 (2021)

23. Krol, O., Porkuian, O., Sokolov, V., Tsankov, P.: Vibration stability of spindle nodes in the zone of tool equipment optimal parameters. Comptes rendus de l'Acade'mie bulgare des Sciences **72**(11), 1546–1556 (2019)
24. Tsankov, P.: Modeling of vertical spindle head for machining center. J. Phys: Conf. Ser. **1553**, 012012 (2020)
25. Ivanov, V., Pavlenko, I., Liaposhchenko, O., Gusak, O., Pavlenko, V.: Determination of contact points between workpiece and fixture elements as a tool for augmented reality in fixture design. Wirel. Netw. **27**(3), 1657–1664 (2021). https://doi.org/10.1007/s11276-019-02026-2
26. Sokolov, V., Porkuian, O., Krol, O., Baturin, Y.: Design calculation of electrohydraulic servo drive for technological equipment. In: Ivanov, V., Trojanowska, J., Pavlenko, I., Zajac, J., Peraković, D. (eds.) DSMIE 2020. LNME, pp. 75–84. Springer, Cham (2020). https://doi.org/10.1007/978-3-030-50794-7_8
27. Sokolov, V., Porkuian, O., Krol, O., Stepanova, O.: Design calculation of automatic rotary motion electrohydraulic drive for technological equipment. In: Ivanov, V., Trojanowska, J., Pavlenko, I., Zajac, J., Peraković, D. (eds.) DSMIE 2021. LNME, pp. 133–142. Springer, Cham (2021). https://doi.org/10.1007/978-3-030-77719-7_14
28. Krol, O., Sokolov, V.: Research of modified gear drive for multioperational machine with increased load capacity. Diagnostyka **21**(3), 87–93 (2020)
29. Krol, O., Sokolov, V.: Research of toothed belt transmission with arched teeth. Diagnostyka **21**(4), 15–22 (2020)
30. Krol, O., Sokolov, V.: Modeling of spindle node dynamics using the spectral analysis method. In: Ivanov, V., Trojanowska, J., Pavlenko, I., Zajac, J., Peraković, D. (eds.) DSMIE 2020. LNME, pp. 35–44. Springer, Cham (2020). https://doi.org/10.1007/978-3-030-50794-7_4

Improving the Productivity of Information Technology for Processing Indirect Measurement Data

Olexander Yenikieiev[1], Dmytro Zakharenkov[1], Magomedemin Gasanov[2], Fatima Yevsyukova[2](✉), Olena Naboka[2], and Andrew Ruzmetov[2]

[1] Donbas State Engineering Academy, 72, Akademicheskaia Street, Kramatorsk 84313, Ukraine
[2] National Technical University "Kharkiv Polytechnic Institute", 2, Kyrpychova Street, Kharkiv 61002, Ukraine
fatimaevsukova@gmail.com

Abstract. The aim of this study is to develop efficient information technologies for processing indirect measurement data in the form of a frequency-modulated signal of the crankshaft speed of an internal combustion engine based on the use of the capabilities of modern software environments. For the first time, a mechanical system with five degrees of freedom is proposed as a deterministic mathematical model of the torque scheme of a power unit. The Laplace Transform is used to solve a system of differential equations of mass motions of a mathematical model under zero initial conditions. On the basis of the method of determinants and methods of the theory of signal graphs, information relationships are established in the form of transfer functions between the torques of cylinders and the signal of fluctuations in the speed of rotation of the first mass of the crankshaft. When analyzing the special points of transfer functions, it is found a satisfactory coincidence of the calculation results of both methods. By comparison, it is found that the method based on the theory of signal graphs provides better calculation performance. It is also established that the determinant method gives a simplified expression of the zero-pole transfer function, which in further studies can be taken into account or not, depending on the type of measurement information signal. Z-transformations of transmission functions are obtained and a scheme for computer modeling of the contribution signal of the first cylinder to the signal of fluctuations in the speed of rotation of the first mass of the crankshaft of the power unit is constructed based on them. In the Mathcad software, a scheme for computer modeling of torque conversion processes with obtaining a measurement information signal is constructed. When compiling it, the possibility of identifying the parameters of a deterministic mathematical model of a power unit based on experimental data of instantaneous speed and adjusting the length of information connections between the contributions of individual cylinders to the fluctuation signal is taken into account. It is established a satisfactory coincidence of the results of computer modeling and experimental data.

Keywords: Hardware · Determinant method · Signal graph theory · Mathematical and computer modeling

D. D. Cioboată (Ed.): ICoRSE 2022, LNNS 534, pp. 80–94, 2023.
https://doi.org/10.1007/978-3-031-15944-2_8

1 Introduction

Modern computer mathematics to increase the productivity of the synthesis of hardware for processing indirect measurement data offers the following integrated software systems: Eureka, Gauss, Derive, Mathcad, Mathematical, Maple V, Matlab, Scilab and others [1, 2]. Their development is based on matrix operations that solve problems of linear algebra. Matulab software with Simulink extension provides significant opportunities to study the dynamics of indirect measurement data processing hardware. The method of frequency characteristics in solving problems of hardware dynamics uses the Fourier transform as a mathematical apparatus. Taking into account a certain number of harmonic components in the processing of source information significantly reduces the productivity of hardware, so this method is used in the analysis of fairly simple transfer functions. The use of graphical integration method slightly increases the productivity of calculations [3]. The construction of hardware is associated with solving the following tasks [4–6]: mathematical modeling of components; study of frequency characteristics of the transfer function; synthesis of the input signal processing device; construction of a computer modeling scheme; establishing uncertainty and speed. These computational actions significantly reduce the productivity of the process of their construction. Thus, the development of new methods of building hardware for processing indirect measurement data, which use the capabilities of modern software environments to ensure better productivity of input information processing, is an urgent scientific and applied task.

2 Literature Review

The issues of constructing hardware tools for monitoring the identity of the operating cycles of diesel generators (DG) based on the processing of indirect measurement data in the form of a signal of uneven rotation of the crankshaft are given sufficient attention in the literature. In [7] the question of interaction of schemes of diagnostics of engines by methods of non-disassembled control is considered, the analysis of existing methods is executed, and the use of the complex approaches increasing reliability of work of engines is offered. In [8], a stochastic mathematical model was developed for studies of the dynamics of changes in the cylinder power of DG. In [9], a method for processing the measurement signal using the capabilities of the Matlab software environment was developed. In order to reduce the impact of random interference on the information signal of the non-uniformity of rotation of the crankshaft of the power unit in the work it is proposed to use a high-pass filter with a final pulse characteristic. In [10], the sensitivity of several criteria for establishing the non-uniformity of the angular velocity of the crankshaft at the maximum pressure of the indicator diagrams of successive operating cycles. Also investigated: coefficient of variation of angular velocity increments over several operating cycles, coefficient of variation of its amplitude values, coefficient of variation of the set of largest maximum and smallest minimum values, criterion of real and kinematic non-uniformity of angular velocity.

In [11], a method for adjusting the parameters of cyclic fuel supply is proposed. The identity of the measuring parameter is established on the basis of processing the signal of non-uniform rotation of the crankshaft. In [12], the expressions of forced oscillations of a discrete mathematical model with several degrees of freedom are obtained, taking into account the matrix of torques. The measuring transducer of amplitude of oscillations gives the chance to establish parameters of a twisting moment, to estimate possibility of occurrence of the resonant phenomena. The line of mathematical expectation of linear regression of degradation of a technical condition of a cranked shaft which allows to establish zones of a satisfactory and unsatisfactory condition is received. In [13], a method for diagnosing the state of a power unit by establishing indicators of unevenness of its operation according to the parameters of vibroacoustic emission signals is proposed. Appropriate hardware has been developed and application software has been implemented. It is established that the use of digital measuring transducers of input signals provides real-time control of indicators.

In [14] the scheme of computer modeling of the sample signal of non-uniformity of crankshaft rotation is constructed and the technique of diagnosing the power unit is offered. In the study [15], the influence of torque unevenness on the dynamic characteristics and power indicators of DG is studied. In the study [16], an expert system is proposed for determining the technical condition of an engine based on the use of a trained computer model. In the work [17], a measuring converter and information technology for processing the fluctuation signal are developed and the results of experimental data for DG 6NVD48UA are presented. In the study [18], a method is proposed to reduce the uncertainty of hardware controls for the fuel supply process based on measurements of the amplitude of fluctuations in the angular velocity of rotation and phase displacements of their extremes relative to the upper dead center of the corresponding cylinder. The lack of methodological foundations for constructing hardware for processing the frequency-modulated signal of the crankshaft speed is their disadvantage.

3 Research Methodology

The aim of the study is to create productive information technologies for processing indirect measurement data in the form of a frequency-modulated signal of the crankshaft speed based on the use of modern software environments.

To achieve this goal the following tasks need to be solved:

- develop a deterministic mathematical model of the DG D964 torque circuit and identify its parameters;
- establish information links in the form of transfer functions between the cylinder torques and the signal of fluctuations in the speed of rotation of the crankshaft using the method of determinants;
- obtain transfer functions using signal graph theory methods;
- investigate the frequency characteristics of torque transmission channels and compare the performance of methods;
- perform an approximation of the powertrain cylinder torques;
- construct computer simulation schemes for the effects of individual cylinder torques on the signal of fluctuations in the speed of rotation of the first mass;
- create a computer simulation scheme for the measurement information signal;
- investigate the performance of hardware monitoring the identity of DG duty cycles.

The authors use the continuous and discrete Laplace Transform as a mathematical research apparatus. The results of theoretical studies of hardware dynamics when processing indirect measurement data are tested using the Mathcad and Matlab software environments with the Simulink extension.

4 Results

4.1 Mathematical Modeling of the Torque Scheme of the Power Unit

When analyzing the structure of DG D964, the authors propose the following: install the primary converter near the first cylinder; provided that vibrations between the masses are taken into account, the torque circuit is presented in the form of a mechanical system that has five degrees of freedom. The dynamic characteristics of a mechanical system that has ten degrees of freedom in the absence of friction are studied in [19, 20]. Accordingly, the dynamics of rotation of cylinder masses of a deterministic mechanical system with five degrees of freedom under the conditions of taking into account friction is represented by the following system of differential equations

$$J_i\phi_i^{//}(t) + \beta\phi_i^{/}(t) - e^{-1}\left[\phi_{i+1}(t) - \phi_i(t)\right] + e^{-1}\left[\phi_i(t) - \phi_{i-1}(t)\right] = M_i(t), \quad (1)$$

where $i = 1,\ 2,\ \ldots\ 5, \phi_i(t)$ – the angle of rotation of the mass, $e = 5.12 \cdot 10^{-7}$ (Nm)$^{-1}$ is the pliability of the bonds between the masses, $J_i = 0.536$ Nm^2 is the moment of inertia of the cylinder mass, $M_i(t)$ – is the torque of the cylinder, $\beta_i = 2.68$ Nms is friction, $J_5 = 0.169$ Nm^2 is the moment of inertia of the fifth mass.

The Laplace Transform under zero initial conditions gives the system of differential Eqs. (1) the following form.

$$\left(Jep^2 + \beta ep + 2\right)\phi_i(p) - \phi_{i+1}(p) - \phi_{i-1}(p) = eM_i(p). \quad (2)$$

Enter the following notation.

$$a = \frac{1}{J_iep^2 + \beta ep + 1},\ b = \frac{e}{J_iep^2 + \beta ep + 1},\ c = \frac{e}{J_iep^2 + \beta ep + 2},$$
$$d = \frac{1}{J_iep^2 + \beta ep + 2},\ f = \frac{1}{J_5ep^2 + \beta ep + 1}. \quad (3)$$

Taking into account these designations, the system of Eqs. (2) takes the following form:

$$\begin{cases} \phi_1 - a\phi_2 = bM_1 \\ \phi_2 - d\phi_3 - d\phi_1 = cM_2 \\ \phi_3 - d\phi_4 - d\phi_2 = cM_3\,. \\ \phi_4 - d\phi_5 - d\phi_3 = cM_4 \\ \phi_5 - f\phi_4 = 0 \end{cases} \quad (4)$$

The procedure for establishing information links between cylinder torques and the fluctuation information signal is performed by two methods. When obtaining the frequency characteristics of torque transmission channels, we use modern software environments. Comparing the research results will allow you to choose the method that provides the best performance.

4.2 Determinant Method

Based on the determinant method, the transfer functions obtained as a result of solving the system of algebraic Eqs. (4) take the following form.

$$\phi_1(p) = \sum_{j=1}^{4} \frac{\Delta_j}{\Delta} M_j(p) = \sum_{j=1}^{4} W_j M_j(p), \tag{5}$$

where $\phi_1(p)$ – Laplace transformations of the signal of fluctuations in the speed of rotation of the first mass; Δ, Δ_j – main and all determinants of Eqs. (4); W_j – transfer functions that establish information relationships between torques and the signal of fluctuations.

Determinants of the system of algebraic Eqs. (4) are established by the authors using the Mathcad software environment. The scripts are shown in Fig. 1 calculation of logarithmic amplitude-frequency characteristics (the Bode plot) of torque transmission paths is performed in the Matlab software environment [21]. Information technology consists of the following actions:

- the command line for specifying expressions (3) has, for example, the following form a = tf ([1], [j * e β * e 1]);
- the transfer functions of the paths are determined by expression (5) and the results of calculating the determinants, which are shown in Fig. 1;
- the Bode Plot of the torque transmission paths is plotted using the bode command (W1, W2, W3, W4). The results of calculating the Bode plot are shown in Fig. 2.

$$\left|\begin{pmatrix} 1 & -a & 0 & 0 & 0 \\ -d & 1 & -d & 0 & 0 \\ 0 & -d & 1 & -d & 0 \\ 0 & 0 & -d & 1 & -d \\ 0 & 0 & 0 & -f & 1 \end{pmatrix}\right| \rightarrow a\cdot d^3 + d^3\cdot f - 2\cdot d^2 - a\cdot d - d\cdot f + a\cdot d^2\cdot f + 1$$

$$\left|\begin{pmatrix} b & -a & 0 & 0 & 0 \\ c & 1 & -d & 0 & 0 \\ c & -d & 1 & -d & 0 \\ c & 0 & -d & 1 & -d \\ 0 & 0 & 0 & -f & 1 \end{pmatrix}\right| \rightarrow b - 2\cdot b\cdot d^2 + a\cdot c + a\cdot c\cdot d - b\cdot d\cdot f + b\cdot d^3\cdot f - a\cdot c\cdot d\cdot f - a\cdot c\cdot d^2\cdot f$$

$$\left|\begin{pmatrix} 1 & b & 0 & 0 & 0 \\ -d & c & -d & 0 & 0 \\ 0 & c & 1 & -d & 0 \\ 0 & c & -d & 1 & -d \\ 0 & 0 & 0 & -f & 1 \end{pmatrix}\right| \rightarrow c - b\cdot d^3 + b\cdot d + c\cdot d - c\cdot d\cdot f - b\cdot d^2\cdot f - c\cdot d^2\cdot f$$

$$\left|\begin{pmatrix} 1 & -a & b & 0 & 0 \\ -d & 1 & c & 0 & 0 \\ 0 & -d & c & -d & 0 \\ 0 & 0 & c & 1 & -d \\ 0 & 0 & 0 & -f & 1 \end{pmatrix}\right| \rightarrow c + b\cdot d^2 + 2\cdot c\cdot d - a\cdot c\cdot d - c\cdot d\cdot f - a\cdot c\cdot d^2 - b\cdot d^3\cdot f - c\cdot d^2\cdot f + a\cdot c\cdot d^2\cdot f$$

$$\left|\begin{pmatrix} 1 & -a & 0 & b & 0 \\ -d & 1 & -d & c & 0 \\ 0 & -d & 1 & c & 0 \\ 0 & 0 & -d & c & -d \\ 0 & 0 & 0 & 0 & 1 \end{pmatrix}\right| \rightarrow c + b\cdot d^3 + c\cdot d - a\cdot c\cdot d - a\cdot c\cdot d^2$$

Fig. 1. Determinant installation scripts

As a result of analyzing the information received, we come up with the following conclusions. The Bode plot has a slope of −20 dB per decade, which corresponds to the presence of an integration circuit in the structure of torque transmission channels, in the frequency band 1050–1400 s^{-1} there are shifters. Methods of the Bode plot correction

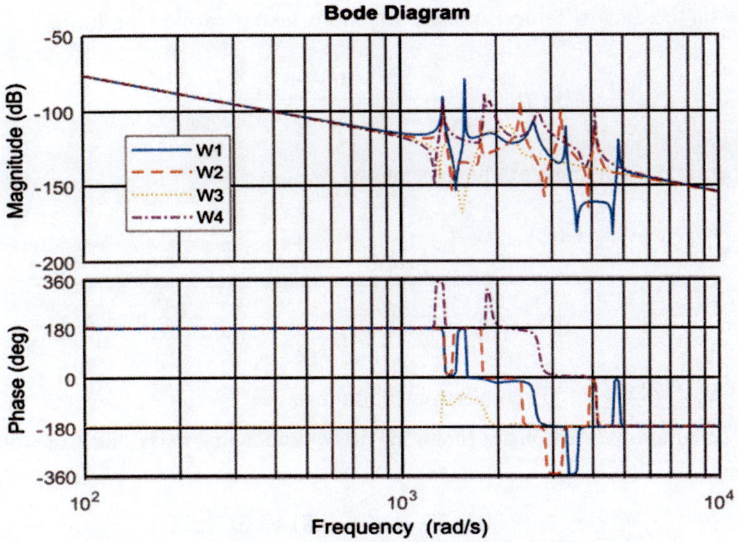

Fig. 2. Bode diagrams obtained using the determinant method

are developed in [21, 22] on the basis of methods of neural network technologies using the computational capabilities of modern software environments.

Representation of transfer functions in the form of a serial connection of elementary circuits is possible due to the establishment of their special points. For these calculations, it is convenient to use the capabilities of the Matlab software environment. When analyzing special points, we adjust the transfer functions. The method of shortening their expressions involves performing the following actions:

- discarding unstable roots;
- repayment of roots whose values are quite close in value;
- discarding second-order roots of smallness, as they affect the beginning of the transition process. In Table 1 shows special points of transfer functions that meet the specified conditions.

Table 1. Special points of transfer functions

Zero				Pole
W_1	W_2	W_3	W_4	W
−4.2 + 1492.7i	−4.0 + 1446.5i	−2.9 + 1327.5i	−2.5 + 1275.1i	0
−4.2 − 1492.7i	−4.0 − 1446.5i	−2.9 − 1327.5i	−2.5 − 1275.1i	−3.3 + 1361.4i
				−3.3 − 1361.4i
				−5.8

Based on the data in Table 1 transfer functions take the following form.

$$W_5(p) = \frac{(p + a_1)^2 + b_1^2}{p\left[(p + a_5)^2 + b_5^2\right](p + a_6)}. \quad (6)$$

After simple mathematical transformations, we have

$$W_5(p) = \frac{k_1\left(\frac{1}{a_1^2+b_1^2}p^2 + \frac{2a_1}{a_1^2+b_1^2}p + 1\right)}{\frac{1}{a_6\left(a_5^2+b_5^2\right)}p^4 + \frac{a_6+2a_5}{a_6\left(a_5^2+b_5^2\right)}p^3 + \frac{2a_5a_6+a_5^2+b_5^2}{a_6\left(a_5^2+b_5^2\right)}p^2 + p}, \quad (7)$$

where $k_1 = \frac{a_1^2+b_1^2}{a_6\left(a_5^2+b_5^2\right)}$ are the conversion coefficients.

After calculations, the transfer function "first cylinder-first mass" takes the following form:

$$W_5(p) = \frac{9.302 \cdot 10^{-8}p^2 + 7.814 \cdot 10^{-7}p + 1}{9.03 \cdot 10^{-8}p^4 + 1.154 \cdot 10^{-6}p^3 + 0.172p^2 + p}. \quad (8)$$

Let's move on to developing a method for establishing information transfer connections between cylinder torques and the frequency-modulated signal of the crankshaft speed based on the methods of signal graph theory.

4.3 Signal Graph Theory Method

It is possible to establish informational transfer connections between the torques and the signal of fluctuations in the speed of rotation of the first mass using the Meson formula. As a result of analyzing the structure of algebraic equations and the organization of system (4), a signal graph is constructed (Fig. 3).

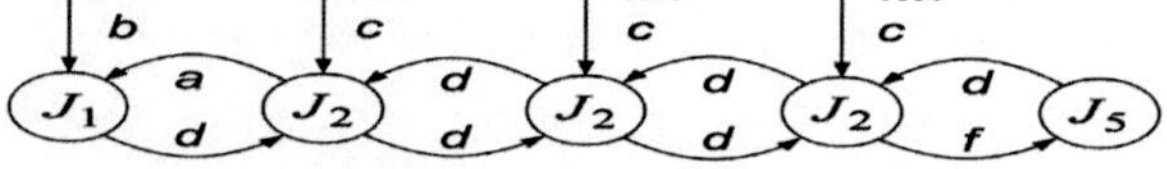

Fig. 3. Signal graph of the system of Eqs. (4)

Based on this graph, using the Meson formula, the determinants are obtained in the following form.

$$\Delta = 1 - ad - df - 2d^2 + ad^3,\ \Delta_1 = b\left(1 - df - 2d^2 + d^3 f\right),$$
$$\Delta_2 = ac\left(1 - d^2 - df\right)\Delta_3 = acd(1 - df)\Delta_4 = acd^2.$$

The results of calculating the Bode plot of torque transmission channels (W6, W7, W8, W9) are shown in Fig. 4. in Table 2 shows the values of special points of transfer functions that meet the above conditions.

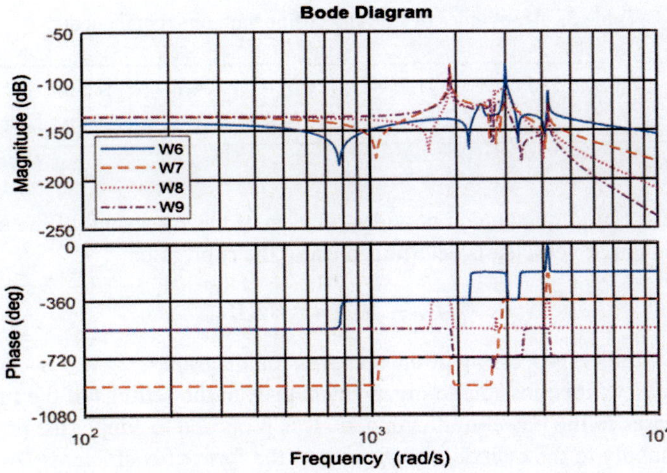

Fig. 4. Bode diagrams obtained using graph theory methods

Table 2. Roots of transfer functions

Zero				Pole
W_6	W_7	W_8	W_9	W
−3.4 + 780.8i	−3.7 + 1052.9i	−4.4 + 1327.5i	−2.5 + 1609.6i	−2.5 + 1909.1i
−3.4 − 780.8i	−3.7 − 1052.9i	−4.4 − 1327.5i	−2.5 − 1609.6i	−2.5 − 1909.1i
				−1139.4

When analyzing the data in Table 2, it is found a match with the previous calculations. Further studies of the transfer functions are performed under the condition that the signal of the measuring information is the fluctuations of the speed of rotation of the first mass of the crankshaft of the power unit. We use the Fourier transform to approximate the torques of the cylinders.

4.4 Torque Approximation

The torques of individual cylinders are obtained as a result of processing experimental pressure data. When setting the torque chart, the indicator and compression chart of the first cylinder are used. The discrete frequency spectrum of the first cylinder's torque from the action of gas forces is shown in Table 3.

Based on the data of this table, the torque of the first cylinder from the action of gas forces is mathematically represented by the following expression.

$$M_1(t) = \sum_{n=1}^{6} (A_n \sin n\Omega t + B_n \cos n\Omega t). \tag{9}$$

Table 3. Amplitudes of sine and cosine harmonic components

A_n, Nm	716.0	641.2	416.5	274.7	180.0	122.1
B_n, Nm	241.5	−18.1	−51.0	−61.5	−79.7	−64.8

The differential torque, which provides rotation of the crankshaft of the power unit with a given angular velocity, is determined using the expression.

$$M_p(t) = M_1(t) - M_2(t), \tag{10}$$

where $M_2(t)$ – torque processing of the compression diagram.

With this view, it is possible to organize changes in the settings of the fuel and air supply processes to the power unit cylinders. It is proposed to supply the processes of fuel and air supply to the cylinders of the unit in the form of coefficients $D_i = 0 \ldots 1$. Based on this, expression (10) takes the following form.

$$M(t) = D_i[M_1(t) - M_2(t)]. \tag{11}$$

The phase delay of cylinders relative to the first one is a multiple of 90° and is calculated taking into account the following sequence of their operation: 1-3-4-2. Accordingly, the representation of the torque of an arbitrary cylinder takes the following form.

$$M_i(t) = D_i e^{-j\Omega\tau_i}[M_1(t) - M_2(t)]. \tag{12}$$

In further research, we use the mathematical apparatus of the Fourier transform. Let us now proceed to the construction of computer simulation schemes for the first-mass rotation velocity fluctuations signal for cylinders.

4.5 Construction of Computer Simulation Schemes

Taking into account the expression (5), the frequency representation of the signal of fluctuations in the speed of rotation of the first mass of the crankshaft DG D964 looks like this.

$$\phi_1(j\omega) = \sum_{i=1}^{4} M_i(j\omega) W_i(j\omega). \tag{13}$$

Accordingly, the transfer function W5 takes the form.

$$W_5(j\omega) = \frac{1 - 9.302 \cdot 10^{-8}\,\omega^2 + 7.814 \cdot 10^{-7}\,j\omega}{1 - 1.154 \cdot 10^{-6}\,\omega^2 - j\left(9.03 \cdot 10^{-8}\,\omega^3 + 0.172\,\omega\right)}. \tag{14}$$

Other transfer functions have a similar appearance. When obtaining a discrete transfer function, expression (8) is decomposed into prime factors.

$$W_{10}(p) = \frac{A_1}{\frac{1}{a_6}p + 1} + \frac{A_2}{\frac{1}{a_5^2 + b_5^2}p^2 + \frac{2a_5}{a_5^2 + b_5^2}p + 1}. \tag{15}$$

Coefficients A_i are defined using the following expressions.

$$A_1 = 9.302 \cdot 10^{-8}\left(a_5^2 + b_5^2\right), \quad A_2 = \left(7.814 \cdot 10^{-7} - \frac{2a_5 A_1}{a_5^2 + b_5^2}\right) a_6.$$

Taking into account the coefficients A_i expression (15) takes the following form.

$$W_{11}(p) = \frac{0.172}{\frac{1}{a_6} p + 1} + \frac{9.713 \cdot 10^{-7}}{\frac{1}{a_5^2 + b_5^2} p^2 + \frac{2a_5}{a_5^2 + b_5^2} p + 1}. \quad (16)$$

We will compare expressions (8) and (16) using the Matlab software. The command line looks like this:

$W5 = tf([9.302 * 10^{-87.814} * 10^{-71}], [5.395 * 10^{-73.561} * 10^{-60.1721}]);$

$W13 = tf([0.172], [0.1721]); W14 = tf([9.713 * 10^{-7}], [5.395 * 10^{-73.561}$

$* 10^{-61}]);$

$W15 = W13 + W14;$ bode(W5, W15).

The comparison results are shown in Fig. 5. As a result of analyzing the Bode plot graphs, we conclude that the transformations are performed on a sufficient level. It is also established that it is advisable to decompose a fraction into prime factors in cases where they are the same: all aperiodic or vibrational. Summation of the transfer function in the presence of aperiodic and vibrational circuits is performed with significant uncertainty. If we sum up several aperiodic chains, there is no uncertainty.

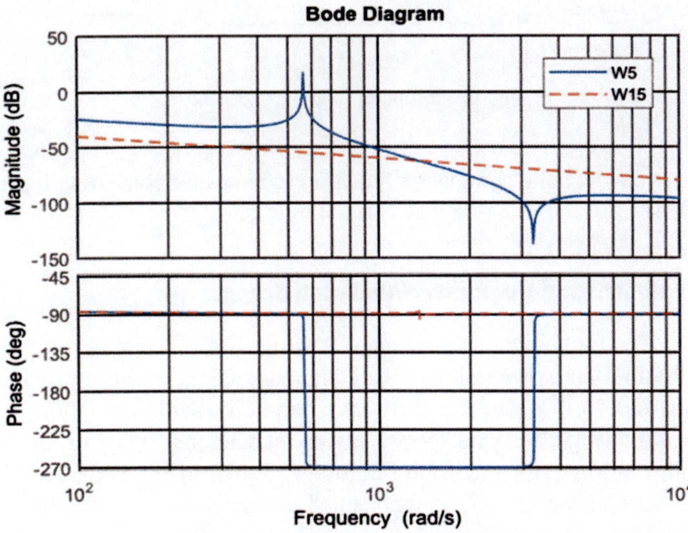

Fig. 5. The Bode plot graphs

To obtain a discrete transfer function, we use expression (8). In accordance with the z-transformation table, this expression takes the following form.

$$W_{16}(z) = \frac{A_1 z}{z - f_1} + \frac{A_2 z f_2 \sin bT}{z^2 - 2z f_2 \cos bT + f_2^2}, \tag{17}$$

where $f_1 = e^{-a_6 T}$, $f_2 = e^{-a_5 T}$, $T = 0.0005\ C$ – quantization period.

After simple mathematical transformations, the discrete transfer function looks like this.

$$W_{17}(z) = \frac{b_1^\circ z + b_2^\circ z^2 + b_3^\circ z^3}{a_0^\circ + a_1^\circ z^1 + a_2^\circ z^2 + a_3^\circ z^3}, \tag{18}$$

where $a_0^\circ = -e^{-Ta_6} e^{-2Ta_5}$, $a_1^\circ = e^{-Ta_5}\left(e^{-Ta_5} + 2e^{-Ta_6} \cos b_5 T\right)$, $a_2^\circ = -\left(2e^{-Ta_5} \cos b_5 T + e^{-Ta_6}\right)$;

$b_1^\circ = e^{-Ta_5}\left(A_1 e^{-Ta_5} - A_2 e^{-Ta_6} \sin b_5 T\right)$, $b_2^\circ = -e^{-Ta_5}(2A_1 \cos b_5 T - A_2 \sin b_5 T)$, $b_3^\circ = A_1$.

The results of calculating the coefficients give expression (18) the following form.

$$W_{17}(z) = \frac{0.172z^3 - 0.343z^2 + 0.171z}{z^3 - 2.549z^2 + 2.544z - 0.994}, \tag{19}$$

Based on the latter expression, a computer simulation scheme for the process of converting the torque of the first cylinder is compiled (Fig. 6).

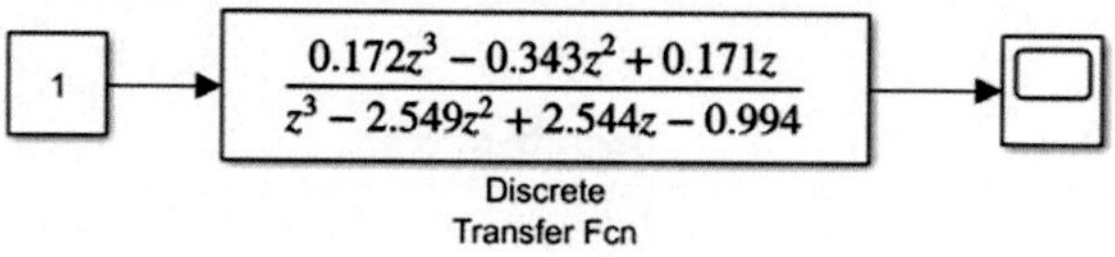

Fig. 6. Computer modeling scheme

Similarly, you can build computer simulation schemes for converting the torques of other cylinders.

4.6 Information Signal Computer Simulation Scheme

To build a scheme for computer modeling of the signal of fluctuations in the speed of rotation of the first mass, we will use the computational capabilities provided by the Mathcad software. In this case, the difference between calculations is that the output signal of the circuit will be a steady process of fluctuations. When using the Matlab software with the Simulink extension, the output signal of the computer simulation circuit has a constant and dynamic component. The very scheme of computer modeling of the signal of fluctuations in the speed of rotation of the first mass is shown in Fig. 7.

Figure 7 shows: $\Omega 1$-signal of fluctuations in the speed of rotation of the first mass of the crankshaft; $\mu 1$ – weight coefficients that are selected as a result of identifying the parameters of the mathematical model of the power unit. This procedure is organized based on the training model method. Identification information technology consists in comparing the signals of fluctuations in the rotation speed of the first mass of the model and experimental data.

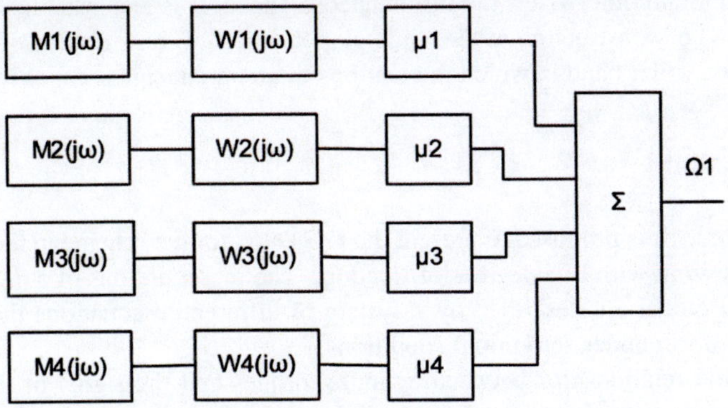

Fig. 7. Scheme of computer modeling in the Mathcad

The results of computer modeling are shown in Fig. 8. The operating speed of the DG D964 crankshaft is 30 s^{-1}.

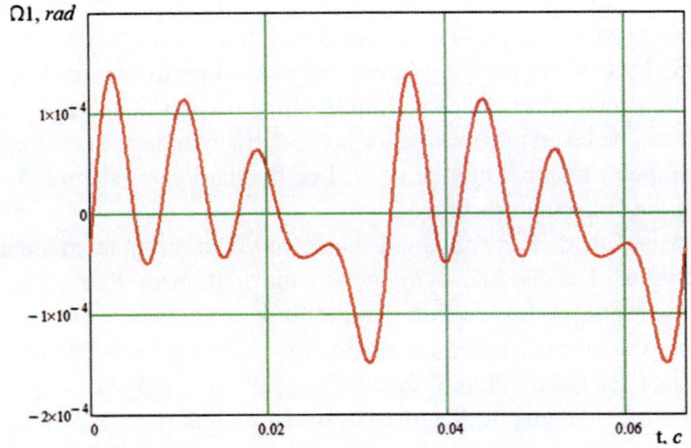

Fig. 8. Results of computer simulation

Analyzing the results of computer modeling allows us to make the following statements:

- the constant component of the discrete frequency spectrum of the fluctuation signal changes with the frequency of the first harmonic;
- the contribution of individual cylinders to the information signal of fluctuations differs in magnitude (maximum – the first, minimum – the fourth).

The second method of determining the transfer functions has less manual labor and provides better performance calculations. It is also established that the first method provides a simplified expression of the zero-pole transfer function, which may or may not be taken into account in further research. If this pole is taken into account, the output signal of the torque transmission channel is the angle of rotation of the crankshaft of the power unit, and otherwise - the instantaneous speed. It is also established that the second method of construction of the Bode plot channels of transmission of torques of cylinders has a wider band in which fluctuations of an output signal are observed.

5 Conclusion

For the first time, it is proposed to present the DG D964 torque scheme in the form of a mechanical system with five degrees of freedom. The mass motions of a deterministic mathematical model are described by a system of differential equations that uses the Laplace Transform under zero initial conditions.

Information relationships between cylinder torques and the signal of fluctuations of the first mass establish transfer functions, for the calculation of which the method of determinants and methods of signal graph theory are used. Logarithmic amplitude-frequency characteristics of torque transmission channels are studied in the Matlab software. A method for simplifying expressions for transfer functions is proposed. As a result of analyzing the special points of Transfer Functions, their satisfactory coincidence is established.

A comparison of methods for determining the transmission functions of torque transmission channels is made and it is established that the second method has a smaller amount of manual labor and provides better calculation performance. It is also established that the first method provides a simplified expression of the zero-pole transfer function, which can be taken into account or not in further studies. Therefore, the output signal of the computer simulation scheme will be fluctuations in the rotation angles of the first mass of the crankshaft or speed.

Discrete transfer functions are obtained, thanks to which computer modeling schemes for the contribution of individual cylinders to the first mass fluctuation signal are constructed in the Matlab software with the Simulink extension.

In the Mathcad software program, a scheme for computer modeling of the signal of fluctuations in the speed of rotation of the first mass of the crankshaft DG D964 is constructed a method for changing the length of information links between the contributions of individual cylinders to the fluctuation signal is proposed to identify the parameters of the mathematical model of the power unit. This procedure is performed on the basis of the method of the trained model, and experimental data on the instantaneous speed of rotation are used. As a result of analyzing the output signal of the computer modeling scheme, a satisfactory uncertainty of the model is established.

It is found that the use of developed methods and information technologies increases the productivity of the hardware construction process.

References

1. Rybin, Yu.K.: Measuring Signal Generators. Signals and Communication Technology. Springer, Cham (2014). https://doi.org/10.1007/978-3-319-02833-0.
2. Klepka, A.: Real time estimation of modal parameters of non-stationary systems using adaptive wavelet filtering and recursive least square algorithm. Diagnostyka **16**(1), 3–8 (2015). ISSN 1641-6414
3. Dorf, R.C.: Modern Control Systems. Richard C. Dorf, Robert H. Bishop, 12th edn. p. cm. Prentice Hall, Addison: Wesley (2010). ISBN-13:978-0-13-602458-3. ISBN-10:0-13-602458-0
4. Kotowski, A.: The method of frequency determination of impulse response components based on cross-correlation vs. fast Fourier transform. Diagnostyka **17**(1), 59–64 (2016). ISSN 1641-6414
5. Chaparro, L.F.: Signals and Systems Using MATLAB. Luis F. Chaparro. p. cm. Elsevier Inc. (2011). ISBN 978-0-12-374716-7
6. Dyakonov V.P.: Maple 9.5/10 v matematike, fizike i obrazovanii [Maple 9.5/10 in mathematics, physics and education] (2009). SOLON-PRESS, Russian. ISBN: 5980032584
7. Bilyk, S., Bozhko, E.: Analysis of methods and methods of diagnosing internal combustion engines by non-assembly control methods. Bulletin of the National Technical University "KhPI". In: New Solutions in Modern Technology, vol. 4, issue number 10, pp. 3–8.NTU "KhPI", Kharkiv (2021). https://doi.org/10.20998/2413-4295.2021.04.01
8. Marchenko, B.G., Myslovich, M.V.: The theory of diagnostics of energy aggregates by deviation of rotating units and its practical application for diesel-electric generators: part 1. modeling the dynamics of cylinder power on the shaft of a diesel-electric generator for the purpose of its diagnostics. Tekhnichna Elektrodynamika **5**, 36–40 (1998)
9. Bodnar, B.E., Ochkasov, O.B., Chernyaev, D.V.: Determination of the method of filtering the signal of non-uniformity of the diesel crankshaft speed. In: VisnykDNUZT Vip, vol. 1, issue number 43, pp. 113–118 (2013)
10. Safronov P.V., Kudinova A.V. Estimation of sensitivity of criteria of non-uniformity of frequency of rotation to non-reproducibility of working cycles in the cylinder of the internal combustion engine in the idling mode. In: Proceedings of the International scientific and technical conference, MADI, 2019, pp. 336–347, Russian (2019)
11. Bashirov, R.M., Insafuddinov, S.Z., Safin, F.R.: Uneven fuel supply in diesels: problems and methods of their solution. In: Izvestiya OGAU, vol. 1, issue number 75, pp. 78–82, Russian (2019)
12. Glushkov, S.P., Glushkov, S.S., Kochergin, V.I., Lebedev, B.O.: Analysis of dynamic characteristics of torsional-oscillatory systems of ship power plants. Mar. Intell. Technol. **2**(40), 59–66 (2018)
13. Bykov, V.V., Mokrushin, D.A., Petrov, A.I.: Diagnostics of the internal combustion engine using the method of vibroacoustic emission. In: Problems and Prospects of Development of the Motor Transport Complex: Proceedings of the International Scientific and Technical Conference, pp. 526–537, Russian (2019)
14. Sivyakov, B.K., Truber, S.S.: Diagnostics of multi-cylinder engines by speed unevenness. In: Vestnik of SGTU, vol. 1, issue number 44, pp. 76–82, Russian (2010)
15. Podrygailo N.M.: The effect of torque unevenness on dynamic and power indicators of internal combustion engines of wheeled vehicles. In: Uchenyie Zapiski KIPU, issue number 38, pp. 18–24, Russian (2013)
16. Savchenko, O.F., Dobrolyubov, I.P.: Modeling of the process of identifying the state of tractor engines. In: Problems of Computational and Applied Mathematics, vol. 4, issue number 6, pp. 4–12, Russian (2016)

17. Pokusayev, M.N., Sibiryakov, K.O., Shevchenko, A.V.: Experimental determination of the degree of irregularity of the collapse of the shaft of the machine-propulsion complex of the ship 1557. In: Vestnik AGTU, vol. 43, pp. 140–144, Russian (2008)
18. Grebennikov, S.A., Grebennikov, A.S., Nikitin, A.V.: Adaptive control of fuel supply by ICE according to indices of uneven rotation of the crankshaft. In: Vestnik SGTU, vol. 2, issue number 71, pp. 80–83, Russian (2013)
19. Enikeev, A.F., Borisenko, A.N., Samsonov, V.P., et al.: Diagnosis of a diesel generator by the deviation in shaft speed. Meas. Tech. **31**, 868–871 (1988). https://doi.org/10.1007/BF00863884
20. Yenikieiev, O., Shcherbak, L.: Information technology for protecting diesel-electric station reliable operation. Tekhnichna Elektrodynamika **4**, 85–91 (2019). https://doi.org/10.15407/techned2019.04.085
21. Yenikieiev O., Yevsiukova F., Prihodko O., Ivanova M., Basova Ye., Gasanov M. Analysis of the frequensy characteristics of the automatic control system of manufacturing process parameters. Acta Technica Naposensis **62**(111), 473–482 (2019). WOS:000489767000015
22. Yenikieiev O., Isikova N., Korotenko Ye., Reshetnyak T. Analysis of characteristics of hardware means for software control of the longitudinal feed of the grinding wheel. Acta Technica Naposensis **63**(11), 149–158 (2020). WOS:000550992100006

Effect of Magnetized Cutting Fluids on Metal Cutting Process

Mardonov Umidjon(✉), Andrey Jeltukhin, Yahyojon Meliboyev, and Baydullayev Azamat

Tashkent State Technical University, Tashkent 100097, Uzbekistan
umidjon.mardonov@tdtu.uz

Abstract. In this paper, the implementation of static magnetic field on lubricating cooling technological condition in the machining process is discussed and its effect on cutting temperature is analyzed experimentally. Moreover, the new construction of UMD-1 magnetizing device is given in the paper. Today, increasing the wear resistance of cutting tools used in metal cutting processes is one of the controversial issues of the manufacturing industry. Using the effect of magnetic field on lubricating cooling technological condition can be a modern solution to this problem. For this purpose, the authors designed a new device for magnetizing flowing lubricating cooling liquids in the machining process. Technical specifications of UMD-1 magnetizing device are also given in the paper.

Keywords: Cutting process · Cutting fluid · Magnetic field · Machining

1 Introduction

In nowadays´ manufacturing industry, the problems of ensuring the quality and durability of parts used in metal cutting process, increasing the strength and service life of cutting tools, the creation of new devices and technologies that ensure high accuracy of workpieces are of particular importance. One of the most important tasks in modern machine building is to increase the wear resistance of the cutting tools that process them in order to increase the size and surface accuracy of the machined parts. In this regard, extensive research is being conducted around the world to increase the stability of cutting tools used in metal cutting process.

Today, the demand for accuracy of manufactured pars is growing day by day. The highest accuracy is achieved by cutting, and the quality of the finished surface is also high. In order to achieve this, it is important to change the properties of cutting fluids used in the cutting process [1]. The scientific works on their physical and mechanical properties plays an important role in achieving the high requirements for the production process.

Many studies in recent years have suggested that the magnetic field (MF) changes the physical and mechanical properties of water [2, 3, 8]. If a liquid passes through a magnetic field, it becomes a magnetized fluid. Han et al. studied the optical properties of water between two strong magnets, and found that the magnetic field changed the

D. D. Cioboată (Ed.): ICoRSE 2022, LNNS 534, pp. 95–104, 2023.
https://doi.org/10.1007/978-3-031-15944-2_9

absorption of infrared light by the water [3]. Holysz found that MF increases water permeability and reduces surface tension [3]. Wang and his team studied the effects of a static magnetic field on a fluid in friction experiments, and the results show that the coefficient of friction in a magnetic field is lower [5]. Cai and other scientists have studied the effects of magnetic fields on the hydrogen bonds of water and talked about the mechanism of molecular dynamic simulation magnetization and their experimental and theoretical models [4].

Scientist in Baghdad F. L. Rashid [5] examined the evaporation rate by placing a 0.5 T magnetic field at different heights (surface, middle, and bottom) of the experimental water. The increase in evaporation time led to an increase in the amount of evaporation, He found that the best place to affect a magnetic field was on the surface of the water and it gave a higher evaporation rate than other parts (6% higher than in the case without a magnetic field). Furthermore, he found no effect when placing the magnetic field at the bottom of the water.

M. C Amiri and A. A. Dadkhah [6] determined whether physical purification of water would result in a decrease in its surface tension. They did their work by measuring purified and untreated water during physical purification. The test results showed that the surface tension of the water is very sensitive to the experimental conditions performed on it in order to say that it is a safe and alternative indicator to study the effect of the magnetic field on the water. They found that the surface tension of the fluid in the sample after one day could be the most optimal indicator to show the physical changes in the sample.

They also found that the surface tension of ordinary and purified water depends on the intensity of the magnetic field. Figure 1 shows the dependence of the number of magnetic field effects on the surface tension of fresh water.

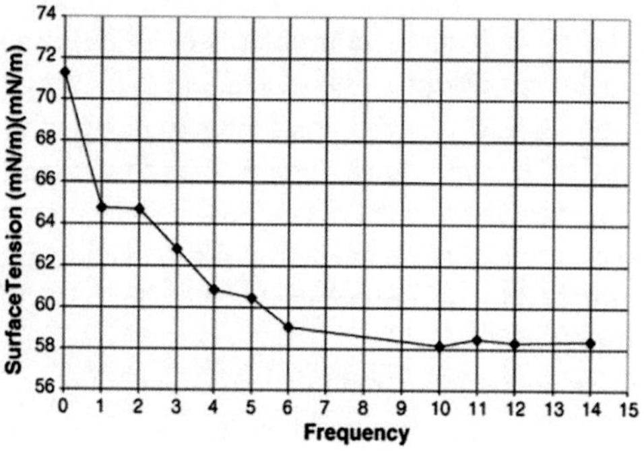

Fig. 1. The relationship between the surface tension of fresh water and the effect of magnetic field intensity on it.

V.N. Tirtigin [7] and others studied the effect of a weak electromagnetic pulsating field on the microflora of a lubricating aqueous lubricant. A hypothesis was made about the effect of an electromagnetic pulse field in a very low frequency range on the microflora. During the study of water-oil emulsion purification, the effect of the lubricating coolant microflora of the electromagnetic pulse field in a very low frequency range was studied.

During operation, the lubricating cooling liquid is contaminated by microorganisms. Signs of microbiological damage to the lubricating coolant include the release of hydrogen sulfide (Class 2 hazard) and other gases with an unpleasant odor. When the lubricating coolant has an unpleasant odor, it is considered unusable even at high technological capacity and requires replacement.

H. Banejad and E. Abdosalehi [9] studied the effect of magnetic field on water hardness. They analyzed 0 T, 0.05 T, 0.075 T, and 0.1 T induced magnetic fields. Three studies and verification of the results using the SAS program revealed that the induction of the magnetic field and the amount of water it affects, as well as their combined changes, have a significant effect on the hardness of the water at 99%.

Youkai Wang, Huinan Wei, and Zhuangwen Li [10] studied the effects of magnetic fields on the physical properties of water. Ordinary water (OW) and 4 different magnetic waters were studied under the same conditions. What they found was that the properties of ordinary water changed after the effect of the magnetic field, the amount of evaporation increased, and the specific temperature and boiling point decreased after magnetization. The change depends on the level of magnetization. In addition, magnetic field strength (MFS) has a strong effect on the magnetization level, the optimal magnetization level is determined when the magnetic field induction is B = 300 mT. The results of their research are very useful in increasing cooling and energy efficiency in industry.

Zhian Liu and his colleagues [11] developed high-tech electromagnetic pulse purification instead of chemical purification of industrially used cutting fluid. For better sterilizing the cooling cutting fluid without harming the environment, some experiments on cleaning the heteratropic bacteria were carried out in a low-pulsating magnetic field. They increased the bacterial loss rate from 7.22% to 20.35% and found that the result was higher when the direction of the magnetic field was parallel to the direction of fluid flow.

From the analysis, it can be concluded that the magnetic field has a significant effect on the cutting process. Another important aspect is that the thermal processing efficiency of the cutting material can also be controlled. To do this, you can use a lubricating coolant during the cutting process to increase the durability. The magnetic field also changes the degree of exposure by changing the physicochemical properties of the lubricating cutting fluid. It is also clear that the effect of a magnetic field on liquids depends on their state of magnetization. In conclusion, modifying the above properties and managing their impact on the cutting process is of great scientific and practical importance.

2 Methods

The accumulated experience in the creation of magnetizing devices for magnetized cutting fluid, their properties, application in the cutting process and the flowing lubricating cooling technological environment show that, in addition to the specific advantages it should be noted that the effect of magnetized cutting fluid with increased cooling capacity on the wear resistance of cutting tool is low when working at small cutting speeds and drilling deep holes. In the magnetization process of flowing lubricating cooling liquids and in the design and construction of a magnetizing device, first of all, the cutting process intended for their use must be taken into account. It is also necessary to pay attention to factors such as the main properties of cutting fluids that affect the cutting process, the method of delivery to the cutting zone, flow rate, cutting mode elements, design solutions of the magnetizing device.

The use of various lubricating and cooling technological conditions in metal cutting process increases the wear resistance of the cutting tool, reduces cutting forces, improves the surface quality of the workpiece, increases the resistance of the product to operational fatigue, increases labor productivity. As a result, product competitiveness increases. Therefore, almost all mechanical processing is carried out using lubricating cooling technological condition. Cutting fluids has an active effect on the friction plastic contact surfaces of the cutting tool.

The main purpose of using cutting fluids is to reduce the cutting temperature, reduce the cutting forces and power consumption, and, consequently, increase the wear resistance of the tool, improve the quality of the machined surface and labor productivity [12]. The use of cutting fluids is especially important when cutting refractory and stainless high alloy steels.

2.1 Magnetization Scheme of a Moving Cutting Fluid

When using magnetized cutting fluid in metal cutting process, a free-flowing method of delivering cutting fluids to the cutting zone was chosen. In order to do the experiment with the effect of cutting fluid on the wear of the cutting tool, it is necessary to develop the technology and scheme of magnetizing of the cutting fluid in flowing condition to the cutting zone.

The upper magnets of the universal magnetizing device UMD-1 (1) are placed at a distance between them relative to the lower magnets. They are separated from each other by long hexagonal screws. Different types of magnetic field strength can be selected to magnetize liquids, and the strength of the magnetic field is controlled by varying the distance between the magnets, or more precisely, we change the height between the upper and lower magnets using hexagonal screws to obtain different magnetic field forces (Fig. 2). The ceramic tube is placed in the middle of the magnets and its diameter can be selected depending on the height between the upper and lower magnets. The polyvinyl chloride pipe passes through the UMD-1 magnetizing device and the magnetic field length is up to a maximum of 350 mm. The liquid that passes through the UMD-1 magnetizing device through the polyvinyl chloride pipe is converted into a magnetized liquid. A scheme of the magnetization process is shown in Fig. 2.

It should be noted that magnetized cutting fluids expand the capabilities of tools and environments to increase production efficiency in mechanical engineering. Magnetized cutting fluids may not be used where conventional cutting fluids have traditionally been used more effectively. However, for some reason, in the technological operations of cutting materials that do not use traditional lubricants, magnetized cutting fluids increase the stability of the cutting tool. In addition, they support to improve the quality of the machined surface, and machining the materials that is difficult to process titanium alloys, corrosion-resistant steels and can prevent accidental breakage of the cutting tool when

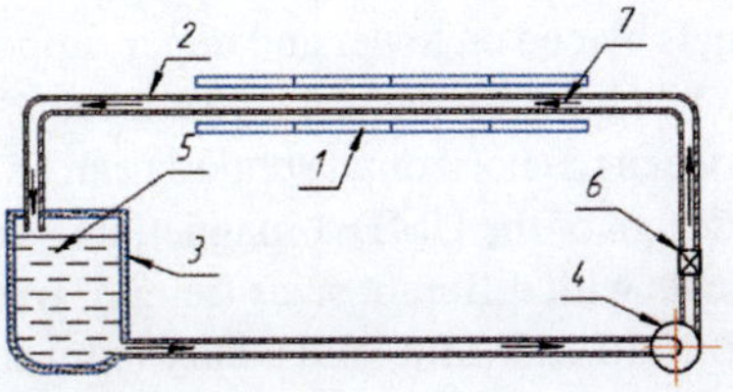

Fig. 2. Magnetization scheme of flowing liquid. 1 – UMD-1 magnetizing device, 2 – Polyvinyl chloride pipe, 3 – container, 4 – pump, 5 – liquid, 6 – tap, 7 – flowing direction of the liquid.

processing raw materials containing other materials. Devices designed for the magnetic cleaning of various liquids during machining process are very important, because changes in the physical properties of the liquid affect the wear resistance of the cutting tool used during machining. The effect of magnetized cutting fluid on the wear resistance of cutting tools helps to increase machining efficiency and save energy.

2.2 Design of a Magnetizing Device to Magnetize Moving Cutting Fluid

It is very important to design a new magnetizing device to magnetize cutting fluids when cutting metals. The universal magnetizing device UMD-1 is designed to magnetize flowing cutting fluids and it is very convenient to apply to any metal cutting machine (turning, milling, grinding, drilling machines). The main purpose of the UMD-1 magnetizing device is to hold the super magnets with different poles and to change the distance between these magnets in order to change the strength of the permanent magnetic field generated between them (Fig. 3.).

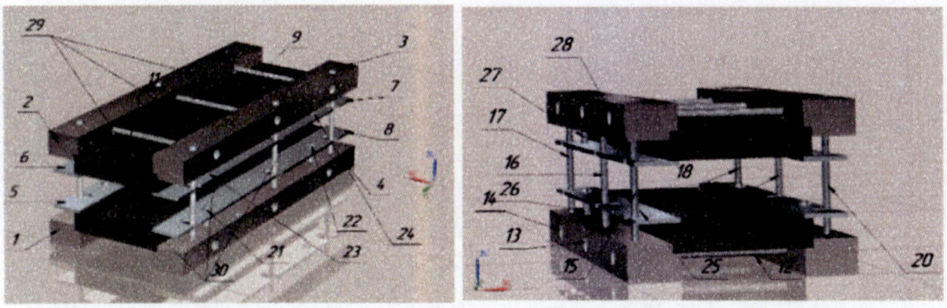

Fig. 3. UMD-1 magnetizing device. 1–4 – supports, 5–8 - magnet holders, 9–14 – horizontal hexagon bolts, 15–20 – vertical hexagon bolts, 21–28 – screws to fix magnet holders, 29, 30 – magnets.

Figure 3 shows that, magnets (29, 30) are placed on lower (1, 4) and upper (2, 3) supports with different poles. And long horizontal hexagon bolts (9, 10, 11, 12, 13, 14) are fixed according to magnets width, then holders(5, 6, 7, 8) are fixed to hold magnets hard by screws(21–28). Holders and supports are developed to change the distance between them. Lower and upper supports have the same construction and the distance between the lower and upper supports can be changed easily by vertical hexagon bolts (15–20). Changing the distance between the lower and upper supports is purposed to change the distance between the magnets placed on lower and upper supports. We can easily change the magnetic field strength by changing the distance between lower and upper magnets. Moreover, the distance between the horizontal supports can be changed according to the sizes of the magnets. The design of the UMD-1 magnetizing device is universal because it is possible to place magnets with different sizes (length, width, height). Magnets that have a minimum size of 20 × 10 × 5 (mm) and a maximum size of 300 × 80 × 30 mm can be placed on UMD-1 magnetizing device. The distance can be changed from 0mm to 50 mm with respect to the sizes of the magnets placed on the supports. Furthermore, UMD-1 magnetizing device is very portable and it is very easy to change its position on cutting machines.

2.3 Measurement of Cutting Temperature During Metal Cutting Process

One of the main tasks of our research work is to analyze the changes in the cutting temperature that generates during the cutting process of metals. The use of the most accurate natural thermocouple method in modern production to measure the temperature in the cutting zone helps to achieve high accuracy. This method allows to obtain accurate results about the average temperature on the surfaces of the tool in contact with the abrasive and the workpiece [13]. Figure 4 shows a schematic diagram of the temperature generated during cutting. The value of the thermal current was recorded using an oscilloscope.

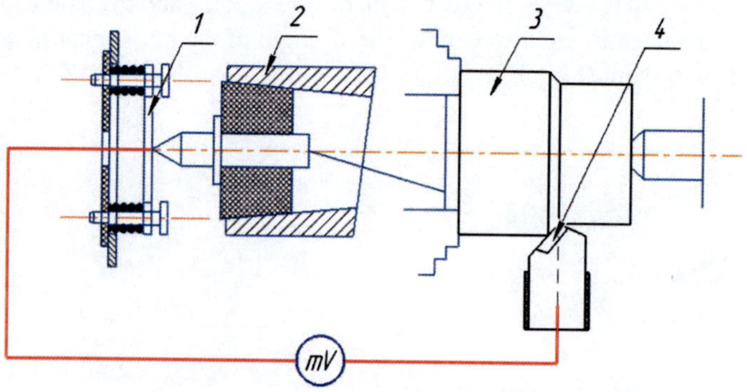

Fig. 4. Scheme of temperature measurement by natural thermocouple method. 1 – current collectors; 2 – spindle; 3 – detail; 4 – cutting tool.

P6M5 high-speed steel cutting tool was used in the experiments. An entire cutting tool was used to ensure the accuracy of the results and the uniformity of the condition (Fig. 5).

A point collector was used in the scheme between the cylindrical raw material being processed and the oscilloscope. A copper conductor mounted on a special plug hole in the back transmits electromotive force to the end of the collector plug. All parts of the collector are made of the same copper to prevent interference. This design of the collector is convenient for accurate delivery of thermocouples in rotating raw materials.

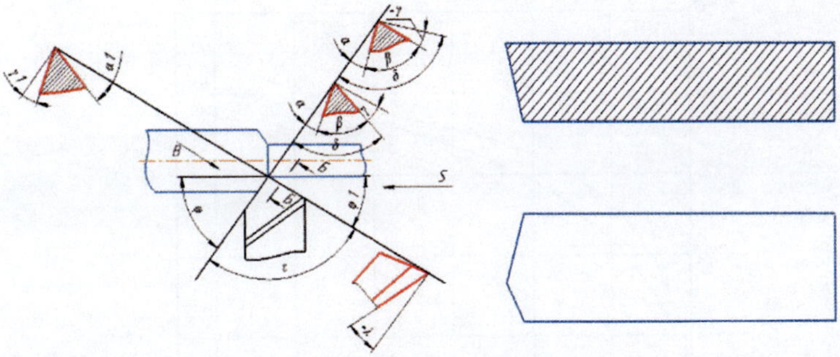

Fig. 5. The whole cutting tool made of P6M5 high-speed steel and its basic geometrical parameters. α – main rear angle, β – the cutting angle if the tool, λ – the angle on the main surface, δ – cutting angle, γ - front angle, φ_1 – auxiliary corner in the plan, φ – main corner in the plan, ε – the planned angle of the cutting edge.

No additional studies have been conducted to investigate the effect of cutting fluids on measurement accuracy when measuring the temperature in the cutting zone using the natural thermocouple method, as the experiments we have performed so far show that the use of cutting fluid in the cutting zone shows that it does not adversely affect measurement accuracy and results [14].

3 Results and Discussions

The condition of the layers of friction surfaces (front and rear surface of the cutting tool, the front surface of the scraper, the surface of the workpiece) is determined by their temperature. Therefore, in the study it is important to have knowledge on the friction laws and wear of the cutting tool, the main parameters of the surface quality of the machined parts, it is important to have information about the temperature of the contact layers [15].

Experimental studies have been conducted to investigate the effect of conventional and magnetized cutting fluids on the cutting temperature when cutting raw materials with a material of steel 45 with P6M5 high-speed steel. Two water-based cutting fluids were selected for the study: chromic (2% aqueous solution of potassium di chromate) and LACTUCA LT 3000 (5% aqueous solution). During the experiments, the dependence of temperature on the cutting speed in different cutting condition was studied, and the results were determined on the basis of this relationship (Fig. 6 and Fig. 7).

It is known that the use of cutting fluids during cutting reduces the caching temperature [16]. Experiments have shown that the drop in cache temperature when using these lubricating fluids in a magnetic cutting process is much higher than when using them in the traditional way. Also, as the cutting speed increases, the effect of liquids on temperature decreases, which is expressed by the following coefficient $K_{\theta v}$.

$$K_{\theta v} = \frac{\theta_{dry}}{\theta_{Cuttingfluid}} \tag{1}$$

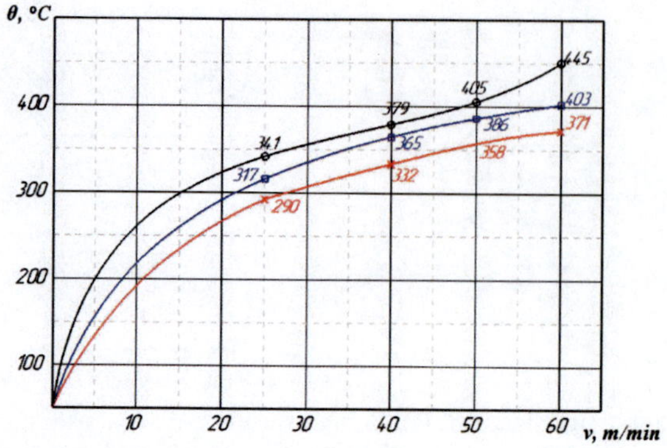

Fig. 6. The dependence of the experimental cutting temperature on the cutting speed when cutting Stal 45 with P6M5 high-speed steel: O – dry, blue line – with traditional cutting fluid (Chromic) x –magnetized cutting fluid (Chromic).

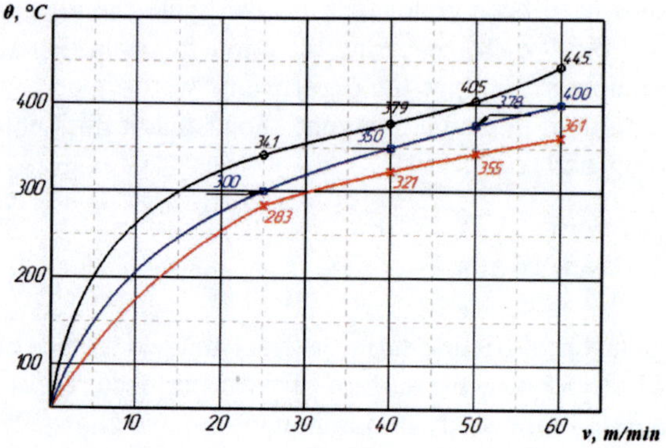

Fig. 7. The dependence of the experimental cutting temperature on the cutting speed when cutting Stal 45 with P6M5 high-speed steel: O – dry, blue line – with traditional cutting fluid (LACTUCA LT 3000) x –magnetized cutting fluid (LACTUCA LT 3000).

To evaluate the effect of magnetized and non-magnetized cutting fluids on the rate of decreasing the cutting temperature at constant cutting speed and different cutting speeds, tables related to the cutting speed of the coefficient $K_{\theta v}$ were developed. The formed dependencies were extrapolated to $K_\theta = 1.01$, and at this value of the coefficient can be assumed that the effect of the cutting fluid on the temperature is almost negligible (Tables 1 and 2).

Based on the results showed in the tables, it can be said that even in experiments conducted in two different fluids, the cutting fluids significantly increased their effect on the reduction of the cutting temperature after magnetization. In the first fluid experiment, this figure increased to 18%, while in the second fluid condition it increased to 23%.

Table 1. Coefficient of impact $K_{\theta v}$ of cutting fluid (chromic) when cutting steel 45 with P6M5 high-speed steel (feed is s = 0.45 mm/rpm).

Cutting condition	Cutting speed v, m/min			
	25	40	50	60
	$K_{\theta v}$			
Cutting fluid (traditional method)	1.07	1.05	1.04	1.1
Magnetized cutting fluid	1.18	1.14	1.13	1.13

Table 2. Coefficient of impact $K_{\theta v}$ of cutting fluid (LACTUCA LT 3000) when cutting steel 45 with P6M5 high-speed steel (feed is s = 0.45 mm/rpm).

Cutting condition	Cutting speed v, m/min			
	25	40	50	60
	$K_{\theta v}$			
Cutting fluid (traditional method)	1.13	1.08	1.07	1.11
Cutting condition	1.20	1.18	1.14	1.23

4 Conclusions

Changes in the cutting temperature when using magnetic and non-magnetic cutting fluids in the metal cutting process of at different cutting speeds were studied, and it was found that the cutting temperature decreased by 10–15% when using magnetized cutting fluids. The obtained results allow to calculate the controllable parameters that are optimal to reduce the wear of the cutting tool when cutting metals on a lathe. A new technology developed for the use of magnetized cutting fluids in metal cutting has shown that it is one of the most promising areas to increase the wear resistance of cutting tools used in machining process.

References

1. Umarov, E., Mardonov, U., Ozodova, Sh.: Analysing the effect of magnetic field on liquids. In: Science, Research, Development Conference 2020, vol. 26/8, pp. 62–66, Berlin, Germany (2020)
2. Umarov, E.O., Mardonov, U.T., Shoazimova, U.: Influence of the magnetic field on the viscosity coefficient of lubricoolant that is used in the cutting process. Int. J. Mechatron. Appl. Mech. **8**(2), 144-149. (2020). https://doi.org/10.17683/ijomam/issue8.50
3. Umarov, E., Mardonov, U., Abdirakhmonov, K., Eshkulov, A., Rakhmatov, B.: Effect of magnetic field on the physical and chemical properties of flowing lubricating cooling liquids used in the manufacturing process. IIUM Eng. J. **22**(2), 327–338 (2021). https://doi.org/10.31436/iiumej.v22i2.1768

4. Holysz, L., Szczes, A., Chibowski, E.: Effects of a static magnetic field on water and electrolyte solutions. J. Colloid Interface Sci. **316**(2), 996 (2007). https://doi.org/10.1016/j.jcis.2007.08.026
5. Wang, Y., Zhang, B., Gong, Z., et al.: The effect of a static magnetic field on the hydrogen bonding in water using frictional experiments. J. Mol. Struct. **1052**(11), 102–104 (2013). https://doi.org/10.1016/j.molstruc.2013.08.021
6. Cai, R., Yang, H., He, J., et al.: The effects of magnetic fields on water molecular hydrogen bonds. J Mol Struct **938**(1–3), 15–19 (2009)
7. Rashid, F.L., Hassan, N.M.: Increasing water evaporation rate by magnetic field, Int. Sci. Invest. J. **2**(9), 61–68 (2013). http://isijournal.info/journals/index.php/ISIJ/article/view/12
8. Amiri, M.C., Dadkhah, A.A.: On reduction in the surface tension of water due to magnetic treatment. Colloids Surf. A **278**(1–3), 252–255 (2006)
9. Tirtigin, V. N., Sobgayda, N.A., Potexa, V.L., Shayxiev, I.G., Makarova, Yu.A.: Vliyanie elektromagnitnogo impulsnogo polya na mikrofloru smazochno-oxlajdayuщey jidkosti [Influence of an electromagnetic pulsed field on the microflora of a cutting fluid]. Vestnik Kazanskogo texnologicheskogo universiteta **16**(3) (2013)
10. Nodir, T., Sarvar, T., Andrey, J., Yahyojon, M.: Mathematical model for calculating heat exchange. In: Cioboată, D.D. (ed.) ICoRSE. LNNS, vol. 305, pp. 243–249. Springer, Cham (2022). https://doi.org/10.1007/978-3-030-83368-8_24
11. Banejad, H., Abdosalehi, E.: The effect of magnetic field on water hardness reducing. In: Water Technology Conference, IWTC, pp. 117–28, May 2009
12. Wang, Y., Wei, H., Li, Z.: Effect of magnetic field on the physical properties of water. Results Phys. **8**(5), 262–267 (2018)
13. Liu, Z., et al.: the sterilization effect of solenoid magnetic field direction on heterotrophic bacteria in circulating cooling water. Procedia Eng. **174**, 1296–1302 (2017). https://doi.org/10.1016/j.proeng.2017.01.274
14. Umarov, T.U., Mardonov, U.T.: General characteristic of technological lubricating cooling liquids in metal cutting process. Int. J. Res. Adv. Eng. Technol. **6**(2), 01–03 (2020)
15. Umidjon, M., Muhammad, T., Andrey, J., Yahyojon, M.: The difference between the effect of electromagnetic and magnetic fields on the viscosity coefficients of cutting fluids used in cutting processes. Int. J. Mechatron. Appl. Mech. **10**(1), 117–122 (2021). https://doi.org/10.17683/ijomam/issue10/v1.14
16. Erkin, U., Umidjon, M., Umida, S.: Application of magnetic field on lubricating cooling technological condition in metal cutting process. In: Cioboată, D.D. (ed.) ICoRSE. LNNS, vol. 305, pp. 100–106. Springer, Cham (2022). https://doi.org/10.1007/978-3-030-83368-8_10
17. Makarov, A.D.: Optimizatsiyaprotsessov rezaniya [Optimization of cutting processes], 1st edn. Mashinostroenie, Moscow (1976)
18. Reznikov, A.N.: Teploobmen pri rezanii i oxlajdenii instrumentov[Heat transfer during cutting and cooling of tools]. 1st edn. Mashgiz (1963)
19. Uljayev, E., Ubaydullaev, U.M., Narzullayev, S., Norboyev, O.N.: Application of expert systems for measuring the humidity of bulk materials. Int. J. Mechatron. Appl. Mech. **9**(1), 131–137 (2021)
20. Uljaev, E., Narzullayev, S., Utkir, U., Shoira, S.: Increasing the accuracy of calibration device for measuring the moisture of bulk materials. In: Cioboată, D.D. (ed.) ICoRSE. LNNS, vol. 305, pp. 204–213. Springer, Cham (2022). https://doi.org/10.1007/978-3-030-83368-8_20

Development of Liquefaction Technology 280X29NL to Increase the Strength and Brittleness of Castings

Saidmakhamadov Nosir[1] (✉) and Karimov Bokhodir[2]

[1] Tashkent State Technical University, Tashkent 100097, Uzbekistan
anvarovichsarvar908@gmail.com
[2] Namangan Engineering – Construction Institute, Namangan, Uzbekistan

Abstract. The chemical composition of the casting material has been developed to increase the ductility of high – chromium cast iron castings in order to increase the service life of high – friction discs of CEMCO and BARMAK crushers, which operate mainly by centrifugal force as well as samples of five grades of sand – clay molds were poured and tested in sand – clay molds without compromising the mechanical properties of the casting.

Keywords: Carbon · Alloys · Abrasion resistance · Corrosion resistance

1 Introduction

According to the results of research conducted by professors and researchers of the Department of "Casting Technologies" of Tashkent State Technical University on the basis of research conducted by NMZ of Navoi Mining and Metallurgical Enterprise, currently engineering, mining, chemical engineering and others. A number of enterprises have developed the technology of casting of high-chromium cast iron in the form of sand – clay, coke and centrifugal castings. The main brands of cast iron are ICh290X28N2, ICh260X17N3G3, ICh290X12M, ICh290X12G5, ICh280X29NL, ICh300X32N2M2TL and others [1].

Nowadays, mechanical engineering requires the use of materials with good mechanical properties, but alloys that provide tensile strength, relative elongation and strength, as well as increased properties such as abrasion resistance, corrosion resistance, heat resistance, and other types of abrasive and aggressive. The details are important to increase the service life of parts in different operating conditions. First of all, it applies to alloys such as iron – carbon alloys, including high – chromium cast iron, the production of which is increasing year by year [2].

Cast iron is an unwanted alloy of the group of iron – carbon alloys with a carbon content higher than 2,14%. The components of the alloy affect its color during fracture: the surface formed by the fracture of a white cast iron sample appears white back to sunlight, which is why white cast iron is composed entirely of cementite. The name of the cementite is Floris Osmond and J. According to Vert's research, iron is a chemical

D. D. Cioboată (Ed.): ICoRSE 2022, LNNS 534, pp. 105–115, 2023.
https://doi.org/10.1007/978-3-031-15944-2_10

compound formed with carbon (Fe_3C) that contains 6,67% carbon, a substance with an orthorombic crystal structure, high hardness. Cementite is present in many steels and cast irons, and their crystalline structure is shown in Fig. 1.

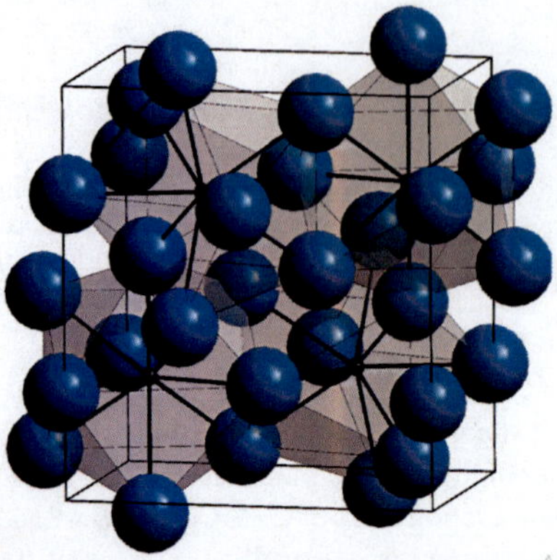

Fig. 1. Orthorombic Fe_3C. Iron atoms are blue.

In addition to iron, cast iron can contain 2,14 to 4,3% carbon (C) and 1 to 3% silicon (Si). Cast iron has become a widely used engineering material due to its brittleness and relatively low melting point, good ductility, easy machining, deformation resistance and good brittleness. It is currently used in parts of the pipes, machinery and automotive industries, for example in the manufacture of cylinders (cylinder blocks and gearboxes). Cast iron is also resistant to corrosion as a result of oxidation.

The oldest cast iron objects date back to the 5th century BC and were found by archaeologists in what is now Jiangsu, China. Cast iron was used in ancient China for war, agriculture, and architecture [3]. During the 15th century French reforms in Burgundy and England, cast iron was used to extract cannons and cannons for the military. The cast iron used to make the balls was produced more than other cast irons [4]. The first iron bridge in Shropshire, England was built in the 1770s by Abraham Darby III. Cast iron was also used in the construction of the buildings.

Cast iron is obtained from ores, which is an iron alloy liquefied in a high – temperature furnace. Cast iron can also be liquefied from ore or secondary shale materials [5]. It is often used in combination with iron, steel, limestone, carbon (coke) and various fluxes for stone removal. By keeping the cast iron in the oven at a high temperature for a certain period of time, harmful additives can reduce phosphorus and sulfur, but in the process the amount of carbon is also burned.

The corrosion resistance of cast iron is mainly provided by carbides with (C, Fe, $Mn)_7$ C_3, $M_{23}C_6$, M_7C_3, M_3C_2 and M_3C structure (C, Fe, $Mn)_7$ C_3. This is because these carbides are 1,5 to 2,0 times harder than cementite carbides. Another complication associated with this is that the amount of chromium in cast iron, which has 3% C for the formation of carbides in the system (Cr, Fe, $Mn)_7C_3$, is in the range of a maximum of 12 to 27% [6–9].

2 Methods

At present, the defects in the disks of CEMCO and BARMAK crushers operating on the basis of centrifugal force in the process of crushing ore in the production conditions of NMZ of Navoi Mining and Metallurgical Plant and the causes of their formation were analyzed (Fig. 2).

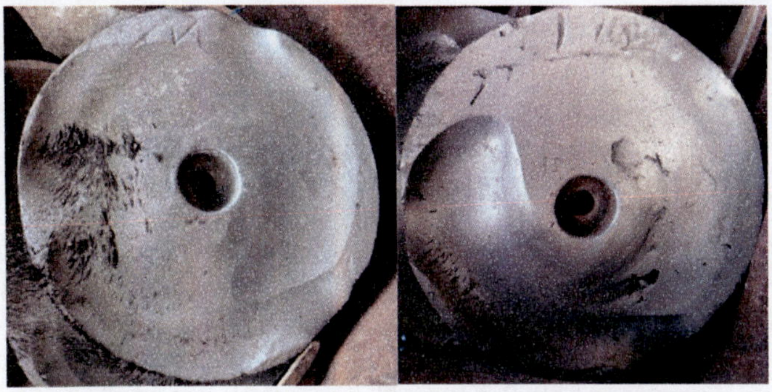

Fig. 2. A view of a disc cast that has become unusable.

In order to increase the service life of the part, changes were made in its chemical composition to ensure durability on the surfaces of parts with a high tendency to corrosion under the influence of strong stress and a high probability of cracking, five samples were developed and results were obtained. Research work of domestic and foreign manufacturers on corrosion – resistant high – chromium cast iron – based cast alloys, as well as research by foreign research institutions and laboratories to extend the service life of high – strength chromium cast discs were analyzed, and casting in molds was recommended (Fig. 3) [10, 11].

Fig. 3. Sand – clay mold drawing for casting.

The chemical composition of the casting material has been developed to increase the ductility of high – chromium cast iron castings in order to increase the service life of CEMCO and BARMAK crushers, which operate mainly under centrifugal force. The chemical composition of the alloy is proposed in Table 1 below [12].

Table 1. Chemical composition of alloys for proposed samples.

Brand	Elements, %									
	C	Si	Mn	Cr	Mo	Ti	Ni	Cu	P	S
280X29NL (Sample 1)	2,8–3,0	1,1–1,2	0,6–0,8	17,2–18,2	1,8–2,0	–	–	–	0,022–0,024	0,007–0,009
280X29NL (Sample 2)	2,8–3,0	1,1–1,2	0,6–0,8	15–16	–	1,31–1,78	–	–	0,02–0,04	0,07–0,09
280X29NL (Sample 3)	2,8–3,0	1,1–1,2	0,6–0,8	14–15	0,8–1,0	–	0,9–1,0	0,9–1,0	0,02–0,04	0,07–0,09
280X29NL (Sample 4)	2,8–3,0	1,8–2,0	0,4–0,5	10–11	–	–	–	–	0,04–0,05	0,04–0,05
280X29NL (Sample 5)	2,7–3,2	0,4–0,8	3,8–4,0	19–20	–	–	–	–	0,05–0,06	0,04–0,05

Sample 1. After coordination, the slag material was heated in an IST – 0,4 furnace to a temperature of 1400–1450 °C, ferroalloys were introduced after the slag was removed, and after holding for 10 min, it was poured into a sand – clay mold.

The internal structure of the alloys obtained by casting was cooled in air after holding for 12 h at a temperature of 1000 °C in a SNOL – 7.2/1100 muffle furnace for heat treatment to increase its physical and mechanical properties (Figs. 4 and 5).

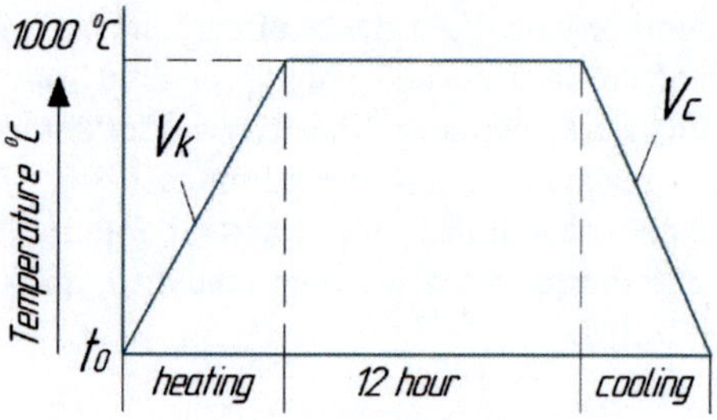

Fig. 4. Sample heat treatment graph.

Fig. 5. SNOL – 7,2/1100 muffle furnace.

Sample 2. After coordination, the slag material was heated in an IST – 0,4 furnace to a temperature of 1420–1430 °C, ferroalloys were introduced after the slag was removed, and after holding for 10 min, it was poured into a sand – clay mold.

The internal structure of the alloys obtained by casting was cooled in air after holding for 1 h at a temperature of 980 °C in a muffle furnace SNOL – 7,2/1100 for heat treatment to improve the physical and mechanical properties (Fig. 6).

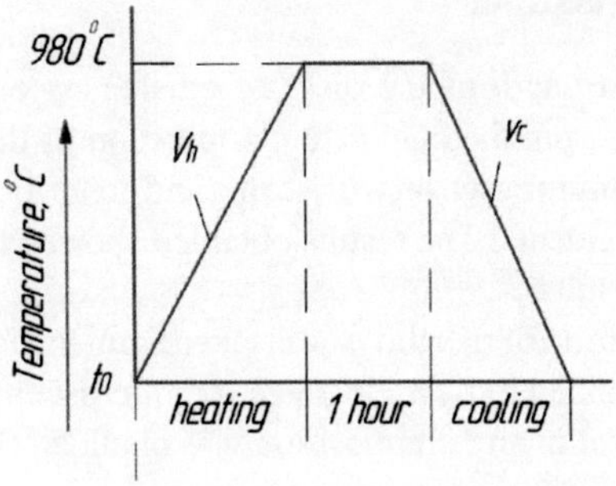

Fig. 6. Sample heat treatment graph.

Sample 3. The slag material was heated in an IST – 0,4 furnace to a temperature of 1390–1400 °C after coordination, ferroalloys were introduced after the slag was removed and after holding for 10 min it was poured into a sand – clay mold.

The internal structure of the alloys obtained by casting was cooled in air after holding for 15 min at a temperature of 850 °C in a SNOL – 7,2/1100 muffle furnace for thermal treatment to increase its physical and mechanical properties (Fig. 7).

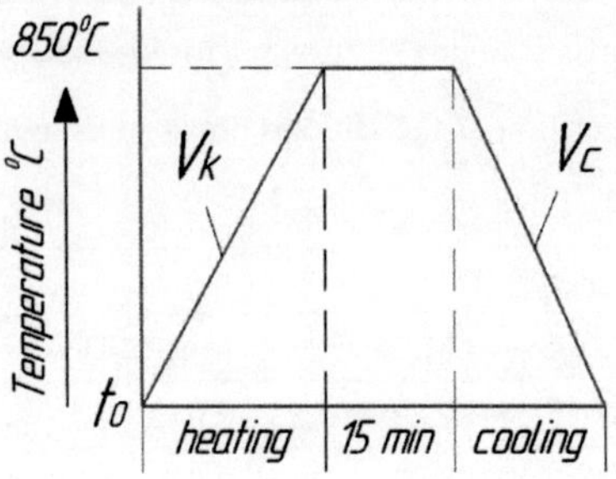

Fig. 7. Sample heat treatment graph.

Sample 4. After coordination, the slag material was heated in an IST – 0,4 furnace to a temperature of 1482 °C, ferroalloys were introduced after the slag was removed, and after holding for 10 min, it was poured into a sand – clay mold.

Sample 5. After coordination, the slag material was heated in an IST – 0,4 furnace to a temperature of 1400–1410 °C, ferroalloys were introduced after the slag was removed, and after holding for 10 min, it was poured into a sand – clay mold.

All five samples were mechanically treated after cooling in a sand – clay mold. The chemical composition of the alloy was determined by electron microscopy scanning sample 1, sample 2, sample 3, heat – treated sample 4 and sample 5 and "SPEKTRO-LAB – 10M". As a result of the study, the hardness of the samples was determined by means of a hardness measuring device brand TK – 2M to determine the hardness of the alloy.

3 Results and Discussions

In order to increase the strength of the discs of crushers operating under high stress from high chromium cast iron, the chemical composition of the shale material for the production of high – chromium cast iron with strong and dendritic structure was increased on the basis of alloying elements. The results obtained showed that research in this area may yield the expected results.

The chemical composition of the alloys was taken from the furnace in the liquid state and all samples were examined by scanning electron microscopy and SPEKTROLAB – 10M equipment, and the following composition was obtained (Fig. 8).

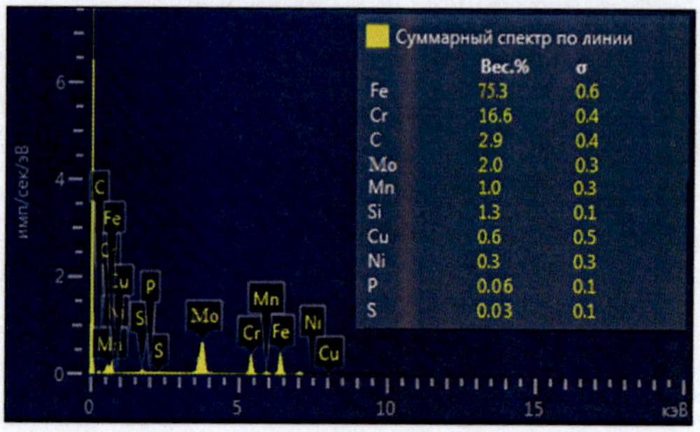

Fig. 8. CEM Zeiss EVO MA 10 Scanning electron microscope sample 1 element analysis image.

Table 2. The chemical composition of the samples obtained using the equipment "SPEKTRO-LAB – 10M".

Brand	Elements, %									
	C	Si	Mn	Cr	Mo	Ti	Ni	Cu	P	S
280X29NL (Sample 1)	2,87	1,18	0,89	16,53	2,07	–	0,63	0,61	0,062	0,036
280X29NL (Sample 2)	2,75	1,87	0,97	14,62	0,03	1,72	0,16	0,96	0,034	0,023
280X29NL (Sample 3)	2,75	0,99	0,67	13,88	0,80	–	0,79	1,05	0,070	0,036
280X29NL (Sample 4)	2,87	1,85	0,58	10,54	0,025	–	0,16	0,81	0,078	0,044
280X29NL (Sample 5)	2,67	0,92	3,10	18,33	0,20	–	0,49	0,48	0,052	0,014

Samples CEM Zeiss EVO MA 10 scanning electron microscope and METAM PB – 23, TK – 2M hardness tester were tested and the following results were obtained (Figs. 9, 10, 11, 12, and 13).

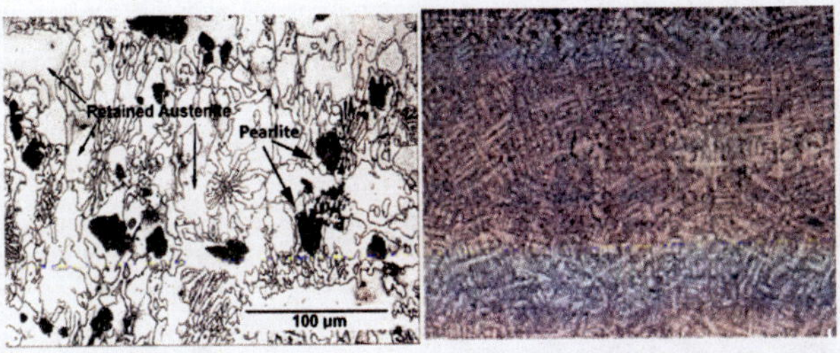

(a) (b)

Fig. 9. Sample 1 CEM Zeiss EVO MA 10 scanning electron microscope 500× times (a), magnification 100× times (b) using METAM PB – 23.

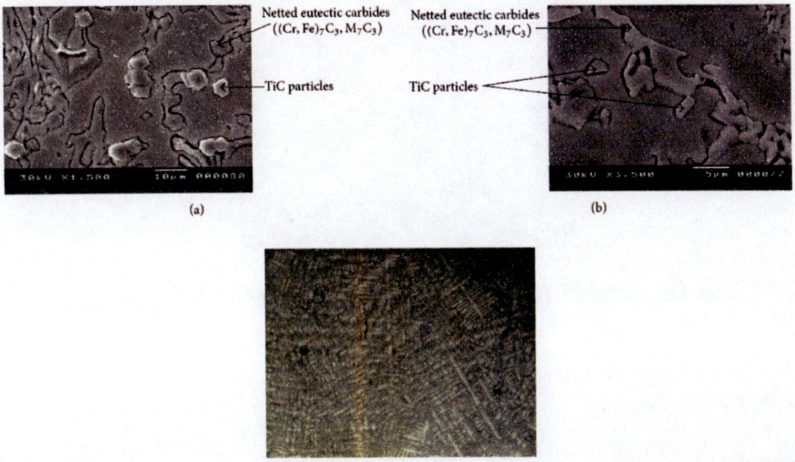

(c)

Fig. 10. Sample 2 CEM Zeiss EVO MA 10 scanning electron microscope magnified 500× times (a, b), 100× times (c) with the help of METAM PB – 23.

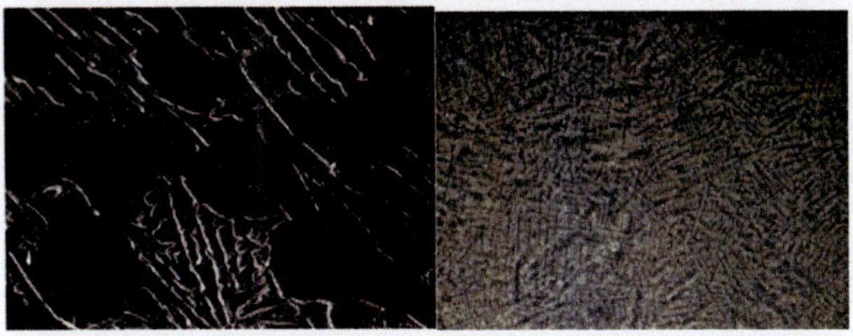

Fig. 11. Sample 3 CEM Zeiss EVO MA 10 scanning electron microscope magnified 100× times (a), 100× times (b) by METAM PB – 23.

Fig. 12. Sample 4 CEM Zeiss EVO MA 10 scanning electron microscope magnified 300× times (a), 100× times (b) using METAM PB – 23.

Fig. 13. Sample 5 100× magnification view using METAM PB – 23.

Table 3. The values in Table 3 were obtained using a TK – 2M hardness tester to determine the hardness of the samples.

Alloy brands	Hardness	
	HB (MPa)	HRC (MPa)
280X29NL (Sample 1)	470–485	46–48
280X29NL (Sample 2)	485–498	48–49
280X29NL (Sample 3)	583–597	58–59
280X29NL (Sample 4)	547–556	54–55
280X29NL (Sample 5)	571–593	56–58

4 Conclusions

As a result of the above experiment, CEMCO and BARMAK crushers, which are made of cast iron from high – chromium cast iron with high bending strength, have been able to increase the service life of the crushing discs. Based on the analysis of the preliminary results obtained, the following conclusions were made:

– there is an opportunity to increase the working resource by 1,3–1,5 times and to develop resource – saving technology in the production of disks;

Sample 1. Diluting the new brand alloy model: 9724, drawing: H – 1939.02B 4 discs were cast. The sand was then removed from the clay mold and cooled in air after 12 h in a SNOL – 7,2/1100 muffle furnace at 1000 °C. As a result, a uniformly distributed dendritic structure was observed on the surface of the sample microvilli, and the hardness index obtained using the TK – 2M hardness tester to determine the hardness of the sample was 46 HRC – 48 HRC;

Sample 2. Diluting the new brand alloy model: 9724, drawing: H – 1939.02B 4 discs were cast. It was then separated from the sand – clay mold and cooled in air after being kept in a SNOL – 7,2/1100 muffle furnace at 980 °C for 1 h. As a result, a uniformly distributed dendritic structure was observed on the surface of the sample microvilli, and the hardness index determined using the TK – 2M hardness tester to determine the hardness of the sample was 48 HRC – 49 HRC.

Sample 3. Diluting a new brand of alloy model: 9724, drawing: H – 1939.02B 3 discs were cast. The SNOL was then separated from the sand – clay mold and cooled in air for 15 min at 850 °C in a 7,2/1100 muffle furnace. As a result, a uniformly distributed dendritic structure was observed on the surface of the sample microvilli, and the hardness index obtained using the TK – 2M hardness tester to determine the hardness of the sample was 58 HRC – 59 HRC.

Sample 4. Dilution of a new brand of alloy model: 9724, drawing: H – 1939.02B 3 discs were liquefied in an induction furnace IST – 0,4 ferroalloys were introduced and poured into a sand – clay mold at a temperature of 1482 °C. As a result, a uniformly distributed dendritic structure was observed on the surface of the sample microvilli, and a hardness index of 55 HRC was determined using a TK – 2M hardness tester to determine the hardness of the sample.

Sample 5. Diluting new brand alloy model: 9724, drawing: H – 1939.02B 4 discs were cast.

As a result, a uniformly distributed dendritic structure was observed on the surface of the sample microvilli, and the hardness index obtained using the TK – 2M hardness tester to determine the hardness of the sample was 56 HRC – 58 HRC.

- new alloys of the brand 280X29NL (samples №1, №2, №3, №4, №5) were developed, which were economically inexpensive and brittle, without changing the chemical composition of high – chromium 280X29NL cast iron and reducing the mechanical properties of the alloy.
- H – 1939.02 B discs are 38% cheaper than the current 280X29NL discs and were introduced to the manufacturer.

References

1. Turakhodjaev, N., Saidmakhamadov, N., Turakhujaeva, S., Akramov, M., Turakhujaeva, A., Turakhodjaeva, F.: Effect of metal crystallation period on product quality. Theoret. Appl. Sci. **11**, 23–31 (2020)

2. Turakhodjaev, N.D., Saidmakhamadov, N.M., Zokirov, R.S., Odilov, F.U., Tashkhodjaeva, K.U.: Analysis of defects in white cast iron. Theoret. Appl. Sci. **6**, 675–682 (2020)
3. Nodir, T., Nosir, S., Shirinkhon, T., Erkin, K., Azizakhon, T., Mukhammadali, A.: Development of technology to increase resistance of high chromium cast iron. Am. J. Eng. Technol. **3**(03), 85–92 (2021)
4. Bekmirzaev, S., Saidmakhamadov, N., Ubaydullaev, M.: Obtaining sand-clay casting". Theory Pract. Modern. Russia **4**(12), 112 (2016)
5. Shirinkhon, T., Azizakhon, T., Nosir, S.: Methods for reducing metal oxidation when melting aluminum alloys. Int. J. Innov. Eng. Res. Technol. **7**(10), 77–82 (2020)
6. Djahongirovich, T.N., Muysinaliyevich, S.N.: Important features of casting systems when casting alloy cast irons in sand-clay molds. ACADEMICIA Int. Multidiscipl. Res. J. **10**(5), 1573–1580
7. Nodir, T., Sherzod, T., Ruslan, Z., Sarvar, T., Azamat, B.: Studying the scientific and technological bases for the processing of dumping copper and aluminum slags. J. Crit. Rev. **7**(11), 441–444 (2020)
8. Nodir, T., Nosir, S., Shokhista, S., Furkat, O., Nozimjon, K., Valida, B.: Development of 280X29Nl alloy liquefaction technology to increase the hardness and corrosion resistance of cast products. Int. J. Mechat. Appl. Mech. **10**(1), 154 (2021)
9. Turakhodjaev, N., Akramov, M., Turakhujaeva, S., Tursunbaev, S., Turakhujaeva, A., Kamalov, J.: Calculation of the heat exchange process for geometric parameters. Int. J. Mechat. Appl. Mech. **9**(1), 90–95 (2021)
10. Sharipovich, K.S., Yusufjonovich, K.B., Yakubjanovich, H.U.: Innovative technologies in the formation of professional skills and abilities of students of technical universities. Int. J. Progr. Sci. Technol. **27**(1), 142–144 (2021)
11. Nodir, T., Sarvar, T., Andrey, J., Yahyojon, M.: Mathematical model for calculating heat exchange. In: Cioboată, D.D. (ed.) ICoRSE 2021. LNNS, vol. 305, pp. 243–249. Springer, Cham (2022). https://doi.org/10.1007/978-3-030-83368-8_24
12. Botirov, A.G., Yusufjonovich, K.B., Makhmudjonovich, T.N.: Planting device. Int. J. Progr. Sci. Technol. **27**(1), 337–341 (2021)
13. Xalimov, S., Karimov, B., Abduraximova, G.: Issledovanie prochnostnix svoystv kompozitsionnix polimernix materialov dlya gazovix ballonov [Study of strength properties of composite polymer materials for gas cylinders]. Nauchnoe znanie sovremennosti **4**, 368–372 (2017)
14. Turakhodjaev, N., Tashbulatov, S., Tursunbaev, S., Kuchkorova, M.: Analysis of technological solutions for reducing the copper concentration in slags from oxygen-flare smelting of copper sulfide concentrates. J. Crit. Rev. **7**(5), 449–452 (2020)
15. Kenjaboev, Sh. Sh., Negmatullaev, S.E.: Obuchenie materialovedeniya kak spesialnix predmetov dlya bakalavrov transportnix napravleniy [Teaching materials science as special subjects for bachelors of transport directions]. In: Sovremennie avtomobilnie materiali i texnologii, pp. 162–166 (SAMIT-2020)
16. Umarov, E., Mardonov, U., Abdirakhmonov, K., Eshkulov, A., Rakhmatov, B.: Effect of magnetic field on the physical and chemical properties of flowing lubricating cooling liquids used in the manufacturing process. IIUM Eng. J. **22**(2), 327–338 (2021). https://doi.org/10.31436/iiumej.v22i2.1768

17. Umidjon, M., Muhammad, T., Andrey, J., Yahyojon M.: The difference between the effect of electromagnetic and magnetic fields on the viscosity coefficients of cutting fluids used in cutting processes. In. J. Mechat. Appl. Mech. **10**(1), 117–122 (2021). https://doi.org/10.17683/ijomam/issue10/v1.14
18. Negmatullaev, S.E., Kenjaboev, Sh.Sh., Bekmirzaev, Sh.B.: Osobennosti mejpredmetnix svyazey pri izuchenii obщeprofessionalnix dissiplin [Features of interdisciplinary connections in the study of general professional disciplines]. in rossiyskie regioni kak sentri razvitiya v sovremennom sotsiokulturnom prostranstve, pp. 71–75 (2020)

Synthesis of the Optimal Structure of the Device for Control and Regulating the Working Gaps of the Picking Apparatus of a Vertical-Spindle Cotton Picking Machine

Uljayev Erkin, Ubaydullaev Utkirjon Murodillaevich, Abdulkhamidov Azizjon Abdulla ugli, and Narzullayev Shohrukh Nurali ugli(✉)

Tashkent State Technical University, Tashkent 100097, Uzbekistan
narzullayevsh1993@gmail.com

Abstract. The paper analyzes the structure of the devices for monitoring and adjusting working gaps picking apparatus device with an automatic operating principle. The mechanism for adjusting the working gap of such a system is equipped with a modern contactless sensor of the working gap width and an electronic unit and other modern blocks. It has been established that the disadvantages of this device are: the lack of the ability to control and regulate additional widths of the working gaps (2-3-4 working slots), which ultimately lead to a decrease in the quality of the harvest and the productivity of the machine. As a result of the analyzes of the compared devices, systems, a new structure of the device for control and regulating the working slot of the picking apparatus was proposed that satisfies the stated requirements and is optimal in terms of the structure of the construction and has wide functionality.

Keywords: Picking apparat · Vertical spindle · Cotton picker

1 Introduction

Various structural, functional, and practical schemes have been developed to control and regulate the working width of the picking apparatus of a vertically spindle cotton picking machine. However, the known works and devices have significant disadvantages related to the structure of the construction, the composition of the blocks, the levels of the element base used, the reliability of the work, etc. indicators. The optimal structure of the device (system) for control and regulating the working width of the picking machine should ensure and meet the following requirements:

- maximum automation of the process of control and regulation of the working gaps cotton picker;
- high reliability of the device operation;
- visibility of information display;
- convenience of adjusting the working width;

D. D. Cioboată (Ed.): ICoRSE 2022, LNNS 534, pp. 116–124, 2023.
https://doi.org/10.1007/978-3-031-15944-2_11

- high accuracy of adjustment of the working width;
- reducing the adjustment time;
- providing improved cotton picker performance;
- wide functionality and informative content, etc.

The development of devices for control and regulating the width of the cotton picker working gaps that meet the specified technical requirements is relevant.

The functional scheme of the first variant of the device [1–3] devices for monitoring and adjusting working gaps contains guides mounted on the frame, each of which is made in the form of two slats mounted on the frame by means of a hinge and movable spring-loaded frames carrying spindle drums. The movable frames are mounted on the guides by means of rollers and have protrusions with rollers that interact with the cams of the rotary shaft. The guide bars have a means of changing the angle of their inclination relative to the longitudinal plane of symmetry of the mechanism, made, for example, in the form of an adjustment screw connecting the end of the bar opposite to the hinge fastening with the frame of the device.

Before the start of picking, the driver, depending on the yield of the field, adjusts the shape of the working slot of the spindle apparatus of the cotton harvester and gives it a conical shape, for which the guide bars of the picking apparatus are moved vertically with the help of adjusting screws.

The disadvantage of the considered structural scheme of control and regulation of the WS is that the means of changing the width of the working slot (the angle of inclination of the slats) is made in the form of an adjustment screw, with which, before the start of each harvest, it adjusts the width of the working slot. In addition, the absence of display units in this scheme for visual control and regulation of the width of the working slot reduces the agrotechnical indicators and productivity of the machine and the insufficient functionality of the control and regulation device.

The structure of the functional scheme of the second variant of the device of the cleaning device additionally contains a sensor mounted on a movable frame, located on the line [5, 6, 21] connecting the centers of the front adjacent drums, a spring-loaded rod-the movable contact of the sensor is in constant mechanical contact with a length-adjustable stop connected to the second movable frame. The sensor consists of a tubular housing with fixed contacts on its inner surface, consisting of a ball, a spring, and a screw. In this case, each contact is electrically connected to the corresponding indicator light. The positions of the width of the working slot are displayed by lighting the corresponding indicator lamp.

The disadvantage of the functional scheme of this device is: the lack of the possibility of continuous control of the size of the working gap; the presence of mechanical contacts of the sensor; the inconvenience of setting the set value i.e., the sensor is calibrated manually before the start of the picking season by changing the length of the stop. All this leads to a decrease in the reliability of the device, the sensor, and the performance of the machine.

Another variant of the structure of the construction of the device of the cleaning device is the control and regulation system made in the semi-automatic principle of action [7, 8, 22, 23]. An electronic tracking system is installed in this device to control the width of the working slot. The tracking system of the working slot adjustment mechanism contains a potentiometric converter-an angular displacement sensor mounted on the rotary shaft of the working width adjustment mechanism, an electronic unit consisting of a sensor system, a comparing element-a resistance bridge, and a two-stage output signal amplifier. The sensor system of the electronic unit consists of quasit-sensor switches, transistors, thyristors, control lights for each value of the working width values), working slots, and other electronic elements.

The disadvantage of the structure of the construction and operation of the devices for monitoring and adjusting working gaps device under consideration is low adjustment accuracy and low reliability of the sensor and the system of the working slot adjustment mechanism itself, as well as the lack of the possibility of constant control of the working slot size, which ultimately leads to a decrease in the quality of picking and productivity of the machine.

The next analyzed structure of the devices for monitoring and adjusting working gaps picking apparatus device is the control and regulation system of the vertical-spindle cotton picking machine [9, 10, 24] with an automatic principle of operation. The working slot adjustment mechanism is equipped with a contactless sensor for the width of the working slot and an electronic unit containing a microcontroller, an alarm, and display unit, a liquid crystal display, a block of matrix touch keyboards, a voltage conversion and stabilization unit, a block of control signal amplifiers. At the same time, photovoltaic sensors are installed inside converging with each other in two tubular housings. The body of the tube of the radiation source and the radiation receiver is fixed to the fixed part of the frame of the cleaning device, connected to the inputs of the electric hydraulic distributor, by means of an electronic unit and an amplifier unit.

The disadvantage of the device under consideration with a mechanism for adjusting the working slot is the lack of the ability to control and regulate additional widths of working gaps (2-3-4 working slots), which ultimately lead to a decrease in the quality of picking and productivity of the machine.

The purpose of the work is to analyze the structure of the construction of known works, according to the developed technical requirement, and to propose an optimal structure for the construction of a control and regulation device for the cotton picker machine, which differs from the known ones by increased agrotechnical indicators and machine performance.

2 Methodology

To eliminate the above shortcomings and meet the developed criteria, a new block diagram is proposed [11–13, 25]. Such a block diagram of the device for control and regulating the width of the cotton picker working gaps is equipped with four contactless photoelectric sensors for control four working gaps and an electronic unit containing a microcontroller, an alarm unit (multi-line) liquid crystal display, a block of matrix keyboards, a block of voltage conversion and stabilization, a multi-channel block of control signal amplifiers for the operation of electromagnetic valves of electric hydraulic distributors, controlling the operation of executive mechanisms that change the width of the working gaps of the picking apparatus.

The device for adjusting the width of the working slot of a vertical-spindle cotton harvester (Fig. 1) contains a group of contactless sensors 1.1, 1.2, 1.3, 1.4 for control the width of the working slots, an electronic unit 2 containing: a microcontroller 3 with extended memory, an alarm, and indication unit 4, a liquid-crystal display 5, a matrix unit 6 of touch keyboards, a block 7 of a voltage converter and stabilization, groups of blocks 8.1, 8.2, 8.3, 8.4 of control signal amplifiers for electromagnetic valves of electro hydro's - limiters 9.1, 9.2, 9.3, 9.4, controlling the operation of executive mechanisms 10.1, 10,2, 10.3, 10.4, changing the width of the working gaps of the cotton picker machines.

The principle of construction of the proposed structural scheme of the device is explained by the scheme of mechanisms for adjusting the width of the working gaps of the vertical spindle cotton harvester, shown in Fig. 1 [14, 20, 21, 26]. The process of control and adjusting the width of the working gaps of the picking machines cotton picker 10.1, 10,2, 10.3, 10.4 is carried out as follows: To control and adjust the width of the working gaps of the spindle picking machines, it is necessary to supply power to the electronic unit 2, unit 7 and a group of blocks. To do this, turn on the toggle switch installed on the housing of the electronic unit 2.

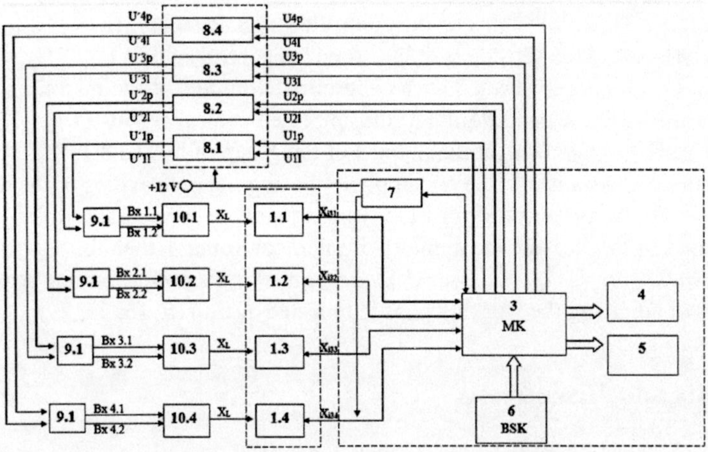

Fig. 1. The principle of constructing the optimal structure of the device for adjusting the width of the working gaps of the cotton picker machine.

Power is supplied to unit 2 for all the listed units. At the same time, the set values of the H_{sv} will be displayed on the upper line of display 5, and the acthal measured values of H_{mv1}, H_{mv2}, H_{mv3}, H_{mv4} corresponding to the values of the width of the working gaps of the spindle devices will be continuously displayed on the first second, third and fourth lines of the display.

To set the required width of the working gaps of the picking machines, the mechanic of the outcome from the state of the agrophone (cotton varieties, bush sizes, the degree of opening of the boxes, and the yield of the field) determines the width of the working slot and enters the value of the selected task (H_{sv}) into the memory of the MK3 using the touch keyboard 6. The value of the selected width of the WS will be displayed on the upper (first) line of display 4 in the form of a decimal code.

From the moment of entering the task into the memory of the microcontroller, control signals are formed at the output of the electronic unit 2, for example, U_{p1}, U_{p2}, U_{p3}, U_{p4}, which are amplified by groups of blocks 8.1, 8.2, 8.3, 8.4 of amplifiers and amplified signals U'_{1p}, U'_{2p}, U'_{3p}, U'_{4p}, are fed to the first inputs, for example, to V_{x1}, V_{x2}, V_{x3}, V_{x4} of electromagnets of electromagnetic hydraulic valves 9.1, 9.2, 9.3, 9.4 depending on the left (right) control signals U_{l1}, U_{l2}, U_{l3}, U_{l4}, (U_{p1}, U_{p2}, U_{p3}, U_{p4}) hydraulic valves, in turn, connect the corresponding pressure lines of the hydraulic system of the machine to the corresponding cavity of the hydraulic valves, which are the actuators of objects 10.1, 10.2, 10.3, 10.4, control and regulation and turn the rotary shafts in the appropriate directions and change the positions of the corresponding groups of blocks of contactless sensors 1.1, 1.2, 1.3, 1.4. Depending on the position of the width of the working gaps of the picking devices, proportional analog signals corresponding to the actual widths of the working gaps of the cotton-picking devices are formed at the corresponding outputs of the group of contactless sensors blocks 1.1, 1.2, 1.3, 1.4. At the same time, groups of blocks of contactless sensors 1.1, 1.2, 1.3, 1.4 continuously monitor changes in the width of the working gaps of objects $10._1$, $10._2$, $10._3$, $10._4$, control and regulation and continue to send controlled signals to the appropriate inputs of the electronic unit 2 (microcontroller 3). If the values of the controlled signals (H_{mv1}, H_{mv2}, H_{mv3}, H_{mv4},) turn out to be equal to the set values of H_{sv}, then the control signals U_{1p}, U_{2p}, U_{3p}, U_{4p}, or U_{1l}, U_{2l}, U_{3l}, U_{4l}, are canceled at the appropriate outputs of the microcontroller 3, which in turn lead to the cancellation of the amplified control signals U'_{1p}, U'_{1p}, U'_{1p}, U'_{1p}, or U'_{1l}, U'_{2l}, U'_{3l}, U'_{4l}, at the outputs of the group of blocks 8.1, 8.2, 8.3, 8.4 of control amplifiers. As a result, the oil supply to the appropriate cavities of the hydraulic valves stops. To change the width of the working slots, it is necessary to re-enter the new set values of the X_{sv} into the memory of microcontroller 3, then the control signals U_{1l}, U_{2l}, U_{3l}, U_{4l}, or U_{1p}, U_{2p}, U_{3p}, U_{4p}, are formed at its corresponding output) and the process of adjusting the working gaps is repeated again [15, 16, 27, 29].

3 Results and Discussions

A comparative analysis of the structure of the construction and functionality of the device for control and regulating the width of the working slits shows that the proposed device for control and adjusting the working slits contains contactless sensors built on the basis of photovoltaic linear displacement converters, which allow to continuously convert the change in the width of the working slits to electrical voltages with high accuracy. A microcontroller is used for operational processing, comparing the output voltages of the working slot width sensors with the set values, and for adjusting the working slots. The comparison of the received information with the set values and adjustments is carried out according to the recorded program in the memory of the microcontroller. The microcontroller converts the processed data into decimal codes, which are displayed in a convenient form on the display.

The comparable structure of the control and regulation device [5, 17, 28, 31–33] is built to control and regulate the working slot of only one cotton picking machine. To control and regulate the four working gaps of the picking machines, it is necessary to make changes to the algorithmic and software parts of the device and it is required to additionally install four amplifiers of control signals of electromagnetic valves corresponding to electric hydraulic distributors. In this device, there is no possibility of operational control and regulation of the working gaps of the picking apparatus, which negatively affect the agrotechnical indicators and productivity of machines. The use of additional blocks in the proposed device is justified by the need for local and continuous and accurate control, and regulation of the four working gaps of the picking machines, and the expansion of the measurement range. The use of independent units: sensors, electronic amplifiers, electric hydraulic distributors, electromagnetic valves, an electronic unit containing a microcontroller with extended memory, as well as a specialized four-row sectional display with simultaneous digital display of the width value of all four working gaps allows you to quickly process information coming from all four working slot width sensors and compare with the set values according to the program recorded in the corresponding memory registers of the microcontroller.

To control and regulate the working slots, local algorithms with appropriate software installed in the memory of the microcontroller are used [18, 19, 30, 31]. At the same time, each working slot of the picking machines, by means of a touch matrix keyboard, is installed separately or at the discretion of the machine operator. In cases of equality of the measured values with the set values, the removal of control signals regulating the widths of the working gaps is automatically provided. The control (regulation) of the working gaps can occur in parallel or separately, depending on the change in the position of the working slots. In addition to these, the microcontroller promptly displays all the processed data and set values in a convenient form for the operator to use.

4 Conclusions

Thus, we can conclude that the proposed structure of the construction and its functionality fully meet the requirements set and is optimal. The device provides operational control and adjustment of the width of the working slot of the picking machine with high accuracy. This is achieved through the use of separate photovoltaic sensors and a microprocessor device operating according to a given algorithm. The given justifications show the optimality of construction and the functionality of the proposed device for control and regulating the width of the cotton-picking machine.

References

1. Sadriddinov, A.S., Abdazimov, A.D., et al.: Positive decision of VNIGPE of 27.02.1995 on the grant of a patent of the Russian Federation for an invention "The mechanism to adjust the working gap vertical-spindle cotton picker machine" under application no 5051399/15 - 032705 from. 07,07,1982 G. Authors KL. MKI AO1 D 46/14. AO1/75/00 (1995)
2. Abdazimov, A.D.: Servo control system of the width of the working gap cleaning machine cotton pickers. Bull. TASHSTU No. **4**, 103–106 (2003)
3. Sadriddinov, A.S., Abdazimov, A.D., Uljaev, E.: Improving the working bodies and controllability of a cotton harvester with the location of harvesting machines behind the tractor drive bridge. Materials of the International Journal. NTC Modern problems of mechanics. Research Institute of Mechanics and SS of the Academy of Sciences of the Republic of Uzbekistan, pp. 273–277 (2009)

4. Uljaev, E., Abdazimov, A.D., Ravutov, Sh.T., Ravutov, S., Tulbaev, F.A.: Hardware supports the system of automatic control and regulation of parameters of mobile objects. Kimeviy tekhnologiya, nazorat va boshkaruv **4**, 29–33 (2012)
5. Abdazimov, A.D.: Tracking system for controlling the width of the working slot of the harvesting machine of the cotton harvesting machine. Bull. TSTU **4**, 103–106 (2003)
6. Uljaev, E., Abdazimov, A.D., Ubaydullaev, U.M., Sherkobilov, S.M., Shodiev, Zh.G., Nosirov, A.M.: Mechanism of adjustment of working slots of vertical-spindle cotton harvesting apparatus//Agencies for Intellectual Property under the Ministry of Justice of the Republic of Uzbekistan. IAP no. 06346 (2020)
7. Abdazimov, A.D., Uljaev, E., Ubaydullaev, U.M., Omonov, N.N.: Fundamentals of automation of control and control of technological parameters of cotton harvesters, TSTU, p. 164 (2014)
8. Uljaev E.U., Ubaydullaev, U.M.: Diagnostic device for changing the airflow velocity in the pneumatic chamber of cotton harvesters. In: Ninth World Conference "Intelligent Industrial Automation Systems", WCIS-2016, October 25–27, Tashkent (2016)
9. Farzane, N.G., et al.: Technological measurements and devices. M.: "Higher School", p. 456 (1989)
10. Cherenkov, V.V.: Industrial devices and automation tools. In: Mashinostroenie, L. (ed.) Handbook, 847p (1987)
11. Shishmarev, V.Yu.: Typical Elements of Automatic Control Systems. Textbook for seed. prof. education/Vladimir Yuryevich Shishmaref. Publisher: Academy, 304p (2004)
12. Babikov, M.A., Kosinsky, A.V.: Elements and devices of automation. High. Sch. Econ. (1975)
13. Uljaev, E., Ubaydullaev, U.M., Tajitdinov, G.T., Narzullaev, S.: Development of criteria for the synthesis of the optimal structure of monitoring and control systems. In: Aliev, R.A., Yusupbekov, N.R., Kachpshik, J., Pedrych, V., Sadikoglu, F.M. (eds.) 11th World Conference "Intelligent Industrial Automation Systems" (WCIS-2020). WCIS 2020. Advances in Intelligent Systems and Computing, vol. 1323. Springer, Cham. https://doi.org/10.1007/978-3-030-68004-6_73
14. Uljaev, E., Ubaidullaev, U.M., Uljaev, Z.E.: Electronic on-board control system for operational and technological parameters of cotton-growing plants. Monthly Sci. Pract. J. Tract. Agric. Mach. **10**, 11–15 (2014)
15. Uljaev, E., Abdazimov, A.D., Ubaidullaev, U.M.: Methodology for diagnosing the probability of trouble-free operation of a cotton harvester. Proc. St. Petersburg State Agrar. Univ. **4**(53), 270–276 (2018)
16. Uljaev, E.U., Abdazimov, A.D., Ubaydullaev, U.M.: Intelligent onboard MPC for monitoring and controlling technological parameters of MTA with tractor TTZ. In: International Scientific and Practical Conference on Technology of the Future: Prospects for the Development of Agricultural machinery, pp. 189–191, Krasnodar (2013)
17. Uljaev, E., Abdazimov, A.D., Ravutov, S., Tulbaev, F.A.: Microprocessor control system and regulation of the working slot of the cotton harvesting apparatus. Probl. Informat. Power Eng. **5**, 48–52 (2011)
18. Uljaev, E., Ubaydullaev, U.M.: Single-channel system of control and regulation of the working slot of the harvesting machine of the cotton harvesting machine. Int. Sci. J. Sci. Educ. Tech. the Kyrgyz-Uzbek Univ. **2**(65), 77–81 (2019)
19. Uljaev, E., Ubaydullaev, U.M.: System of control and regulation of working slots of a four-row vertical-spindle harvesting machine. Republican scientific and technical conference "Problems and prospects of development of innovative equipment and technologies", TSTU, Tashkent (2019)

20. Uljaev, E., Ravutov, Sh.T., Ubaydullaev, U.M.: Remote control device for controlling the contact uniformity of brush pullers on the surface of the spindle of a cotton harvesting machine. In: 1st International Conference on Energy, Civil and Agricultural Engineering 2020, IOP Conference Series: Earth and Environmental Science 614 012139, Tashkent, October 14–16, 2020. https://doi.org/10.1088/1755-1315/614/1/012139
21. Sadriddinov, A., Tulaev, B., Sotqinov, E., Abdazimov, A., Mirpulatov, R., Shaykhov, I.: Certificate of authorship. No. 799690 USSR, Mechanism for adjusting the working gap of a vertical-spindle cotton picker. (Tashkent Order of Friendship of Peoples Polytechnic Institute named after A.R. Biruni). - 2760804; declared 04/27/1979; publ. 01/01/1981
22. Sadriddinov, A., Khasanov, P., Tashkhuzhaev, T., Kiselev, O., Abdazimov, A.: Certificate of authorship. No. 1301345 USSR, Mechanism for adjusting the working slot of a vertical-spindle cotton picker. (Tashkent Order of Friendship of Peoples Polytechnic Institute named after A. R. Biruni). - 3961275; declared 07/30/1985; publ. 12/8/1986
23. Sadriddinov, A., Abdazimov, A., Tulyaev, A., Tashkhodzhaev, T., Mirzaev, A.: Certificate of authorship. No. 1169560 USSR, Cotton picker. (Tashkent Order of Friendship of Peoples Polytechnic Institute named after A. R. Biruni). - 3562514; declared 03/10/1983; publ. 1.04.1985
24. Sadriddinov, A., Rashidov, N., Tashkhodzhaev, T., Abdazimov, A., Avazov, Sh.: Certificate of authorship. No. 988225 USSR, Mechanism for adjusting the working gap of a vertical-spindle cotton picker. (Tashkent Order of Friendship of Peoples Polytechnic Institute named after A. R. Biruni). - 3310979; declared 2.07.1981; publ. 09/14/1982
25. Rashidov, N., Sadriddinov, A., Abdazimov, A., Sotqinov, E.: Certificate of authorship. No. 917768 USSR, Mechanism for adjusting the working slot of a vertical-spindle cotton picker. (Tashkent Order of Friendship of Peoples Polytechnic Institute named after A. R. Biruni). - 2940417; declared 06/11/1980; publ. 12/7/1981
26. Sadriddinov, A., Abdazimov, A., Shakamalov, A., Bratchikov, A.: Certificate of authorship. No. 1115676 USSR, Mechanism for adjusting the working gap of a vertical-spindle cotton picker. (Tashkent Order of Friendship of Peoples Polytechnic Institute named after A. R. Biruni). - 3549279; declared 02/7/1983; publ. 1.06.1984
27. Sadriddinov, A., Tulyanov, A., Yakubov, Kh., Abdazimov, A.: Certificate of authorship. No. 1147276 USSR, Cotton picker drum. (Tashkent Order of Friendship of Peoples Polytechnic Institute named after A. R. Biruni). – 3687129; declared 01/06/1984; publ. 1.12.1984
28. Sadriddinov, A., Abdazimov, A., Tulyaev, A., Karimov, Sh., Turgunov, U.: Certificate of authorship. No. 1576015 USSR, Cotton picker. (Tashkent Polytechnic Institute named after A. R. Biruni). – 4475038; declared 08/18/1988; publ. 03/08/1990
29. Uljaev, E., Abdazimov, A., Ubaydullaev, U., Ravutov, Sh., Omonov, N., Erkinov, S., Nosirov, A., Saidov, S.: Certificate of authorship. Republic of Uzbekistan No. IAP 06346/Mechanism for adjusting the working gap of a vertical-spindle cotton picker. (Toshkent State University named after Islam Karimov) announced on 12 December 2017; publ. 12/31/2020 Bulletin No. 12
30. Uljaev, E., Ubaydullaev, U.M., Abdulxamidov, A.A.: Selection of methods and sensors for monitoring width changes between moving objects, International Conference on Scientific Collection «International Conference»: Science, Education, Innovation: Topical Issues And Modern Aspects, May 11–12, 2020
31. Uljaev, E., Ubaydullaev, U.M., Abdulxamidov, A.A.: Analysis of the current state of automation of control and regulation of the width of the working slots of the cotton harvesting machine with a vertical spindle. In: International Conference on Scientific Collection «International Conference»: Challenges in Science Of Nowadays, April 4–5, 2021
32. Uljayev, E., Ubaydullaev, U.M., Narzullayev, Sh.N., Norboyev, O.N.: Application of expert systems for measuring the humidity of bulk materials. Int. J. Mechat. Appl. Mech. 9(1), 131–137 (2021). https://doi.org/10.17683/ijomam/issue9.19

33. Plakhtiev, P., Denmukhammadiev, A., Gaziev, G: Information universal non-contact transducers of control and management systems. IOP Conf. Ser. Mater. Sci. Eng. **869** (2020) 022034, https://doi.org/10.1088/1757-899X/883/1/012182
34. Umidjon, M., Muhammad, T., Andrey, J., Yahyojon, M.: The difference between the effect of electromagnetic and magnetic fields on the viscosity coefficients of cutting fluids used in cutting processes. Int. J. Mechat. Appl. Mechan. **10**(1), 117–122 (2021). https://doi.org/10.17683/ijomam/issue10/v1.14
35. Uljaev, E., Narzullayev, S., Utkir, U., Shoira, S.: Increasing the accuracy of calibration device for measuring the moisture of bulk materials. In: Cioboată, D.D. (ed.) ICoRSE 2021. LNNS, vol. 305, pp. 204–213. Springer, Cham (2022). https://doi.org/10.1007/978-3-030-83368-8_20

Obtaining Liquid Hydrocarbons by Processing of Natural and Associated Petroleum Gas in a Flow Reactor with a Non-equilibrium Electric Discharge

Alimbabaeva Zulkhumar Latipovna[1(✉)], Yakubov Lazizkhan Ergashkhanovich[2], and Narimov Dilshodjon Shukhratovich[1]

[1] Branch of the Russian State University of Oil and Gas named after I.M. Gubkin, Tashkent, Uzbekistan
dimonbizon@bk.ru

[2] Almalyk branch of the Tashkent State Technical University named after Islam Karimov, Tashkent, Uzbekistan

Abstract. In this article, we consider one of the new methods for producing liquid hydrocarbons by processing natural or associated petroleum gas in a flow reactor with a nonequilibrium electric discharge. The essence of this method is that in this reactor, a flow of natural gas of atmospheric pressure is exposed to a pulsed volume discharge. This discharge is initiated by a pulsed electric beam.

Keywords: Natural and associated gas · Pulsed volume discharge · Reactor

1 Introduction

In this paper, a new method in the field of GTL technologies (gas to liquid) is studied and considered, namely, the production of liquid hydrocarbons by processing natural or associated petroleum gas in a flow reactor with a non-equilibrium electric discharge. The essence of this method lies in the fact that a natural gas flow at atmospheric pressure is exposed to a pulsed volumetric discharge in the reactor. This discharge is initiated by a pulsed electric beam.

The decisive success in the development of this method is the absence of synthesis gas as an intermediate necessary for the production of synthetic products. It's no secret that the stage of obtaining synthesis gas in the GTL technology is the most complex and capital-intensive stage, it accounts for 60–70% of all production costs.

Currently, there are many methods and methods for the conversion of hydrocarbon-containing gases into liquid products. For these methods, chemical, electrochemical and plasma-chemical technologies are usually used.

The traditional way to obtain liquid products from natural gas (through its conversion) is the chemical Fischer-Tropsch process. This process is carried out in several stages, or rather includes the stages of obtaining synthesis gas from a mixture of natural gas

D. D. Cioboată (Ed.): ICoRSE 2022, LNNS 534, pp. 125–129, 2023.
https://doi.org/10.1007/978-3-031-15944-2_12

with such oxidizing agents as O_2, H_2O, CO_2. And after receiving the synthesis gas, the conversion into liquid products is carried out:

$$CO + H_2 \rightarrow CH_3OH \rightarrow \text{liquid hydrocarbons (mobil process)}$$

The conditions for the implementation of the traditional method or the chemical Fischer-Tropsch process are high pressure (several tens of atmospheres) and high temperature (about 900 °C).

In addition to this traditional method, there are many other methods and varieties of technology for obtaining liquid products from natural gas and associated petroleum gases. But, if we conduct a comparative analysis, we can see that all modifications of these technologies have the following common disadvantages:

1. All of them are carried out under high pressure and high temperature.
2. To ensure these parameters (high pressure and temperature), large-sized and metal-intensive structures are created. This, in turn, makes it difficult to use them in remote areas and places with poor infrastructure.
3. At all stages of the synthesis, the presence of a catalyst is necessary. This factor also creates additional problems associated with regeneration, as well as significant economic losses.

There are three types of the Fischer-Tropsch process:

- with a fixed catalyst bed (the process of the company "ARGE");
- with a fluidized or pseudo-transport bed of catalyst particles (Kellog-Synthol process);
- with a stationary catalyst suspended in high-boiling products (the process of the firms "Research Inc." and "Standard Oil").

The synthesis of hydrocarbons with a fixed catalyst bed is carried out in tubular reactors. The catalyst is placed in tubes with an inner diameter of 50 mm and a length of 12 m. Each reactor contains ~2000 such tubes. The process uses 8 reactors, 5 of which operate at a temperature of 220–225 °C, and the rest at 320 °C. The system maintains a pressure of 1.7–2.5 MPa.

The disadvantages of the Fischer-Tropsch fixed bed process are:

- low heat transfer in the catalyst layer;
- ifficulties with temperature regulation; the organization of a large recycle of the gas flow is required;
- clogging of tubes with carbonaceous compounds or waxy products.

The essence and innovation of this work lie in the fact that it is possible to obtain liquid hydrocarbon products from natural gas and associated petroleum gas by selective vibrational excitation of source gas molecules in the reactor (Fig. 1 and Fig. 2), as well as the implementation of chain processes in non-equilibrium plasma of a pulsed gas volume discharge high pressure, which is initiated by an electron beam.

As a result, we get a dimer R2, while the effective conversion of gaseous hydrocarbons (HC) is ensured by the reverse reaction of the formation of light RH molecules, that is, using this method, we obtain a high conversion of light HC in this reactor, and to be more precise, up to 90% and at In this way, we obtain low energy consumption for the production of liquid hydrocarbons (approximately 1 kWh/kg). Using these indicators, one can undoubtedly say that the activation energy in this method is significantly less than in the conditions of a barrier discharge.

Flow reactions (pressure reactions, constant reactions, microchemical reactions) are being used more and more in scientific research because of their productivity and efficiency. Optimized reaction conditions can be used to produce more products or can be stored in a database.

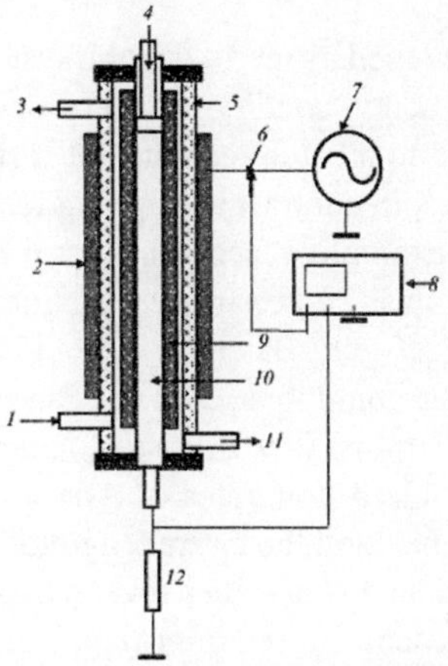

Fig. 1. Schematic design of a flow reactor with a non-equilibrium discharge: 1, 3 - gas inlet and outlet, 2 - external electrode, 4, 11 - inlet and outlet water, 5 - dielectric barrier (pyrex), 6 - high-voltage probe, 7 – high voltage source, 8 – two-channel digital oscilloscope, 9 – porous hydrophilic material, 10 – inner electrode, 12 - resistor (100 Ω).

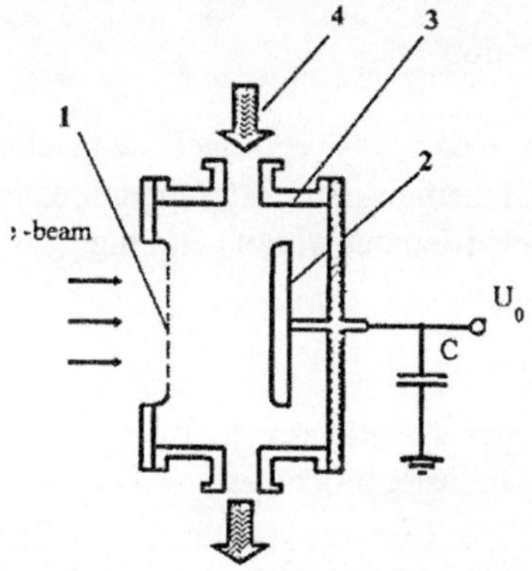

Fig. 2. Schematic description of the working chamber of the reactor. 1-mesh electrode; 2- solid electrode; 3- combustion chamber; 4-gas supply; C- capacitor.

Flow reactions are carried out in a very small amount of material (microflow). Microflow allows precise monitoring, control, and analysis of liquids from picoliter to microliter.

2 Differences in a Flow Reactor from a Reactor Periodic Work

To optimize the reaction conditions for a classical reactor with power is a rather laborious and lengthy process. Leaking reactors make it possible to do this several times faster and with a very low cost of reagents. The conducted studies have shown that the time of the reaction in a periodic mode is compared with the time of the reaction proceeding under the same conditions.

The response time can be reduced by increasing the system pressure (using a pressure module); The flow synthesis reaction conditions can be optimized in a flow reactor and the results obtained are applied to a batch reactor. But it should be noted that periodic reactions can vary several times, in contrast to ongoing reactions. Not all reactions can be carried out in a stream (for example, reactions to form a solid product). Not all flow reactions can be recreated in a flask (for example, reactions with the initiation of some cycles). The BFB (Batched-Electricity-Batched) system is a fast and efficient method for formulating and optimizing conditions. The mixing of the components of series microorganisms occurs due to laminar flow diffusion, and not in a turbulent, as in batch reactors (sausages). At the standard flow rates of Africa reactors, the mixing process takes place over a 10 mm channel, and the entire length of the reactor is about 1 m.

The efficiency of heat exchange is not only several times better than batch reactors but also makes it easy to maintain the required temperature in the chemical process occurring in endothermic reactions.

3 Benefits of Reaction Flow

Conducting a reaction stream has several advantages over traditional methods:

1. Management and automation

Flow reactors allow you to perfectly reproduce the reaction conditions in a running chemical process and control them, n-p, time, temperature, the ratio of mixed reagents, exothermic reactions can be carried out without cooling.

2. Speed up the reaction

Due to the easier increase in system pressure (up to 300 psi), reactions can occur in ultra-precise form, resulting in faster processes.

3. Synthesis and analysis of the combination.

The reaction stream is supplied with a reactor liquid crystal chromatography (LCX) module. The flow after the reaction receives a receiving module, from which a 5 μl sample is selected and fed to the dilution module, from which the sample already enters the chromatography system.

4. Efficiency

Leak reactors operate in very small quantities during the development of raw materials, the volume of the starting material can only be 100 μl for the reaction.

Summarizing all of the above, we note that the advantage of this method and technology is a significant reduction in energy costs for obtaining liquid products in connection with the implementation of the chain process and an increase in plant productivity compared to known analogs, as well as the fact that the process of obtaining liquid hydrocarbons is carried out without high temperature and pressure, which in turn reduces both the metal consumption of the structure and the dimensions of the equipment itself.

4 Conclusions

The method of processing natural gas into liquid products relates to the field of oil and gas processing and can be used at petrochemical enterprises, oil and gas fields for processing light gaseous hydrocarbons (natural and associated petroleum gases) into liquid products, and single-stage synthesis of hydrocarbon oils. The invention concerns a method for processing natural gas into liquid hydrocarbons in a flow reactor with a nonequilibrium electric discharge, while the flow of natural gas of atmospheric pressure in the reactor is exposed to a pulsed volumetric discharge, which is initiated by a pulsed electron beam. The technical result is a reduction in energy costs for producing liquid hydrocarbons and an increase in reactor performance.

References

1. Vosmerikova, L.N., Vosmerikov, A.V.: Oil refining and petrochemistry No. 3 (2011)
2. Rozovsky, A.Y.: The main ways of processing methane and synthesis gas, (1999)
3. Aryutonov, V.S., Krylov, O.V.: Oxidative transformations of methane (1998)
4. Mordkovich, V.Z., Sineva, L.V., Kulchakovskaya, E.V., Asalieva, E.Y.: Four generations of liquid fuel technology based on Fischer-Tropsch synthesis. Catal. Ind. **5**(15), 23–45 (2015)
5. Luque, R., Speight, J.: Gasification for Synthetic Fuel Production.1st edn. 348p. Woodhead Publishing, Sawston (2014)
6. Okhatrina, V.S.: International experience in the production of synthetic liquid fuels using GTL technology and prospects for its development. Probl. Modern Econ. **4**, 114–116 (2012)
7. Mirgayazov, I.I., Abdullin, A.I.: GTL industry: state and prospects. Bull. Kazan Technol. Univ., pp. 253–257 (2014)
8. Matkovsky, P.E., Sedov, I.V., Savchenko, V.I., Yarullin, R.S.: Technologies of synthesis gas production and processing. Gas Chem. 77–84 (2011)
9. Karimov, S., Mamirov, S., Khabibullayeva, I.A., Bektemirov, B., Khusanov, N.: Friction and wear processes in tribotechnical system. Int. J. Mechat. Appl. Mech. **10**(1), 204–208 (2021)
10. Bektemirov, B.S., Ulashov, J.Z., Akhmedov, A.K., Gopirov, M.M.: Types of advanced cutting tool materials and their properties. Euro-Asia Conf. **5**(1), 260–262 (2021)
11. Assessing a Coal-to-Liquids Fuel Industry in the United States. Research brief. https://www.rand.org/pubs/research_briefs/RB9342/index1.html. Copyright, 2008 RAND Corporation
12. Henderson, R., Rodvel, R.M., Kharji, A.: Modification of refineries for processing unconventional heavy oils. Oil Gas Technol. **1**, 67–73 (2006)

Research of Technological Modes of Production of Small Diameter Rods from Niobium

Tilavov Yunus Suvonovich[1], Urokov Kamoliddin Khushvakt ugli[1], and Bektemirov Begali Shuhrat ugli[2](✉)

[1] Karshi State University, Karshi, Uzbekistan
[2] Tashkent State Technical University, Tashkent, Uzbekistan
begalibektemirov94@gmail.com

Abstract. This article examines the study of technological modes of production of small-diameter niobium rods, which allows to obtain wires from niobium and its alloys with high values of strength and plastic characteristics in both deformed and annealed states, compared with the mechanical characteristics of niobium. The microstructure of niobium rods has been studied. It can be seen that rods made of niobium and its alloy with a diameter of 20–40 mm have high mechanical properties.

Keywords: Rod · Wire · Pressing · Heat treatment · Deformation · Ingot

1 Introduction

Pressed niobium billets with a diameter of 20 mm were subjected to rotary forging for 8 transitions at $t_H = 20$ °C and rods with a diameter of 11 mm were obtained according to the following modes [1]:

Table 1. Rotary forging modes for pressed niobium billets

Transitions	Rod size, mm Diameter of initial blank	Compression, %	Feed rate, m/min
1	20		Manually
2	17	11	
3	16	23	
4	14	32	
5	11	27	
6	9	22	
7	8	23	
8	7	24	
	6		

D. D. Cioboată (Ed.): ICoRSE 2022, LNNS 534, pp. 130–138, 2023.
https://doi.org/10.1007/978-3-031-15944-2_13

The analysis of the microstructure of the samples (Fig. 1) showed that after the first and second passes, the microstructure (Fig. 1a, b) is fine-grained, deformed, after the first pass in the niobium workpiece, the deformation spreads over the entire cross-section of the sample. After the third pass, additional deformation led to the appearance of recrystallization over the entire cross-section of the sample, and after the fourth pass, additional grain growth occurred, i.e., collective recrystallization occurred. There were no signs of cracks and stratification. The quality of the bars is satisfactory.

At the same time, this technology for the production of niobium rods is inefficient because it is time-consuming and multi-cyclical: for the production of diam rods. 4–6 mm requires 10–12 transitions.

2 Research Results and Discussion

Rotary Forging of niobium is carried out manually, which is harmful to maintenance personnel. All this makes it necessary to find ways to improve the production technology of niobium rods with the replacement of the manual rotary forging process with more efficient pressure treatment processes.

The author conducted comprehensive research on the development of technology for obtaining rods with a diameter of 6 mm from niobium using long-range rolling.

For this purpose, in relation to the DUO 230 single-celled rolling mill existing at the OHMZ GIREDMET plant (Podolsk), a system of calibers has been developed: -rhombic, oval and round 8 calibers are placed on one pair of rolls (Fig. 2), on the other pair of rolls −16 calibers. Calibration of rolls for rolling niobium rods was carried out using the methodology, algorithm and calculation program developed at the Department of Plastic Deformation of special steels and alloys of MISIS for rolling long metal, which are based on the results of generalization of studies of calibration rolls of long mills operating in Russian factories.

According to this mathematical model of varietal rolling, the production of a varietal profile is presented in the form of separate stages, at each of which the strip is rolled in a certain system of calibers. In this case, the strip obtained from the previous pair of calibers is the initial one for the next system of calibers in such a way that the condition is met:

$$\mu = \prod^{n} \mu_1 \tag{1}$$

where n- number of gauge pairs.

μ_1- the coefficient of extraction in a pair of calibers (1, 1,2,..., n).

At the same time, the energy-power parameters and the possibility of capturing metal by rolls must meet the following conditions:

$$P_1\langle[p]_j, M_{nj}\langle[M_n]_j, \alpha_{3j}\langle[\alpha_3]_j$$

Here $P_1\langle[p]_j, M_{nj}\langle[M_n]_j, \alpha_{3j}\langle[\alpha_3]_j$ - The actual and permissible rolling forces and moments, as well as the angles of metal capture by the rolls in the corresponding passages. The program can work both in dialog and in batch mode.

With the use of the specified software package, the calculation of the calibration of rolls and technological parameters of niobium rolling on the single-cell mill 230 OHMZ GIREDMET was performed.

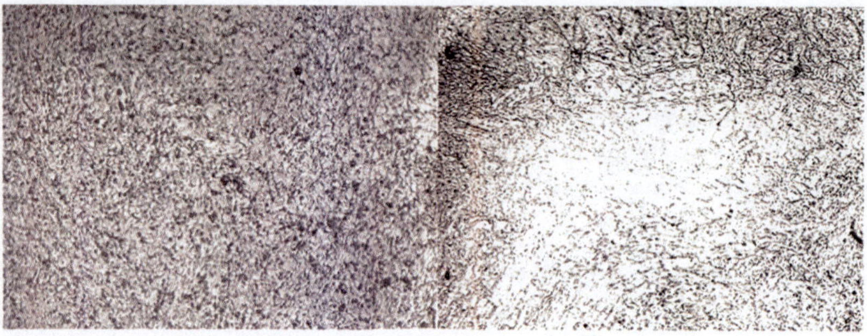

a) b)

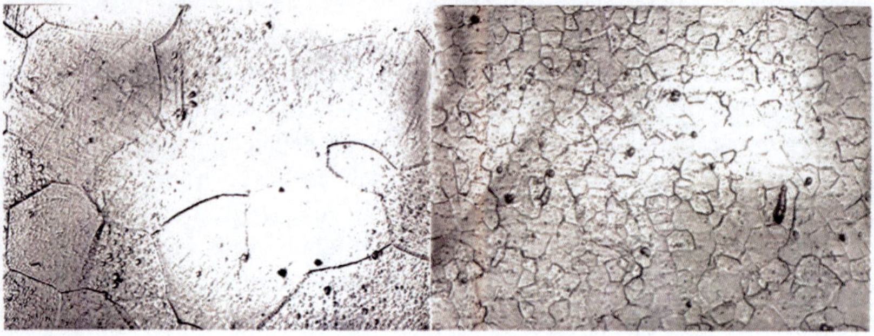

c) d)

Fig. 1. Microstructure of niobium billet by rotational forging processes. ×250. a-blank ø 17 mm; b-blank ø 16 mm; c-blank ø 14 mm; d-blank ø 11 mm.

For rolling niobium rods, a system of rhombus-rhombus exhaust calibers with angles at the top of 95–102° is adopted as the main one. This is due, firstly, to the limited possibilities of metal capture by rolls during cold rolling of niobium due to the low coefficient of friction.

Secondly, rolling is carried out manually without wiring due to the absence of any mechanization on the mill and difficulties in installing roller fittings on the mill. Therefore, the system of exhaust gauges should ensure reliable stability of the strip in caliber and good metal capture by rolls. The selected rhombus-rhombus calibre system meets such requirements for specific conditions to the maximum extent.

Rhombic calibers are made with a flat top, which made it possible to maximize the possibilities of the mill and prevent overflow of calibers and associated sunsets. The hoods per pass were 1.12–1.25, which made it possible to obtain a rod with a diameter of 6 mm in 22 passes from a square billet with a cross-sectional dimension of 30 × 30 mm. At the same time, the size of the capture angles did not exceed 14–15°, and the filling of the calibers is in the range of 0.98–1.0 (Table 1).

The rolling of the pressed blanks was carried out without preheating them. There was no sticking of niobium on the rolls. From the data in Table 2, it can be seen that the pressures have a comparatively low value of 320–380 Pa. The analysis of the microstructure of rolled niobium rods (Fig. 3) and the quality of their surface allows us to conclude that the rods rolled on rolls of this calibration were of good quality and did not require additional machining, which allowed further processing of the rods by drawing, eliminating rotary forging. Pilot-industrial rolling of niobium rods has shown that the plasticity resource of the metal allows deformation without the use of intermediate heat treatment of niobium.

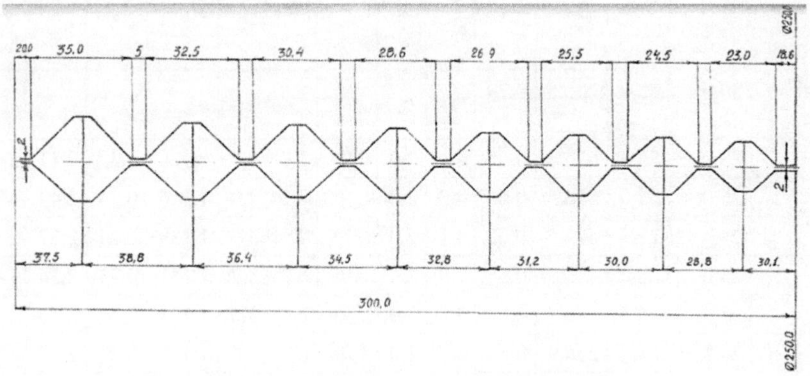

Mounting of gauges on rolls. / set № 1/.

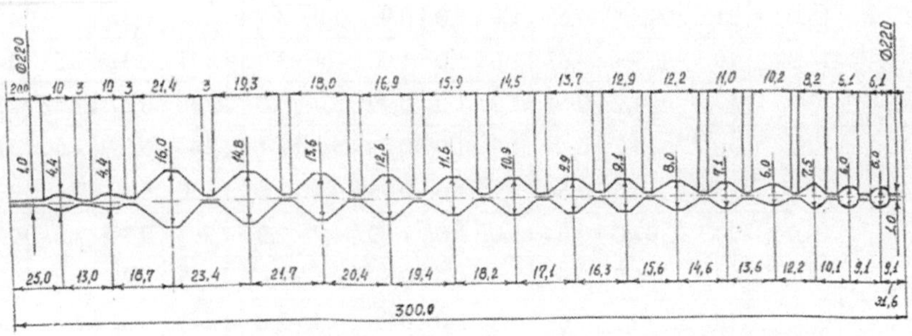

Fig. 2. Mounting of gauges on rolls /set № 2/.

This technology for the production of niobium rods using the rolling process has significant advantages over the one previously used at the OHMZ GIREDMET plant, since it excludes the process of rotary forging with all the disadvantages inherent in this process, and is more productive.

In 1993, this technology was successfully implemented in industry at the OHMZ plant in Podolsk.

The microstructure of the niobium billet according to the processing of longitudinal rolling. ×250. a-blank after the second pass octahedron, b-blank after the third pass octahedron, c-blank after the fifth pass octahedron.

Table 2. Calibration parameters of rolls for rolling a niobium circle with a diameter of 6 mm.

NN caliber	Caliber height, N, mm	Gauge width, N, MM	Cross-sectional area, F, mm^2	Gauge shape coefficient, K_F	Blunting of the strip, m, mm	The gap between the rolls, S, mm	Gauge fill factor, $\varkappa$	Compression, <h, mm	Broadening, Δb, mm	Extraction coefficient, *M*	Relative compression, *E*	The broadening indicator, $\Delta B/\Delta h$	Capture angle, α, degree
Initial blank	30,0	30,0	706,5	0,79									
1	28,0	34,3	626,5	0,65	4,4	2,0	0,98	2,0	4,3	1,128	0,22	2,17	14,0
2	26,0	32,0	546,5	0,66	4,3	2,0	0,98	8,3	4,0	1,146	0,24	2,47	15,6
3	245	29,9	479,6	0,65	4,1	2,0	0,98	7,5	3,9	1,139	0,24	2,52	14,7
4	23,0	28,2	424,3	0,65	3,9	2,0	0,99	6,9	3,7	1,131	0,23	2,54	14,1
5	21,8	26,6	378,2	0,65	3,8	2,0	0,99	6,4	3,6	1,122	0,23	2,56	13,6
6	20,5	25,4	336,9	0,65	3,5	2,0	1,00	6,1	3,6	1,123	0,23	2,59	13,2
7	19,0	24,3	298,6	0,65	3,6	2,0	0,99	6,4	3,8	1,128	0,25	2,59	13,5
8	17,5	23,0	258,4	0,64	3,2	2,0	1,00	6,8	4,0	1,155	0,28	2,58	13,8
9	16,0	21,2	213,8	0,63	1,9	1,0	0,99	7,0	3,7	1,209	0,32	2,54	15,0
10	14,8	19,3	175,1	0,63	1,0	1,0	1,00	6,4	3,3	1,221	0,32	2,51	14,4
11	13,6	17,8	154,2	0,64	2,0	1,0	0,99	5,7	3,0	1,136	0,27	2,52	13,5
12	12,6	16,7	134,2	0,64	1,9	1,0	0,99	5,2	3,1	1,149	0,29	2,60	12,8
13	11,6	15,6	116,6	0,64	1,9	1,0	0,98	5,1	3,0	1,150	0,30	2,60	12,6
14	10,9	14,5	97,0	0,64	1,0	1,0	1,00	4,7	2,9	1,203	0,34	2,61	12,2
15	9,9	13,5	86,8	0,65	1,8	1,0	0,99	4,6	2,6	1,118	0,28	2,57	12,0
16	9,1	12,7	75,7	0,65	1,8	1,0	0,98	4,4	2,8	1,147	0,32	2,63	11,8
17	8,0	12,1	63,7	0,66	1,7	1,0	0,99	4,7	3,0	1,189	0,37	2,63	12,1
18	7,1	10,9	51,9	0,67	1,6	1,0	0,99	5,0	2,9	1,226	0,40	2,59	12,4
19	6,0	10,1	41,4	0,69	1,6	1,0	0,99	4,9	3,0	1,255	0,44	2,60	12,3
20	7,5	8,2	36,5	0,60	1,5	1,0	1,00	2,6	2,2	1,134	0,35	2,85	9,6
21	4,4	9,0	31,5	0,79	2,2	1,0	0,90	1,7	2,9	1,158	0,42	1,72	9,9
22	6,0	5,0	28,3	0,79	2,6	1,0	0,98	3,0	1,6	1,115	0,34	1,52	9,7

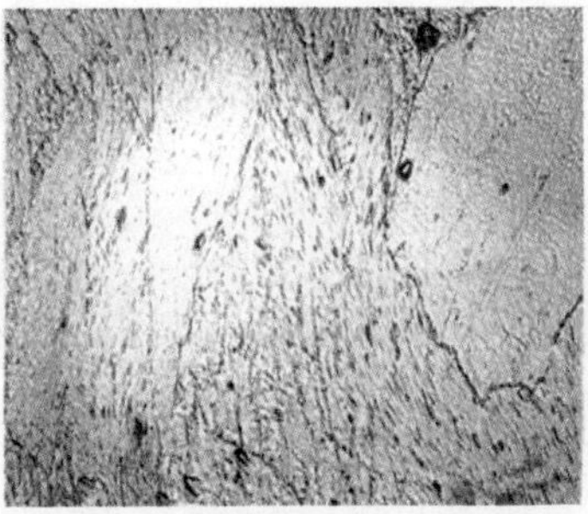

a- the blank after the seventh pass is an octahedron

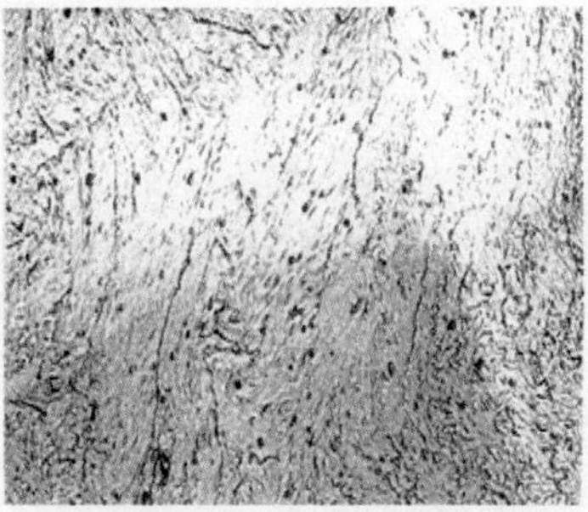

b- the blank after the ninth pass is an octahedron

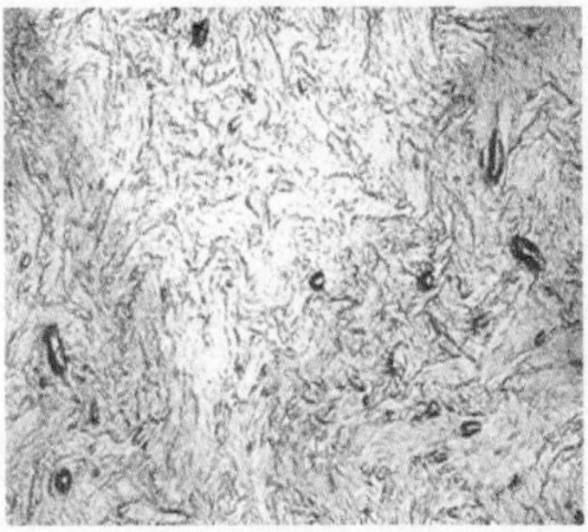

c- the blank after the eleventh pass is an octahedron

d- the blank after the thirteenth pass is an octahedron

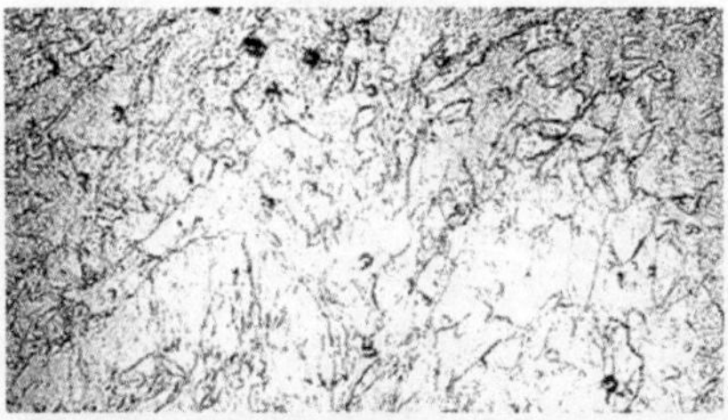

e- the blank after the fifteenth pass is an octahedron

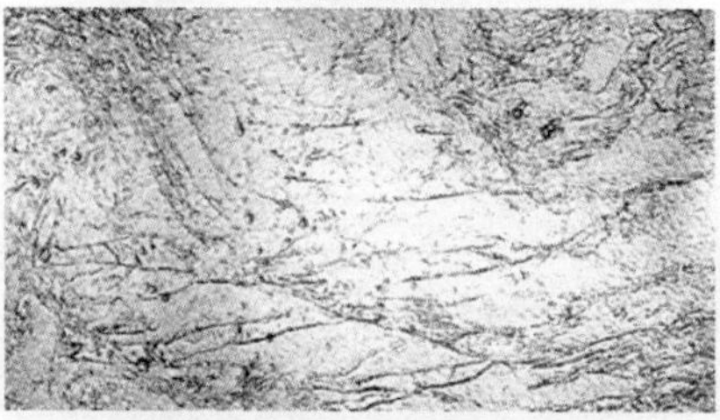

f- the blank after the seventeenth pass is an octahedron

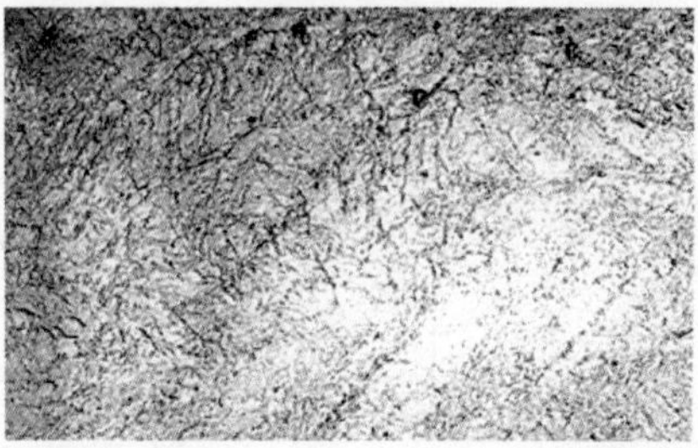

h- blank after the nineteenth pass octahedron

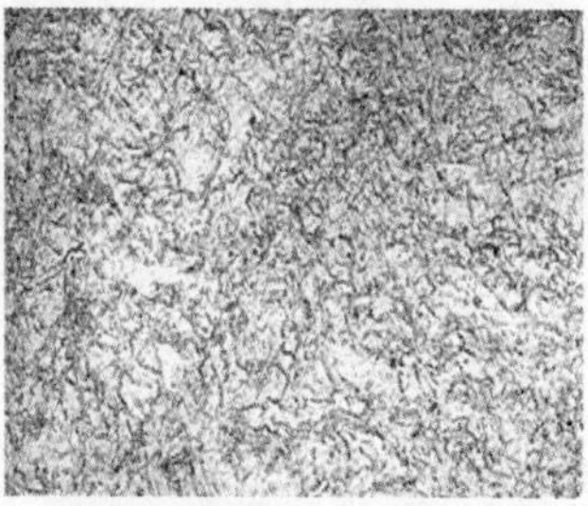

g- the blank after the twenty-first pass is an octahedron ø 6 mm

Fig. 3. Microstructure of niobium billet by longitudinal rolling processes.

Table 3. Indicators of the process of rolling niobium rods diameter 6 mm.

NN caliber	Temperature $t_H = 20$ °C		The length of the deformation focus, l_d, mm	Cross-sectional area, F, mm^2	The number of revolutions of the rolls, n rpm	Average pressure, P_{cp}, N/mm^2	Effort, P, N	Torque, M_{kp}, KNm	Specific energy consumption
1	32,7	32,7	24,0	824,7	23,0	352	29,1	0,90	3,14
2	45,6	45,6	24,9	748,9	23,0	342	25,6	0,70	3,28
3	58,2	58,2	24,4	686,4	23,0	347	23,8	0,70	3,28
4	70,1	70,0	23,4	620,5	23,0	352	21,8	0,60	3,26
5	81,3	81,3	22,5	563,9	23,0	357	20,1	0,50	3,26
6	92,4	92,4	22,2	529,7	23,0	357	18,9	0,50	3,35
7	103,9	103,8	22,3	508,4	23,0	357	18,1	0,50	3,61
8	116,7	-	23,3	503,3	23,0	346	17,4	0,50	4,04
9	39,2	39,2	21,8	453,4	23,0	323	14,6	0,40	4,32
10	57,7	57,7	20,5	380,7	23,0	317	12,1	0,30	4,30
11	71,6	71,5	18,5	301,9	23,0	355	10,7	0,20	3,84
12	86,1	86,0	18,5	296,1	23,0	352	10,4	0,20	4,22
13	100,2	100,0	18,6	273,4	23,0	353	9,7	0,20	4,41
14	115,4	114,9	18,2	253,3	23,0	331	8,4	0,20	4,41
15	125,6	125,0	16,5	205,0	23,0	370	7,6	0,10	4,24
16	137,3	136,4	17,5	208,7	23,0	361	7,5	0,10	4,87
17	150,2	148,9	17,9	207,3	23,0	347	7,2	0,10	5,51
18	163,8	161,8	18,5	193,9	23,0	338	6,5	0,10	6,09
19	177,1	173,9	18,5	178,8	23,0	335	6,0	0,10	6,75
20	182,1	178,4	15,8	126,7	23,0	379	4,8	0,10	5,77
21	183,8	183,6	16,2	146,6	23,0	386	5,7	0,10	7,28
22	190,9	-	16,1	94,8	23,0	389	3,7	0,10	5,70

3 Conclusions

Comprehensive studies of technological schemes and pressure treatment processes have been carried out:

- small diameter bars made of sintered molybdenum stacks: rotary forging and longitudinal rolling on mills with multi-roll calibers;
- bars of small diameter made of molybdenum and niobium ingots using hot and high-temperature pressing and forging in crimps as primary processing, followed by processing of workpieces by rotary forging, long rolling, forging on radial forging machines.

As a result of the analysis of the features of processing sintered molybdenum in the form of stacks by rotary forging and rolling on mills with multi-roll calibers, it was found that the most effective scheme for the production of small-diameter rods is a scheme using the process of longitudinal rolling on multicellular mills with multi-roll calibers. Calibration schemes and thermomechanical modes of rolling molybdenum bars developed in relation to the MK-230 MOZTMITS industrial mill made it possible to eliminate manual rotary forging, reduce the number of heating, preheating, annealing and pressure treatment operations by $5 \cdots 10$ times, improve the surface quality of bars and increase KIM by reducing metal losses from stratification and carbon monoxide.

As a result of the comparative analysis of the features of the processes of primary pressure treatment of ingots of refractory metals, it was found that the process of high-temperature hydro-pressing is one of the effective ways of deformation of refractory metals, which ensures the production of bars of dia. $20 \cdots 40$ mm of high quality with increased mechanical properties by $15 \cdots 20\%$ compared to conventional pressing.

A new highly efficient process and thermo-mechanical modes of forging on a radial crimping machine have been developed to produce small-diameter rods from refractory metals, providing improved product quality, increased equipment productivity and improved working conditions.

References

1. Tilavov, Y.S.: Razrabotke i vnedrenie maloothodnoi tehnologii proizvodstva prutkov iz molibdena i niobiya [Development and implementation of low-waste technology for the production of molybdenum and niobium rods]. – Diss. na soiskanie uchyonei stepeni kandidata tehnicheskih nauk, p. 285 (1994)
2. Kolikov, A.P., Tilavov, Y.S., et al: Matematicheskaya model' termonapryajennogosostyaniya tehnologicheskogo instrumenta pri kovke trudnodeformiruemyh materialov [Mathematical model of the thermally stressed state of a technological tool during forging of hard-to-deform materials]. Dep. VINITI, No. 5 (1994)
3. Kolikov, A.P., Levickiy, I.A., Tilavov, Y.S. i dr.: Matematicheskaya model' teploobmena i termonapryajennogosostyaniya v sisteme zagotovka – rabochii instrument pri obrabotke tugo plavkih metallov [Mathematical model of heat exchange and thermal stress in the billet - working tool system in the processing of tight fusible metals]. Izv. Vuzov. CHernaya metallurgiya, № 9 (1994)

4. Mihailov, V.G., Novojonov, G.I., Tilavov, Y.S., et al: Issledovanie processa vysokotemperaturnogo gidropressovaniya tugoplavkih metallov [Investigation of the process of high-temperature hydro-pressing of refractory metals]. Izvestiya vuzov «CHernaya metallurgiya» №1, pp. 42–44 (1994)
5. Vardiyashvili, A.A., Tilavov, Y.S., Urokov, K.Kh.: Analiz rezultatov rascheta parametrov temperaturnogo i termonapryajennogo sostoyaniya v sisteme zagotovka-instrument pri radial'noi kovke [Analysis of the results of calculating the parameters of the temperature and thermally stressed state in the workpiece-tool system during radial forging]. Nauchno–tehnicheskii jurnal FerPI **24**(5), 213–216 (2020)
6. Tilavov, Y.S., Urokov, K.Kh., Elmurodov, N.S.: Raschet temperaturnogo polya v zagotovkah pri deformacii na radialno - objimnyh mashinah [Calculation of the temperature field in workpieces during deformation on radial crimping machines]. Innovaciontehnologiyalarilmii-tehnik jurnal. Qar MII, pp. 140–146 (2021)
7. Tilavov, Y.S., Vardiyashvili, A.A., Urokov, K.Kh.: Issledovanie tehnologicheskih shem i processov proizvodstva prutkov malogo diametra [Research of technological schemes and processes for the production of small diameter rods]. Nauchno–tehnicheskii jurnal FerPI **25**(5), 191–195 (2021)
8. Karimov, S., Mamirov, S., Khabibullayeva, I.A., Bektemirov, B., Khusanov, N.: Friction and wear processes in tribotechnical system. Int. J. Mechat. Appl. Mech. **10**(1), 204–208 (2021)
9. Bektemirov, B.S., Ulashov, J.Z., Akhmedov, A.K., Gopirov, M.M.: Types of advanced cutting tool materials and their properties. Euro-Asia Conf. **5**(1), 260–262 (2021)
10. Izotov, V.M., Yarcev, V.G., Voroncov, V.K., et al.: Sovershenstvovanie tehnologii rotacionnoi kovki tugoplavkih metallov [Improving the technology of rotary forging of refractory metals]. CNII cvetmet ekonomiki informacii. Obrabotka cvetnyh metallov: Obzor.informac (1978)
11. Kolikov, A.P., Tilavov, Y.S., Negoda, V.V.: Analiz tehnologicheskih shem proizvodstva polufabrikatov malogo diametra iz splavov tugoplavkih metallov [Analysis of technological schemes for the production of semi-finished products of small diameter from refractory metal alloys]. Izv.vuzov.CHernaya metallurgiya №. 3, pp. 85–87 (1993)
12. Umarov, E., Mardonov, U., Abdirakhmonov, K., Eshkulov, A., Rakhmatov, B.: Effect of magnetic field on the physical and chemical properties of flowing lubricating cooling liquids used in the manufacturing process. IIUM Eng. J. **22**(2), 327–338 (2021). https://doi.org/10.31436/iiumej.v22i2.1768
13. Voroncov, V.K., Poluhin, P.I., Belevitin, V.A., Brinza, V.I.: Eksperimental'nye metody mehaniki deformiruemyh tverdyh tel/tehnologicheskie zadachi obrabotki davleniem [Experimental methods of mechanics of deformable solids /technological problems of pressure treatment], Metallurgiya, p. 480 (1990)

Geometrical Model for Tool Wear Assessment in the Processing of Reinforced Composite

Gennadii Khavin(✉) and How Zhiwen

National Technical University "KhPI", Kharkiv, Ukraine
gennadii.khavin@gmail.com

Abstract. A theoretical study of tool wear assessment during machining of reinforced polymer composite materials is presented. Based on the known experimental data on turning, drilling and milling of polymer composites, published in the open press, a theoretical model for assessing tool wear was proposed. This model is based on an analytical description in tool geometry change over time. One of the most effective, simple and accurate methods for determining tool wear is to weigh it before and after machining. The resulting absolute weight loss of the tool falls on the cutting edges, which in the process of contact interaction change their shape due to the material removal under conditions of force loading and thermal heating. When machining reinforced composites, the wear along the flank surface is taken as the main criterion for tool wear. This is due to the specifics of the composite material structure and the nature of the destruction in contact with the tool. Thus, having a reliable analytical apparatus that predicts the shape and nature in the shape of the tool cutting edges change and the value of the weight loss of these cutting edges, one can try to determine the distribution of the lost weight along the geometric tool shape and, most importantly, along the tool flank surface. In other words, the task is to link the tool weight loss during cutting with a change in the amount of wear along the tool flank surface. To implement this approach, a geometric model is proposed.

Keywords: Composites mechanical processing · Tool weight loss · Geometric model · Microgeometry of the cutting edge · Flank wear

1 Introduction

In the last few decades, there has been a steady upward trend in the consumption of composite materials (CM) in various industries. At the same time, the quantitative growth in the consumption of composite various types is accompanied by the expansion of products range made from them. The latter circumstance requires the development of methods for their processing and, first of all, those based on mechanical cutting of composites. Despite the well-established techniques and the developed scientific and technical base for the implementation of these processes, the cutting of composites has a number of specific features due to the heterogeneity of their structure. The main one is the high tool abrasive wear by the reinforcing elements, which leads to a decrease in tool life.

D. D. Cioboată (Ed.): ICoRSE 2022, LNNS 534, pp. 139–145, 2023.
https://doi.org/10.1007/978-3-031-15944-2_14

Intensive tool wear is unrecoverable, determined by the high abrasive properties of the filler. In addition, the relative position of the tool cutting edge and the angle of composite reinforcement play an important role here. High tool wear leads not only to deterioration in the quality of the machined surface, but also to an increase in the cost of operations in general.

Mechanical processing of CM leads to the appearance of treated surface specific defects, such as delamination, pulling out of fibers, cracking. The appearance of these defects is directly related to the cutting tool sharpness. Therefore, tool wear during the cutting process largely determines the quality of processing. There is a need to create mathematical models for predicting its guaranteed operation without costly and lengthy experiments. The task of predicting tool life for a given quality of the machined surface is relevant and of great practical value.

2 Literature Review

A very large number of publications are devoted to the study of the wear phenomenon during various types of CM processing, in which, first of all, the experimental studies obtained by the authors were discussed. An analysis of the results made it possible to formulate a qualitative, and in some cases, a quantitative picture of tool wear due to a change in the cutting edge, including the dynamics of changes in the cutting edge micro geometry during cutting. The analysis of the obtained results is reflected in numerous review publications, in which the main attention is paid to ensuring defect-free processing [1]. The main approach to the practical use of tools made of tungsten carbide and high speed steel is the rational choice of initial sharpening angles and optimization of processing modes. At the same time, the main attention was paid to methods and analysis of tool flank wear, detection, measurement and processing of signals using intelligent decision-making methods [2].

A systematic study of the tool wear process as a physical phenomenon made it possible to draw a number of important conclusions. Tool wear actually occurs on the flank surface and its value on the rake surface is insignificant [3]. On the other hand, there is a wear mechanism due to the rounding of the cutting edge, and an approach to the quantitative prediction of wear has been proposed [4]. Based on this model, analytical methods created on modeling the micro geometry of the cutting tool tip have been developed [5]. This paper presents an analysis of tool wear characteristics in five micro geometry parameters used the straight line-ellipse-straight line approach, which can also describe a developing asymmetric cutting edge. Based on experiments on orthogonal machining with different fiber orientations and tool geometry, it was concluded that the micro geometry changes with increasing cutting length, while it is possible to detect an increase in the resulting friction length and elastic compression height, which has a significant effect on the force component of the process. An experimental study of the tool wear process made of tungsten carbide during orthogonal cutting of carbon fiber reinforced plastic (CFRP) is presented in [6]. To describe the form of tool cutting part wear, the scheme proposed by the authors earlier with five parameters [5] is used. It is shown that the loss of material due to wear differs in the tool contact area and strongly depends on the flank angle and the reinforcing fibers orientation angle, and the distortion of the rake angles is negligible. The magnitude of the elastic squeeze is especially pronounced when processing carbon fiber reinforced plastics with an inclination angle of the reinforcing fibers of 30° and 60°. This spring action tends to increase as tool tip wear increases.

The influence of the sharpened cutting edge geometry on the tool wear characteristics, technological forces and the thermal field of the inserts during turning operations is described in [7], where the proposed special micro geometry of the cutting edge, which leads to a significant increase in the productivity of the cutting tool and an increase in its durability. The resource maps of various tool profiles are given.

In [8], a series of experiments on drilling CFRP plates using carbide drills is presented. A study was made of the correlation between tool wear, cutting forces, temperature and hole surface quality. Particular attention is paid to the evaluation of the surface quality (the number of uncut fibers) depending on the temperature in the destruction zone. For this, a special method for measuring temperature was developed.

The work [9] presents experimental and theoretical studies on the study of cutting reinforced polymer composites with and without the imposition of vibrations. It has been shown that in a conventional cutting process, the chips generated by a juvenile cutting edge are larger than that of a worn cutting edge. It follows from the analysis of experimental data that tool wear not only changes the nature of chip formation and, as a consequence, the magnitude of the cutting force, but also the entire mechanism of material removal and surface quality.

The influence of the cutting edge radius, sharpening and feed angles per tooth, on technological forces and the appearance of tool vibrations is presented in [10]. An experimental study on the study of chip formation was carried out for micro mills. In [11], an experimental study of a standard double-fluted carbide drill wear during high-speed drilling (up to 15,000 rpm) of woven carbon reinforced plastics with an epoxy matrix is presented. The interdependence of the wear process and surface damage of the composite is shown. The main manifestation of tool wear is the appearance of chipping and abrasion. Friction of sheared graphite fibers and tungsten carbide grains against a soft epoxy matrix causes an increase in temperature and, as a result, increased flank wear. It has been experimentally established that at the beginning of the tool operation, the wear of the cutting edges flank surface predominates, and then carbon formation appears, which is accompanied by abrasion. The consequence of this is an increase in delamination and an increase in the roughness of the treated surface. A correlation has been established between tool wear, delamination damage, and surface roughness.

The issues of predicting tool performance under conditions of intense wear were considered in [12] for face milling of fiberglass. When analysing the results obtained, an fuzzy adaptive neural network (ANFIS) is used, which makes it possible to predict the degree of tool wear depending on the spindle speed. The results obtained showed the high accuracy and efficiency of ANFIS in assessing wear under non-linear conditions. The use of neural network technology to analyse the effect of tool wear on the value of axial force is presented in [13]. Traditional technological inputs (spindle speed and feed) and drill point angle were used in an experimental study of drilling with an uncoated carbide tool.

The use of carbide tools with adaptable geometry and wear-resistant diamond coating for drilling CFRP is presented in [14]. The diamond layer thickness determines the radius of the cutting edge and, ultimately, tool life. A comparison is made between the methods of processing the cutting edge by traditional sandblasting and laser processing. It is shown that laser processing gives a cutting edge radius of 3–4 μ, which provides a higher quality of the exit hole.

The article [15] presents the results of a study of a carbide cutting tool wear during the CFRP milling using dry air cooling. The experiments were carried out for various cutting speeds, feeds and depths of cut using air with a temperature of −10 °C and a flow rate of 4.1 m/s. There was a reduction in tool wear under chilled conditions compared to operation without air cooling. The main factor affecting tool life is the feed rate. The use of chilled air in CFRP machining increases the life of uncoated carbide cutting tools compared to uncooled machining.

3 Research Methodology

The main purpose of the study is to use the method of weighing a tool before and after its use in machining to determine the degree of wear. The weight loss of the tool cutting edges is uneven. They change their geometric shape over time. The processing of composite materials is accompanied by intense wear along the flank surface and an insignificant change in the rake angle due to wear along the rake surface. Therefore, there is a relationship between the total weight loss of the tool tip and flank wear, which is accompanied by a change in the clearance angle. The task is to relate the tool weight loss in during of work with the value of its wear along the flank surface.

As a working hypothesis, it was assumed that in the process of wear, the tip of the initially sharpened tool shifts along its rake surface, i.e. there is a constant displacement of the original top of the sharpened tool along the rake surface, Fig. 1. In this case, it is assumed that there is no change in the value of the rake angle. Such an assumption became possible from the analysis of the cutter worn tip profile presented in [4].

In the presented formulation, there is also an inverse problem, when it is necessary to determine the total change in tool geometry and weight loss from the existing value of wear along the tool flank surface. In addition, having an apparatus for geometrically changing the area of the tool tip (in fact, weight loss), it is possible to calculate the amount of wear along the flank surface and determine the tool life under given machining conditions, i.e. for a material with a given orientation of the hardening fibers and technological parameters of cutting.

Thus, the actual wear of the tool during its operation can be represented as the difference in the tool tip surface areas, calculated for two consecutive moments of time, multiplied by the width of the cutting edge and the tool material density.

In the machining of composite materials, the standards of different countries provide for a limit value for flank wear, which usually does not exceed 0.3 mm on average (for example ISO 36851993 for cemented carbides and high speed steel). Upon reaching this value, during the cutting process, the appearance of machining defects is observed, and the tool itself must be replaced and reground.

In fact, two practical formulations of the problem follow from the formulated problem statement. First task is to calculate the change in its shape and, first of all, wear on the

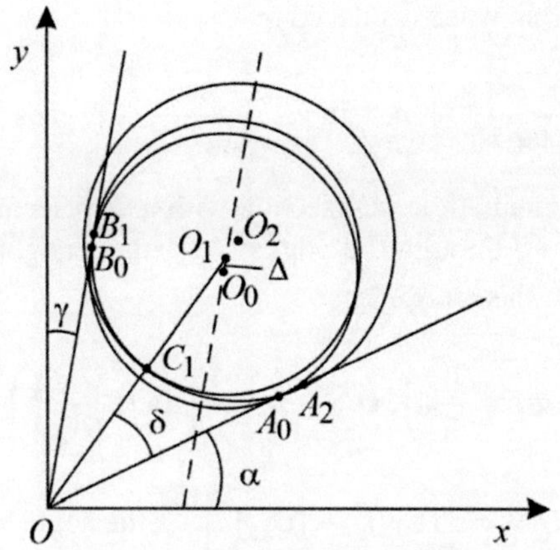

Fig. 1. Worn profile geometric model

flank surface from a given weight loss of the tool. The second task is to determine the expected weight loss for a given value of flank wear and hence tool life.

Considering that the wear rate changes with time, and with the appearance of wear, the stress state in the contact changes, the tool tip heats up, the friction conditions (friction coefficient) change, both formulations have a pronounced nonlinearity. To solve such problems, it is advisable to use step-by-step algorithms that take into account the hereditary change in the shape of the cutting tool tip.

4 Results

Consider the tool tip wear over some time of its operation, starting with its sharp state and some worn out. We will assume that the initial geometrical parameters of the tool tip cutting part are known, namely, Fig. 1: flank angle α, rake angle γ and cutting edge rounding radius þ. The touch points of the flank – *A*, and rake – *B* edges to the conditional rounding circle of radius þ can be calculated from the ratios:

$$A_0\left(\frac{\rho \cdot \cos\alpha}{\tan\beta}, \frac{\rho \cdot \sin\alpha}{\tan\beta}\right); B_0\left(\frac{\rho \cdot \sin\gamma}{\tan\beta}, \frac{\rho \cdot \cos\gamma}{\tan\beta}\right),$$

where $\beta = \left[\pi/2 - (\alpha + \gamma)\right]$.

The coordinates of the initial circle O_0 centre of the rounded cutting edge of radius þ are equal

$$O_0\left(\frac{\rho \cdot \cos(\alpha+\beta)}{\sin\beta}, \frac{\rho \cdot \sin(\alpha+\beta)}{\sin\beta}\right) = O_0\left(O_{0x}, O_{0y}\right).$$

The circle O_1 centre of radius þ displaced by Δ as a result of wear has the coordinates

$$O_1\left(\frac{\rho \cdot \cos(\alpha+\beta)}{\sin\beta} + \Delta \cdot \sin\gamma, \frac{\rho \cdot \sin(\alpha+\beta)}{\sin\beta} + \Delta \cdot \cos\gamma\right) = O_1\left(O_{1x}, O_{1y}\right).$$

The contact point circle when displaced by Δ

$$B_1\left(\frac{\rho \cdot \sin\gamma}{\tan\beta} + \Delta \cdot \sin\gamma, \frac{\rho \cdot \cos\gamma}{\tan\beta} + \Delta \cdot \cos\gamma\right).$$

The point of the original circle intersection with a displacement of Δ due to wear and a straight line passing through the points OO_1 (the straight line has an angle of inclination $arctg\frac{O_{1y}}{O_{1x}}$) has the coordinates

$$C_1\left(O_{1x} - \rho \cdot cos\left(arctg\frac{O_{1y}}{O_{1x}}\right), O_{1y} - \rho \cdot sin\left(arctg\frac{O_{1y}}{O_{1x}}\right)\right) = C_1\left(C_{1x}, C_{1y}\right).$$

Straight line length $OC_1 = \left[(C_{1x})^2 + \left(C_{1y}\right)^2\right]^{1/2}$, the angle δ between the segment OC_1 and the conditional straight line passing through the points O and A at an angle α (flank surface) is equal to $\delta = \text{arctg}\ (O_{1y}/O_{1x}) - \alpha$. The value of the radius ρ_2 of the inscribed circle formed as a result of wear is $\rho_2 = OC_1 \cdot \sin\delta/(1 - \sin\delta)$, and the centre of this circle has the coordinates

$$O_2((OC_1 + \rho_2) \cdot \cos(\alpha + \delta)), ((OC_1 + \rho_2) \cdot \sin(\alpha + \delta)) = O_2\left(O_{2x}, O_{2y}\right).$$

The contact point of the circle with the flank surface initial inclination line (at an angle α), corresponding to the new position after wear, can be determined from the expression

$$A_2((OC_1 + \rho_2) \cdot \cos\delta \cdot \cos\alpha, (OC_1 + \rho_2) \cdot \cos\delta \cdot \sin\alpha) = A_2\left(A_{2x}, A_{2y}\right).$$

Then the wear cutter tip surface area can be determined as the difference between the area of the cutter tip initial rounding and the area of intersection of circular areas with centres O_0 and O_2.

In this case, the controlled amount of wear along the flank surface h_z can be calculated as the projection of the circles contact points on the x axis, describing the shape of the cutter tip to the conditional straight OA_2 flank surface passing at an angle α.

5 Conclusions

A simple one-parameter geometric model is presented that allows calculating the change in the tool tip area and determining the change in flank angle. After determining the area of the worn surface, knowing the width of the cutting edge and the tool material density, it is possible to determine the tool weight loss during cutting, which can also be used as a criterion for technological wear, as well as the amount of wear of the flank surface. For the numerical implementation of the formulated wear problem in time, a simple geometric model is proposed. The implementation of the approach is based on the current wear calculation, taking into account the already accumulated during the tool operation time. This allows, by simple weighing, technical control of cutting edge wear and prediction of tool life.

References

1. Lokesh, K.S., Pinto, T., Ramachandra, C.G.: Effect of tool wear and machinability studies on polymer composites; a review. Int. J. Eng. Inf. Syst. **1**(5), 99–110 (2017). Hal-01571294f
2. Siddhpura, A., Paurobally, R.: A review of flank wear prediction methods for tool condition monitoring in a turning process. Int. J Adv. Manuf. Technol. **65**, 371–393 (2013). https://doi.org/10.1007/s00170-012-4177-1
3. Hrechuk, A., Bushlya, V., Saoubi, M.R., et al.: Experimental investigations into tool wear of drilling CFRP. Procedia Manuf. **25**, 294–301 (2018). https://doi.org/10.1016/j.promfg.2018.06.086
4. Faraz, A., Biermann, D., Weinert, K.: Cutting edge rounding: an innovative tool wear criterion in drilling CFRP composite laminates. Int. J. Mach. Tools Manuf **49**, 1185–1196 (2009). https://doi.org/10.1016/j.ijmachtools.2009.08.002
5. Seeholzer, L., Voss, R., Grossenbacher, F., et al.: Fundamental analysis of the cutting edge micro-geometry in orthogonal machining of the unidirectional Carbon Fibre Reinforced Plastics (CFRP). Procedia CIRP **7**, 379–382 (2018). https://doi.org/10.1016/j.procir.2018.09.040
6. Seeholzer, L., Kneubühler, F., Grossenbacher, F., Wegener, K.: Tool wear and spring back analysis in orthogonal machining unidirectional CFRP with respect to tool geometry and fibre orientation. Int. J. Adv. Manuf. Technol. **115**(9–10), 2905–2928 (2021). https://doi.org/10.1007/s00170-021-07222-z
7. Denkena, B., Lucas, A., Bassett, E.: Effects of the cutting edge microgeometry on tool wear and its thermomechanical load. CIRP Ann. Manuf. Technol. **60**, 73–76 (2011). https://doi.org/10.1016/j.cirp.2011.03.098
8. Ramireza, C., Poulachona, G., Rossia, F., et al.: Tool wear monitoring and hole surface quality during CFRP drilling. Procedia CIRP **13**, 163–168 (2014). https://doi.org/10.1016/j.procir.2014.04.028
9. Xu, W., Zhang, L.: Tool wear and its effect on the surface integrity in the machining of fibre reinforced polymer composites. Compos. Struct. **188**, 257–265 (2018). https://doi.org/10.1016/j.compstruct.2018.01.018
10. Biermann, D., Baschin, A.: Influence of cutting edge geometry and cutting edge radius on the stability of micromilling processes. Prod. Eng. Res. Devel. **3**, 375–380 (2009). https://doi.org/10.1007/s11740-009-0188-7
11. Rawat, S., Attia, H.: Wear mechanisms and tool life management of WC–Co drills during dry high speed drilling of woven carbon fibre composites. Wear **267**, 1022–1030 (2009). https://doi.org/10.1016/j.wear.2009.01.031
12. Azmi, A.I., Lind, R.J.T., Bhattacharyya, D.: Fuzzy logic predictive model of tool wear in end milling glass fibre reinforced polymer composites. Adv. Mater. Res. **214**, 329–333 (2011). https://doi.org/10.4028/www.scientific.net/AMR.214.329
13. Feito, N., Muñoz-Sánchez, A., Díaz-Álvarez, A., et al.: Analysis of the machinability of carbon fiber composite materials in function of tool wear and cutting parameters using the artificial neural network approach. Materials **12**, 27–47 (2019). https://doi.org/10.3390/ma12172747
14. Voss, R., Henerichs, M., Capricano, G., et al.: Post-coating treatment of cutting edge for drilling carbon fibre reinforced plastics. Procedia CIRP **46**, 161–164 (2016). https://doi.org/10.1016/j.procir.2016.03.173
15. Nor Khairusshimaa, M.K., Sharifahb, I.S.S.: Study on tool wear during milling CFRP under dry and chilled air machining. Procedia Eng. **184**, 506–517 (2017). https://doi.org/10.1016/j.proeng.2017.04.121

Assessment of Methods for Fan Blades Arrangement for Static Balancing of the Fan of a Turbofan Engine

Anna Stefanova[1], Georgi Georgiev[2], and Vladimir Serbezov[1](✉)

[1] Department of Aeronautics, Technical University of Sofia, Sofia, Bulgaria
vserbezov@tu-sofia.bg

[2] Institute for Nuclear Research and Nuclear Energy, Bulgarian Academy of Sciences, Sofia, Bulgaria

Abstract. The fan blades of a high by-pass ratio turbofan engines require frequent maintenance actions due to erosion and foreign object damage. To keep the fan disk imbalance within limits when repairing or replacing fan blades a fan blades rearrangement procedure must be applied. In this work various methods of fan blades (re)arrangement are suggested and evaluated by applying them on large number of randomly generated sets of fan blades for a typical commercial turbo fan engine. A special attention is payed to the subsequent minimization of the resultant moment of imbalance by applying all to all blade switch recursive algorithms. The optimum values of the recursive algorithm parameters are determined. The results of the investigation show that the resultant moment of imbalance of the fan can be significantly reduced, compared to the standard methods used in the most of the airplane maintenance and repair organizations.

Keywords: Turbofan engine · Fan balancing · Iterative balancing algorithm

1 Introduction

The fan blades of a modern high by-pass turbofan engine during aircraft operation are highly exposed to erosion and foreign object damage, resulting in relatively frequent replacement of damaged blades. For this reason the turbofan engines are designed with provision for easy replacement without disassembling the engine and without removing it from the aircraft (on-wing maintenance). On every such event it is necessary to find suitable arrangement of the fan blades around the fan disc so that the resultant moment of imbalance, generated by fan blade moment weights, is acceptably small. This is a combinatorial problem and it has no analytic solution. Although there are many software solutions to the problem, in the airline and maintenance and repair organizations (MRO) practice there is still a wide use of manual solution methods, provided as last means by the engine manufacturers. These methods are time consuming and do not guarantee a good practical solution. In this context the development and assessment of relatively simple and reliable fan balancing algorithms continues to be a challenging topic. The

D. D. Cioboată (Ed.): ICoRSE 2022, LNNS 534, pp. 146–159, 2023.
https://doi.org/10.1007/978-3-031-15944-2_15

motivation of this study was to evaluate and propose such fan balancing algorithms with aim at practical applications in the airline and MRO industry.

The fan balancing problem itself is a subcase of the turbine balancing problem (TBP), which is well studied in the literature. There are many mathematical approaches to get fine enough approximate solutions of the problem. Methods developed include Monte Carlo method applied by Mosevich [7], formulation as a Quadratic Assignment Problem (QAP) by Amroune et al. [2], by Mason and Rönnqvist [5] and by Pitsoulis et al. [9]. It should be mentioned that the QAP has multidisciplinary applications, the research in this area is ongoing and new results continue to emerge, e.g. Misevicius, and Verene [6]. General influence coefficient algorithms were used by Yu [11], convex optimization by Li et al. [4], Ant Colony Optimization by Piskin et al. [8]. The application of stochastic optimization, iterative random search and genetic algorithms was studied by Zhai and Gong [12, October]. Heuristic algorithms were proposed by Amiouny et al. [1] and by Fathi and Ginjupalli [3]. Another tendency in the TBP studies is to account for factors other than the blade's imbalance, such as the concentricity of the disk [10].

An important issue is the evaluation of the performance of the proposed balance algorithms. The common approach, used by virtually all of the researchers is the approbation of the algorithms over a large sample of randomly generated blade sets with certain standard deviation of the mass moment of the blades.

In this work some of the methods listed were applied for the fan of the CFM56-5B turbofan engine. Some new methods were suggested that weren't found in the literature. A new implementation in the algorithm for static balancing was also made by developing an iterative procedure for minimization of the resultant moment of imbalance. The different methods for fan blades sorting and efficiency of the minimizing procedures were compared based on a great number of randomly generated fan blade sets.

2 Problem Definition and Methodology

2.1 The Fan Balancing Problem

The fan balancing problem is well defined in the work of Amiouny et al. (2000). It can be stated as a combinatorial problem of sequencing the blades around the fan disk so that the resultant imbalance about the axis of rotation is beyond a certain limit. The magnitude of the resultant imbalance mim for a fan with n blades can be expressed as follows:

$$m_{im} = \sqrt{\left(\sum_{i=1}^{n} (m_{\omega i}.\sin(i.\Delta_\theta))\right)^2 + \left(\sum_{i=1}^{n} (m_{\omega i}.\cos(i.\Delta_\theta))\right)^2} \quad (1)$$

where $m_{\omega i}$ is the moment weight of the blade located in the i-th position on the fan disk and Δ_θ is the angular step between two adjacent blade slots.

As an exemplary fan for this study was chosen the fan of the CFM56-5B engine for the Airbus A320 family aircraft. The fan has 36 blades. Operational requirements to the static balance of the fan, listed in the Aircraft Maintenance Manual (AMM) are that the resultant mass moment of imbalance should be less than 80 g.cm and that the difference of the mass moment between the opposite blades should be less than 200 g.cm.

As far as the number of combinations for n blades is n!, e.g. for 36 blades it is 36! ≈ 3.72*10^{41} combinations, the straightforward solution is not possible. This requires the application of heuristic algorithms for blade arrangement.

2.2 Methodology for the Arrangement Methods Assessment

The algorithms for blade arrangement are implemented for CFM56-5B and are compared based on randomly generated "virtual" fan blade sets. All calculation procedures are developed in the Python programming language.

To perform a statistically valid comparison between the different algorithms a large amount of fan blade sets is required. For this purpose a random blade pair generator is developed. The random blade pair generator generates fan blade sets in pairs with moment weights following normal distribution with parameters mean moment weight $\mu = 16000$ g.cm and standard deviation $\sigma = 3000$ g.cm. These parameters are estimated by analyzing a big sample of real fan blades for CFM56-5B, obtained from sources in the MRO industry. The fan blades for each set are generated in pairs with maximum difference between blade moment weights in each pair 200 g.cm. A total of 500000 random fan blade sets were generated (Fig. 1) and used in the analysis that follows.

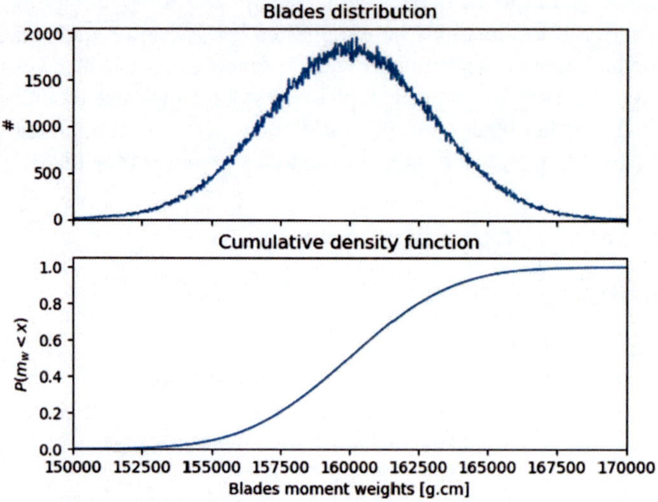

Fig. 1. Gaussian distribution and cumulative distribution function for 500000 randomly generated values with an average value of 16 000 g cm and standard deviation $\sigma = 3000$ g.cm.

3 Description of the Algorithms for Initial Fan Blades Arrangement

The proposed fan blades arrangement algorithms can be applied to jet engines with fans, constructed from even number fan blades. In all algorithm descriptions the following notations are used: *n* is the number of fan blades and $m_{\omega 1} \ldots m_{\omega n}$ are the moment weights of the blades. The positions of the fan blades around the fan disc are numbered from 1 to n counter-clockwise, beginning from position 12 o'clock. All of the figures show the fan blades distribution for the jet engine CFM56-5B.

A total of 11 algorithms for initial fan blades arrangement are examined in this work. As a reference the AMM manual procedure for fan mapping is included. Five of the algorithms are taken from Amiouny et al. (2000). Lastly five of the methods are proposed by the authors of this work. The list of the algorithms is presented in Table 1. A detailed description is provided only for the algorithms proposed by the authors of this work.

Table 1. List of the examined algorithms for initial fan blades arrangement.

No	Algorithm name	Source	Blade pairs ordering principle
1	Symmetrical mapping about the line 12 h–6 h decreasing	Proposed by the authors	Moment weights
2	Symmetrical mapping about the line 12 h–6 h alternating	Proposed by the authors	Moment weights
3	One way mapping decreasing	Proposed by the authors	Moment weights
4	Double beam heavy light adjacent decreasing	Amiouny et al. (2000)	Moment weights
5	Double beam heavy light adjacent alternating	Amiouny et al. (2000)	Moment weights
6	Quadraple beam heavy light adjacent decreasing	Amiouny et al. (2000)	Moment weights
7	Quadraple beam forward reverse decreasing	Amiouny et al. (2000)	Moment weights
8	AMM manual procedure	Airbus A320 AMM	Moment weights
9	Delta sort dynamic	Proposed by the authors	Difference of the moment weights
10	Ordinal pairing	Amiouny et al. (2000)	Difference of the moment weights
11	Delta sort decreasing	Proposed by the authors	Difference of the moment weights

3.1 Algorithms Based on Blade Pairs Ordered by Moment Weights

The fan blades are sorted by their moment weights in decreasing order. Each two successive blades are put together in pair k_i, so that the first pair k_1 is formed by the blades with moment weights m_{w1} and m_{w2}. The blades from each pair are placed diametrically opposite to one another.

Symmetrical Mapping About the Line 12 h–6 h Decreasing. The first blade pair k_1 is placed on positions 1 and n/2 + 1 with the heavier fan blade in the first position for the pair. The second pair k_2 is placed on positions 2 and n/2 + 2 with the lighter fan blade in the first position for the pair. The third pair k_3 is placed on positions n and n/2 with the heavier fan blade in the first position for the pair. The remaining blade pairs are placed in the same manner symmetrically about the line 12 h–6 h with alternating position of the heavier blade in each pair. Example of this sorting procedure is shown on Fig. 2.

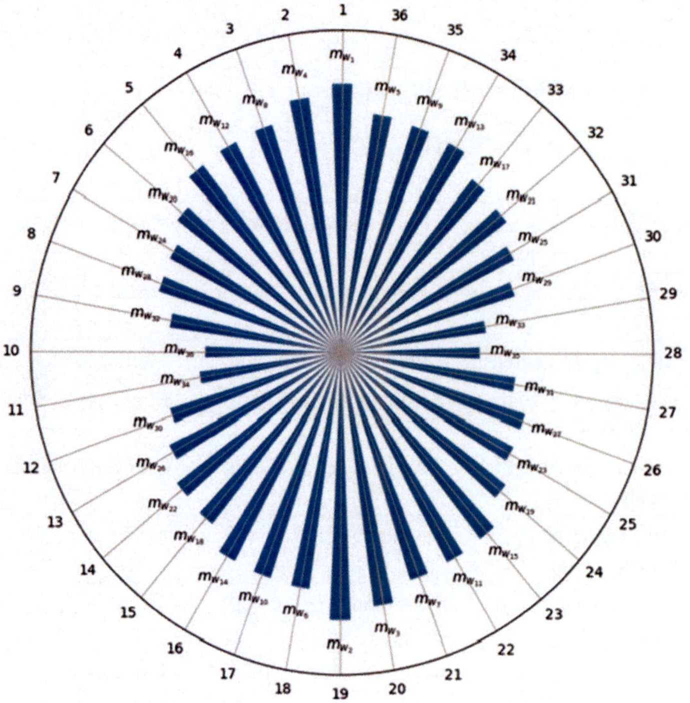

Fig. 2. Symmetrical mapping about the line 12 h–6 h decreasing

Symmetrical Mapping About the Line 12 h–6 h Alternating. The first blade pair k_1 is placed on positions 1 and $\frac{n}{2}+1$. Next to it, on positions 2 and $\frac{n}{2}+2$ is placed the last blade pair $k_{\frac{n}{2}}$. On positions n and $\frac{n}{2}$ are placed the blades from the second to last pair $k_{\frac{n}{2}-1}$. . For the "heavy" pairs (from k_1 to k_9) the heavier fan blade is in the first position for the pair, and for the "light" pairs (from k_{10} to k_{18}) the lighter fan blade is in the first position for the pair. The goal of this sorting method is to surround each heavy blade with two light and each light blade with two heavy. Example of this sorting procedure is shown on Fig. 3.

One Way Mapping Decreasing. The blade pairs are placed around the fan disk sorted by their moment weights in decreasing order. The first blade pair is placed on positions 1 and $\frac{n}{2}+1$, the second – in positions 2 and $\frac{n}{2}+2$ and so on. The heavier blade is always put in the first position for the pair. Example of this sorting method is shown on Fig. 4.

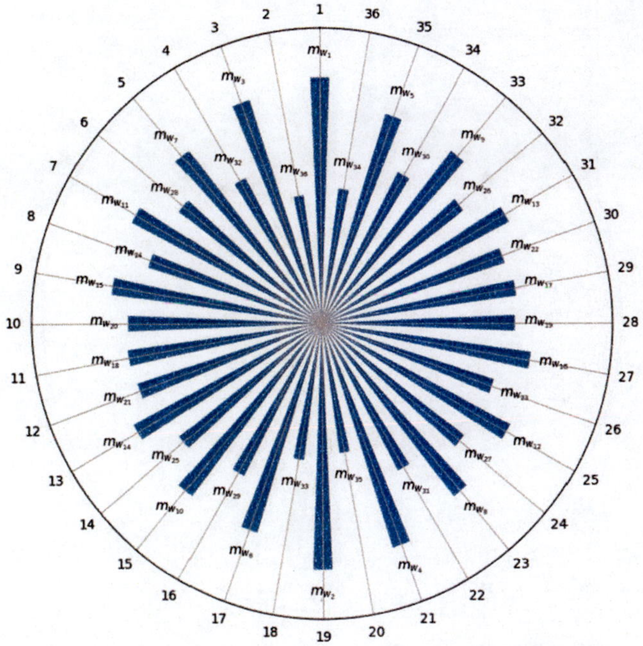

Fig. 3. Symmetrical mapping about the line 12 h–6 h alternating

3.2 Algorithms Based on Blade Pairs Ordered by the Difference of Their Moment Weights

The fan blades are sorted by their moment weights in decreasing order. Each two successive blades m_{w_i} and $m_{w_{i+1}}$ are put together in a pair k_i and the difference between the blades moment weights in each pair is calculated as $\Delta_i = m_{w_i} - m_{w_{i+1}}$, where Δ_i is the difference between the moment weights, m_{w_i} and $m_{w_{i+1}}$ are the moment weights of the fan blades in the pair. The blade pairs are sorted and enumerated by the value of Δ_i in decreasing order so that for blade pair k_1 the difference between the moment weights is Δ_1, for blade pair k_2 respectively Δ_2, ..., for blade pair $k_{\frac{n}{2}}$ - $\Delta_{\frac{n}{2}}$ and $\Delta_1 > \Delta_2 > \ldots > \Delta_{\frac{n}{2}}$.

Delta Sort Dynamic. This method is developed as a computer algorithm which determines the places of each blade pair except from the first one k_1, which is placed on positions 1 and $\frac{n}{2}$. The algorithm places the blades from the next pair k_2 on each of the available positions and calculates the resultant moment of imbalance. Blade pair k_2 is assigned to the positions where it generates minimal resultant moment of imbalance. The positions for the rest of the blade pairs are determined the same way. Example of this sorting method is shown on Fig. 5.

Delta Sort Decreasing. The idea of this method is to obtain the worst possible solution, in order to use it as a starting point for the recursive minimization procedure.

The blade pairs are placed around the fan disk by the difference between the moment weights in each pair in decreasing order. The first blade pair is placed on positions 1 and $\frac{n}{2} + 1$, the second – in positions 2 and $\frac{n}{2} + 2$ and so on. The heavier blade is always put in the first position for the pair. Example of this sorting method is shown on Fig. 6.

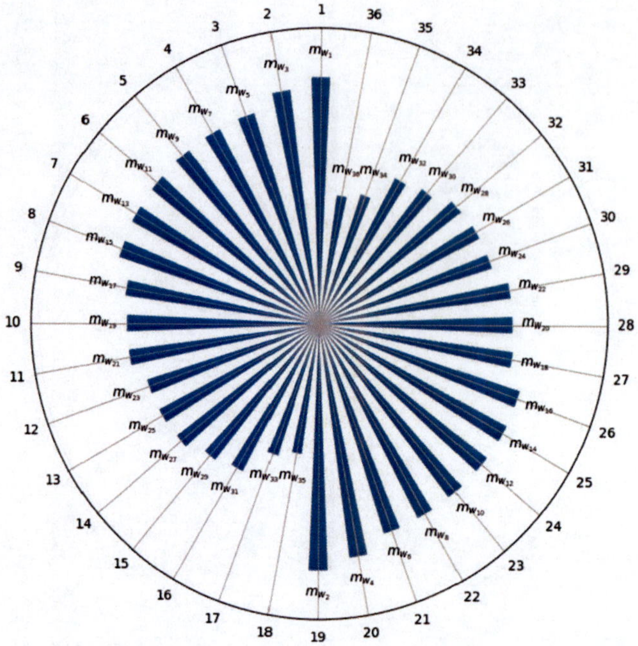

Fig. 4. One way mapping decreasing

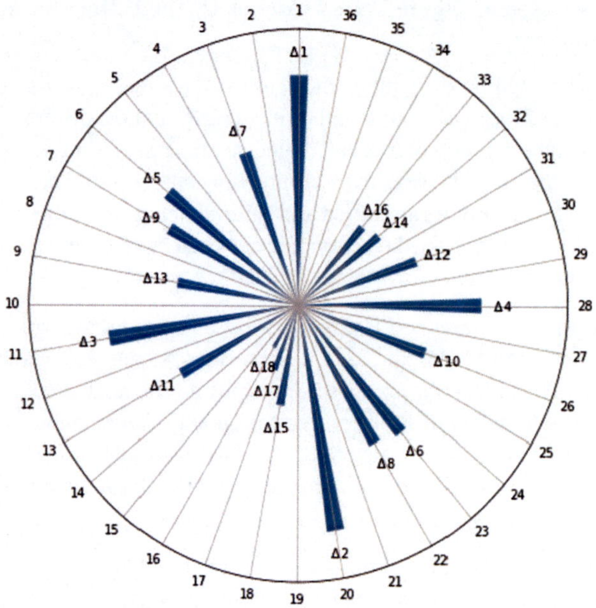

Fig. 5. Delta sort dynamic mapping

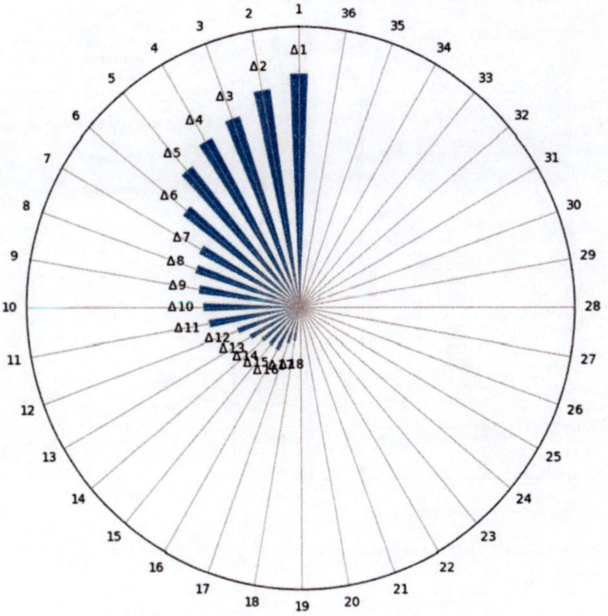

Fig. 6. Delta sort decreasing mapping

4 Initial Comparison of Fan Blades Arrangement Algorithms

The algorithms for blade arrangement are implemented for CFM56-5B. Each algorithm is applied on a sample of 10^5 randomly generated fan blades sets and each time the resultant moment of imbalance of the fan is calculated. The distribution of the resultant moments of imbalance for each arrangement algorithm is then represented as a histogram and the histograms of all of the algorithms are compared with each other. The results are shown on Fig. 7.

The following inferences can be drawn from Fig. 7:

- Delta sort dynamic algorithm is the most effective in minimizing the resultant moment of imbalance with most probable value of $m_{im} < 80$ g.cm in almost all of the cases.
- The second most effective algorithm is Ordinal pairing (Amiouny et al. 2000)
- The least effective algorithms for fan blades arrangement are 'Symmetrical mapping about the line 12–6 h decreasing', 'One way mapping decreasing' and 'Delta sort decreasing'.
- The rest of the algorithms, including the AMM procedure, have virtually identical performance, with less than 30% of the solutions not meeting the imbalance limit of 80 g.cm
- Algorithms based on ordering the blade pairs by their moment weight difference are the most effective (except for 'Delta sort decreasing' which is purposefully designed to deliver the worst result) due to the fact that the resulting moment of imbalance is actually the vector sum of the deltas.

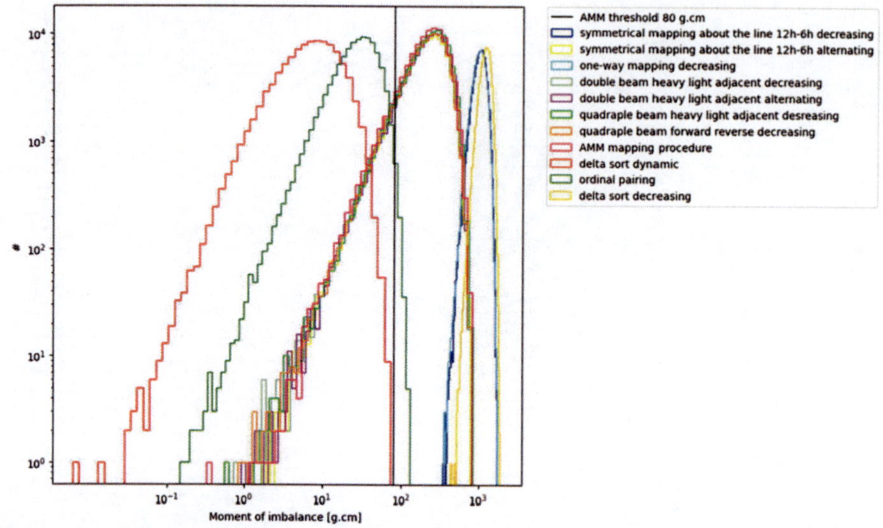

Fig. 7. Comparison of the distributions of the resultant imbalance of the examined algorithms for initial fan blades arrangement.

5 Subsequent Imbalance Minimization

After analyzing the results from the arrangement algorithms it is becoming clear that subsequent minimization of the resultant moment of imbalance is needed for almost all of them in order to satisfy the AMM requirement m_{im} <80 g.cm. Even when this requirement is satisfied the resultant moment of imbalance can be further minimized, probably resulting in improved fan performance.

5.1 Minimization Target for the Resultant Moment of Imbalance

The values of the fan blade moment weights are given by the manufacturer with accuracy of 1 g.cm. Let's consider a fan composed of 36 fan blades with equal moment weights up to the uncertainty of 1 g.cm. The resultant moment of imbalance of such a fan will be due to the rounding error of the fan blades moment weights. Sample of 500000 fans were custom generated and their resultant moments of imbalance were calculated in order to get the distribution of the error in the resultant moments caused by the rounding error of the blade moment weights. This distribution is shown on Fig. 8 where the blue line shows the distribution of the error and the orange line – the cumulative density function (CDF) of the distribution. The CDF shows the probability that some m_{im} from the horizontal axis will take value less than or equal to a value from the vertical right axis.
From the CDF it can be seen that less than 5% of the resultant moments of imbalance due to the rounding error in fan blades moment weights are less than 0.39 g.cm. This is the value that was adopted as a minimization target.

5.2 Iterative Procedure for Subsequent Minimization of the Resultant Moment of Imbalance

The iterative procedure for subsequent minimization is a computer algorithm, developed by the authors, based on the manual imbalance minimization procedure described in the AMM. The idea is to swap each pair of the fan set with each of the remaining pairs in order to improve the fan blades arrangement by achieving a smaller value of the resultant moment of imbalance.

The iterative algorithm is based on a recursive function which creates a list of arrangements in increasing order of the value of the resultant moment of imbalance and calls itself for each of the arrangements from the list. This is done until no further swapping of the blade pairs improves the resultant moment of imbalance.

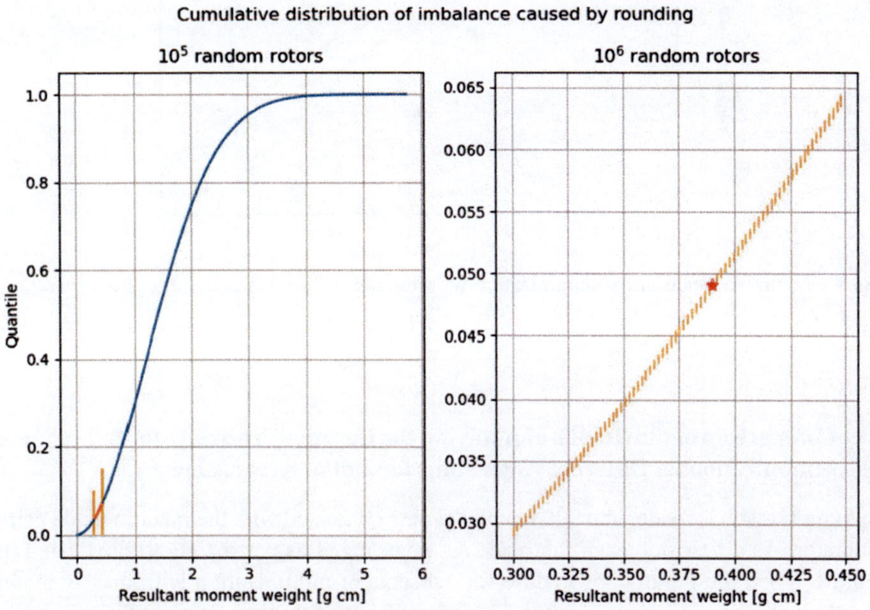

Fig. 8. Distribution of the error in m_{im} caused by the rounding error in the fan blades moment weights, with closer look at the cumulative density function between 0.03 and 0.065 quantiles.

Going through all the possible states that improve the initial arrangement could take a lot of computational time. That is why the iterative procedure can be customized with the following parameters:

– Number of branches – this parameter sets how many fan blades arrangements that improve the initial ones are saved in the lists for further iterations.
– Maximal depth – this parameter sets how many times the recursive function should call itself on a single branch. Once the maximal depth is reached, the iterative procedure for minimization returns the current best arrangement of the fan blades without further swapping of the blades pairs and continues on other branches if any.

- Minimization target – once the iterative procedure reaches a fan blade arrangement with resultant moment of imbalance equal to or smaller than the minimization target, the procedure ends and returns this arrangement.

A sample diagram illustrating the improvement of the resultant moment of imbalance during the iterations of the recursive algorithm is shown on Fig. 9.

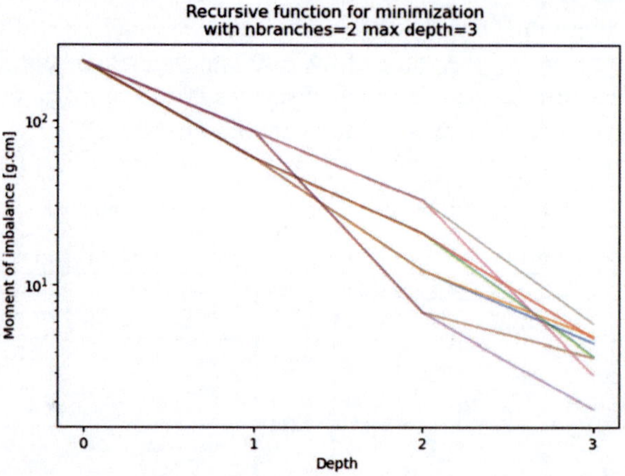

Fig. 9. Recursive function for minimization with number of branches equal to 2 and maximal depth of 3

5.3 Comparison of the Results of Applying the Iterative Procedure for Subsequent Minimization on Different Algorithms for Initial Arrangement

This comparison is made with 900 fan blades sets generated with the random blade pairs generator. As a first step each fan blade set is arranged as per the algorithms listed in Table 1. Then the iterative procedure for subsequent minimization with one, two and three branches is applied on each set. There is no limit on the max depth of the recursion. The minimization target for the iterative procedure is set to 0.39 g.cm. The comparison between the resultant moments of imbalance after the subsequent minimization for each method is shown on Fig. 10 as violin plots. For each method there are three violin-like figures, showing the distribution of the resultant moment of imbalance after the subsequent minimizations.

From Fig. 10 can be seen that only for methods 1, 3 and 11, corresponding to arrangement algorithms 'Symmetrical mapping about the line 12–6 h decreasing', 'One way mapping decreasing' and 'Delta sort decreasing', the applying of the iterative procedure for subsequent minimization with two branches always satisfies the minimization target. Based on that, it can be concluded that these three algorithms are the best initial arrangement for the iterative procedure for subsequent minimization.

On the other hand it is evident that all of the methods satisfy the AMM imbalance limit of 80 g.cm even if the recursive function is used with only one branch. It is also interesting to note that the initial arrangement methods with best performance have the worst potential for subsequent minimization.

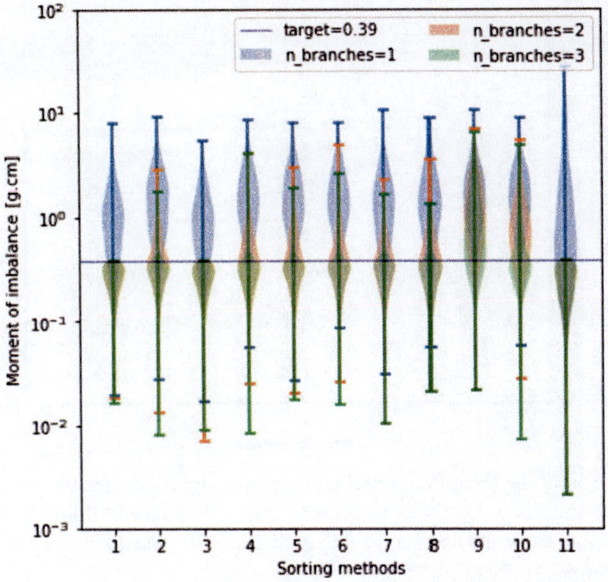

Fig. 10. Comparison of the results of applying the iterative procedure for subsequent minimization. The numbers on the horizontal axis refer to algorithm numbering in Table 1. The horizontal line on 0.39 g.cm shows the minimization target. The violin figures represent the results after the subsequent minimization: blue - one branch; orange - two branches; green - three branches.

5.4 Comparison of the Three Best Initial Arrangement Algorithms for the Iterative Procedure for Subsequent Minimization

To find out the most suitable algorithm for the iterative procedure for minimization a comparison is made based on 50000 random generated fan blade sets. Each of the three arrangement algorithms is applied on each set and then the resultant moments of imbalance is reduced with the iterative procedure for subsequent minimization with two branches. This comparison is shown on Fig. 11. The distribution of the resultant moments of unbalance after the subsequent minimization for each algorithm is shown again in violin plots. Once again it is evident that the 'worst' initial arrangement algorithm 'Delta sort decreasing' is the most appropriate for subsequent recursive imbalance minimization.

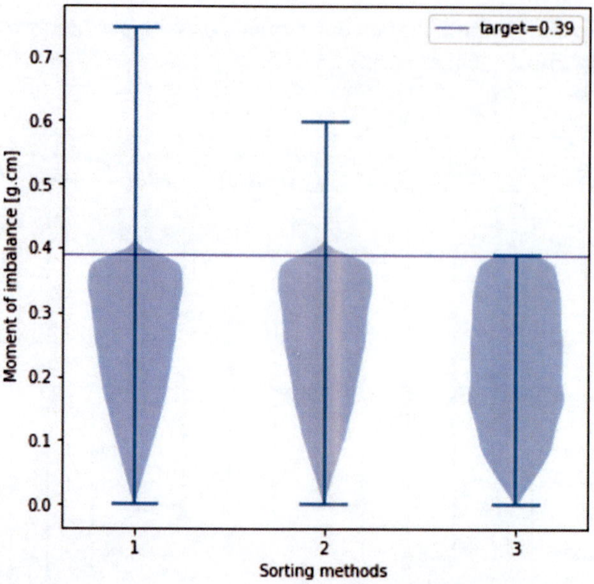

Fig. 11. Comparison of the three best initial arrangement algorithms for the iterative procedure for subsequent minimization. On the horizontal axis are the numbers of the three methods: 1 - Symmetrical mapping about the line 12–6 h decreasing, 2 - One way mapping decreasing and 3 - Delta sort decreasing.

6 Conclusions

In this work different approaches for static balance of fans of turbofan engines were examined against a large number of computer simulated blade sets of a particular engine model, namely CFM56-5B.

Firstly 11 algorithms for initial fan blades arrangement were tested. The results show that 'fixed rule' algorithms satisfy the balance accuracy only in part of the cases. Algorithms based on 'dynamic' principle, e.g. the 'delta sort dynamic', with calculation of the optimal position of each next pair of blades to be placed, show much better performance, but still in some of the cases the solution falls close to the AMM imbalance limit. It should be noted that there was a group of three 'worst' fixer rule algorithms that did not satisfy the balance target at all.

Secondly the method for subsequent balance improvement by pair swapping was developed into an iterative computation procedure and tested with all of the initial arrangement algorithms. The results showed that by using iterative subsequent imbalance minimization procedures the resultant moment of imbalance of the fan can be minimized to marginal values, much below the limits stated by the manufacturer. The value of such minimization for the prolongation of the turbofan engines life and minimization of maintenance costs could be an object of further studies.

As expected the subsequent imbalance minimization algorithm performed best, when applied on the results of the 'worst' initial arrangement algorithms and the performance was much poorer for the 'best' initial arrangement algorithms. This is explained with the suggestion that the 'worst' initial arrangement solutions stay faraway from local minimums of the balance solutions space, providing better conditions for the subsequent imbalance minimization algorithm to find deeper depressions in the solutions space.

Finally it can be concluded that despite the turbine balance problem has no analytic solutions and the finding of the absolute minimum of the imbalance is impractical, there are quite simple computational methods, easily realizable on contemporary computer systems, which can provide fine enough solutions. Although in this work the balance methods were examined for the fan of the CFM56-5 turbofan engine, they can be easily adapted for any compressor or turbine wheel with even number of blades.

Acknowledgment. This work is supported by the European Regional Development Fund within the Operational Program "Science and Education for Smart Growth 2014–2020" under the Project CoE "National center of mechatronics and clean technologies" BG05M2OP001-1.001-0008.

References

1. Amiouny, S.V., Bartholdi, J.J., III., Vande Vate, J.H.: Heuristics for balancing turbine fans. Oper. Res. **48**(4), 591–602 (2000)
2. Amroune, S., Belaadi, A., Menasri, N., Zaoui, M., Mohamad, B., Amin, H.: New approach for computer-aided static balancing of turbines rotors. Diagnostyka **20** (2019)
3. Fathi, Y., Ginjupalli, K.K.: A mathematical model and a heuristic procedure for the turbine balancing problem. Eur. J. Oper. Res. **65**(3), 336–342 (1993)
4. Li, G., Lin, Z., Allaire, P.E.: Robust optimal balancing of high-speed machinery using convex optimization. J. Vibrat. Acoust. **130**(3) (2008)
5. Mason, A., Rönnqvist, M.: Solution methods for the balancing of jet turbines. Comput. Oper. Res. **24**(2), 153–167 (1997)
6. Misevičius, A., Verenė, D.: A hybrid genetic-hierarchical algorithm for the quadratic assignment problem. Entropy **23**(1), 108 (2021)
7. Mosevich, J.: Balancing hydraulic turbine runners—a discrete combinatorial optimization problem. Eur. J. Oper. Res. **26**(2), 202–204 (1986)
8. Piskin, A., Aktas, H.E., Topal, A., Turan, O., Baklacioglu, T.: Rotor balancing with turbine blade assembly using ant colony optimization for aero-engine applications. Int. J. Turbo Jet-Engines **1**(ahead-of-print) (2017)
9. Pitsoulis, L.S., Pardalos, P.M., Hearn, D.W.: Approximate solutions to the turbine balancing problem. Eur. J. Oper. Res. **130**(1), 147–155 (2001)
10. Sun, C., Xiao, P., Wang, X., Liu, Y.: Blade sorting method for unbalance minimization of an aeroengine concentric rotor. Symmetry **13**(5), 832 (2021)
11. Yu, X.: General influence coefficient algorithm in balancing of rotating machinery. Int. J. Rotat. Mach. **10** (2004)
12. Zhai, W., Gong, W.B.: Optimal blade placement for large turbofan balancing. In: Proceedings of the Fourth International Conference on Computer Integrated Manufacturing and Automation Technology, pp. 261–266. IEEE, October 1994

Analysis and Synthesis of Mobile Portable Machine Tools Layouts

Ihor Yakovenko(✉), Dmitry Shepeliev, Vladislav Sharlay, Alexander Permyakov, Serhii Slipchenko, and Yurii Havryliuk

National Technical University "Kharkiv Polytechnic Institute", 2, Kyrpychova Street, Kharkiv 61002, Ukraine
igor.dych59@gmail.com

Abstract. The research touches upon modern trends in the machine tool industry, which are associated with a constant increase of the portable metal-cutting machine's demand to perform repair work or modernize the structures of large-sized equipment, the dismantling of which is impossible or not economically feasible. The increase in demand for such equipment makes it necessary to reduce the design and production time for this type of machine tool, with unconditional observance of quality parameters.

In the process of research, the existing world practice of designing, manufacturing, and operating equipment of this class was used, as well as the tasks for technical re-equipment of the energy industry.

The paper highlights and analyzes the main features of the layout of portable machine tools formulates the main classification features and proposes a description of a hierarchical classification model, which is based on the methods of discriminant analysis. As the main classification features of layout solutions, possible options for implementing the kinematics of shaping the machined surface, as well as basing and fixing a portable machine on the processing object are considered. Further refinement of the classification for each hierarchical branch is associated with the geometric parameters of the surface, parameters of the cutting process are not the purpose of this work and will be considered in the next publication.

This research is aimed at typing design layout solutions and, in the future, optimizing the design and production process using the principles of aggregation and modular layout of portable machines.

Keywords: Portable machine tools · Layouts · Classification · Repair · Large equipment · Principles of aggregation

1 Introduction

Particular attention in modern machine tool building is paid to the creation of special portable machine tools. This is due, first of all, to the need to modernize and repair large units, assemblies and systems, the dismantling of which is difficult, requires significant costs or is impossible in principle. As an example of such equipment, one can consider turbines and systems for their operational support of power plants, main pipelines and

D. D. Cioboată (Ed.): ICoRSE 2022, LNNS 534, pp. 160–171, 2023.
https://doi.org/10.1007/978-3-031-15944-2_16

gas pipelines, mining equipment, etc. The service life of such equipment is quite long, however, during operation, it becomes necessary to repair and, often, modernize it based on scientific achievements in a particular area. And, if earlier such work was mainly carried out manually using a universal tool, now special portable machines are being used more and more often. This approach allows to significantly improve the quality of work performed, on the one hand, and on the other hand, significantly reduce the time for modernization and repair.

Portable machine tools are mainly designed to perform a specific technological operation, so the layout solutions for such equipment should make the most use of normalized components and assemblies, which will significantly reduce the time for designing and manufacturing equipment. This approach is most typical for the aggregate principle of constructing layout diagrams of metal-cutting machine tools.

On the other hand, the development of IT technologies implies the widespread use of computer-aided design of technological equipment based on aggregate layout solutions, which significantly reduces the design time due to the use of ready-made layout schemes, which mainly include normalized components and assemblies with minimal use of original components and elements for a portable machine.

Reducing the time of modernization, repair and restoration of large-scale equipment (requiring technological impact with chip removal) with the rejection of dismantling and transportation of the product is determined by the widespread use of mobile portable machines. A promising direction for solving this problem can be the development and subsequent technical implementation of equipment layout diagrams, which is based on the principles of aggregation. With this approach, the main emphasis is on the use of unified components and assemblies, and the design and manufacture of original elements takes much less time and, accordingly, costs.

2 Literature Review

One of the most important issues by designing the metal-cutting tools is the question of describing the layout of the future machine. At the same time, the designer needs to take into account a lot of issues: the kinematics of shaping, the relationship of nodes and their spatial arrangement, the system for basing and fixing the workpiece, etc.

Many scientists have tried to facilitate the task of designing general layouts of metal-cutting equipment. The most complete description of the layout of machine tools are described in [1]. Based on the totality of the entire set of movements (rotational and translational), a set of principal machine layouts is determined. This approach creates tremendous opportunities for the development, analysis and optimization of machine tool layouts, however, it is rather complicated and redundant in the case of the synthesis of special equipment for performing one technological operation.

A simpler approach to describing the structures of machine tools is proposed in [2]. This International Standard describes the machine coordinate system relating to the primary movements of individual numerically controlled assemblies and their associated movements. The machine coordinate system is used to determine the coordinates of a moving tool (or a point in the workspace or drawing) in relation to a stationary workpiece. The shaping movement of the cutting tool itself is not taken into account for some layouts, since there is no direct movement mechanism (typical for the rotational movement of the spindle). For simplicity, most of this International Standard is written in terms appropriate to all types of machine tools. This standard was last revised in 2017 and the version is confirmed to be current. However, the proposed standard is so general in nature that it is practically difficult to take into account the layout of portable machine tools or does not give a complete picture of its structure.

The work of H. Goebel [3] considers the principles of synthesis of layout solutions for special multi-operation machine tools built according to the aggregate principle, which are used to perform metalworking processes with a high concentration of operations aimed at processing one specific part or group of parts. Obviously, this approach is largely redundant for the synthesis of the structure of portable machine tools, which usually perform one technological operation.

In [4], a layout of a portable machine for processing is proposed, which contains only a few elements (power plant-converter-guides-tool-part). Such a simplified approach does not allow coding a variety of layouts of mobile portable equipment and using it in the process of designing this type of equipment.

The article [5] discusses the methodology for modeling the architecture of machine tools, taking into account their rigidity and dynamic characteristics, which is important for portable machine tools. However, the questions of classification options for layout solutions are not considered.

In [6], a system for automated design of modular reconfigurable machine tools for processing a group of parts is considered. The proposed method consists of three successive steps in the formation of the machine layout: module identification, module definition and the final synthesis of the layout. The choice of the module and specification of the characteristics is carried out from the corresponding libraries, which are formed on the basis of the kinematics of possible movements of the nodes. A similar approach to designing layouts of modular machine tools is described in [7], while in the process of choosing a layout solution, a possible variant of a power plant is taken into account, which provides the kinematics of the selected module (pneumatic, hydraulic, electric). However, these works do not take into account the design features of portable machine tools, which significantly affect the final layout.

In [8], there considered a system that makes it possible to assemble machines from the same type of modules with different possibilities of movement and degrees of freedom of such movement to ensure an unlimited use of the possibility of rotation of two neighboring modules. However, in addition to the considered advantages, such a system is redundant for options for synthesizing the layout structures of portable machine tools.

Articles [9, 10] discuss the design of specific portable machine tools for processing parts of turbine units of power plants. The works [11–13] consider the designs of portable machine tools, which involve the use of an aggregate-modular principle of building a layout in the design process. However, in these works there is no system of generalizations and classification of possible options for layout solutions for this type of machine tools, which is necessary to automate the process of synthesizing the structure of portable machine tools.

Thus, the analysis of existing methods showed that in order to design layouts of portable machine tools on the principles of modular construction, it is necessary to develop a system for classifying and describing layouts that would allow, on the one hand, to simply characterize its structure, and on the other hand, to take into account the features and various technological requirements as much as possible. to such machine.

To do this, it is necessary to analyze the most common processing objects for such machines, the technological processing parameters and the existing designs of portable machine tools from the world's leading manufacturers that implement these parameters.

3 Research Methodology

The main purpose of this study is to improve the quality and reduce the complexity of the design and production of mobile portable metal-cutting machines to perform technological operations in the repair and modernization of non-dismantled units and assemblies.

To achieve this goal, the following tasks were set and solved:

- an analysis of the main technological operations and processing methods that are performed during the repair and modernization of large-sized equipment was carried out;
- an analysis of the existing layout solutions of portable machine tools that implement the considered technological operations was carried out;
- on the basis of the analysis carried out, a classification and a system for describing the layout solutions of portable machine tools were developed;
- a formalized mathematical model for the synthesis of layout solutions for portable machine tools was developed.

In the course of the study, the authors considered the issues of the regularity of classification and subsequent synthesis of layout diagrams of portable machine tools based on unified components and assemblies, with minimal use of original components and elements, to ensure the necessary principles of basing and relative movement of machine elements relative to the processing object.

Portable mobile machines can perform an almost complete list of technological operations of machining, but there are limitations in providing complex multi-coordinate kinematics. However, such operations are practically not used to restore the operability of non-dismantled units, therefore, such machines are not considered in this paper for the analysis of layout schemes.

Figure 1 shows the layout of a classic metal cutting machine and a mobile portable machine.

Basic elements of the machine design are marked in gray, the nodes that provide tool movement are green; nodes that provide movement of the fixture with the workpiece - yellow; the workpiece locating system and the processing object itself - blue and dark blue.

The movements of nodes and mechanisms are considered in the classical coordinate system. Possible linear movements of the mechanism along the corresponding axis are indicated by the letter L followed by the letter of the axis (x, y, z). Rotational movements are denoted by the letter R followed by the letter of the axis.

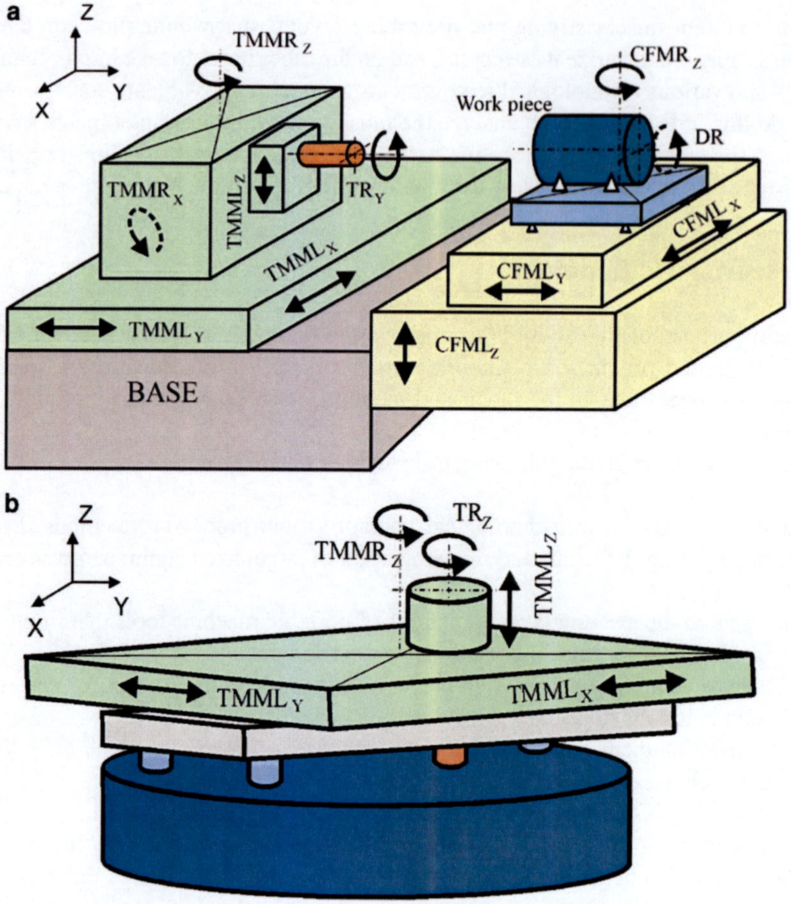

Fig. 1. Description of the layout of metal cutting equipment: a) - a universal machine of a traditional layout; b) - portable mobile machine (Symbols: Tool movement mechanism-TMM; Clamp fixed moving mechanism CFM; Divider mechanism–D; Tool spindle -T)

Compared to classic stationary mobile portable machines have several fundamental differences.

Firstly, unlike machines of a stationary layout, the workpiece is not delivered to the machine, but the machine is transported to the workpiece.

Secondly, the workpiece is not based and fixed on the machine, but the machine is based and fixed on an existing processing object, along with the tool and the locating system.

Thirdly, a portable mobile machine almost always performs one specific operation, unlike universal machines and stationary modular machines.

Fourthly, since a portable machine is mounted directly on the workpiece, there is a number of requirements for the layout of the machine and the design of its units and assemblies (simplicity of transportation, assembly and dismantling of units with the obligatory condition to ensure high rigidity and, accordingly, the required accuracy of processing).

As can be seen from Fig. 1, for a portable machine, there are no systems and nodes for moving the setting and clamping device, since the locating system itself, in this case, is the base / bed / base for installing the nodes and mechanisms for moving the tool.

In addition to these features, a number of other factors must be taken into account while processing the design implementation. A variety of designs and sizes of processing objects requires taking into account the methods of locating and mounting the machine directly at the place of operation. So, for example, the processing of flanges is possible with an internal and external option for locating and attaching the machine to the processing object. However, consideration of the issues of locating the machine and its subsequent fixing is in itself a complex research task and is not included in the objectives of this study.

From the considered design features of portable machine tools, it is obvious that the machine design must be delivered to the place of operation in a disassembled block form and the machine is already installed on site. That is, the design of a portable machine should be easy to install and maintain, its units should, if possible, have a minimum weight, high rigidity and reliability.

All these features of the layout of portable machines determine their design implementation. Most of these machines are designed to perform fairly simple operations with an efficient cutting tool. Figure 2 shows the most frequently implemented technological operations on the examples of leading manufacturers of portable machine tools.

4 Results

The analysis of the designs of portable machine tools manufactured by the world's leading manufacturers of this class of equipment: CLIMAX, SERCO, ENERPAC, MIRAGE, PROTEM, etc., and also taking into account the real characteristics of the processed surfaces, technological characteristics of the equipment and operating conditions, made it possible to identify the characteristic features layouts of portable machines. On the basis of these features, many possible variants of the kinematics of the shaping of the machined surfaces were formed. For further classification of layout solutions, the most common classification methods were analyzed - hierarchical and faceted.

Facing (with external and internal basing of the machine).

The authors settled on the hierarchical classification method due to its naturalness and large informative capacity. The use of faceted classification carries a high redundancy of groupings and is somewhat difficult in the formation of a database of layout solutions for portable machines.

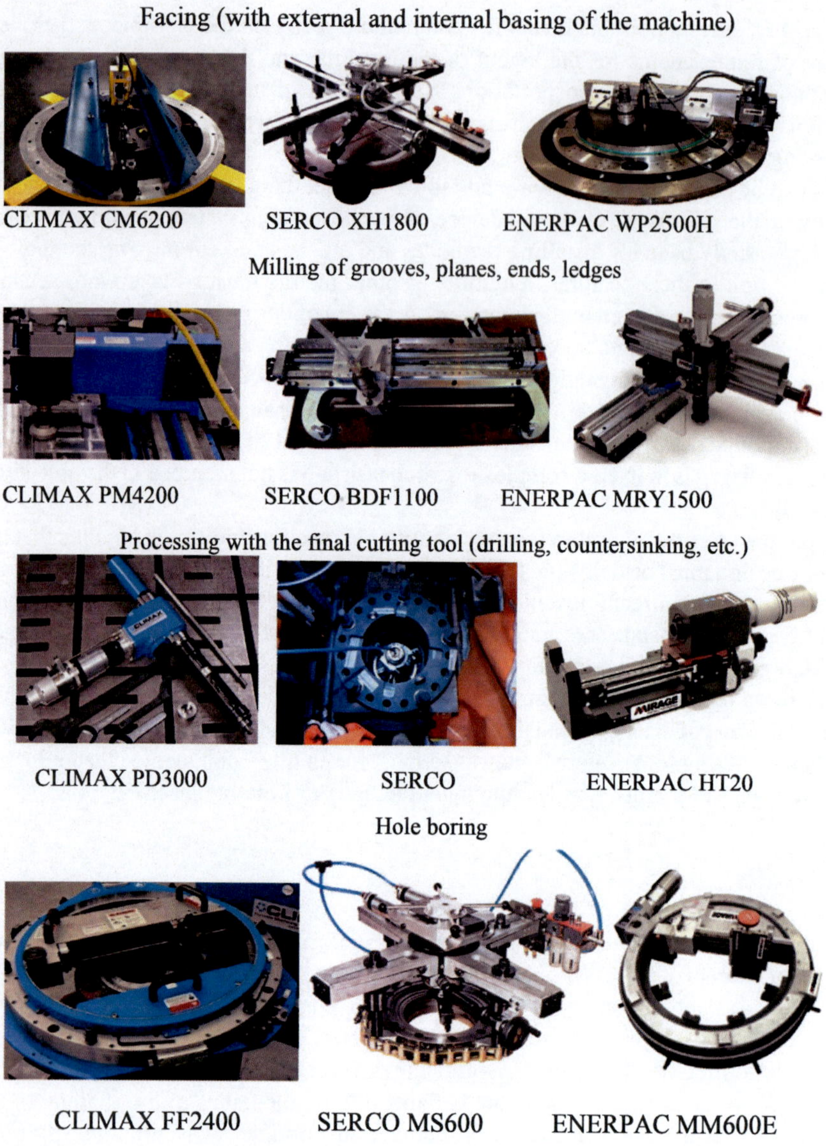

Fig. 2. Typical technological operations and their implementation by portable machines on the example of the world's leading manufacturers

This method is characterized by fulfillment of the following conditions:

- the union of subsets of classification groupings mi of the same hierarchy level gives the initial set of objects M;
- the intersection of classification groupings mi of the same hierarchy level gives a zero subset 0, i.e.

$$\cup_i m_i = M, \forall_i = 1, I \quad (1)$$

$$\cap_i m_i = 0, \forall_i = 1, I \quad (2)$$

Figure 3 shows a fragment of the hierarchical classification of portable machine tools. The upper level of classification (classes) is determined by the kinematics of the shaping of the treated surface (Fig. 4). Each class contains groups, which are determined by the portable machine basing system on the processing object. The locating system is hierarchically brought to a higher level than power units, since the possibility of using one or another type of drive that will provide a given kinematic scheme largely depends on the locating scheme and fixing the machine on the processing object. The basing system is a determining factor in the design of the original elements of a portable machine: the base (bed), basing, auxiliary and fixing elements and assemblies. Accordingly, the groups include subgroups, the classification feature of which is the types and models of power plants that provide the necessary shaping movements for a given class.

Thus, it is possible to describe the layout of a portable machine tool and the tool movement system for performing a technological operation using fairly simple formulas that fully give the idea of the structure and subsequent design implementation of equipment of this class.

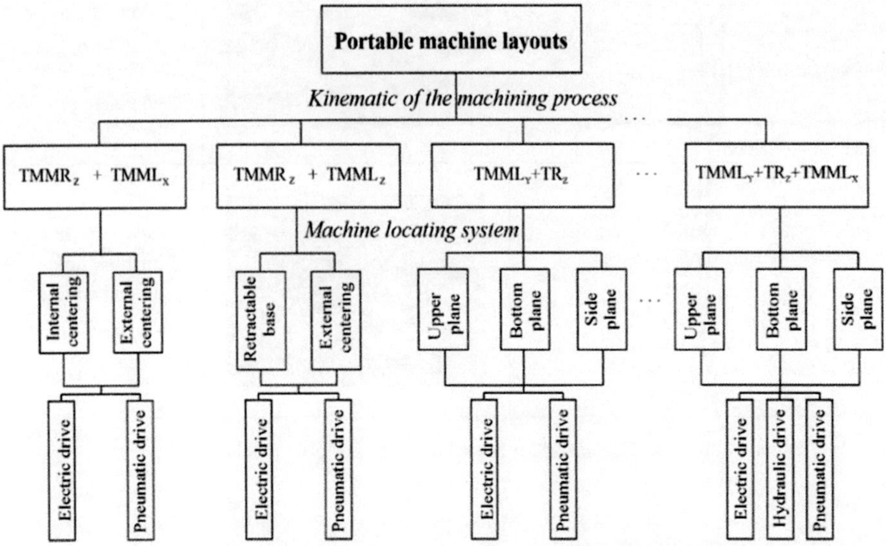

Fig. 3. Classification system for portable machine tool layouts

Scheme and formula	Example of a machine	Scope of use
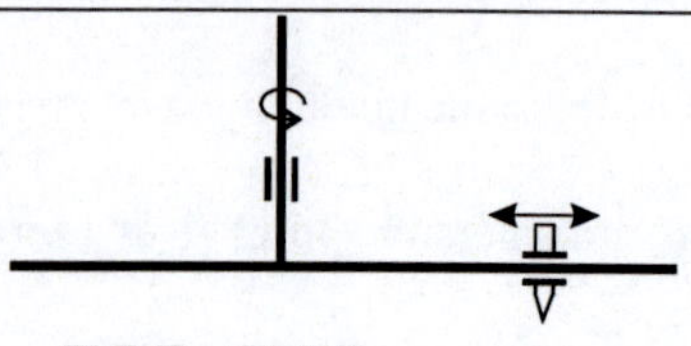 $TMMR_Z$+$TMML_X$		Processing of end faces, diametrical planes, grooves, compact flanges.
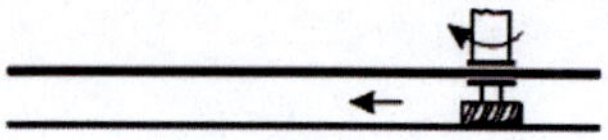 $TMML_Y$+TR_Z	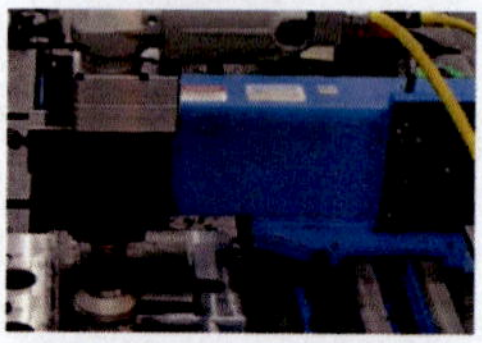	Processing of planes, grooves of a rectilinear profile.
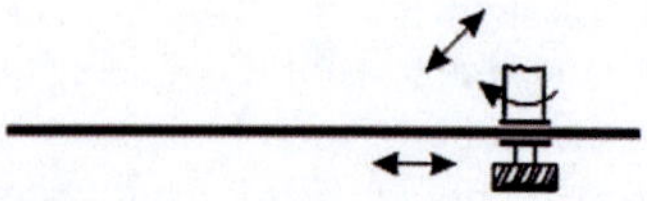 $TMML_Y$+TR_Z+$TMML_X$		Processing of planes, grooves of a free profile.
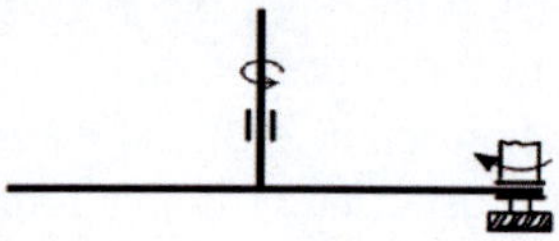 $TMMR_Z$+ TR_Z		Milling of end faces, diametrical planes, ledges.
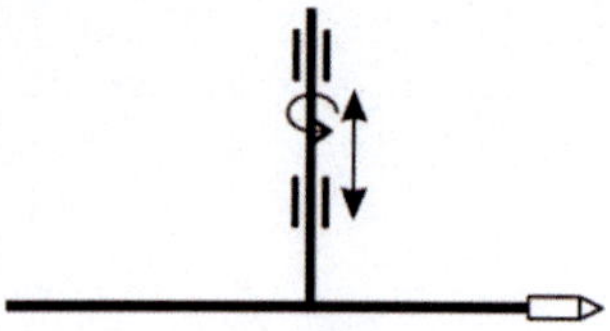 $TMMR_Z$+ $TMML_Z$	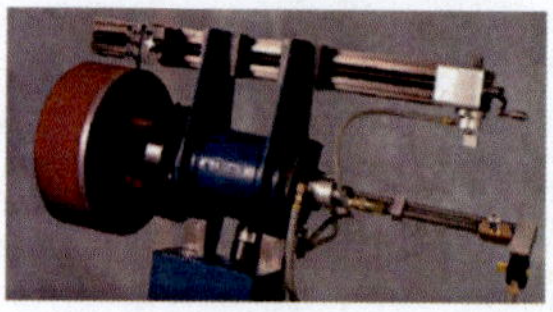	Processing (boring) of openings of big diameters.
 $TMMRL_Z$		Finishing holes with end cutting tools.

Fig. 4. Classes of layout schemes of portable machines

Based on the analysis of basic technological operations, kinematics of formation and classification of layout schemes of portable machines, a formalized mathematical model for designing the layout structure of such equipment was proposed. This model is based on the principles of aggregation with the maximum use of unified components and structural elements.

Modular design based on unified components and elements can significantly reduce the time of design and production of portable machines, use units of third-party manufacturers, which are mass-produced and therefore have high quality and much lower price than their own production, with the same technical characteristics.

This approach is also due to the fact that two main schemes of operation of such equipment are actually used.

The first is the acquisition of the machine in the property and, often, with the involvement of specialists from the supplier of equipment for repair and restoration work. In this case, after performing the necessary work, the machine becomes useless, as such repairs are very rare (frequency can reach decades).

The second scheme involves the lease of equipment and the invitation of specialists from the company that manufactures the equipment to perform the necessary repair and restoration work. The equipment remains the property of the manufacturer. In this case, it is possible to reuse unified units and elements for other machines.

Obviously, the second scheme for the customer's company is more preferable from an economic point of view.

The enlarged mathematical model of the synthesis of the structures of portable machine layouts is presented in the form of a HIPO diagram (Fig. 5). The notations are presented in the form of variable structures for the further programming process.

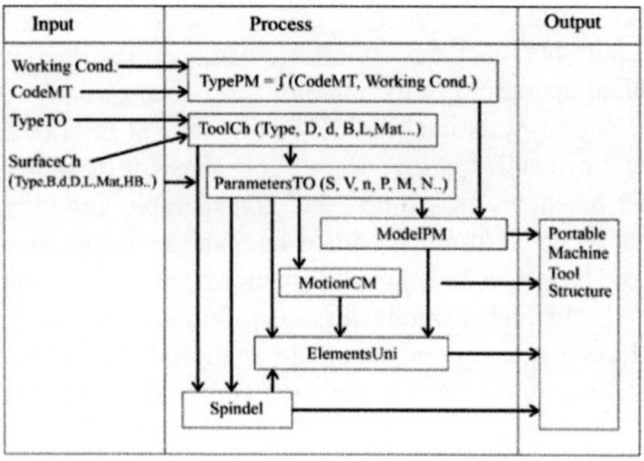

Fig. 5. HIPO diagram of synthesis of structures of configurations of portable machines

Symbols: Working conditions - WorkingCond; Machine tool layout code - CodeMT; Type power machine - TypePM; Model of power machine - ModelPM; Surface characteristics - SurfaceCh; Type of technological operation - TypeTO; Parameters of technological operation - ParametersTO; Tool characteristics - ToolCh; Spindle - Spindle;

Unified elements - ElementsUni; Motion conversion mechanism - MotionCM; Structure of the portable machine Tool - PortableMTstructure.

Initially, based on the operating conditions and the layout code of the machine, the type of portable machine is selected, for which other layout characteristics will be formed in the future. In parallel, the method of technological impact on the object of processing (technological operation) is selected on the basis of technical and dimensional characteristics (material, its physical and mechanical properties, shape and size of the treated surface, structural rigidity, etc.). The cutting tool is chosen for the selected method and its characteristics are set.

The next step is to determine the characteristics of the selected technological operation (cutting speed, feed, speed, power characteristics, etc.). This stage can be performed both automatically on the basis of regulatory materials from the database, and manually by a technologist with the subsequent entry of the necessary information into the system.

The obtained technological information and the previously selected version of the machine layout allow to organize the choice of the model of power units, unified mechanisms of motion conversion, spindle group and some other unified elements of the design of the portable machine. The choice can be made on several criteria (cost, size, weight, etc.) depending on the need to provide the most effective solution. Usually, value is accepted as a criterion for selection, and the rest act as a system of constraints. The search can be performed both on the internal database, which includes units available in production or at the nearest partners, and on the extended database, which includes known serial suppliers of the necessary equipment (especially for pneumatic motors and pneumatic motors).

5 Conclusions

Layout schemes of portable machines are established and classified, which implement the main technological operations in the modernization, repair and restoration of large parts, dismantling and transportation of which is impractical and / or unprofitable. The basic approaches and principles of creation of formulas for the description of layout schemes of portable machines are offered that considerably simplifies a synthesis of layout decisions and a choice of units, unified knots and constructive elements which realize the synthesized layout. In the future, this reduces the time of design and production of such equipment. A mathematical model for the synthesis of layout solutions for mobile portable machines based on the principles of aggregation with maximum use of unified units, assemblies and structural elements of such equipment is proposed and partially implemented.

References

1. Vragov Yu.D. Analysis of the layouts of metal-cutting machines: (Fundamentals of components). - M.: Mashinostroenie, 1978. - 208 p., ill. (in Russian)
2. ISO 841:2001 Industrial automation systems and integration. Numerical control of machines. Coordinate system and motion nomenclature. 28p.

3. Gebel H. Layout of modular machines and automatic lines. - M.: Mashgiz, 1959. - 189p. (in Russian)
4. Shukla, Sharad and Pandey, Ankit and Anasune, Kartik and Gond, Mohit and Chauhan, Parth, Design of Portable Light Weighing Metal Cutting Machine (April 6, 2019). Available at SSRN: https://ssrn.com/abstract=3373785 or http://dx.doi.org/https://doi.org/10.2139/ssrn.3373785
5. M Lajili, H. Chanal, B. C. Bouzgarrou Machine tool architecture selection at the preliminary design stage: application to hard material machining / Journal of Computational Design and Engineering, 2020, 7(5), 657–667. https://doi.org/10.1093/jcde/qwaa049
6. Zhengyi Xu, Fengfeng Xi, Lilan Liu, Li Chen A Method for Design of Modular Reconfigurable Machine Tools. /Machines 2017, 5, 5; doi:https://doi.org/10.3390/machines5010005
7. Majija, N., Mpofu, K., Modungwa, D.: Conceptual Development of Modular Machine Tools for Reconfigurable Manufacturing Systems. In: Azevedo, A. (ed.) Advances in Sustainable and Competitive Manufacturing Systems. LNME, pp. 467–477. Springer, Heidelberg (2013). https://doi.org/10.1007/978-3-319-00557-7_38
8. M. Hajduk, L. Koukolova The Structural of the Basic Module for Construction of Modular Machines. / Applied Mechanics and Materials Vol. 613, 2014, pp. 434–440. https://doi.org/10.4028/www.scientific.net/AMM.613.434
9. Permyakov A.A., Ischenko M.G., Shepeliev D.K. Layouts of mobile machines for the repair of non-dismantling parts and assemblies of turbine units / / Cutting and tools in technological systems: Intern. sci.-tech. Sat. / Under the total. ed. prof. A. N. Shelkovy, editorial board.: prof. A. I. Grabchenko (responsible editor) and others - Kharkiv : NTU "KhPI", 2018. – Ed. 88, (101). pp. 166–171. (in Russian)
10. Permyakov A.A., Ischenko G.I. Processing of large parts of turbine units using portable machines // Modern engineering and innovative technologies, Germany. - Ed 15. - 2021. - pp.17–26. (ISSN 2567–5273) (in Ukraine)
11. Yakovenko I.E., Ischenko M.G., Balayan B.A., Logvin V.E., Mileeva O.O. Design of portable machines for modernization and repair of dimensional parts. // XIX MMNTK "Mechanical Engineering through the eyes of young people: progressive ideas - science - production", 25–26 November 2020, - Sumy: Sumy State University, 2020. - pp.144–145. (in Ukrainian)
12. Yakovenko I.E., Mileeva O.O., Ischenko M.G. Aggregate-modular principle in the design of portable machine layouts. // XIV International scientific-practical conference of undergraduates and graduate students "Theoretical and practical research of young scientists" (December 1–04, 2020): conference proceedings / ed. Prof. E.I. Sokol. - Kharkiv: NTU "KhPI", 2020. - p.491. (in Russian)
13. Yakovenko I.E., Sharlay V.R. Analysis of portable machine layout schemes // XV International scientific-practical conference of undergraduates and graduate students "Theoretical and practical research of young scientists" (December 1–03, 2021): conference proceedings / ed. prof. E.I. Sokol. - Kharkiv: NTU "KhPI", 2021. - p.439. (in Ukrainian)
14. Mirage (Enerpac) Home page. https://www.enerpac.com/en-us/products/USPortableMachineTools#redirect, last accessed 2022/03/01
15. Serco. Home page. https://www.serco-tools.com/products, last accessed 2022/03/01
16. Tomco. Home page. https://www.tomco.ca/tomco-industrial-ltd/machine-tool/lathes, last accessed 2022/03/01
17. Climax. Home page. https://www.climaxportable.com, last accessed 2022/03/01
18. Protem. Home page. https://www.protemusa.com, last accessed 2022/03/01

Creation of a Superhydrophilic Surface with Anti-icing Properties for X18H10T Stainless Steel Using a Nanosecond Laser

Sergey Dobrotvorskiy[1(✉)], Yevheniia Basova[1], Ludmila Dobrovolska[1], Viktor Popov[2], and Abou Samra Youssef Mounif[1]

[1] National Technical University "Kharkiv Polytechnic Institute", 2, Kyrpychova Street, Kharkiv 61002, Ukraine
sergiy.dobrotvorskyy@khpi.edu.ua
[2] JSC "FED", 132, Sumska Street, Kharkiv 61023, Ukraine

Abstract. The process of laser texturing and the creation of periodic surface nanostructures is a promising method for changing the wettability of a metal surface, on which micro-/nano-rough structures are created to change the contact angle between the textured surface and a liquid drop.

In this work, several models of linear microstructures were textured with a nanosecond laser on the 12X18H10T stainless steel surface, which is of interest to the aviation, aerospace, and food industries.

Superhydrophobic and superhydrophilic stainless steel surfaces were obtained. Using the atmospheric aging method, hydrophilic to superhydrophobic transformation takes 20–27 days and depends on processing parameters. The two-temperature model of laser action was considered. Based on this model, it was substantiated that the aging mechanism is due to the relaxation of periodic non-equilibrium surface structures and nanoclusters created as a result of the appearance of a deformation hardening wave. The results of the study of anti-icing properties in contact with a humid atmosphere and water were presented. Low adhesion of water droplets to the steel surface at low temperatures was demonstrated. Attention was focused on the resistance of surfaces to the loads accompanying freeze/thaw cycling, indicating a good potential for coating surfaces for outdoor use.

Keywords: Superhydrophobic surface · Laser processing · Super high-speed deformation mechanism · LIPSS · AISI 321

1 Introduction

Stainless steel has found a very wide application in a variety of fields. They are used in aviation, mechanical engineering and instrumentation, the food industry, the chemical industry, the automotive industry, construction, and household appliances. Often stainless steels are designed for outdoor use, while in cold weather their performance can be severely affected by icing on work surfaces. The problem of preventing icing

D. D. Cioboată (Ed.): ICoRSE 2022, LNNS 534, pp. 172–184, 2023.
https://doi.org/10.1007/978-3-031-15944-2_17

and eliminating its influence on the operation of various equipment is the central topic of many studies [1–5]. Among the promising approaches to solving this problem, it is worth highlighting a strategy based on the use of superhydrophobic surfaces. Such a strategy can provide continuous passive de-icing, and unlike active methods, it does not require the use of de-icing fluids or thermal energy to prevent freezing and de-icing.

2 Literature Review

From an analysis of current research in this area, it has been found that there are several approaches to creating hydrophobic surfaces on stainless steel. The first approach to creating such surfaces is based on laser texturing followed by silane processing.

In article [3], the authors showed that superhydrophobicity is achieved on workpieces made of AISI 4130 steel using a surface nanostructuring process based on a nanosecond laser. After that, the surface with a laser texture was additionally processed by immersion in a chlorosilane reagent for a certain period. The result of the experiment in this paper [3, 6] is based on the fact that metals and many metal oxides can strongly adsorb silanes when a chelating functionality such as a diamine or dicarboxylate is present. Article [6] also noted the possibility of using an organofunctional silane with reactivity. However, it should be noted that the stability of such a surface can be maintained under limited operating conditions. Therefore, many studies are aimed at the direct laser formation of hydrophobic surfaces by femtosecond, picosecond, and nanosecond pulses. The prospects of such an approach for anti-icing of aviation materials are substantiated in the articles by Vercillo V. et al. [7] and Volpe A. et al. [8]. Gaddam A. et al. [9] presented a study where nanosecond laser processing is used to fabricate hydrophobic 17-4 PH stainless steel surfaces with microscale and submicron structures.

In the work [9], the AISI 430 stainless steel laser with an average power of 5 W, a pulse duration of 310 fs, was used to create periodic surface structures for anti-icing purposes. An analysis of the studies showed that both fs, ps, and ns pulses make a unique contribution to surface texture.

Simões J.G.A.B. et al. in their work [10] studied the creation of periodically surface structures on the AISI 304 stainless steel surface using a 100 ns nanosecond pulsed laser. Among the results of the study, the high efficiency of the nanolaser compared to femto- and picosecond lasers was noted.

In almost all the considered works on the creation of Laser-induced periodic surface structures (LIPSS), it was found that hydrophilic surfaces are formed immediately after irradiation, and only after a long time from 17 to 50 days can they become hydrophobic. However, a unified view on the mechanism of aging and the formation of hydrophobicity has not yet been formed.

This work aims to try to fill in some of the problems in this area.

3 Materials and Methods

To prepare for the experiment, stainless steel plates 12X18H10T (similar to AISI 321) were cut. The thickness of the experimental samples was 1 mm, and the overall dimensions of the plates were 35 × 55 mm. The surface in the delivered condition had a smooth, mirror-like surface characteristic of steel. The chemical composition of steel (in weight%): Fe 67.0; Cr 17.0′19.0; Ni 9.0′11.0; Mn < 2.0; Si < 0.8; Ti 0.6 = 0.8.

To obtain the polymodal surface roughness required to achieve the superhydrophobic state, IR laser surface processing was used. In the studies, a laser setup with an IR ytterbium fiber laser (wavelength 1064 nm) and a BM 2500+ biaxial galvanometric scanner was used. Laser processing was carried out in ambient conditions with a humidity of 40–50% and a temperature of 20–25 °C. This study used laser exposure with a pulse duration of 100 ns, a repetition rate of 20–100 kHz, and peak power of up to 1 mJ in the TEM_{00} mode. The samples were scanned with a raster at a linear speed of up to 400 mm/s with a parallel line density of up to 1.25 μm. Previously, for aluminum alloys, it was found that the laser's number of passes through the same places on the sample surface affects the chemical phase composition and surface morphology due to different times and temperatures of the molten state of the surface layers. To study the effect of multiple melting and crystallization processes on the state of the stainless-steel surface, samples were studied that were subjected to either single-pass laser processing or three-pass laser processing on one of the wafer sections at a minimum laser radiation intensity (on all samples, the upper right section was always subjected to multi-pass laser processing). The surface of the steel immediately after laser treatment is mainly superhydrophilic, as evidenced by the complete retention of water droplets on the surface of the plates, even if it is in a vertical position (Fig. 1). On the untreated part of the surface (the extreme section on the left), on the contrary, a partial runoff of the drop along the vertical wall is observed, that is, it can be characterized as simply hydrophilic. On Fig. 1 shows the results of the processing experiment in 11 areas with different scanning parameters. On the left is the raw base area of the surface. The fact of superhydrophilicity can be explained by the fact that the surface of the samples after irradiation with a nanolaser contains both micro-and nanosized elements, including nanoclusters. Therefore, the surface composition, including micro-and nanosized austenite crystallites, corresponds to a material with high surface energy [9, 11].

4 Results and Discussion

For further research, out of 11 processing modes, 4 most characterizing different surface morphology were selected. Areas with the following energy were processed: area 1 (Fig. 2) was processed with an energy value of 0.12 MJ (3 passes), Section 2 (Fig. 2) was processed with an energy value of 0.2 MJ, Section 3 (Fig. 2) was processed with an energy value of 0.25 MJ, Section 4 (Fig. 2) was processed with an energy value of 0.3 MJ. Longitudinal scanning speed 100 mm/s. Transverse reaming speed 30 mm/min. On the left is the untreated base area Section 5 (Fig. 2, 3, 4, 5, 6, 7 and 8).

Fig. 1. Retention of water droplets on the surface of a vertically located steel sample immediately after laser processing

The change in the adsorption state of the surface was evaluated periodically for 45 days. Major changes were noted within 25–27 days and are presented below.

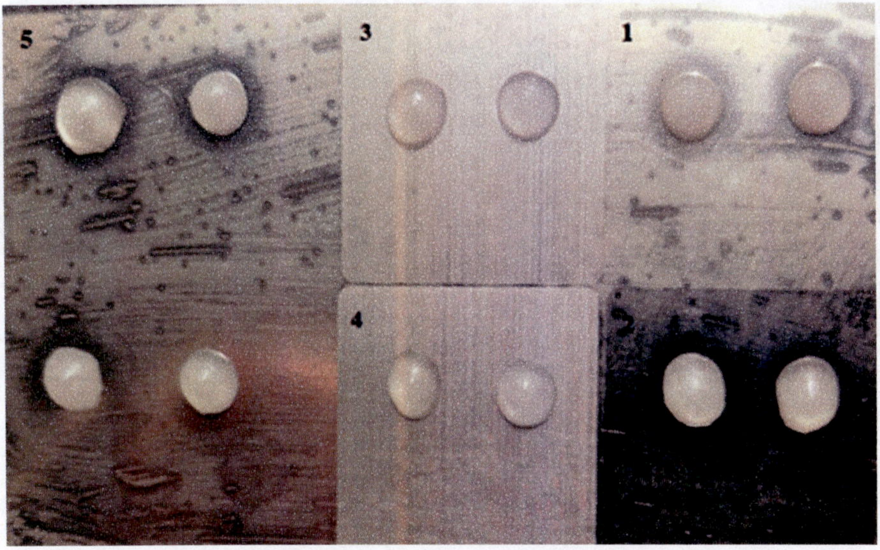

Fig. 2. Frozen water drops on the surface of a steel sample 12X18H10T

4.1 Freezing and Melting of Water Droplets

Figure 2 shows a photograph of the initial surface with drops after freezing to a temperature of −18 °C, and Fig. 3 shows the result of the defrosting process at a temperature of +18 °C with a vertical plate. Faster melting and dripping of drops in areas with intense laser processing (middle areas, Fig. 2, areas 3–4) were observed. The runoff starts time on the base surface (Fig. 2, area 5) and the surface with the minimum processing energy (Fig. 2, area 1) practically does not differ and exceeds the drop start time on the middle sections (Fig. 2, area 3–4) by 50–60 s.

It should also be noted that the droplet flow rate in the middle section (Fig. 2, area 3–4) is 4–5 times higher than the droplet velocity on the untreated surface (Fig. 2, area 5). The start time and the flow rate in the area with an intermediate intensity of treatment have intermediate values (Fig. 2, area 2).

Fig. 3. Melting and dripping of water droplets from the surface of the sample during thawing

An analysis of the results obtained indicates that the adsorption capacity of the untreated surface remained hydrophilic. The surface area treated with minimum energy (Fig. 3, area 1), despite the multi-pass, also showed in this case a rather weak change in the direction of increasing hydrophobicity. Conversely, areas with intense laser treatment showed a significant increase in hydrophobic properties (Fig. 3, areas 3–4). This can be explained by the fact that as the ice melts, a large contact angle is formed near the droplet, as a result of which its rolling speed becomes very high.

4.2 Snow Melting

An important result is also the study of the behavior of snow melting on the surface of stainless steel 12X18H10T in the temperature range close to zero. This becomes possible when an aircraft takes off or lands in the presence of snowfall.

Steel samples were cooled to a temperature of 0–2 °C and covered with a layer of snow (Fig. 4), and then moved to conditions with a temperature of 3–4 °C, and the dynamics of snow melting were recorded (Fig. 5).

Fig. 4. 12X18H10T steel plate with a layer of snow on the surface

An analysis of the results shows that the highest melting rate is observed in the area with the maximum processing energy (Fig. 4, area 4), which has a large meltwater coverage and lacks the structure of melting snow. At this point, snow is still present in other areas. This may be due to the highly developed surface topography and larger contact area due to the absence of an air gap at the initial stage of melting.

4.3 Evaporation of Water at Low Temperatures

Further studies were aimed at determining the rate of water evaporation from the surface of the same samples of steel plates at temperatures of 2–4 °C, i.e., under conditions close to the onset of freezing (Fig. 6).

The temperature during the experiment was in the range of 2–4 C. The process of evaporation of meltwater was monitored for 24 h. With air humidity of 70%, we obtained the following results: area 3–4 (Fig. 6) - water evaporation took 8–9 h; area 2 (Fig. 6) - water evaporation took 11–12 h, area 1 (Fig. 6) - water evaporation took 15 h.

It should be noted once again that the experiments were carried out in natural conditions with high humidity. From the comparative results shown in Fig. 6, it can be seen that in the area of intense laser processing on hydrophobic areas (Fig. 6, area 2–4), the

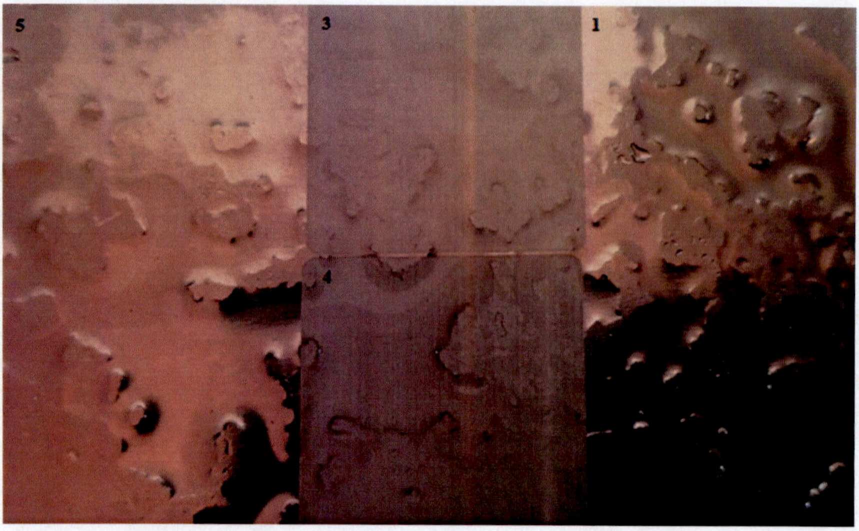

Fig. 5. Melting snow on the surface of steel 12X18H10T

Fig. 6. The result of the study of water evaporation at low temperatures

evaporation of drops occurred much earlier than on the untreated surface (Fig. 6, area 5). The area with multi-pass treatment (Fig. 6, area 1) also outperforms the base sections in terms of evaporation rate, which indicates the presence of hydrophobicity, in contrast to the experiment with freezing (Fig. 2). To clarify the change in the adsorption properties of the studied samples, the surface was sprayed with the surfactant 2-methilisothizol (Henkel). Figure 7 was shown the dynamics of water droplet spreading in the presence of surfactants.

As expected per the previous stages of the study, the droplet spreading rate on the hydrophobic areas and the spreading size are significantly less than on the untreated surface. This is also direct evidence of the possibility of creating hydrophobic surfaces on stainless steel using nanosecond laser radiation. The increased rate of droplet evaporation from hydrophobic areas is probably due to the larger droplet-free surface due to large wetting angles.

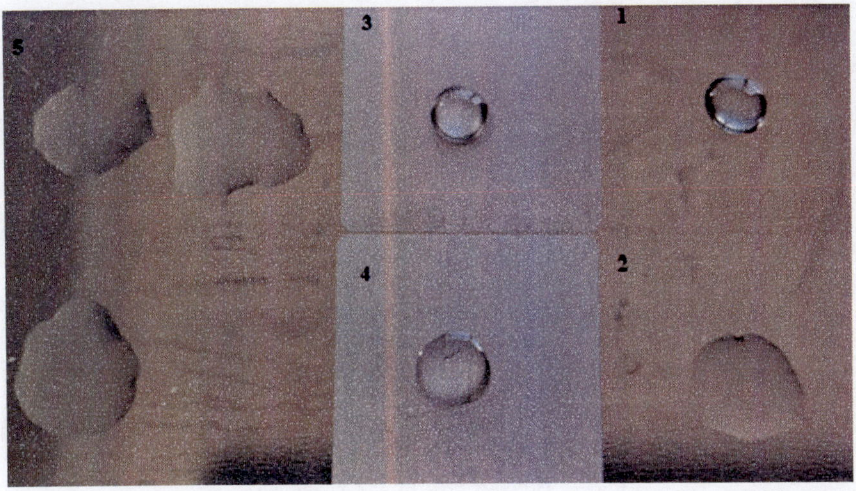

Fig. 7. Spreading of water droplets on the surface in the presence of surfactants (initial stage)

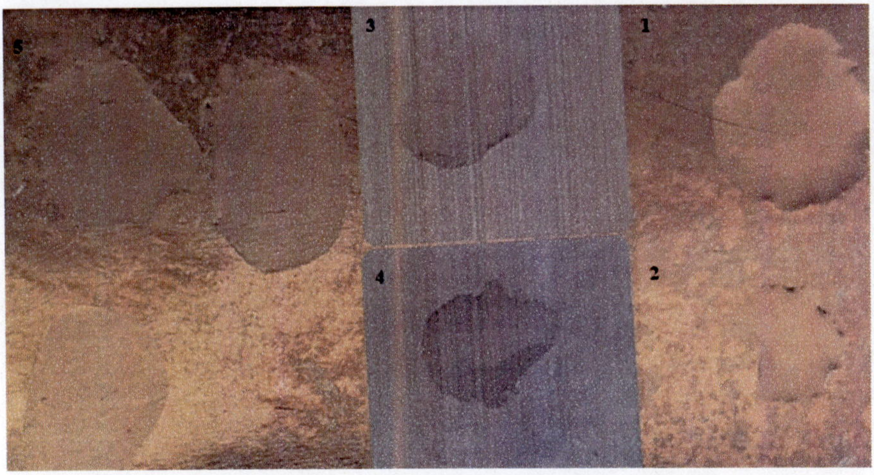

Fig. 8. Spreading of droplets in the presence of surfactants (finishing stage)

4.4 Droplets Flow Around Each Other

It should be noted that the movement of drops along the hydrophobic surface with a change in the angle of inclination of the plate differs from the movement along the base surface. The movement of drops along the base surface is characterized by randomness in direction. Drops of water, when approaching each other, try to merge. A fundamentally different picture is observed on a hydrophobic surface. Drops, as a rule, move along a clear trajectory and try not to merge, but go around each other (Fig. 9, area 4).

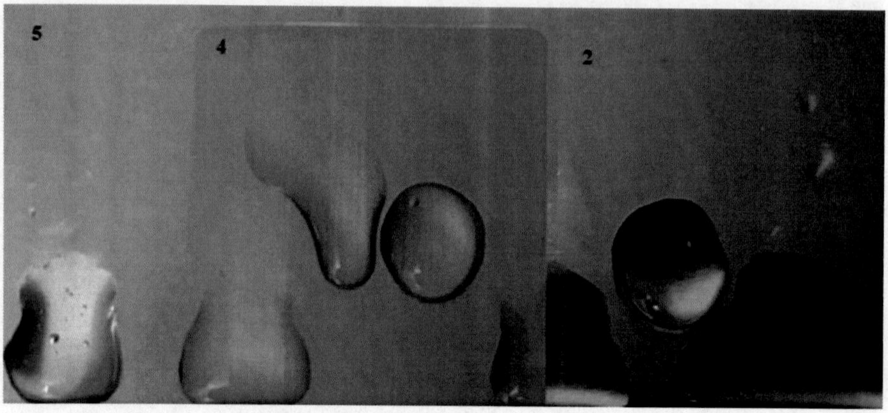

Fig. 9. Droplets flow around each other

4.5 On the Mechanism of Formation of a Hydrophobic Surface

Various models of hydrophilic and hydrophobic surfaces were summarized in the paper [1]. The properties of a hydrophobic surface with LIPSS structures are usually considered based on the Cassie-Baxter model. In the case of superhydrophobicity, the droplet no longer infiltrates the surface roughness, but a kind of combined phase contact is established between the solid-liquid and gas-liquid interfaces. So, Cassie and Baxter introduced a new relationship for the apparent contact angle [12]:

$$cos\theta* = f_1 cos\theta_1 + f_2 cos\theta_2 \tag{1}$$

where θ_1 and θ_2 are, respectively, the young contact angles of the liquid droplet on the solid and gas bearings, while f_1 and f_2 are the percentages of the solid-liquid and gas-liquid contact area, with $f_1 + f_2 = 1$.

However, within the framework of the Cassie-Baxter model, it is rather difficult to explain the features of the formation and behavior of a hydrophobic surface under the action of laser radiation, since it mainly takes into account the roughness geometry and the presence of an air gap inside the roughness. Therefore, let us consider in more detail the processes that may be responsible for the creation of hydrophobicity.

Two related questions need to be answered:

1. First question: Why does hydrophobicity appear only after some long time and depends on the intensity of laser radiation, scanning speed, and scanning step?
2. Second question. Why is it possible to create hydrophobic surfaces on steels under the action of the ultrashort nanosecond, picosecond, and femtosecond laser pulses?

The problem of the aging of steel in this class is not so obvious. Under conventional thermal heating methods, these steels do not harden due to the absence of corresponding austenite transformations and therefore they are not characterized by the presence of an aging process with time under atmospheric conditions. However, it is a well-known fact that the aging period of hardened steels with classical hardening methods is 14–17 days.

Apparently, in the case of exposure to short-pulse laser radiation, a process occurs that is somewhat reminiscent of the hardening of the surface, followed by the aging process. To explain this, it is necessary to consider and compare the processes of surface modification during laser pulses in ultrashort time ranges. The first important fact is that hydrophobic surfaces can be obtained in all these ranges. It is known that the shape of pulses in TEM_{00} modes has a shape close to triangular, but with a steeper leading edge of the energy contribution. In the case of femtosecond laser pulses, the mechanism of cold laser ablation of materials is realized. The physics of the ablation process is explained based on a two-temperature model [13, 14]. Initially, the absorption of laser radiation occurs on collectives of free electrons, and the electron temperature is much higher than the lattice temperature. The so-called electron melting and the formation of nanocluster structures take place in the surface layer [15]. A shock wave is superimposed on these processes during the ablative removal of material. The Boltzmann melting temperature is not realized during processing. The process of surface modification by a nanosecond laser pulse, in contrast to the femtosecond one, allows the presence of the classical process of melting and evaporation of matter in the final part of the pulse. The processes occurring during the rise of the leading edge of the nanosecond pulse are similar to the processes during femtosecond irradiation. The removal of the matter is accompanied by the creation of a high-intensity shock wave. During rapid isochoric heating of a surface layer of a substance with a thickness of about several hundred nanometers, megabar pressures arise in the surface layer itself, which lead to the appearance of shock waves of very high amplitude and short duration. Their distinctive feature is a record-breaking small localization in space. The characteristic localization scale of such a shock wave in a substance is several hundreds of nanometers.

Experiments show that the dynamic strength of nanolayers under such short loads increases hundreds of times, approaching the limiting, atomic strength of the substance. That is, to break the substance very quickly, you need to apply a much greater effort. Here, other mechanisms for the occurrence of plastic deformations and fractures operate in comparison with a conventional explosive shock wave [15].

This allows us to conclude that in all ranges of ultrashort laser exposure, the common thing is the appearance of a hardening deformation wave with the formation of nonequilibrium structures, including nanostructures and nanoclusters. Thus, we can conclude that during the "aging" of the steel surface, relaxation of some of these structures occurs over time. It is reasonable to associate the observed decrease in the contact angle with a deformation change in the structural (phonon) component of the disjoining pressure. For hydrophobic surfaces, the phonon component of the disjoining pressure contributes to an increase in the wetting angle. It should also be noted that the presence of Boltzmann heating and the melting process in the nanosecond range does not make a decisive contribution to the formation of hydrophobic properties. However, it is important that the emission of electrons from the surface due to multiphoton absorption and expansion of the plasma can lead to ionization of atmospheric oxygen and various degrees of oxidation of the elements that make up the steel [16–19]. The color of oxides of iron, chromium, and other elements, in combination with the "structural color" generated by laser-induced structures, creates the color gamut of the coloring of various parts of the samples (Fig. 1). Since the sliding of water droplets over dielectric and hydrophobic surfaces occurs with the participation of droplet electrification [20], the consequence of this may be the ability of a droplet to accumulate an electric charge on its surface. As a result, when the drops slide (Fig. 9), the electric fields of the drops on the hydrophobic surface are mutually repelled, and the drops tend to flow around each other. When the surface is tilted, the advancing and receding fronts form different angles with the surface. Based on this, Eq. (1) must be considered depending on the "aging" time (t):

$$cos\theta(t)* = K_1(t) \cdot K_2(t) \cdot (f_1 cos\theta_1 + f_2 cos\theta_2) \quad (2)$$

where K_1 – strain aging factor, K_2 – coefficient taking into account the change in the chemical composition of the surface.

In connection with the course of the deformation process of "aging" and the formation of static electricity, a time dependence should also be introduced into the equation for the sliding of a drop on an inclined surface [1]. Then the angle α characterizing the moving drop is determined depending on the advancing and receding contact angles Q_A end Q_R (3):

$$sin\alpha(t) = K_1(t) \cdot K_2(t) \cdot K_3(t) \cdot \left(\frac{G \cdot R \cdot k}{m \cdot g}\right) \cdot (cosQ_R - sinQ_A) \quad (3)$$

where $K_3(t)$ is the coefficient taking into account the influence of electrostatic charge, G is the surface tension, R and k are length scale and shape constants for describing the droplet profile (R is generally taken to be the drop radius, while k is an empirical parameter obtained by fitting the experimental data), respectively, m is the mass of the droplet, and g is gravitational.

Further studies are required to determine the numerical values of the new coefficients.

5 Conclusions

Nanosecond pulsed lasers can be used to create superhydrophobic surfaces using raster scanning, provided that rational parameters are chosen. The hydrophobicity parameters are very sensitive to the laser processing parameters. As in other cases of laser generation of superhydrophobic surfaces, there is a transition time after laser processing for the surface to become hydrophobic. The transition time is highly dependent on the laser processing parameters used. It is shown that the mechanism of "aging" is based not on a thermal mechanism, but a superhigh-speed deformation mechanism for the formation of nonequilibrium nanostructures and nanoclusters, followed by their slow partial relaxation. As a result of this rearrangement, the surfaces began to acquire anti-icing properties. This expands the possibilities of using steels at low temperatures without the formation of ice on their surface. In general, the anti-icing phenomenon based on LIPSS is a complex synergy of various physical and chemical mechanisms and requires further research, especially related to the change in parameters over time.

Acknowledgments. The general approach was developed within the research project "Development of a methodology for optimal design and manufacture of highly efficient, highly reliable turbomachines, taking into account various operating modes" (No. 0121U107511).

References

1. Volpe, A., Gaudiuso, C., Ancona, A.: Laser fabrication of anti-icing surfaces: a review. Materials (Basel) **13**(24), 5692 (2020). https://doi.org/10.3390/ma13245692
2. He, H., Guo, Z.: Superhydrophobic materials used for anti-icing theory, application, and development. iSciense **24**(11), 103357 (2021). https://doi.org/10.1016/j.isci.2021.103357
3. Cao, L., Jones, A.K., Sikka, V.K., et al.: Anti-icing superhydrophobic coatings. Langmuir **25**(21), 12444–12448 (2009). https://doi.org/10.1021/la902882b
4. Farhadi, S., Farzaneh, M., Kulinich, S.A.: Anti-icing performance of superhydrophobic surfaces. Appl. Surf. Sci. **257**(14), 6264–6269 (2011). https://doi.org/10.1016/j.apsusc.2011.02.057
5. Ruan, M., Li, W., Wang, B., et al.: Preparation and anti-icing behavior of superhydrophobic surfaces on aluminum alloy substrates. Langmuir **29**(27), 8482–8491 (2013). https://doi.org/10.1021/la400979d
6. Gelest: Hydrophobicity, Hydrophilicity, and Silane Surface Modification. https://technical.gelest.com. Accessed 07 April 2022
7. Vercillo, V., Tonnicchia, S., Romano, J.-M., et al.: Design rules for laser-treated icephobic metallic surfaces for aeronautic applications. Adv. Funct. Mater. **30**(16), 1910268 (2020). https://doi.org/10.1002/adfm.201910268
8. Volpe, A., Gaudiuso, C., Di Venere, L., Licciulli, F., Giordano, F., Ancona, A.: Direct femtosecond laser fabrication of superhydrophobic aluminum alloy surfaces with anti-icing properties. Coatings **10**, 587 (2020). https://doi.org/10.3390/coatings10060587
9. Gaddam, A., Sharma, H., Karkantonis, T., Dimov, S.: Anti-icing properties of femtosecond laser-induced nano and multiscale topographies. Appl. Surf. Sci. **552**, 149443 (2021). https://doi.org/10.1016/j.apsusc.2021.149443
10. Simões, J.G.A.B., Riva, R., Miyakaw, W.: High-speed Laser-Induced Periodic Surface Structures (LIPSS) generation on stainless steel surface using a nanosecond pulsed laser. Surf. Coat. Technol. **344**, 423–432 (2018). https://doi.org/10.1016/j.surfcoat.2018.03.052
11. Kolobov, Yu., Zhidkov, M., Ligachev, S., et al.: The influence of laser radiation of emtosecond duration on the structure and properties of surface layers of steel 12Kh18N10T. Composit. Nanostruct. **8**(4), 223–235 (2016). (in Russian)
12. Cassie, A.B.D., Baxter, S.: Wettability of porous surfaces. Trans. Faraday Soc. **40**, 546–551 (1944)
13. Mikhail, N.: Libenson: Nonequilibrium heating and cooling of metals under action of supershort laser pulse. In: Proceedings of SPIE 4423, Nonresonant Laser-Matter Interaction (NLMI-10) (2001). https://doi.org/10.1117/12.431201
14. Anisimov, S.I., Lukyanchuk, B.S.: Selected problems of the theory of laser ablation [Izbrannye zadachi teorii lazernoj ablyacii]: Reviews of Current Problems. UFN **172**(3), 301–333 (2002). https://doi.org/10.3367/UFNr.0172.200203b.0301. (in Russian)

15. Popov, V.: The concept and principles of designing parts of aggregate constructions, and their effective machining by a cutting tool with nanostructures. Dissertation.... doctor of technical sciences. Kharkiv. Code and name of the specialty 05.03.07 processes of physical and technical processing. Special Council D64.062.04 of the National Aerospace University named after M.E. Zhukovsky "Kharkov Aviation Institute" (2021)
16. Sotnikov, V.T., Dobrotvorsky, S.S., Zapechelnyuk, E.F.: Threshold effects in the emission of charged particles from the surface under the influence of laser radiation [Porogovye effekty v emissii zaryazhennyh chastic s poverhnosti pod vozdejstviem lazernogo izlucheniya]. Ukrainian Phys. J. **11**(5), 744–748 (1986). [in Russian]
17. Sotnikov, V.T., Dobrotvorsky, S.S., Zapechelnyuk, E.F., Dobrotvorskaya, M.V.: Self-oscillations of the emission of charged particles from the surface of optical transparent media with nonlinear absorption of laser radiation [Avtokolebaniya emissii zaryazhennyh chastic s poverhnosti opticheskih prozrachnyh sred pri nelinejnom pogloshchenii lazernogo izlucheniya]. Surface. Phys. Chem. Mech. **9**, 103–110 (1986). (in Russian)
18. Sotnikov, V.T., Dobrotvorsky, S.S., Zhuk, V.A.: Influence of oxygen and water vapor on thermally stimulated exoelectronic emission of zinc selenide [Vliyanie kisloroda i parov vody na termostimulirovannuyu ekzoelektronnuyu emissiyu selenida cinka]. Ukrainian Phys. J. **11**, 48–54 (1984). (in Russian)
19. Sotnikov, V.T., Dobrotvorsky, S.S., Zapechelnyuk, E.F., Dobrotvorskaya, M.V., Galiy, P.V.: The role of thermal and electronic excitations in changing the stoichiometry of the surface of alkali halide crystals under the influence of concentrated energy flows [Rol' teplovyh i elektronnyh vozbuzhdenij v izmenenii stekhiometrii poverhnosti shchelochno-galoidnyh kristallov pri vozdejstvii koncentrirovannyh potokov energii]. Solid State Phys. **28**(7), 2254–225 (1986). (in Russian)
20. Li, X., Bista, P., Stetten, A.Z., et al.: Spontaneous charging affects the motion of sliding drops. Nat. Phys. (2022). https://doi.org/10.1038/s41567-022-01563-6

Theoretical and Experimental Research on Step Coverage Optimization for Integrated Microstructures of Thin Films

Georgeta Ionascu[1], Elena Manea[2], Ileana Cernica[2], and Edgar Moraru[1(✉)]

[1] Politehnica University of Bucharest, 313, Splaiul Independentei, 060042 Bucharest, Romania
edgar.moraru@upb.ro

[2] National Institute for Research and Development in Microtechnologies, 126A, Erou Iancu Nicolae Street, 077190 Bucharest, Romania

Abstract. A reliable deposition process of thin films must guarantee their material homogeneity and properties. In this context, the metallization of integrated microstructures needs to be performed specific requirements, among which those related to the conformity, respectively to the way in which layers uniformly cover the existing unevenness/steps on the surface of the substrate. This paper deals with studying and experimentation of conformity of metal films on steps, being demonstrated that the geometry of placement of silicon wafers in the evaporation system and the step profile influence the cracks developing. A mathematical model of vacuum evaporation process, validated through numerical simulation and experiment is given. The experiments have consisted of some planarization films (thermal grown SiO_2 and deposited, undoped and phosphorous doped, SiO_2 using APCVD - a process of Atmospheric Pressure Chemical Vapor Deposition) and Al metallization. Results of slope and roughness measurements, and SEM investigations are presented. Based on them, interesting conclusions are inferred regarding the optimization of working parameters for a uniform covering of the substrates with irregularities.

Keywords: Conformal step coating · Vacuum evaporation · Metallization · Planarization · CVD silicon dioxide film · SEM examination

1 Introduction

Metallization performing for integrated microstructures raises some problems, as optimization of planarization and metal conformity on steps are. At the deposition time, the surface profile is determined through a high number of previous technological stages, whose succession forms an irregular relief.

The properties of the evaporated thin metal films (resistivity; characteristics of etching; properties of diffusion, surface roughness, and resistance to electromigration) are established by their microstructure and thickness uniformity of the layer, which are influenced by two main effects:

- the vapor incident flux shadowing. If the film is deposited over topography, certain regions of the film will receive less flux than others, resulting in a non-uniform film

D. D. Cioboată (Ed.): ICoRSE 2022, LNNS 534, pp. 185–202, 2023.
https://doi.org/10.1007/978-3-031-15944-2_18

thickness. This effect is strongly dependent on the angular distribution of the incoming adatom flux. The distribution function of the vapor flux depends at its turn on the geometry of the evaporation system (the geometry and the emission characteristics of the vapor source, the shape of the deposition supports, the relative movement between the parts/wafers (substrates) and the source), the position and the orientation of the substrates on the support.

Step coverage on trenches in typical physical vapor deposition (PVD) techniques is illustrated in Fig. 1:

(a) The incident flux is not directional and has an angular distribution (e.g., vacuum evaporation [1]). Due to the geometrical shadowing, the film is preferentially deposited to the top corners and the coverage is poor especially at the bottom corners of the trench. This results in "overhang" at the top corners and "void" at the bottom corners.
(b) The incident flux is directional and normally incident to the substrate (e.g., by sputter deposition process). However, it approaches to the sidewalls of the trench at high oblique angles. Due to the shadowing effect, film on the sidewalls is in the form of columnar, porous structure [2]. In addition, due to the atomic attraction forces, the incident atoms are preferentially deposited to the top corners of the trench (atomic shadowing). This again gives rise to overhangs at the top corners and voids at the bottom corners.

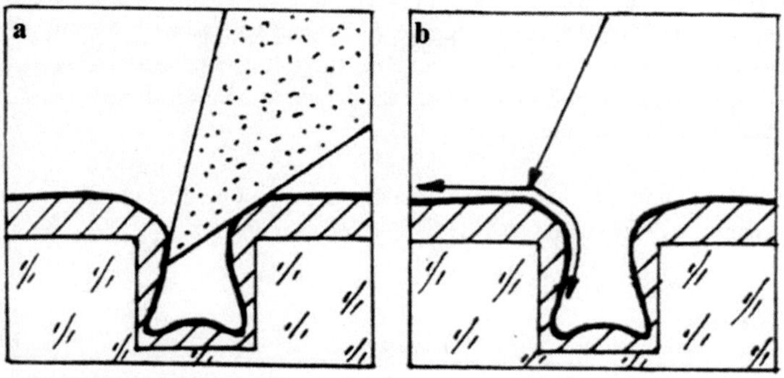

Fig. 1. a – shadowing of vapor flux due to substrate surface topography (contacts, steps, trenches etc.); b) – adatoms mobility on surface (the adatoms of growing film are subjected to a surface diffusion and have a characteristic diffusion length).

- The evolution of the film profile is also strongly determined by the mobility of the adatoms (absorbed atoms) after they strike the growing film. The diffusion length is a function of the ratio between substrate temperature and melting point of the material being deposited. Increased mobility of the atoms in the growing film on a hot substrate leads to a reduction of film surface area and a filling of the cracks caused by the shadow effect. At low temperatures of the substrate, the migration of the atoms onto the surface is negligible and the shadow effect prevails.

The emission characteristics of the real (practical use) vapor sources can be approached with the ones represented by two types of ideal source: the point source (with spherical symmetry) and the directed surface source of small area.

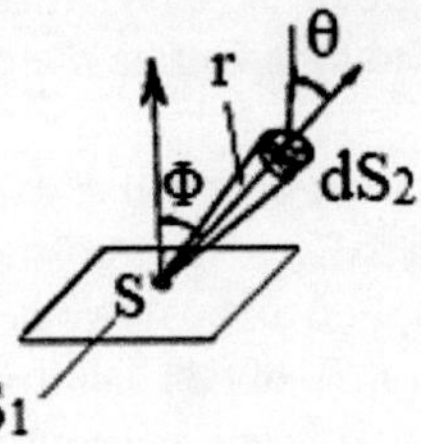

Fig. 2. Evaporation from a point source or a directed surface source of small area.

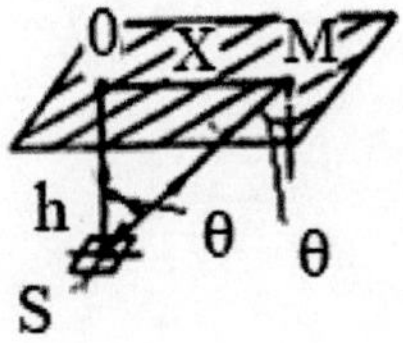

Fig. 3. Evaporation from a small area surface source onto a parallel plane substrate.

As it is seen in Fig. 2, the receiving surface dS_2 is inclined at an angle θ related to the medium direction of the vapor flux, emitted from the source dS_1.

The thickness of the film condensed on the elementary surface dS_2 is given by the expression:

$$t = \frac{m}{4\pi\rho} \cdot \frac{\cos\theta}{r^2}, \text{ in the case of the point source,} \quad (1)$$

$$t = \frac{m}{\pi\rho} \cdot \frac{\cos\Phi \cdot \cos\theta}{r^2}, \text{ in the case of the surface source,} \quad (2)$$

where m and ρ are the total mass and the density of the deposited material, respectively; Φ represents the angle between the direction of the vapor flux and the normal to the plane emitting source.

In the case of a plane receiving surface (substrate), the film thickness at a certain point M (Fig. 3) is according to the relations (1) and (2):

$$t = \frac{m}{4\pi\rho} \cdot \frac{1}{h^2\left[1 + \left(\frac{x}{h}\right)^2\right]^{\frac{3}{2}}}, \text{ point source,} \quad (3)$$

$$t = \frac{m}{\pi\rho} \cdot \frac{1}{h^2\left[1 + \left(\frac{x}{h}\right)^2\right]^2}, \text{ surface source parallel to the substrate,} \quad (4)$$

$$(\cos\theta = h/r \text{ and } \Phi = \theta)$$

For both types of sources, it is obviously difficult to obtain thin films with a uniform thickness when substrates are situated on a plane support. It can be observed that thickness decreases as the distance x to the source center increases.

The uniformity of the deposition thickness on plane substrates can be improved either by putting them into a single or planetary rotational motion or by use a rotating mask whose shape is designated to compensate the directional emitting characteristics of the source [3–6].

Let us consider a surface source of small area – dS_1 (Fig. 4a) and assume that evaporation takes place according to Knudsen's cosine law. This is a gas-kinetic analogy to Lambert's law for radiation, applied to the movement of the molecular/atomic beams, which are obtained in high vacuum. To obtain a uniform thickness of the deposited film, it is necessary that the substrates – dS_2 and the source – dS_1, too, be disposed onto the surface of a sphere, called Knudsen's sphere. In this case:

$$cos\theta = \cos\Phi = \frac{r}{2r_0} \tag{5}$$

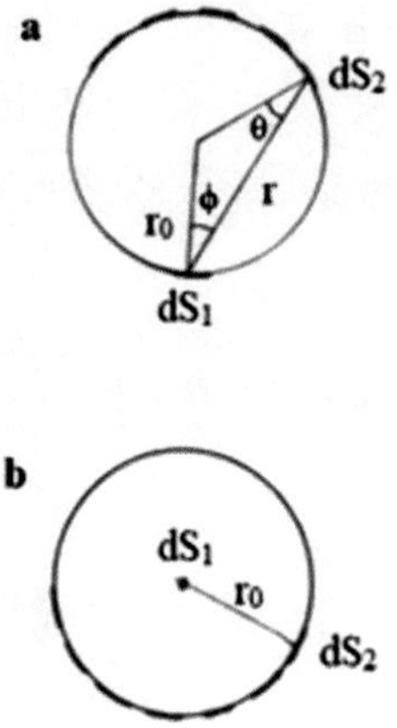

Fig. 4. Ideal deposition system by evaporation: a – evaporation onto the inside of a sphere from a surface source situated on circumference; b – as in the previous figure, but with a point source at the centre of the sphere.

The deposition rate is constant for any point of the sphere, and we obtain a uniform deposit with such an arrangement, that is:

$$t = \frac{m}{\pi\rho} \cdot \frac{1}{4r_0^2} \tag{6}$$

To perform thin films of uniform thickness by using a point source dS_1 (Fig. 4b), the substrates dS_2 must be placed on a concentric sphere with the source and the radius of the sphere must be large compared to the highest linear dimension of the substrates. In this case:

$$t = \frac{m}{4\pi\rho} \cdot \frac{1}{r_0^2} \tag{7}$$

If the substrates are situated on a plane support, the placement of this support must be made at a very large distance, at semi-infinite, relative to the source, to minimize the ratio x/h from the relations (3) and (4).

So, the finite plane surface of the substrate could be considered as a part of the association (concentric) sphere. But this solution reduces very much the deposition rate corresponding to a certain evaporation rate.

Based on these findings, the supports using planetary deposition systems in form of spherical calotte (Fig. 5) are widely spread.

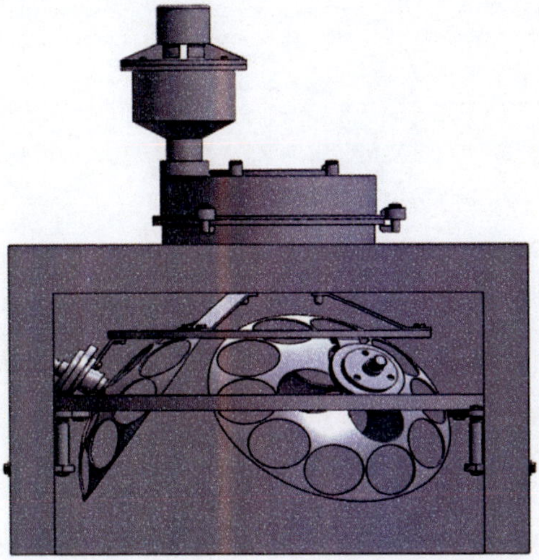

Fig. 5. 3D model in Solid Works of a planetary deposition system.

In an oblique angle physical vapor deposition (OAPVD) [7], the incident flux approaches the rotating substrate at an angle θ from the surface normal. The θ is set to small values, for example, ~15° for a 2:1 aspect ratio (ratio between the height and the minimum dimension in plane), and ~7° for a 4:1 aspect ratio trench such that the whole sidewall and the bottom corner is exposed to the incident flux by avoiding shadowing effect.

By this way, the step coverage especially on the sidewalls and the bottom corners can be greatly enhanced. In addition, the film on these regions becomes denser and the thickness is uniform. The developed programs allow the simulation of surface profile and film microstructure using the adsorption and thermodynamics theories as regards condensation and formation of thin films from gaseous phase [8–11].

Exploring the surface over a specific length of diffusion, these theories place the deposition species in positions characterized by a minimum of potential energy. This supposes a great mobility of adatoms, which means depositions at high temperatures of the substrate. The deposition of the refractory metals by cathodic sputtering or chemical vapor deposition (CVD) is such an example [12].

It is preferred that the metallization with aluminum by evaporation be made onto not heated substrates, because the particles generated inside the deposition bell jar can affect the finest interconnections. The layer adhesion is ensured, in this case, by a careful cleaning of the silicon wafers before introducing them into the bell jar and, also, by using of electron beam gun to evaporate the metal or by using of metallization multi-layer (e.g., Al-Si, Al-Ti).

In addition, the thermal treatment of sintering applied after deposition, to obtain good contacts between the metal interconnections and the silicon, leads also to the crystallite raising in the layer, the improvement of the crystalline structure and an effect of filling of the cracks produced by the step non-uniform coverage.

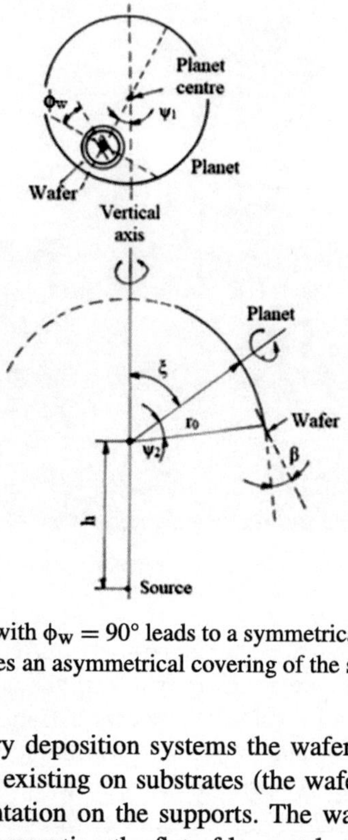

Fig. 6. The wafer orientation with $\phi_W = 90°$ leads to a symmetrical covering of the step, and $\phi_W = 0°$ and $\phi_W = 180°$ determines an asymmetrical covering of the step

Because in the planetary deposition systems the wafers do not rotate around the own axes, Fig. 6, the steps existing on substrates (the wafers) are differently covered as a function of their orientation on the supports. The wafer orientation is obtained through the primary flat representing the flat of longest length with technological role of mechanical orientation in the working devices. The steps parallel to the radius of planetary support are symmetrically covered (A steps), while the steps perpendicular to the radius are asymmetrically covered (B steps), the deposited films presenting deep cracks, as it will see below.

The cracks in Al metallization (the most widespread conducting material) of the integrated circuits and MEMS devices, due to the improper coverage of the steps by the metal layer, are an important source of failure, as shown by the reliability tests of these devices. Because these cracks cannot be observed with optical microscopy, scanning electron microscopy (SEM) is required. It has been shown that the shape of these cracks in the metal films is related to the geometry of the wafer-source placement in the deposition equipment and to the profile of the substrate step.

To cover the steps in an electron beam evaporation installation, the authors have developed a model for evaluating the profile of the deposited layer, using vector calculation, aiming to optimize the configuration of the planetary system-wafers-source and the profile of the step by numerical simulation.

The aluminum and dielectric films have been investigated by SEM to assess the coverage of the steps performed by wet chemical etching of undoped and doped silicon dioxide, obtained by various CVD methods, and to validate the proposed calculation model. Interesting conclusions have inferred about the factors that influence the aspect and quality of the deposited dielectric and metallic films.

2 Calculation Model for Step Covering by Deposition of Thin Films

The geometry of the considered planetary system is presented in Fig. 7. It is supposed that the deposited layer is homogeneous in terms of composition, has a uniform density, does not have an internal structure, and presents a condensation coefficient equal to the unit (for the cold substrates it can be neglected the mobility of adatoms during the growth of the thin film). It is assumed that the planetary rotates at a uniform speed, high enough that the profile of the deposited layer is independent of both the rotational speed and the initial position ψ_1 (Fig. 6). Since the rotation around the central axis of the planetary device does not change the evaporation equations (because the geometry of the system materializes Knudsen's "sphere"), only the rotation of the planets around its own axes is considered in the calculation. For ease of calculation, it is assumed that the planetary is fixed and that the source rotates around the axis of the planet. In Fig. 7b there are presented the geometry of the substrate and the rotating source at radius R in position A. Distances R', r, W and L are also represented in Fig. 7. The steps on the substrate surface are shown in detail in Fig. 7b.

The following vectors are defined according to Fig. 6: $\overline{i}, \overline{j}, \overline{k}$ – vectors parallel to the axes x, y and z.

$$\overline{N} = -\overline{k} \tag{8}$$

– normal to the plane described by the rotating source

$$\overline{N_1} = -(R \cdot cos\omega) \cdot \overline{i} - (Rsin\omega) \cdot \overline{j} - W\overline{k} \tag{9}$$

– distance h in the direction of normal to the source

$$\overline{J} = -(Rcos\omega - r) \cdot \overline{i} - (Rsin\omega) \cdot \overline{j} - L\overline{k} \tag{10}$$

– source-wafer distance

$$\overline{n} = -(sin\beta) \cdot \overline{i} + (cos\beta) \cdot \overline{k} \tag{11}$$

– normal on the wafer

$$\overline{n_1} = -(cos\beta) \cdot \overline{i} - (sin\beta) \cdot \overline{k} \tag{12}$$

– normal on the B outside step

$$\overline{n_1'} = (cos\beta) \cdot \overline{i} + (sin\beta) \cdot \overline{k} \tag{13}$$

– normal on the B inside step

$$\overline{n_2} = \overline{j} \tag{14}$$

– normal on the A step

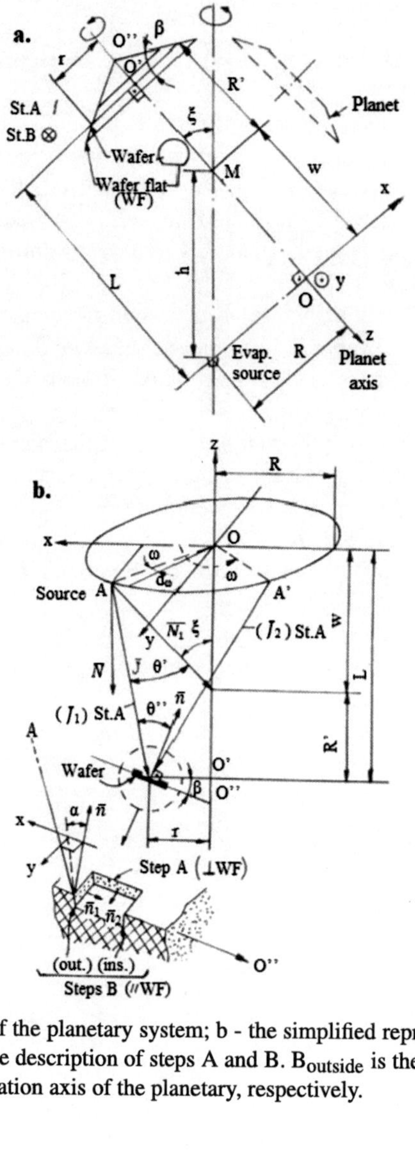

Fig. 7. a - the geometry of the planetary system; b - the simplified representation of the source-wafer relative position. The description of steps A and B. $B_{outside}$ is the farthest step and B_{inside} is the closer step to the rotation axis of the planetary, respectively.

The vapor flux should be calculated as a function of the incidence angle, α - defined as the angle between the normal to surface of the wafer and the projection of vector J in the xOz plane, for step B, and as the angle between the Oz axis and the projection of vector J in the yOz plane for step A. According to the Eq. (2):

$$I(\omega) \approx \frac{cos\theta' \cdot cos\theta''}{|\overline{J}|^2} \tag{15}$$

represents the intensity of the vapor flux in the normal direction at the wafer surface, without taking into account the shadowing effect introduced by the wafer topography. By introducing α angle, the shadowing effect will be taken into account and I(α) is obtained – the deposition rate in the direction of the Oz axis with respect to:

$$I(\alpha)d\alpha = I(\omega)d\omega \tag{16}$$

which represents the amount of material that arrives in the unit of time, per unit area of the wafer, in the angular range dα. I(α) can also be written as the product of three functions, which are to be expressed as a function of the angle α, as follows:

$$I(\alpha) = f_1 \cdot f_2 \cdot f_3 \tag{17}$$

where:

$$f_1 = (cos\theta')/|\overline{J}|^2 \tag{18}$$

$$f_2 = d\omega/d\alpha \tag{19}$$

$$f_3 = cos\theta'' \tag{20}$$

Taking into account the defined vectors, and for the $B_{outside}$ step:

$$\begin{aligned} f_1 = \frac{\overline{J} \cdot \overline{N_1}}{|\overline{J}|^3 \cdot |\overline{N}_1|} &= \frac{R}{L}[1 + \frac{W}{R}\frac{L}{R} - (\frac{r}{R})^2 - \frac{rL}{R^2}tg(\alpha - \beta)] \\ &\cdot [1 + (\frac{W}{R})^2]^{-1/2} \cdot L^{-2} \cdot [1 + (\frac{R}{L})^2 - (\frac{r}{L})^2 - 2\frac{r}{L}tg(\alpha - \beta)]^{-3/2} \end{aligned} \tag{21}$$

$$f_2 = \frac{d\omega}{d\alpha} = -\frac{L}{R}cos^{-2}(\alpha - \beta)\{1 - \left[\frac{r}{R} + \frac{L}{R}tg(\alpha - \beta)\right]^2\}^{-1/2} \tag{22}$$

$$f_3 = \frac{\overline{J}}{|\overline{J}|}\frac{\overline{n}}{|\overline{n}|} = L\big[tg(\alpha - \beta)sin\beta - cos\beta\big] \cdot [R^2 - r^2 + L^2 - 2rLtg(\alpha - \beta)]^{-1/2} \tag{23}$$

After the deposition of the metal in the time interval ΔT, a certain point of the substrate, of x_i and z_i coordinates, will move to a new position with the coordinates $x_i + \Delta x_i$, $z_i + \Delta z_i$. Amounts $\Delta x_i(\alpha)$ and $\Delta z_i(\alpha)$ are determined by the relations:

$$\Delta z_i(\alpha) = I(\alpha) \tag{24}$$

$$\Delta x_i(\alpha) = \Delta z_i(\alpha) tg(\alpha - \beta) \tag{25}$$

The total displacements of the x_i and z_i coordinates, due to the deposition under all angles α between α_{min} and α_{max}, are:

$$\Delta x_i = \int_{\alpha_{min}}^{\alpha_{max}} \Delta x_i(\alpha) d\alpha \tag{26}$$

$$\Delta z_i = \int_{\alpha_{min}}^{\alpha_{max}} \Delta z_i(\alpha) d\alpha \tag{27}$$

For B_{inside} step, the calculation is made with the same relations, but the values of α change, because it is determined in different coordinate points.

In the case of step A, for each value of α, there are two values of the vector modulus representing the source-wafer distance, $|\overline{J}|$, which must be taken into account in the calculation of $I(\alpha)$.

3 Numerical Simulation of Step Covering

By using the mathematical model from above, the distribution of vapor flux as a function of the incidence angle has been determined, considering a Balzers 550 vacuum evaporation plant endowed with planetary support and an electron beam evaporation source. The graphical representations of the distribution function for steps A and B are given in Fig. 8. The deposition rate is lower at the base of steps due to the shadowing effect. This effect, as well as the increased maxima of the vapor flux distribution function towards its limits, lead to deep cracks, especially for the B steps. At points where the shadow effect is not felt, the deposited layers increase at the nominal deposition rate.

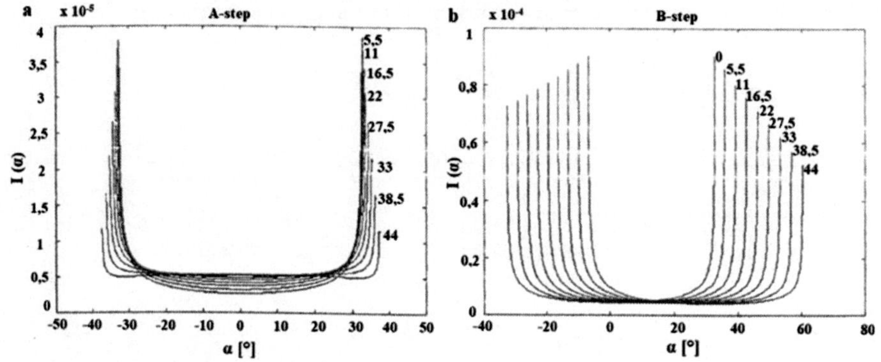

Fig. 8. The vapor flux distribution functions in the case of planetary evaporator for step A (a) and step B (b); $\beta = 0°\ldots44°$.

By using the calculation program, whose block diagram is shown in Fig. 9, and the programming environment MATLAB, the layer profile has been drawn for the geometric

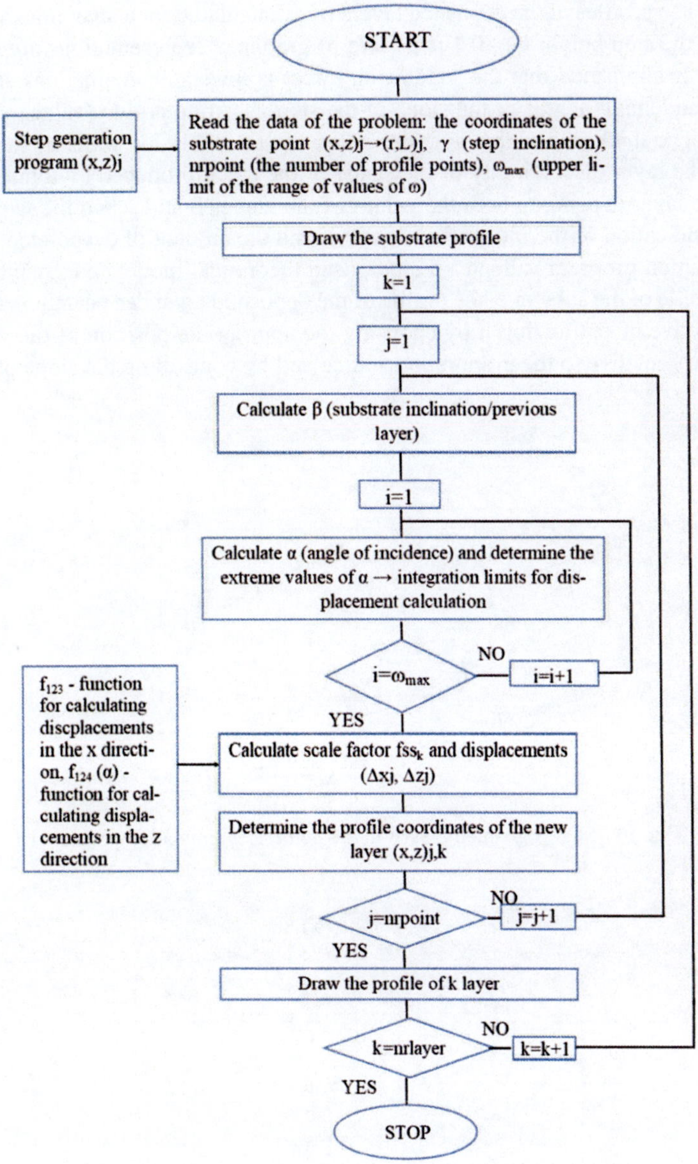

Fig. 9. Block diagram of the numerical simulation program.

data of the BAK 550 vacuum evaporation plant. The wafers were considered to be placed at the top ($\beta = 5°$), centre ($\beta = 10°$ and $15°$) and bottom ($\beta = 30°$) of the planet.

The profiles in Fig. 10 are ideal, but $\gamma > 0$ corresponds to the profile of the steps in a photoresist layer, for selective metal deposition by the lift-off method, and $\gamma \leq 0$ corresponds to the profile of the steps etched by wet chemical etching of the usual dielectric layers (undoped and doped silicon dioxide), in case of non-selective deposition of metals. The profiles of the deposited layers were calculated for a layer thickness equal to 1/10 of the step height i.e., 0.1 μm. Two of graphical representations are shown in Fig. 11. It is confirmed that the shadowing effect is lower for inclinations small β of the wafer and small angles of the slope of the step-substrate profile (values γ strongly negative). The thickness of the layer deposited on the inner side walls of steps B, the aspect of the layer discontinuity or the shape of the crack produced by thinning at the base of the layer depend on both the values of the angles β and γ and the width of the window (indication of the integration density), and the amount of evaporated material. The calculation program validates the proposed theoretical model and, by introducing the real profile of the substrate, the profile of the deposited layer can be anticipated, with the perspective of optimizing it by choosing the appropriate position of the wafers on the planetary relative to the evaporation source and by controlling the slope of the step profile.

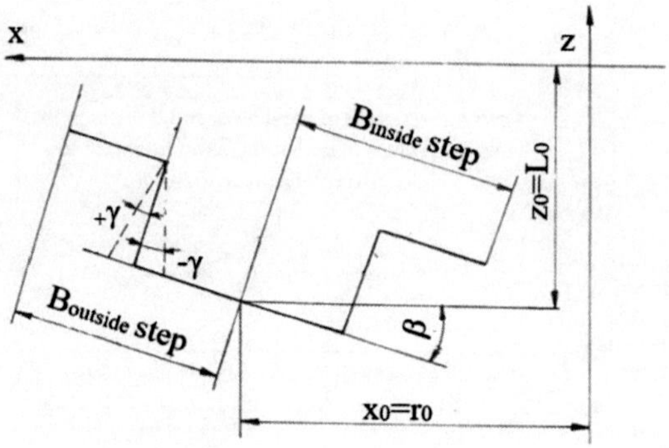

Fig. 10. Step B geometry, used in the numerical simulation program.

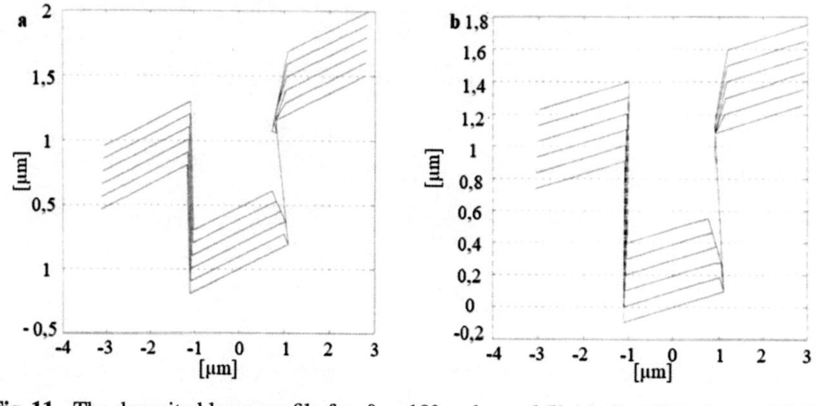

Fig. 11. The deposited layer profile for: β = 10° and γ = 6.5° (a); β = 5° and γ = 6.5° (b).

4 Experimental Research

Undoped and doped SiO_2 films are amorphous, so that their wet chemical etching is an isotropic process with profile inclination angles of about 80°–90°, usually. To obtain a small base angle, the control of profile inclination can be performed by several methods.

Following the evaluation and comparison of the etched profiles in thermal SiO_2 films and APCVD (Atmospheric Pressure Chemical Vapor Deposition) undoped/phosphorous doped SiO_2 films (PSG or Phospho-Silicate Glass), as well as their influence on the aluminum metallization conformity, the authors used two methods in experiments:

- *etching in BHF* ("Buffered" HF with ammonium fluoride (NH_4F)). Addition of NH_4F avoids the exhaustion of fluor ions and so, the change of reaction velocity. The reproducibility of chemical attack rises appreciably. At the same time, the presence of NH_4F diminishes HF attack on photoresist mask and, consequently, the solution selectivity is very high (almost 100) relative to the Si substrate. Both thermally grown SiO_2 films and deposited SiO_2 films can be etched in BHF solution. But the etch rate of the deposited oxides is much higher than the thermal oxides, which made interesting the application of the next method;
- *the double layers technique.* If a layer is deposited onto another layer, and the second layer possesses a higher etch rate, the lateral etch rate becomes superior to the vertical one, resulting in a profile with a smaller inclination. The top surface film can be a vapor phase deposited doped or undoped oxide.

Table 1. Structure of the dielectric and metal films under study

Substrate	Lapped (100) Si wafers, initial roughness $R_a \approx 0{,}2\ \mu m$		
Dielectric films	No	Type	Thickness [nm]
	1	Thermally grown SiO_2	1400 ± 50
	2	Thermally grown SiO_2	150
		APCVD undoped SiO_2	1300 ± 50
	3	APCVD undoped SiO_2	900 ± 10
		APCVD 4% wt. P doped SiO_2	500 ± 10
	4	APCVD undoped SiO_2	900 ± 10
		APCVD 11% wt. P doped SiO_2	500 ± 10
Metal film	Vacuum evaporated Al 1100 ± 100 nm, e-gun process		

For use as insulating, sacrificial, passivating or planarizing layers (after "flow"), the studied oxide layers have similar thicknesses and compositions both for the technology of micromechanical structures and in the manufacture of integrated circuits. This ensures the compatibility of the two technologies and is important in the process of integrating the microstructures.

3" (100) Si wafers with various resistivities were used. To simulate the worst case, the wafers were lapped on the polished face (so, we had a high roughness of the surface before starting the process sequence). Succession and thickness of the deposited films involved in our experiment are centralized in Table 1.

The technological process consists of the following stages:

1. initial cleaning of the wafers,
2. thermal processes (thermal oxidation),
3. APCVD processes,
4. standard photolithography process through M1 mask to open contact windows having square geometry of 10 × 10 μm to 50 × 50 μm in the oxide layer,
5. oxide wet etching,
6. photoresist removing,
7. cleaning again,
8. Al film deposition,
9. photolithography process through M2 mask for metal lines (interconnections) of 20 ÷ 50 μm in length and 7 ÷ 20 μm in width,
10. metal wet etching,
11. photoresist removing,
12. cleaning (followed by the usual rinse and dry steps),
13. sintering,
14. chips cutting – off from the wafers,
15. SEM examination.

Table 2. Values of the profile slope and the surface roughness for the investigated oxide films

No	Type of oxide film	Medium slope of the profile [°]	Surface roughness Ra parameter [μm]
1	Thermal SiO_2	11,8	0,11
2	CVD undoped SiO_2/ thermal SiO_2	9,6	0,08
3	PSG (P4%)/CVD undoped SiO_2	6,2	0,41
4	PSG (P11%)/CVD undoped SiO_2	4,8	0,24

To avoid the electrical charging of dielectric areas during SEM examination, the deposition by vacuum evaporation of a very thin double layer of graphite + Ag (100 Å–200 Å total thickness) was performed.

Observations:

1. For composed oxide films, between thermal and deposited SiO_2 no cleaning process was done.
2. For composed APCVD films, the doped and undoped SiO_2 were deposited in the same run to preserve the good stability of interfaces.
3. All the wafers with APCVD SiO_2 were reflowed using a standard process (O_2, 1000 °C, 30 min).

The slope of the steps in the windows and roughness surface oxide were determined before deposition of Al film using an "alpha-step" profilometer. The obtained results are presented in Table 2.

An improvement of the initial surface quality of wafers was observed for thermal SiO_2 and CVD undoped SiO_2/thermal SiO_2 films. This can be explained by making the layers with Si substrate participation (consuming) during the oxidation process, resulting in a better filling of the surface irregularities.

A worsening of the surface quality was obtained for PSG (P 4%)/CVD undoped SiO_2 films, which means that the layers "follow" and even "amplify" slowly the initial surface irregularity. Rising the content of phosphorous (P 11%), a surface roughness comparable with the initial one was observed.

These observations lead to the conclusions that (composed) films containing thermal SiO_2 have a denser structure than doped oxides PSG/CVD undoped SiO_2 films, reflected also in bigger slopes of the etched steps. The smallest slope appears at the oxide layer with the highest content of doping element (P11%), which presented the highest etching rate.

The high lateral chemical attack (under the mask) at the deposited oxides surface (not the ones thermally grown), especially the doped oxides, can be due to the weak adherence of the photoresist during the etching process, when an adherence promoter (HMDS) lacks.

An optimal solution has been inferred if the following conditions are met: to be obtained a minimum lateral chemical attack (under the photoresist mask); to minimize the slope angle changes; the final slope (nearby the Si) must have a minimum angle. Such a slope of 4.1°, nearby the silicon wafer, was obtained for a composed oxide film, as follows: thermal SiO_2 + APCVD undoped SiO_2 + APCVD 4% wt. P doped SiO_2 + APCVD 11% wt. P doped SiO_2.

Another aspect we have studied is the conformity of metal covering on steps. SEM investigations have revealed that the thickness uniformity of the metal layer on steps is influenced by the wafer position on planet (support) and the step slope. A better covering is obtained for smaller slopes and for wafers placed in the central zone of planet.

By induced tensions, the non-uniformity of thickness of the metal layer can produce its crack. The high level of tensile tensions in a thermally grown oxide can be another cause of the cracks found in the metal layer.

The best quality of the surface was observed at the sample with CVD undoped SiO_2/thermal SiO_2, where the initial irregularities of wafer surface almost disappear. At the sample with PSG (P 11%)/CVD undoped SiO_2, the etched steps in this layer are embedded in the surface microrelief, because of the phosphosilicate glass (PSG) flowing. All these observations are exemplified in the Figs. 12, 13, 14 and 15.

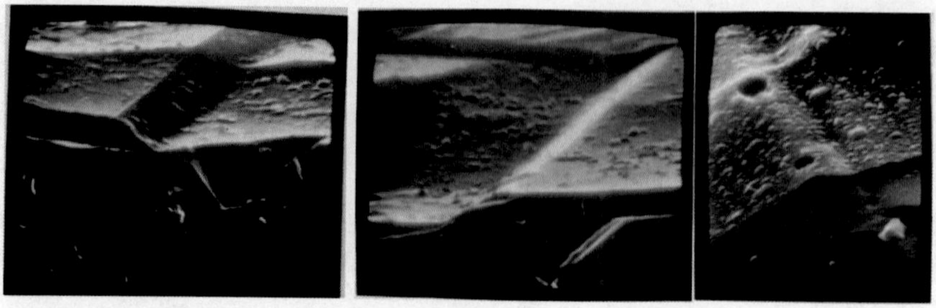

a b c

Fig. 12. SEM images of the sample with thermal SiO_2 and Al film, step slope 31°, placed at bottom of the planet: a – step B_{inside}, magnification 1631; b – step $B_{outside}$, magnification 1631; c – rotated sample with highlighting of cracks in the metal layer, magnification 1720.

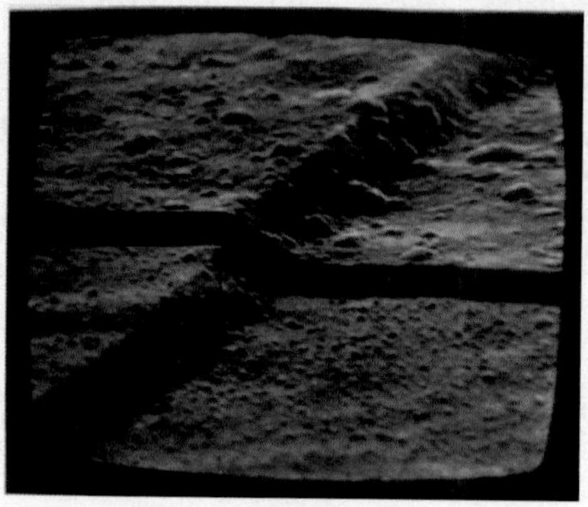

Fig. 13. SEM image of the sample with CVD undoped SiO_2/thermal SiO_2 and Al film, step slope 29°, placed at central zone of the planet, magnification 1525.

Fig. 14. SEM image of the sample with PSG (P4%)/CVD undoped SiO_2 and Al film, step slope 22°, placed at central zone of the planet, magnification 1525.

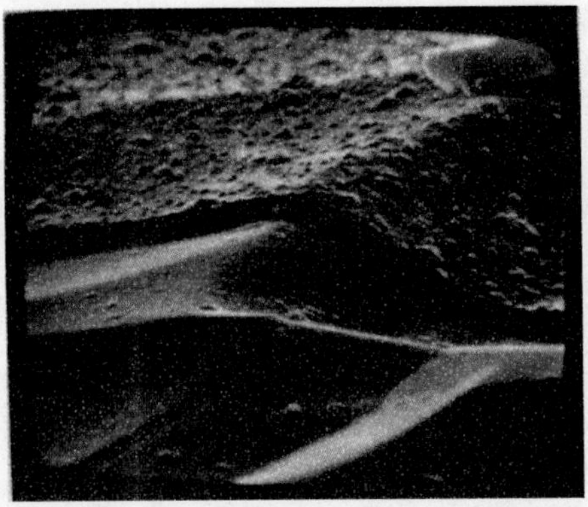

Fig. 15. SEM image of the sample with PSG (P11%)/CVD undoped SiO_2 and Al film, step slope 15°, placed at central zone of the planet, magnification 1525.

5 Conclusions

In this paper, the authors have developed a mathematical model for covering through vacuum evaporation the steps of a substrate (silicon wafer), placed on a planetary support. The theoretical model has been validated by numerical simulation and experiments of aluminum metallization of silicon wafers presenting etched steps in SiO2 layer, single or double; the step slope has been changed considering the nature, thickness, and succession of the oxide films. The best surface quality was observed to the sample of CVD undoped SiO2/thermal SiO2, where the initial irregularities of the wafer surface almost disappear. The descending order of the slope of steps is maintained: thermal oxide↓ undoped deposited oxide↓ doped deposited oxides↓. To the sample of PSG (11%), due to the phosphosilicate glass (PSG) flowing, it was noticed that the etched steps were buried in the microrelief of the surface.

For use as dielectric, sacrificial, passivating or planarizing layers, the studied oxide films have similar thicknesses and compositions both for the technology of micromechanical structures and in the manufacture of integrated circuits. This ensures the compatibility of the two technologies and is important in the process of integrating the microstructures.

The examination of the step profile and of metal lines has been performed by using the scanning electron microscopy (SEM) and the profilometer method. An optimal solution has been inferred if the following conditions are met: to be obtained a minimum lateral etching (under the photoresist mask); to minimize the slope angle changes; the final slope must have a minimum angle. By introducing the real profile of the substrate, the profile of the deposited layer can be anticipated, with the perspective of optimizing its conformity by choosing the appropriate position of the wafers on the planetary relative to the evaporation source and by controlling the slope of the step profile.

References

1. Ionascu, G.: Utilization of the Thin Film Structures Technologies in Precision Engineering and Mechatronics (in Romanian). Printech Publishing House, Bucharest (2004)

2. Smy, T., Tait, R.N., Westra, K.L., Brett, M.J.: Simulation of density variation and step coverage for via metallization. In: Proceedings of the Sixth International IEEE VLSI Multilevel Interconnection Conference 1989, pp. 292–298 (1989)
3. Ionascu, G., Manea, E., Gavrila, R., Moraru, E.: Technologies for thin layers on ceramics substrate. In: Cioboată, D.D. (ed.) ICoRSE 2021. LNNS, vol. 305, pp. 250–265. Springer, Cham (2022). https://doi.org/10.1007/978-3-030-83368-8_25
4. Ionascu, G., Manea, E., Gavrila, R., Moraru E.: Perspectives on the use of thin films technologies in precision mechanics and mechatronics. Int. J. Mechatron. Appl. Mech. **II**, 204–208 (2020)
5. Stevenson, I.C.: Optimizing Source Location for Control of Thickness Uniformity, Denton Vacuum, LLC - Moorestown, NJ, USA. https://citeseerx.ist.psu.edu/viewdoc/download?doi=10.1.1.593.8225&rep=rep1&type=pdf
6. Wang, F., Crocker, R., Faber, R.: Simulation of precision evaporation coaters - beyond thickness uniformity, (SVC) Society of Vacuum Coaters. In: 53rd Annual Technical Conference Proceedings, 17–22 April 2010, Orlando, FL, USA (2010)
7. Karabacak, T., Lu, T.M., Barthel, J.R.: Enhanced Coverage of Thin Films on Patterned Substrates by Oblique Angle PVD, patent no. 7244670 (2007)
8. Barranco, A., Borras, A., Gonzalez-Elipe, A.R., Palmero, A.: Perspectives on oblique angle deposition of thin films: from fundamentals to devices. Progress Mater. Sci. **76**, 59–153 (2016). https://doi.org/10.1016/J.PMATSCI.2015.06.003
9. Wang, B., et al.: Simulation and optimization of film thickness uniformity in physical vapor deposition. Coatings **8**(9), 325 (2018). https://doi.org/10.3390/coatings8090325
10. Ji, L., Kimb, J.-K., Ji, Q., Leung, K.-N., Chen, Y., Gough, R.A.: Conformal Deep Trench Coating with both Conducting and Insulating Materials (2006). https://www.osti.gov/servlets/purl/890653-IwLxQR/
11. Panjan, P., Drnovšek, A., Gselman, P., Cekada, M., Panjan, M.: Review of growth defects in thin films prepared by PVD techniques. Coatings **10**, 447 (2018). https://doi.org/10.3390/coatings10050447
12. Morosanu, C.E.: Thin Films by Chemical Vapour Deposition, vol. 7. Thin Films Science and Technology. Elsevier (1990)

Innovative Technology to Combat Sars-Cov Using a Finely Dispersed Catching Medium and Microwave Energy

Borys A. Aleksenko[1], Sergey Dobrotvorskiy[1], Yevheniia Basova[1(✉)], Yevgen Sokol[1], Milan Edl[2], and Ludmila Dobrovolska[1]

[1] National Technical University «Kharkiv Polytechnic Institute», Kharkov, Ukraine
Yevheniia.Basova@khpi.edu.ua
[2] University of West Bohemiadisabled, Plzen, Czech Republic

Abstract. Due to the current acute problem of combating the SARS-CoV coronavirus, our team has proposed an innovative technology to combat the virus in closed or ventilated rooms. The developed design of ventilation equipment ensures the inactivation of coronavirus by thermal exposure of sufficient duration. The virus is destroyed outside the human body, so sterilization is preventive. Capturing the virus from the airflow and its retention with subsequent disinfection occurs using a finely dispersed catching medium, using the effect of coagulation of the medium vapors, its coalescence, and intense heating. The use of highly efficient heating technology using microwave energy allows sterilizing the virus with minimal energy consumption. Unlike virus disinfection technologies developed in the world using ultraviolet radiation, the technology we offer involves a long-term deactivating thermal effect on the virus, which ensures a high degree of disinfection.

Keywords: SARS-CoV coronavirus · Disinfection · Ventilation · Microwave energy

1 Introduction

Today, when humanity is faced with the problem of large-scale epidemics, such as the coronavirus epidemic, the direction of the fight against the global virus threat must move to an interdisciplinary level. In addition to virologists, physicians, and epidemiologists, the field of mechanical engineering is called upon to make the most important contribution to solving this urgent problem by developing and improving disinfection technologies and creating modern equipment for the destruction of viral infections. The task of machine builders, in this case, is to create safe, efficient, economical, environmentally friendly, and reliable equipment, which must use the most effective disinfection technology.

Researchers are currently investigating various disinfection technologies to avoid chemical spraying of the disinfectant. This is due to the high cost, the laboriousness of

D. D. Cioboată (Ed.): ICoRSE 2022, LNNS 534, pp. 203–214, 2023.
https://doi.org/10.1007/978-3-031-15944-2_19

processing, and the long duration of the spray disinfection process. Therefore, the attention of scientists are attracted by the technologies of virus inactivation using ultraviolet LED radiation at different wavelengths or frequencies [1], the destruction of viruses by acoustic vibrations [2–5], and the like. Particular attention is paid to the method of destroying coronavirus outside the human body using high temperatures. Contrary to the fact that this issue has not yet been fully studied by scientists of the World Health Organization, scientists [6] propose a method for the destruction of coronavirus using heat treatment. At the same time, as indicated in our previous articles [7–11], the use of microwave energy as a heat source is a modern trend in the development of technological processes of heat treatment. Also, the role of microwaves in the sterilization of various types of materials is attracting increasing attention. Studies [12–14] indicate that the thermal microwave effect is much stronger than a simple thermal effect.

Based on the results of research by virologists [6] and the knowledge about the propagation of microwave radiation energy obtained by us in previous studies [15], our team is developing an innovative technology that involves a long-term deactivating thermal microwave effect on the virus, which provides a high degree of disinfection.

2 Research Problem

The disinfection process serves to destroy the virus in the air mass. To do this, the virus must be effectively isolated, fixed, and subjected to microwave heat exposure for a certain period of time. At the same time, the energy impact must be highly effective and uniform in space and time. The achieved temperature level and exposure time on the virus should guarantee its inactivation. According to research [6], for the virus SARS-CoV-2, this exposure must last at least 15 min at a temperature of at least 92 °C.

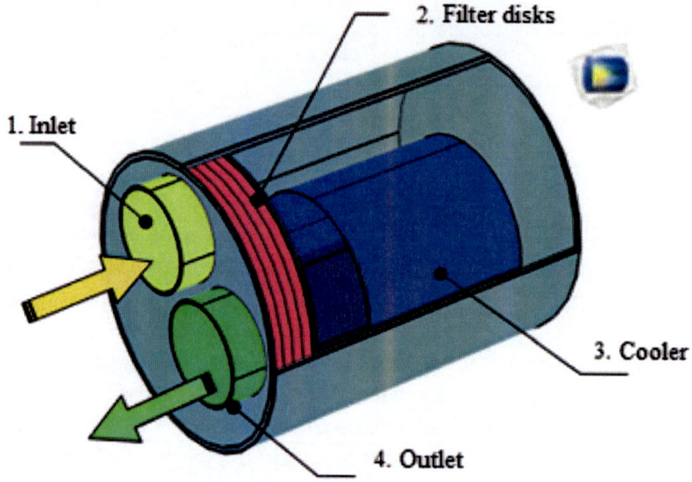

Fig. 1. The equipment for microwave air disinfection. Design and general appearance.

The technology we propose involves several stages.

In the first stage, the air is saturated with moisture vapor. At the same time, fine dispersed water particles, dissolved in the air stream, serve to trap the virus from the air by coalescence with viral particles of the nuclear droplet phase (>30 μm), and coarse droplet phase (<1 μm) and bacterial dust.

In the second stage, the air passes through a set of rotating disc fiber filters. Unlike a sieve in which only those particles larger than the holes are retained, the particle retention efficiency of the fibrous structure is high regardless of the particle size. The airflow passes between the fibers. Particles that touch the surface of the fiber are removed from the flow and are firmly held by the fiber due to intermolecular forces. As the temperature rises and under the influence of microwave radiation, the energy imparted to the particles leads to their detachment and secondary transfer with the flowing air stream. To recapture particles to increase the time of disinfection exposure, a set of several disc fibrous filters is used, which are installed in series.

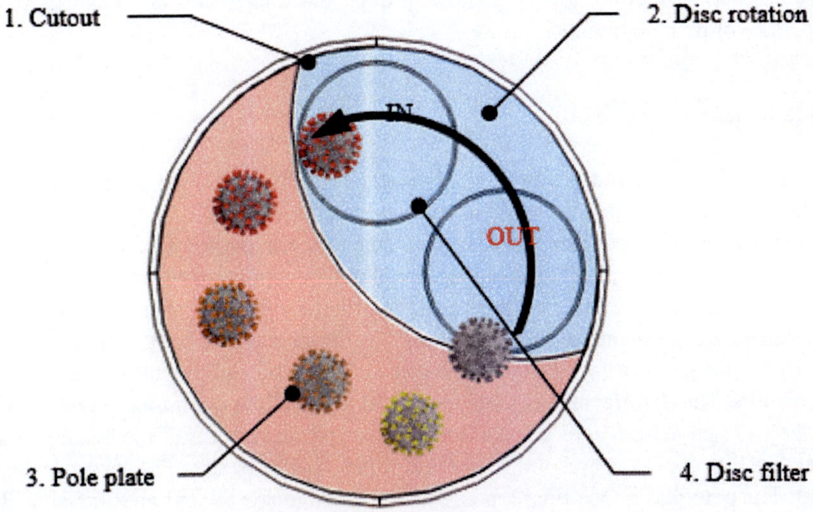

Fig. 2. The disc filter and the pole plate.

In the third stage, the particles held by the fibrous structure of the disc filters are subjected to a uniform and prolonged exposure to microwave heating. This assumes that all areas of the filter volume must be energized. This eliminates the occurrence of residual zones in which the virus will not be disinfected. This requirement makes it impossible to use one or a group of magnetrons for the disinfection of a large volume of filter material. The problem is the limited depth of penetration of wave energy into the volume of the material and the uneven distribution of the electromagnetic field strength level in it. It is proposed to solve this problem by applying the pole plate technology that we have previously studied and published [15].

At the final stage, the disinfected air is cooled in a radiator and passes through a set of disk filters in the opposite direction. This causes a local decrease in temperature in those sectors of the disks through which the outgoing air passes at a given time. It also helps clean the filters' fibrous structure from solid contaminants and moisture. The direction of rotation of the disc filters is chosen in such a way that the areas cooled by the outgoing airflow thereafter meet with the incoming airflow. Reducing the temperature of the fibrous structure increases the efficiency of capturing viral particles from the incoming air stream.

The design of disinfecting equipment that uses the presented technology should provide:

- uniform laminar airflow through the entire volume of the filter disks;
- uniform energy impact on the disk sector which is disinfecting for a sufficient period [6];
- effective cooling of the disinfected disk sector.

The presented work is devoted to the study of the processes of exposure to microwave energy on filter discs, the flow of airflow through the filter discs, and changes in the temperature of their segments.

3 Materials and Methods

The study is carried out using a mathematical multiphysics computer model of disinfecting equipment that uses the technology we offer.

Models have been built to study the movement of the purge airflow through a set of rotating disc fibrous filters, the intensity of high-frequency exposure to disc filters, and to assess the uniformity of temperature distribution in their volume. Modeling is performed using the Komsol Multiphysics ^{tm} software environment. The calculation was carried out using a full-scale mathematical model of equipment with a capacity of 1 m^3/min. The disinfectant equipment model is a hollow cylindrical housing. The cylinder is closed at both ends with bottoms. In one of the bottoms of the housing, there is inlet (see Fig. 1, pos. 1) and outlet (see Fig. 1, pos. 4) air nozzles. The case of material is steel. The pole plates (see Fig. 2, pos. 3) are located inside the cylinder housing. The pole plate is a flat disk with an elliptical cutout. Cutouts in the pole plates (see Fig. 2, pos. 1) are located opposite the inlet and outlet pipes and designed for unobstructed incoming and outgoing airflow. Fiber filters are placed between the pole plates. Filters are disks made of fibrous material. Thus, the volume of the filter material is divided by plates into several cylindrical volumes. The axis of disc filters is the axis of their rotation and coincides with the axis of the cylindrical housing (see Fig. 1, pos. 2).

During the disinfection process, air containing viral particles is passed through a set of rotating disk fiber filters. The scavenging air moves (see Fig. 1) along the axis of rotation of the disk filters. The physical properties of the fibrous filter material are given in the table (Table 1).

Passing through the filters, the incoming air saturated with viral microparticles carries the virus into the layer of the fibrous structure. Due to the axial rotation of the disk filters, the layer of filter material with the virus particles retained by the fibrous structure moves into the gap between the pole plates, where it is exposed to a uniform long-term exposure to microwave radiation and high temperature.

Table 1. Filter material physical properties.

Parameter	Dimension	Value
Density	kg/m^3	500
Relative permeability	1	1
Relative permittivity	1	4,1
Electric conductivity	S/m	0
Heat capacity	Cp	330

The filtered air passes through a cooling radiator (see Fig. 1, pos. 3), where it is cooled to a temperature of +10 °C. After that, the air passes through that sector of the filter discs, which will then be introduced into the incoming air stream during the process of rotation of the disc. The outgoing air moves in the opposite direction to the incoming air. Due to this, the current sector of the filter discs is cooled down before starting to participate in the process of filtering viral microparticles, which increases the efficiency of the filter. Also, the flow of outgoing air contributes to the removal of mechanical impurities from the fibrous structure. The repeated passage of air through the filter further increases the reliability of the disinfection process.

Linear dimensions of the model (see Fig. 3) are given in the table (Table 2).

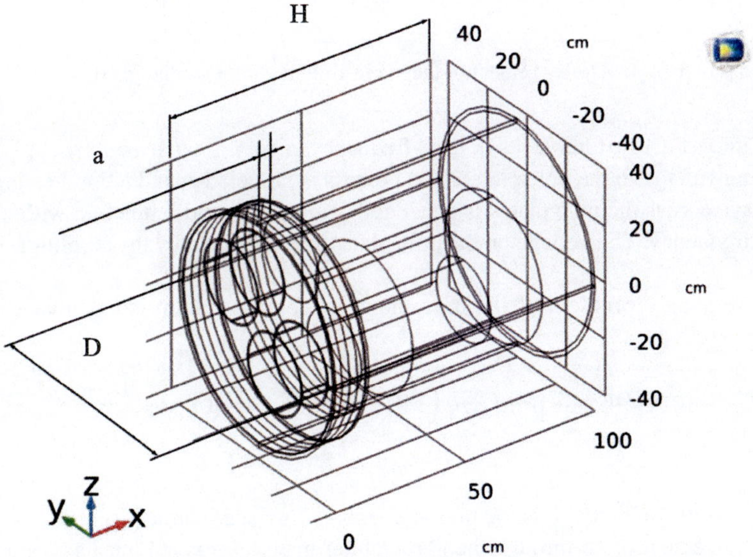

Fig. 3. Linear dimensions of the equipment for air disinfection and the distance between the pole plates.

Table 2. Linear dimensions of the equipment for microwave air disinfection.

Index	Size	Dimension, mm
H	Housing lenght	1000
D	Housing diameter	800
a	Distance between the pole plates	6

The consumption of purge air throughout the route through the device (see Fig. 4) does not exceed 1.2 m/s. Therefore, the flow rate is calculated as laminar according to the Eq. (1–4).

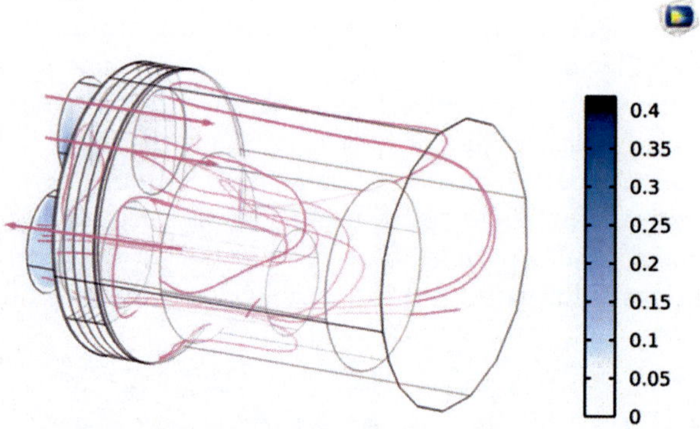

Fig. 4. The movement of the purge air in the filter media, [m/s].

The impact on the filter disks of high-frequency energy is carried out by applying an alternating voltage to the pole plates. The process is modeled as induction heating. The AC voltage across the pole plates is generated using a sinusoidal function with a given angular frequency. The voltage on the pole plates is determined by the amplitude of this function.

The stationary problem of the movement of the air medium is calculated by the equation [16, 17]:

$$\rho(u \cdot \nabla)u = \nabla\left[-pl + \mu\left(\nabla u + (\nabla u)^T\right) - \frac{2}{3}\mu(\nabla \cdot u)l\right] + F \tag{1}$$

$$\nabla \cdot (\rho u) = 0 \tag{2}$$

where: u - is the velocity, p- is the pressure, and μ - is the dynamic viscosity.

The movement of air through the fibrous filter material was modeled as a motion in a porous medium and was calculated by the equation:

$$\frac{1}{\epsilon_p}\rho(\mathrm{u} \cdot \nabla)\mathrm{u}\frac{1}{\epsilon_p} == \nabla\left[-\mathrm{pl} + \mu\frac{1}{\epsilon_p}(\nabla\mathrm{u} + (\nabla\mathrm{u})^{\mathrm{T}}) - \frac{2}{3}\mu\frac{1}{\epsilon_p}(\nabla \cdot \mathrm{u})\mathrm{l}\right] - \\ -\left(\mu k^{-1} + \beta_F|\mathrm{u}|\frac{\theta_m}{\epsilon_p^2}\right)\mathrm{u} + F \tag{3}$$

$$\nabla \cdot (\rho u) = Q_m \tag{4}$$

where: θ_m- the value of porosity of the material.

The dynamic heat transfer problem was calculated by the equation:

$$\rho C_p \mathrm{u} \cdot \nabla \mathrm{T} + \nabla \cdot q = Q + Q_{vd} \tag{5}$$

$$q = -k_{eff} \nabla \mathrm{T} \tag{6}$$

where: Cp - heat capacity value, T- temperature, Q - quantity of heat.

Heat transfer in a porous medium was calculated as:

$$k_{eff} = \theta_p k_p + (1 - \theta_p) k + k_{disp} \tag{7}$$

where: θ_ρ- is the volume fraction of the material, k_ρ- is the thermal conductivity.

The influence of purge air flow on thermal processes, taking into account the rotation of filter discs domains, is calculated as dynamic heat transfer model through the following equations:

$$\rho C_p \frac{\partial T}{\partial t} + \rho C_p v_{trans} \cdot \nabla T + \nabla \cdot q = Q + Q_{ted} \tag{8}$$

where ρ is the density, C_p– heat capacity, q– heat flux, T– temperature,Q_e – heat sources and sinks.

The initial temperature of the all domains is 20 °C. The purge air also has an initial temperature of 20 °C.

The electric field generated in the fibrous filter material volume is calculated with equations:

$$\nabla \times ({\mu_\gamma}^{-1} \nabla \times E) - {k_0}^2 \left(\varepsilon_\gamma - \frac{j\sigma}{\omega \varepsilon_0} \right) E = 0 \tag{9}$$

$$\nabla \cdot J = Q_{\mathrm{j,v}} \tag{10}$$

$$J = \sigma E + j\omega D + J_e \tag{11}$$

$$\omega = 2\pi f \tag{12}$$

where μ_γ is the complex relative permeability; E – complex amplitude representing an oscillating electric field, ε_γ – complex relative permittivity, ω – circular (angular) frequency, k_0 – phase constant of free space, σ – electrical conductivity, j – is an imaginary unit, J – externally generated current density, D – electric displacement, $Q_{\mathrm{j,v}}$ –charge, f – frequency.

The heating of the filter material due to the effect of alternating voltage (see Fig. 5) is calculated as:

$$\rho C_p u \cdot \nabla T = \nabla \cdot (k \nabla T) + Q_e \tag{13}$$

$$Q_e = J \cdot E \tag{14}$$

The force, which is affects on the water molecule is calculated as:

$$\overline{F} = \overline{\rho} \cdot \overline{E} \tag{15}$$

$$F = \Sigma_i \frac{\partial \overline{E}}{\partial x_i} \rho_i \tag{16}$$

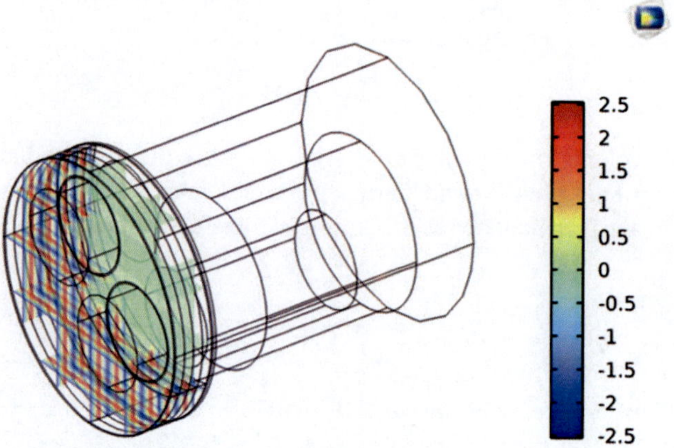

Fig. 5. Variable electrical potential in the space between the pole plates, [V].

$$E = \frac{U}{d} \tag{17}$$

$$E = -\nabla V \tag{18}$$

where $\overline{F}$ is the electric displacement, $\overline{\rho}$ – dipole moment, U – potential difference, d – distance between the pole plates, V – electric potential.

The force, which is affects on the water molecule is calculated as:

$$E = -\nabla V \tag{19}$$

where $\overline{F}$ is the electric displacement, $\overline{\rho}$ – dipole moment, U – potential difference, d – distance between the pole plates, V– electric potential.

The air capacity of the equipment was set through the air flow rate, based on the equation:

$$Q_a = 60\pi \cdot v\left(\frac{d}{2}\right) \tag{20}$$

where Q_a is the air consumption, v – air speed, d – inlet pipeline diameter.

The calculations were carried out in the time interval from 0 to 600 min, with step 10 min.

4 Results and Discussion

Calculation of the computer model shows that the proposed design provides effective uniform heating of the filter material to guarantee the disinfection of viruses for the whole required time [6], during these segments are in the area between the pole plates (see Fig. 6, 7).

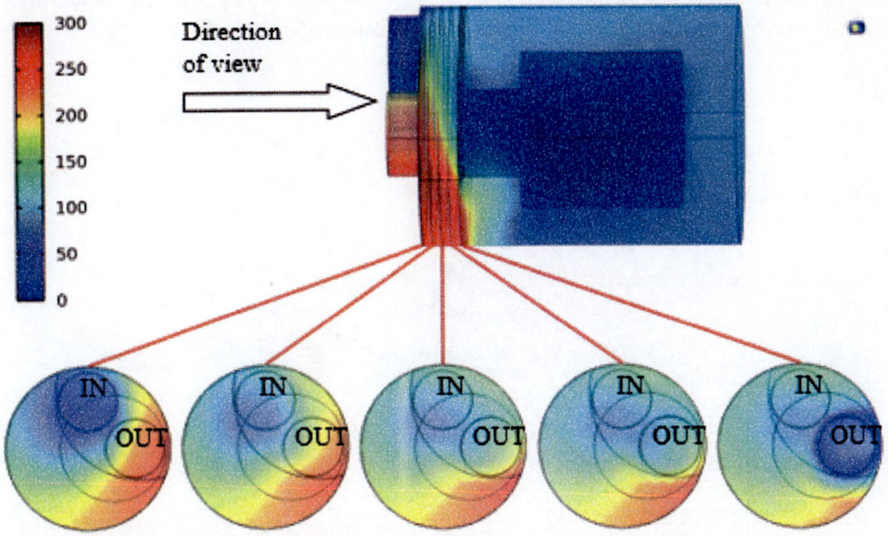

Fig. 6. Temperature distribution inside the disc filter material under the influence of microwave energy and airflow, [°C].

On the basis of comparative experiments, that we carried out on a mathematical model, we obtained the optimal values for the parameters of the studied equipment (Table 3).

Table 3. Optimal values for the parameters of the equipment.

Parameter	Dimension	Value
Angular speed of rotation	rpm	0.03
Air inflow speed	m/s	0.2
Electric potential	V	3
Cooler temperature	deg C	10

The calculation of the computer model shows the complex influence of such factors as the angular speed of rotation of the filter disks, the speed of the airflow, and the voltage on the pole plates on the ongoing disinfection process.

So, at an increased speed of rotation of the filter discs, the outgoing air does not have time to cool the filter surface (see Fig. 8, Pos. A). As a result, the viral particle-retaining medium does not effectively linger on the fibrous structure heated above 100 °C. This leads to the fact that viral particles quickly pass through the disinfectant device and are not exposed to microwaves. In case of excess voltage on the pole plates, the filters overheat over the required value (see Fig. 8, Pos. B). With reduced voltage, the required disinfection temperature [6] is not reached.

The computer model we built showed the technical feasibility of creating equipment for microwave air disinfection and the practical feasibility of its use.

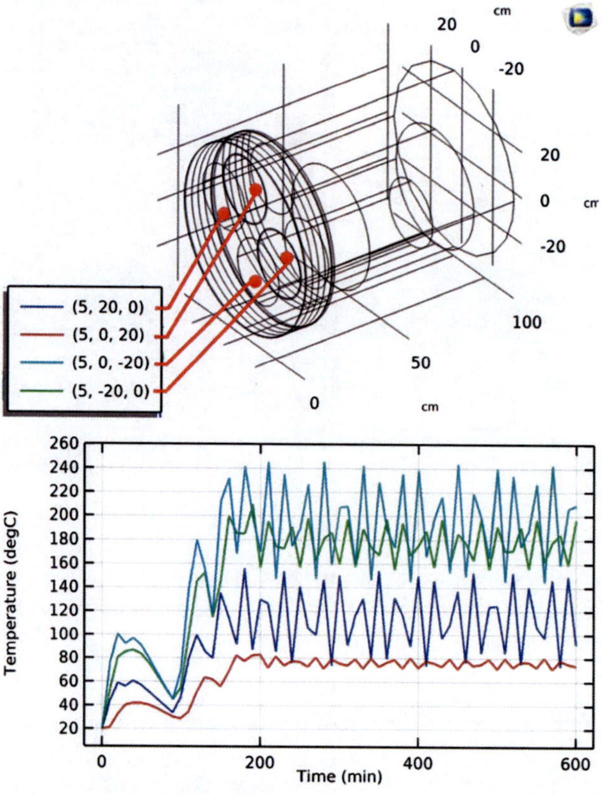

Fig. 7. Temperature distribution inside the disc filter material at four reference points with specified coordinates (X, Y, Z) inside the filter media, [°C].

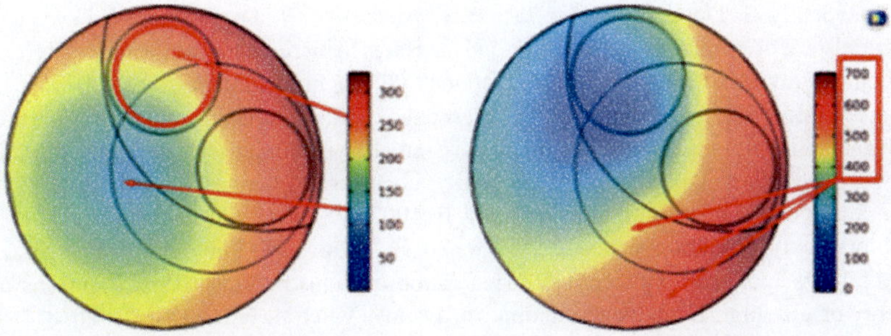

Fig. 8. Temperature distribution inside the disc filter material under the influence of microwave energy and airflow, [°C].

5 Conclusions

The paper presents an innovative technology for air disinfection using a finely dispersed trapping medium and highly efficient microwave heating technology. The possibility of scaling the proposed technology, its efficiency, and its economy is shown.

It is described that the effect of microwaves on the sectors of filtering disks with the help of pole plates makes it possible to evenly distribute energy in the volume of these sectors.

The influence on the course of the disinfection process of such factors as the frequency and voltage on the pole plates, as well as the angular speed of rotation of the filter discs, is considered.

The study shows the applicability of the proposed technology in the process of creating disinfectant equipment both as part of the ventilation systems of buildings and operating autonomously.

References

1. Gerchman, Y., Mamane, H., Friedman, N., Mandelboim, M.: UV-LED disinfection of coronavirus: wavelength effect. J. Photochem. Photobiol. B Biol. **212**, 112044 (2020). ISSN 1011-1344, https://doi.org/10.1016/j.jphotobiol.2020.112044
2. Wierzbicki, T., Li, W., Liu, Y., Zhu, J.: Effect of receptors on the resonant and transient harmonic vibrations of Coronavirus. J. Mech. Phys. Solids. **150**, 104369 (2021). https://doi.org/10.1016/j.jmps.2021.104369. Epub 2021 Feb 18. PMID: 33623172; PMCID: PMC7890278
3. Balandin, A., Fonoberov, V.: Vibrational modes of nano-template viruses. J. Biomed. Nanotechnol. **1**, 90–95 (2005). https://doi.org/10.1166/jbn.2005.005
4. Yang, S.C., et al.: Efficient structure resonance energy transfer from microwaves to confined acoustic vibrations in viruses. Sci. Rep. **9**(5), 18030 (2015). https://doi.org/10.1038/srep18030.PMID:26647655;PMCID:PMC4673452
5. Liu, T.-M., et al.: Microwave resonant absorption of viruses through dipolar coupling with confined acoustic vibrations. Appl. Phys. Lett. **94**, 043902 (2009). https://doi.org/10.1063/1.3074371
6. Jiang, Y., et al.: Sub-second heat inactivation of coronavirus using a betacoronavirusmodel. Biotechnol. Bioeng. **118**, 2067–2075 (2021). https://doi.org/10.1002/bit.27720

7. Dobrotvorskiy, S., Dobrovolska, L., Basova, Y., Aleksenko, B.: Diagnostics of uneven heating of the adsorbent by microwave radiation and measures to increase the uniformity of its heating. Periodica Polytech. Chem. Eng. **63**(4), 640–649 (2019). https://doi.org/10.3311/PPch.13234
8. Dobrotvorskiy, S.S., Dobrovolska, L.G., Aleksenko, B.A.: Computer simulation of the process of regenerating the adsorbent using microwave radiation in compressed air dryers. In: Hamrol, A., Ciszak, O., Legutko, S., Jurczyk, M. (eds.) Advances in Manufacturing. LNME, pp. 511–519. Springer, Cham (2018). https://doi.org/10.1007/978-3-319-68619-6_49
9. Dobrotvorskiy, S., Dobrovolska, L., Aleksenko, B., Basova, Y.: The use of waveguides with internal dissectors in the process of regeneration of industrial adsorbents by means of the energy of ultrahigh-frequency radiation. In: Ivanov, V., et al. (eds.) DSMIE 2018. LNME, pp. 433–442. Springer, Cham (2019). https://doi.org/10.1007/978-3-319-93587-4_45
10. Dobrotvorskiy, S., Aleksenko, B., Dobrovolska, L., Basova, Y.: Effect of the application of microwave energy on the regeneration of the adsorbent. Acta Polytechnica **58**(4), 217–225 (2018). https://doi.org/10.14311/AP.2018.58.0217
11. Dobrotvorskiy, S., Dobrovolska, L., Basova, Y., Aleksenko, B.: Particulars of adsorbent regeneration with the use of microwave energy. Acta Polytechnica **59**(1), 12–23 (2019). https://doi.org/10.14311/AP.2019.59.0012
12. Kakita, Y., Kashige, N., Murata, K., Kuroiwa, A., Funatsu, M., Watanabe, K.: Inactivation of Lactobacillus bacteriophage PL-1 by microwave irradiation. Microbiol. Immunol. **39**(8), 571–576 (1995). https://doi.org/10.1111/j.1348-0421.1995.tb02244.x. PMID: 7494495
13. Barbora, A., Minnes, R.: Targeted antiviral treatment using non-ionizing radiation therapy for SARS-CoV-2 and viral pandemics preparedness: technique, methods and practical notes for clinical application. PLoS ONE **16**(5), e0251780 (2021). https://doi.org/10.1371/journal.pone.0251780
14. Wang, A., Cheng, N., Liou, Y., Lin, K.: Inactivation of bacteriophage by microwave irradiation. J. Exp. Microbiol. Immunol. **1**, 9–18 (2001). Corpus ID: 53380108
15. Dobrotvorskiy, S., Borys, A., Yepifanov, V., Basova, Y., Dobrovolska, L., Popov, V.: The absorbents nanoporous structures regeneration for industrial dryers by microwave energy. In: Cioboată, D.D. (ed.) ICoRSE 2021. LNNS, vol. 305, pp. 8–22. Springer, Cham (2022). https://doi.org/10.1007/978-3-030-83368-8_2
16. RF Module User's Guide. COMSOL© (2018). https://doc.comsol.com/5.4/doc/com.comsol.help.rf/RFModuleUsersGuide.pdf. Accessed 29 Feb 2018
17. COMSOL User's Guide. COMSOL© (2018). https://extras.csc.fi/math/comsol/3.4/doc/multiphysics/wwhelp/wwhimpl/js/html/wwhelp.htm?context=multiphysics&file=html_guideintro.5.1.htm. Accessed 22 Aug 2018

Comparative Analysis of Anchorage Length for Rebars in Reinforced Concrete Structures

Julian Kasharaj(✉), Igli Kondi, and Irakli Premti

Faculty of Civil Engineering, Polytechnic University of Tirana, Tirana, Albania
jkasharaj69@yahoo.ca

Abstract. In this work, the authors present the findings obtained from the analysis of calculating the rebars anchorage length in reinforced concrete structures. The calculation of anchorage length is performed according to Albanian Normative (AN) and Eurocode 2 (EC 2). Furthermore, is made a comparison between the analysis results. In the beginning a presentation is made with the theoretical solution of the problem and after that the comparison is based on numerical solution. The conclusions are followed from the recommendations given in the end of this work.

Keywords: Albanian normative · Eurocodes · Anchorage length for rebars in reinforced concrete structures

1 Introduction

The design of anchorage length for rebars in reinforced concrete elements is a very delicate and important problem. The anchorage length must ensure the transmission of stresses from concrete to rebars and vice versa, avoiding in any case the detachment (separation) of rebars from concrete. When the anchorage length is calculated correctly, experiments have shown [1] that first the rebars breaks (destroys) rather than detaches (separates) from concrete. The cohesion force between concrete and rebars is larger than the rebars own tensile force capacity (Fig. 1).

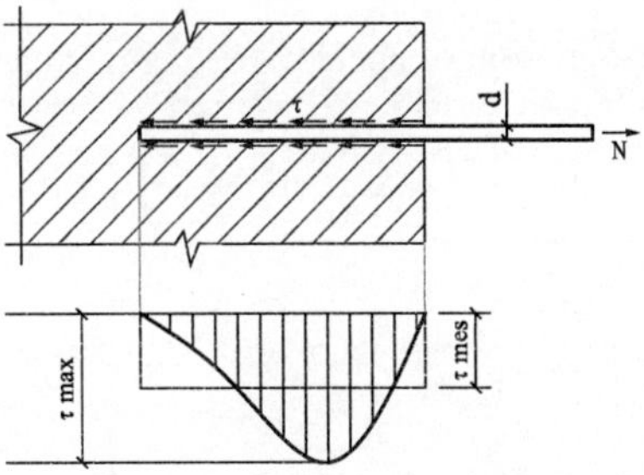

Fig. 1. The diagram of rebars stresses [1]

D. D. Cioboată (Ed.): ICoRSE 2022, LNNS 534, pp. 215–221, 2023.
https://doi.org/10.1007/978-3-031-15944-2_20

Cohesive connection between concrete and rebars occurs during concrete hardening process, which reduces the volume and tightens around the rebar. Cohesive connection between concrete and rebars depends on:

- Quantity of cement in concrete. If quantity of cement increases, the cohesive connection between concrete and rebars also increases.
- Age of concrete. Over time the cohesive connection between concrete and rebars improves.
- Quality realization of the element. The higher the quality, the better the vibration of concrete, better the connection between concrete and rebars.
- Concrete class. With the increase of the concrete class, the connection between concrete and rebars improves.
- Lateral surface of rebar. Deformed bar related better with concrete.
- The connection between concrete and rebars in circular cross section is better than in square cross section rebars.
- In tension rebars the connection between concrete and rebars is weaker than in compression rebars. This is explained with out of plane strain (expansion) of rebars in compression [1].
- At low temperatures the connection between concrete and rebars is reduced, due to contraction of rebar.

1.1 Symbols

The symbols between {…} belong to the Eurocodes. Only the symbols that differ are shown.

- l_{an} – anchorage length; $\{l_{bd}\}$
- τ_{mes} – design value of ultimate bond stress; $\{f_{bd}\}$
- R_{sn} – characteristic yield strength of reinforcement; $\{f_{yk}\}$
- R_s – design yield strength of reinforcement; $\{f_{yd}\}$
- d – diameter of rebar; $\{\varphi\}$
- R_b – design value of concrete compressive strength; $\{f_{cd}\}$

1.2 Calculation of Anchorage Length According to Albanian Normative [8]

According to AN [8] for calculation of anchorage length is used the expression:

$$l_{an} = \left(\omega_{an}\frac{R_s}{R_b} + \Delta\lambda_{an}\right)d \tag{1}$$

It is required:

$$l_{an} = \lambda_{an} \cdot d \tag{2}$$

$$l_{an} \geq l^*_{an} \tag{3}$$

The value of ω_{an}, $\Delta\lambda_{an}$, λ_{an} and l^*_{an} – are taken in the table below [8].

Table 1. The value of ω_{an}, $\Delta\lambda_{an}$, λ_{an} and l_{an}* [8]

NR	Gjendja e nderur e armaturës dhe kushtet e mberthimit	ϖ_{an}	$\Delta\lambda_{an}$	λ_{an}	l^{*}_{an} (mm) [minimale]
1	Mbërthimi (inkastrimi) i armaturës së tërhequr në betonin e tërhequr	0.7	11.0	20.0	250mm
2	Mbërthimi i armaturës së shtypur apo të tërhequr në betonin e shtypur	0.5	8.0	12.0	200mm

The first row concerns to anchorage tensile rebars. The second row to anchorage compressive rebars.

1.3 Calculation of Anchorage Length According to EC 2 [10]

Ways of rebar anchorage are shown in the picture 8.1 of EC 2 [10]. The design values of ultimate bond stress are defined by the expression:

$$f_{bd} = 2.25 \cdot \eta_1 \cdot \eta_2 \cdot f_{ctd} \tag{4}$$

fctd – design tensile strength of concrete.

η1 – coefficient related to bond conditions, as well as the position of rebar during the casting.

Based on the picture 8.2 to EC 2 [10], η1 = 1 (good bond condition), η1 = 0.7 for all other cases.

η2 – coefficient related to rebar diameter.

$$\eta 2 = 1.0 \text{ for } \varphi \leq 32 \text{ mm}, \ \eta 2 = (132 - \varphi)/100 \text{ for } \varphi > 32 \text{ mm} \tag{5}$$

If stresses in rebar are σ_{sd}, then the basic anchorage length $l_{b,rqd}$ is:

$$l_{b,rdq} = (\varphi/4) \cdot (\sigma_{sd}/f_{bd}) \tag{6}$$

Usually is accepted $\sigma_{sd} = f_{yd}$.

The design anchorage length, l_{bd}, is defined by the expression:

$$l_{bd} = \alpha_1 \cdot \alpha_2 \cdot \alpha_3 \cdot \alpha_4 \cdot \alpha_5 \cdot l_{b,rdq} \geq l_{b,min} \tag{7}$$

Coefficients $\alpha_1, \alpha_2, \alpha_3, \alpha_4, \alpha_5$ are taken from Table 8.2 of EC 2 [10].

Minimum anchorage length:

$$l_{b,min} = \max\left(0.3 \cdot l_{b,rqd}; 10\varphi; 100 \text{ mm}\right) - \text{for tensile rebars} \tag{8}$$

$$L_{b,min} = \max\left(0.6 \cdot l_{b,rqd}; \ 10\varphi; \ 100 \text{ mm}\right) - \text{for Compressive Rebars} \tag{9}$$

Some faster approximations can also be performed. Based on the picture 8.1 of EC 2 [10], the role of anchorage length can be played by equivalent length, $l_{b,eq}$.

For cases b) and d):

$$l_{b,eq} = \alpha_1 \cdot l_{b,rqd} \tag{10}$$

For the case e):

$$l_{b,eq} = \alpha_4 \cdot l_{b,rqd} \tag{11}$$

1.4 Applied Example 1. Calculation of Anchorage Length According to Albanian Normative [8]

Concrete class B30 (cubic resistance), $R_b = 160$ daN/cm^2. $R_s = 4348$ daN/cm^2.

For rebars in tension based on expression (1) and Table 1 determine $l_{an} = 30d$. With expressions (2) and (3) determine $l_{an,min1} = 20d$ and $l_{an,min2} = 250$ mm. If $d = 16$ mm, then $l_{an} = 480$ mm, $l_{an,min1} = 320$ mm. Eventually $l_{an,min} = 320$ mm and $l_{an} = 480$ mm.

For rebars in compression based on expression (1) and Table 1 determine $l_{an} = 21.6d$. With expression (3) determine $l_{an,min1} = 12d$ and $l_{an,min2} = 200$ mm. If $d = 16$ mm, then $l_{an} = 346$ mm, $l_{an,min1} = 192$ mm. Eventually $l_{an,min} = 200$ mm and $l_{an} = 346$ mm.

1.5 Applied Example 1. Calculation of Anchorage Length According to EC 2 [10]

Concrete class C 25/30, $f_{cd} = 141.6$ daN/cm^2, $f_{ctd} = 12$ daN/cm^2. $f_{yd} = 4348$ daN/cm^2. Good bond condition, $\eta_1 = 1$. $\varphi = 16$ mm ≤ 32 mm, $\eta_2 = 1$. $f_{bd} = 27$ daN/cm^2.

For rebars in tension: Not straight rebar. Cover to reinforcement $c = 3.5$ cm $< 3\varphi = 4.8$ cm, $\alpha_1 = \alpha_2 = 1$. $\alpha_3 = \alpha_4 = \alpha_5 = 1$. $l_{b,rqd} = 40.25\varphi = 644$ mm $= l_{bd}$. While from expression (8) $l_{b,min} = \max(12.07\varphi; 10\varphi; 100$ mm$)$ or $l_{b,min} = \max(194$ mm; 160 mm; 100 mm$)$. Eventually $l_{b,min} = 194$ mm and $l_{bd} = 644$ mm.

For rebars in compression: Not straight rebar, $\alpha_1 = \alpha_2 = \alpha_3 = \alpha_4 = \alpha_5 = 1$. $l_{b,rqd} = 40.25\varphi = 644$ mm $= l_{bd}$. Meanwhile by the expression (9) $l_{b,min} = \max(24.14\varphi; 10\varphi; 100$ mm$)$ or $l_{b,min} = \max(388$ mm; 160 mm; 100 mm$)$. Eventually $l_{b,min} = 388$ mm and $l_{bd} = 644$ mm.

2 Comparative Analysis

If we analyze the expression (1), as well as the expressions (4) to (7), results that even according to Albanian Normative and according to Eurocodes, in generalized form, the anchorage length can be expressed:

$$\text{Anchorage length} = k \cdot \text{rebar diameter} \tag{12}$$

According to AN:

$$k = \left(\omega_{an}\frac{R_s}{R_b} + \Delta\lambda_{an}\right) \tag{13}$$

$$k = \alpha_1 \cdot \alpha_2 \cdot \alpha_3 \cdot \alpha_4 \cdot \alpha_5 \cdot \frac{f_{yd}}{9 \cdot \eta_1 \cdot \eta_2 \cdot f_{ctd}} \tag{14}$$

According to EC 2:

According to AN "k" depends on tensile strength, design strength, rebar diameter and by concrete class . According to EC 2 "k" depends on bond condition between concrete and rebar, the position of rebar during the casting, from rebar diameter, from rebar shape (straight or not straight), by tightening the concrete with transverse reinforcement, from concrete class, tensile rebar strength, from cover to reinforcement.

The following graphs show "k" from the tensile strength of rebar relation and from the concrete class, where "k" is calculated according to AN and EC 2, for rebars in tension and in compression. To make the comparison of "k" according to AN and EC 2, is accepted not a straight rebar, as even AN takes this type of rebar. It is also accepted a good bond condition between concrete and rebar, as AN does not take this factor into account. The rebar diameter is accepted $\varphi \leq 32$ mm, as the most usable diameter and $c_d > 3\varphi$. As a consequence of all the above admissions it results (Figs. 2, 3 and 4):

$$\eta_1 = \eta_2 = \alpha_1 = \alpha_2 = \alpha_3 = \alpha_4 = \alpha_5 = 1 \tag{15}$$

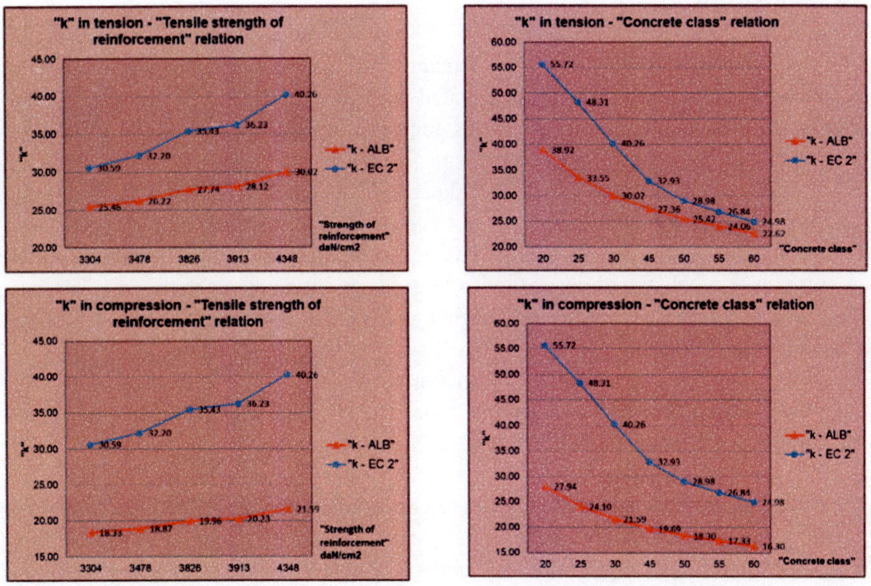

Fig. 2. Comparison of "k" for rebars in tension and in compression, according to AN and EC 2

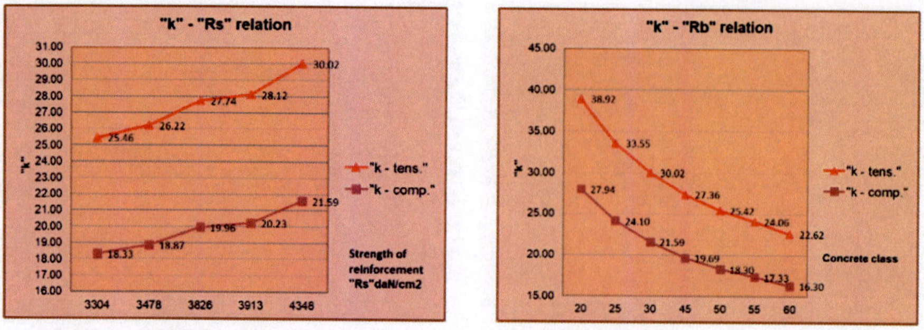

Fig. 3. Comparison of "k" for rebars in tension and in compression according to AN

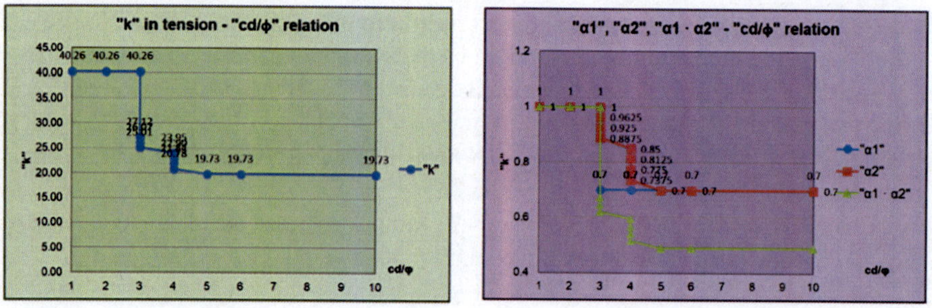

Fig. 4. "k", "α1", "α2", "α1· α2" from "cd/φ" relation

3 Conclusions

- The anchorage length for rebars in tension and rebars in compression calculated according to AN is less than that calculated according to EC 2.
- Increasing the tensile strength of rebars, increases the anchorage length of rebars in tension calculated according to both norms. Increasing strength 1.31 times, increases 1.17 times the anchorage length calculated according to AN and 1.31 times that calculated according to EC 2.
- Increasing the tensile strength of rebars, increases the anchorage length of rebars in compression calculated according to both norms. Increasing strength 1.31 times, increases 1.17 times the anchorage length calculated according to AN and 1.31 times that calculated according to EC 2.
- Increasing concrete class, reduces the anchorage length of rebars in tension calculated according to both norms. Increasing 3 times the concrete class, reduces 1.72 times anchorage length calculated according to AN and 2.23 times that calculated according to EC 2.
- Increasing concrete class, reduces the anchorage length of rebars in compression calculated according to both norms. Increasing 3 times the concrete class, reduces 1.72 times anchorage length calculated according to AN and 2.23 times that calculated according to EC 2. See the graphs in Picture 1 above.
- According to AN the anchorage length of rebars in tension is larger than that of rebars in compression. See picture 2. According to EC 2 the anchorage length of rebars in

tension is equal to or less than that of rebars in compression. It all depends on "α" coefficients explained in Table 8.2 of EN 1992–1-1. ($\alpha_1 \cdot \alpha_2 \cdot \alpha_3 \cdot \alpha_4 \cdot \alpha_5$) for rebars in tension $\leq$ ($\alpha_1 \cdot \alpha_2 \cdot \alpha_3 \cdot \alpha_4 \cdot \alpha_5$) for rebars in compression.

- The graphs in picture 3 express "k" (for rebars in tension), "α_1", "α_2", "$\alpha_1 \cdot \alpha_2$" - "c_d/φ" relation, where c_d is the cover to reinforcement.

References

1. Negovani, K., Verdho, N.: Teoria e ndërtimeve prej betoni të armuar, Vëllimi I-rë, Tiranë (1973)
2. Negovani, K., Verdho, N.: Konstruksione prej betoni të armuar, Vëllimi II-të, Tiranë (1975)
3. Ministria E Ndërtimit e Republikës së Shqipërisë - Kushte teknike të projektimit, Librat 1 deri 23, Tiranë (1978)
4. MINISTRIA E NDËRTIMIT E REPUBLIKËS SË SHQIPËRISË - Kusht teknik projektimi për ndërtimet antisizmike KTP - N.2 - 89, Tiranë (1989)
5. Verdho, N., Mukli, G.: Shembulla numerikë në konstruksionet prej betoni të armuar, Vëllimi I-rë, Tiranë (1980)
6. Verdho N., Mukli G. - Shembulla numerikë në konstruksionet prej betoni të armuar, Vëllimi II-të, Tiranë, 1981
7. Verdho, N., Mukli, G.: Konstruksione prej betoni të armuar, Pjesa I-rë, Tiranë (1996)
8. I.S.T.N.: Kusht teknik i projektimit të konstruksioneve b.a. me metodën e gjendjeve kufitare, K.T.P. – N.30 (1991)
9. I.S.T.N.: Kusht teknik mbi simbolet në b.a., K.T.P. – N.28 (1987)
10. Eurocode 2 – Design of concrete structures – Part 1-1: General rules and rules for buildings (EN 1992-1-1)

Towards the Improvement of Yard Management Systems (YMS) Using Radio Frequency Identification (RFID)

Volodymyr Alieksieiev[1](✉), Valentyn Kovalenko[2], Vsevolod Stryzhak[2], Ivan Varchenko[2], Mariana Stryzhak[2], Bernhard Heiden[3], and Bianca Tonino-Heiden[4]

[1] Leibniz University Hannover, Welfengarten 1, 30167 Hannover, Germany
v.alieksieiev@stud.uni-hannover.de
[2] National Technical University "Kharkiv Polytechnic Institute", Kyrpychova 2, Kharkiv 61002, Ukraine
[3] Carinthia University of Applied Sciences, Europastrasse 4, 9524 Villach, Austria
[4] University of Graz, Heinrichstrasse 26/V, 30167 Graz, Austria

Abstract. In the last years strict quarantine regulations and the associated supply chain disturbances have changed the usual business processes, especially increasing logistics costs. In a pandemic environment, the rapid trend to online sales, and with it, the increase of turnover, have forced companies, especially logistics service providers, to look for solutions, which allow fast and accurate recognition of goods and establish the continuous control of material flow in all steps of the supply chain. One of the major problems in this field represents the lack of integration of yard, warehouse and transport management systems (YMS, WMS and TMS correspondingly), that results in schedule disturbances and uncontrolled material flow inside of a yard of logistics service providers. The goal of our work is to improve the YMS, obtaining new data about the status of the material flow object inside the yard of logistics complexes, using a universal RFID-based logistics complex model, which allows better integration of YMS with the WMS and TMS. Firstly, we provide a theoretical background of YMS and set up of the universal RFID-based logistics complex model. Then the model is mirrored in the software package AnyLogic, where three possible scenarios of transport vehicle identification are shown. Finally, the results and outlook of possible future research are presented.

Keywords: Logistics · Yard management · YMS · RFID

1 Introduction

The development of digital means of goods identification and information processing opens up new possibilities for the management and optimization of material flows and increases the relevance of this task. A large number of publications are devoted to the optimization of the movement of individual material handling machines within enterprises (e.g. [1–3]). However, the results of these studies do not provide a holistic picture of

D. D. Cioboată (Ed.): ICoRSE 2022, LNNS 534, pp. 222–232, 2023.
https://doi.org/10.1007/978-3-031-15944-2_21

the technical improving of complex material flow systems, such as factories, or logistics distribution centers. To make it possible to manage the complex material flows, different (software-based) management systems are implemented in logistics companies (cf. [4]). The main IT-based systems for controlling of material flows are the Warehouse Management System (WMS), the Transport Management System (TMS) [5] and the Yard Management System (YMS).

The WMS represents a system to manage and control of stock keeping processes. The key functions of such systems are controlling of all processes in the warehouse from the goods receiving to the goods shipping. In addition to this, depending on the specific of certain warehouses, additional functions can be included in the WMS (e.g. for dealing with special kinds of goods). [5] Many research works are devoted to the implementing and improving of the WMS (see e.g. [4, 6–8]).

As the demand on costs and transport optimization has been becoming more critical for enterprises in the last years, the role of the TMS as an effective instrument for planning and control of transport processes has become crucial for achieving of high logistics performance. The important functional areas of the TMS comprise among others transport order management, routes optimization, transport service management, etc. [5]. The exploration of different applications and perspectives of the TMS are performed, e.g. in [9] and [10].

Since the complexity of yard flows in the logistics service providers companies has been increasing in the last years, driven by the enlarging the volume of delivery orders, e.g. in the retail area (e.g. in the time of pandemic), companies are looking for solutions, that enable them to optimize the management and control of yard flows. This can result in the increase of turnover and, with it, of profits. Such a control of complex yard systems, that is the case, e.g. in different logistics service providers companies, can be realized with the YMS. This management system can be seen as a bridging between TMS and WMS, and is hence, highly important for achieving of high logistics performance and increasing of the revenue. For this reason, the improving of the YMS is in the focus of this paper.

Since the increasing of revenue in logistics service providers companies as a rule depends on the corresponding turnover, the technical solutions for its increasing are also addressed in this work. According to many research works, that will be described in the next section, it can be achieved, e.g. by using Radio Frequency Identification (RFID)-technology, which is based on the radio wave transmission between the transponder, where the information about the good is written, and the reader. The RFID has following advantages in comparison to another popular identification technology, the barcode technology. Among others, the advantages of the RFID are simultaneous scanning of several items, full automation and bigger distance from transponder to reader are possible as well as rewriteability, etc. [11]. In addition to this, the RFID enables fully contactless identification of goods, which can be an additional advantage, e.g. in time of pandemic.

Based on the aforementioned arguments, this work primarily focuses on the improving of the YMS, obtaining new data about material flows using RFID-technology. The main idea of this paper is to bridge the gap between technical and managerial challenges while implementing RFID for yard management processes, using the universal model of the RFID-based YMS. To reach it, the following structure of the paper was developed: first, the state of the art of the YMS solutions and RFID-applications as well as RFID-based YMS solutions are analyzed in Sect. 2 to identify the research lack, based on available works. Further in Sect. 3 the scalable model of the universal RFID-based YMS of warehouse complex is proposed to present an application independent solution. This system is then mirrored to the virtual model for the purpose of evaluating its economic potential and bottleneck revealing, using the software package *AnyLogic*, that is described in Sect. 4. Finally, the summary and outlook of further research is given in Sect. 5.

2 Literature Review: RFID and YMS

In [12] it was mentioned that the on-time collection of logistics information and detection of errors in the supply of materials play a crucial role in modern production systems. An approach for obtaining multi-parameter logistics information in production logistics and error detection from a large amount of RFID data is proposed. The proposed solution takes into account multi-attributes, random processes and different units of logistic states. In [13] the authors use the method of graphs to evaluate the logistics processes in production. In [14], a solution to the problem of storing containers in the yard of a warehouse that already contains other containers based on fuzzy logic was proposed. The system is designed to reduce the number of reloading operations with unknown container storage time. The algorithm also reduces container retrieval time, but it requires more data.

[15] describes a container YMS based on RFID technology, which includes the use of active and passive tags to track control points of cargo containers for the purpose of cargo security. Three tags are used. The first and second are passive tags that contain information to identify the truck and container. The third active one is for storing information about the route. When a container arrives in the yard, the monitoring system periodically interrogates the tags in order to track unauthorized opening of the container or its movement. The focus of this paper is on describing the system structure, tag polling commands, etc. The questions of optimal placement of containers in yard, elimination of chaos and management errors are not solved by this system. [16] presents the concept of a cargo container sorting management in the YMS. The system is intended for use in rail transport. The developed real-time sorting optimization system is described, but no data are given about hardware components with which this system can be implemented on real objects. In [17] the high relevance of optimization of processes in the yard is noted. The article considers such areas of yard management: yard crane management, yard vehicle management, yard space management. However, as in [16], the hardware component basis for implementation is not described. [18] explores a container YMS based on the use of active RFID tags. This work describes the hardware features of using RFID tags in container identification, however the optimization of logistics processes is not considered. Similar studies can be found in the other works, such as [19] and [20], where the processes of optimization of yard logistics without reference to the hardware base are considered. A comprehensive solution to optimize yard management processes using RFID technology requires thus additional research.

3 Scalable Model of Universal RFID-Based YMS of Warehouse Logistics Complex

The concept of the scalable RFID-based YMS of the warehouse logistics complex, presented in this paper (in the following – logistics complex) is based on the concept of the scalable logistic cell, presented in [21]. The scalability of this cell is understood as the system property, which enables applying size or/and volume change. In addition to this, the border of the material, energy and information flow (G_1) in such a system is strongly connected to the border of the cell (G_2) (see Fig. 1). In other words, "the bigger is the Material-Energy-Information (MEI)-flow, the bigger is the "input surface" and with it tendentiously the cell" [21].

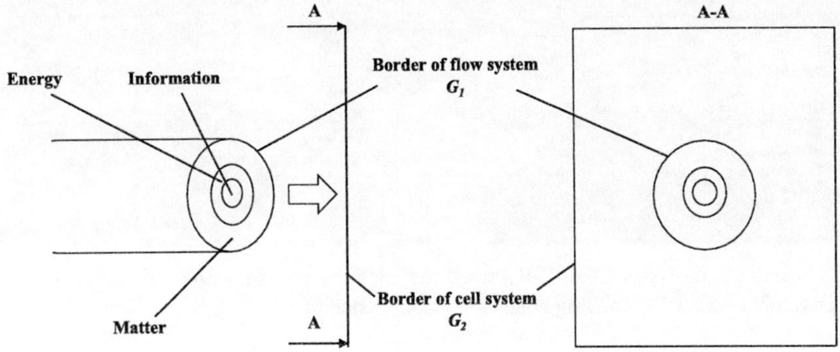

Fig. 1. Dependence between borders of incoming flow and logistic cell in scalable logistic cell [21]

The logistics complex model is formed from the flow of information about each stage of material flow. Thereby the logistics complex yard is broken down into study zones, which are scaled and unified to an adaptive model. Scalability of the model allows us to change the amount of information depending on conditions, and universality enables us to highlight those zones that are inherent in the modern warehouse. To summarize, we highlight the universality of the logistics complex yard in 4 zones (see Fig. 2):

1. Zone of vehicle entry (checkpoint);
2. Unloading zone (receiving of goods);
3. Loading zone (loading of goods into a vehicle);
4. Zone of vehicle exit (checkpoint).

Therefore, we present the model as a four operating zones system: "Entry-Unloading-Loading-Exit" (EULE). This universality allows us to systematize the information flow, that supports the corresponding material flow. Also, we can allocate that part of the information about the object of the material flow, which is the most relevant in our particular case. In other words, universality allows to consider both the full cycle of material flows and a particular case of given conditions.

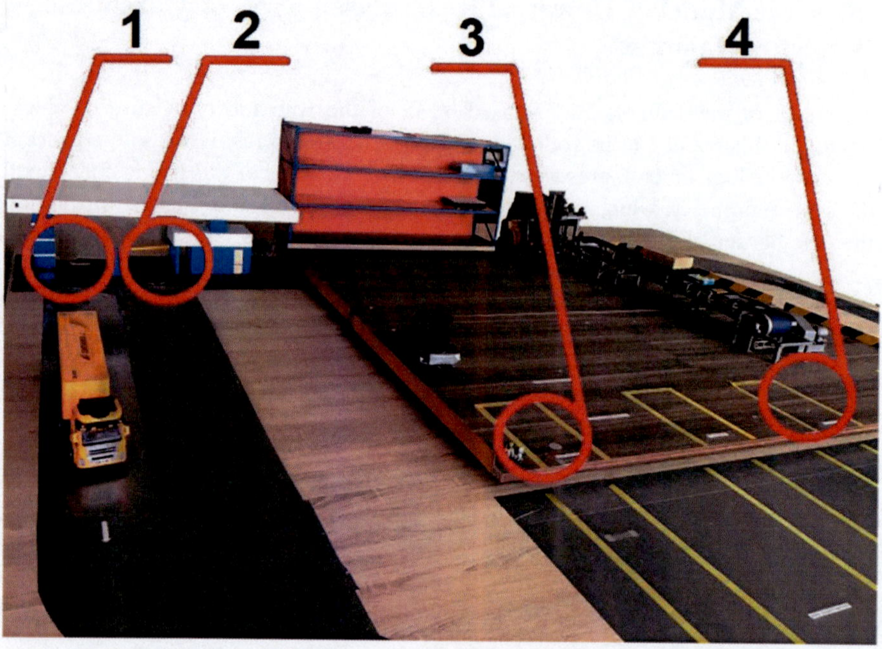

Fig. 2. Scalable prototype of the RFID-based logistics complex: 1 - Zone of vehicle entry, 2 - Zone of vehicle exit, 3 - Unloading zone, 4 - Loading zone

The main function of the above-mentioned prototype is the simulation of goods and transport identification processes using an RFID-based hardware system, which is integrated into the YMS. A typical process scheme of this prototype can be as follows: A vehicle enters the territory of the warehouse at the checkpoint. After entering it goes to the unloading ramp, where the unloading is being processed. After unloading, the goods are transferred to the storage rack and are being stored. For picking orders, the conveyor is used, which transports ready-to-dispatch goods to the loading ramp, where they are loaded in vehicles and dispatched.

For the realization of the aforementioned RFID-based yard management RFID-transponders have to be installed: (1) inside or outside of the vehicle to transmit the information about the shipping company, the goods, the destination, etc., and (2) inside of the goods packages to transmit the information about the package content, the producer, the destination, etc. of certain goods. RFID-readers, which have to collect the corresponding information have to be installed: (1) in the checkpoint (to identify incoming and outgoing vehicles), in loading and unloading points (to establish a reliable control of leaving and incoming material flows) as well as in the storage racks (to establish the bin location warehousing).

The automation of such a logistics complex model, using RFID-technology, has the goal to reduce the time for goods identification procedure, to avoid bottlenecks and to increase the turnover. With achieving these goals, logistics costs can be optimized and a more precise material flow control can be established. To show these potential economical and managerial advantages, the simulation of such a model in the software package AnyLogic is performed and described in the following Section.

4 Simulation of Scenarios in AnyLogic

There is a large variety of software applications, which work with both visual and analytical process modeling. However, not every software can provide a fully-fledged fusion and joint functioning of visual and analytical components. Another important criterion for the evaluation of the software is the usability spectrum and library of design elements.

After analyzing the existing options, we have chosen the simulation software *AnyLogic* [22]. This software has a free version for educational institutions and combines visual and analytical modeling in algorithmic, 2D and 3D modes. In this work the 3D simulation of the in the previous Section described model of the logistics complex is presented, for which the software version *AnyLogic 8.7.2* was used. The virtual 3D model allows us to change the arrangement, location, and type of equipment in the warehouse to reduce delays of goods in a particular area of the warehouse, and with it, to improve yard logistics of the presented logistics complex. It also avoids future errors in the building of the full-scale model. The virtual model enables it to monitor the movement of goods visually (see Fig. 3) and represents the corresponding algorithm for the entire material flow (see Fig. 4).

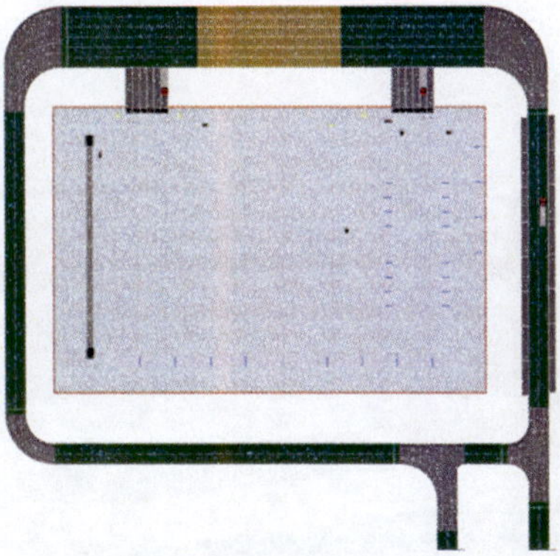

Fig. 3. 2D model of the simulated logistics complex in AnyLogic

Using the virtual model, different scenarios during relatively long time periods can be simulated. It enables us to forecast and plan logistics costs as well as to analyze possible risks or disturbances. Looking at the existing software functionality, the purpose of this virtual model is to analyze the work principles of the RFID-based YMS of the logistics complex described in the previous Section, compare possible technical solutions, and visualize the yard processes in the 2D- and 3D-model.

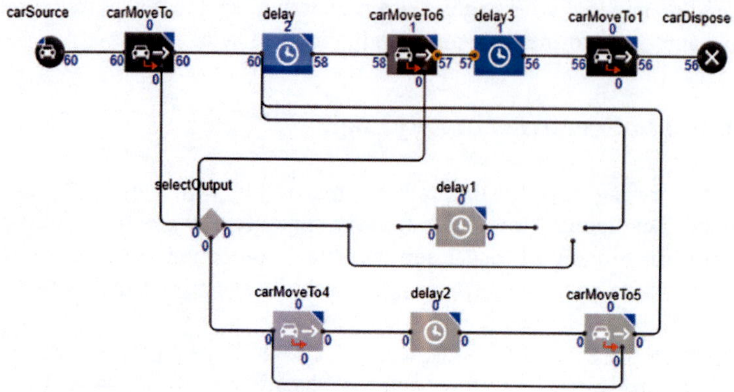

Fig. 4. Analytical model of the simulated logistics complex in AnyLogic

The virtual model, presented in the Fig. 5, consists of five key stages:

1. the arrival of a vehicle into the territory of the logistics complex;
2. unloading of goods;
3. movement of goods through the warehouse;
4. loading of goods in a vehicle;
5. leaving of the warehouse by a vehicle.

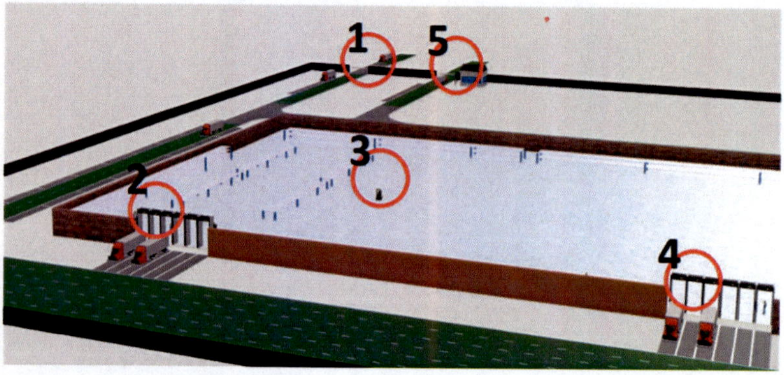

Fig. 5. Stages of material flow through the logistics complex in AnyLogic: 1 - the arrival of a vehicle into the territory of the logistics complex, 2 - unloading of goods, 3 - movement of goods through the warehouse, 4 - loading of goods in a vehicle, 5 - leaving of the warehouse by a vehicle

To analyze the model, we introduce the parameter "*Time on-route*". This parameter includes all stages of material flow and idle time of the vehicle at the territory of the logistics complex. Considering the time on-route, we explore three scenarios of vehicle movement with and without goods at the territory of a logistics complex (see Fig. 6):

- "Classic": Identification of goods and vehicles is being done manually. Arrived goods are identified and accounted for by scanning the barcode of each item;
- "RFID": Identification of goods is being done automatically, using RFID-technology: the information about goods is transferred from the transponder to the reader, which is connected with the database.
- "RFID-": Identification of goods is done automatically, using RFID-technology, however under the condition of reader breakage at the stage of goods unloading.

Simulating these three scenarios, we obtain graphs of vehicle movement (see Fig. 6). As a vehicle the truck was chosen for simulation. The initial data are set in the form of the frequency of the vehicles' arrival, which arrive according to the triangular law. This law in AnyLogic indicates that an event will occur in a random order from minimum to maximum, taking into account the shifting to the middle area. The intensity corresponds in this case to 50 cars in 2.5 days (60 h).

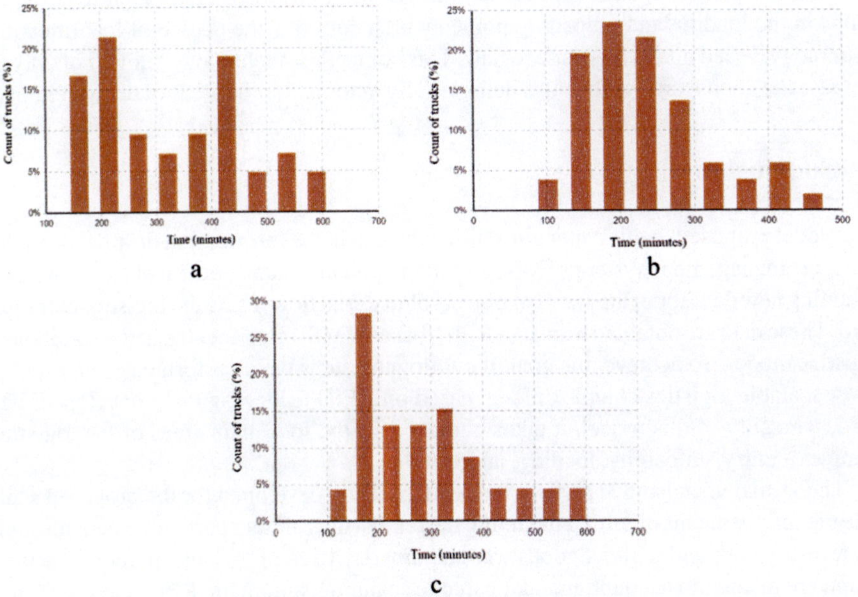

Fig. 6. On-route time of trucks at the territory of the logistics complex in AnyLogic: a – scenario "Classic", b – scenario "RFID", c – scenario "RFID-"

The model also considers the time of identification of goods and machines, as well as the time of unloading and loading of goods. Besides, we enter the downtime in the parking zone in case there is no free terminal for unloading or loading. We also consider the difference in the time of identification of goods when using barcode and RFID technology to compare them. The resulting comparison graphs are summarized in Fig. 7, where all three scenarios for the movement of trucks are given, taking into account the minimum, average, and maximum time in percentage.

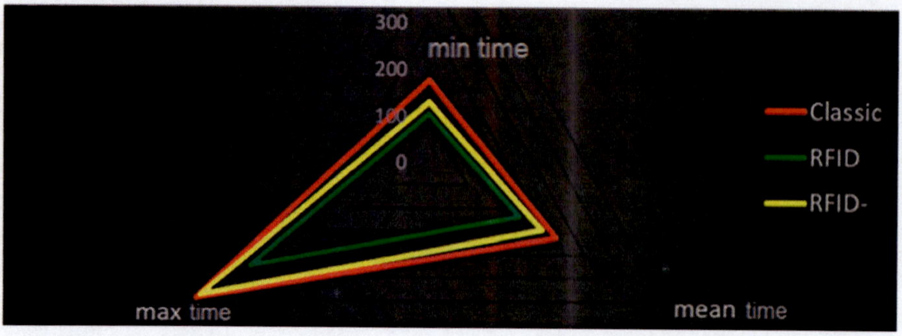

Fig. 7. A comparison chart of three scenarios in AnyLogic

To summarize the results of the simulation, we can conclude that the RFID technology has a significant time advantage over the classic barcode technology. On average, based on the above presented simulation results RFID allows for 28% faster identification of goods in the loading and unloading points, which confirms the choice of identification technology, based on the literature review. This contactless technology allows not only to increase the goods turnover but also helps to fully automate some areas of the warehouse.

5 Conclusions

The model proposed in this paper aims filling the gap in the information flow between the transport management system (TMS) and the warehouse management system (WMS), obtaining new data about the status of objects of material flow inside the logistics complex yard. These data are obtained using the RFID-based YMS, implemented in the warehouse logistics model. To achieve this goal, the following steps were performed:

A scalable logistics complex model based on RFID technology was developed. For the universality of this model, it takes into account the four main areas of the logistics complex - entry, unloading, loading, and exit.

The virtual visual and analytical models have been developed for the proposed scalable model. By means of virtual modeling the regularities of transport movement through the territory with and without goods were obtained. Different variants of goods identification are taken into account: manual barcode scanning, automatic RFID identification, automatic RFID identification under the condition of the reader failure at the stage of goods unloading. It was determined that the yard management process based on RFID technology can reduce the identification time compared with traditional methods. In the simulation presented in this work, this reduction is up to 28%. At the same time, both maximum and minimum values of identification time by different intensity of material flows are reduced.

A distinctive feature of this study is that the proposed model identifies not only goods, but also transport, which is important for tracking its current status and can be useful for all stakeholders in the supply chain - the manufacturer, the transport company and the warehouse management.

Future research can be conducted, e.g. towards the implementation of artificial intelligence control in such a logistics complex model for automated decision making in the YMS and costs reduction for installation and maintenance of RFID-based YMS.

References

1. Perig, A.V., Stadnik, A.N., Deriglazov, A.I., Podlesny, S.V.: 3 DOF spherical pendulum oscillations with a uniform slewing pivot center and a small angle assumption. Shock. Vib. (2014). https://doi.org/10.1155/2014/203709
2. Arabasi, S., Masoud, Z.: Simultaneous travel and hoist maneuver input shaping control using frequency modulation. Shock. Vib. (2017). https://doi.org/10.1155/2017/5703820
3. Grigorov, O., Druzhynin, E., Anishchenko, G., Strizhak, M., Strizhak, V.: Analysis of various approaches to modeling of dynamics of lifting-transport vehicles. IJET (2018). https://doi.org/10.14419/ijet.v7i4.3.19553
4. Andiyappillai, N., Prakash, T.: Implementing Warehouse Management Systems in Logistics: A Case Study (2019). https://doi.org/10.5281/ZENODO.2576011
5. Albrecht, W.: Aufgabenbereiche von IT-Systemen in der Logistik. In: Wehking, K.-H. (ed.) Technisches Handbuch Logistik. Fördertechnik, Materialfluss, Intralogistik, pp. 11–70. Morgan Kaufmann, Berlin, Heidelberg (2020)
6. Anđelković, A., Radosavljević, M.: Improving order-picking process through implementation of warehouse management system. Strategic Manage. (2018). https://doi.org/10.5937/StraMan1801003A
7. Hamdy, W., Mostafa, N., Elawady, H.: Towards a smart warehouse management system. In: Proceedings of the International Conference on Industrial Engineering and Operations Management, 27–29 Sept 2018, pp. 2555–2563. IEOM Society International, Southfield, Michigan, USA (2018)
8. Tamás, P., Dobos, P., Illés, B.: Examination of improvement possibilities in warehouse management systems (2017)
9. Griffis, S.E., Goldsby, T.J.: Transportation management systems: an exploration of progress and future prospects. JOTM (2007). https://doi.org/10.22237/jotm/1175385780
10. Agarwal, V., Sharma, S.: IoT based smart transport management system. In: Luhach, A.K., Jat, D.S., Bin Ghazali, K.H., Gao, X.-Z., Lingras, P. (eds.) ICAICR 2020. CCIS, vol. 1394, pp. 207–216. Springer, Singapore (2021). https://doi.org/10.1007/978-981-16-3653-0_17
11. Martin, H.: Informationslogistik. In: Nagl, A. (ed.) Transport- und Lagerlogistik, pp. 504–529. Springer Fachmedien Wiesbaden, Wiesbaden (2014)
12. Cao, X., Li, T., Wang, Q.: RFID-based multi-attribute logistics information processing and anomaly mining in production logistics. Int. J. Prod. Res. (2019). https://doi.org/10.1080/00207543.2018.1526421
13. Feng, S., Han, G., Zhang, F., He, B., Hui, J.: An RFID data matrices-based evaluation method for process logistics state. Math. Probl. Eng. (2019). https://doi.org/10.1155/2019/1638031
14. Abbas, A., Al-Bazi, A., Palade, V.: A constrained fuzzy knowledge-based system for the management of container yard operations. Int. J. Fuzzy Syst. **20**(4), 1205–1223 (2018). https://doi.org/10.1007/s40815-018-0448-9
15. Park, J., Oh, S., Cheong, T., Lee, Y.: Freight container yard management system with electronic seal technology. In: 2006 IEEE International Conference on Industrial Informatics, 16.08.2006 - 18.08.2006, pp. 67–72. IEEE, Singapore (2006). https://doi.org/10.1109/INDIN.2006.275719

16. Dimitrov, L., Purgic, S., Tomov, P., Todorova, M.: Approach for development of real-time marshalling yard management system. In: 2018 International Conference on High Technology for Sustainable Development (HiTech), 11.06.2018 - 14.06.2018, pp. 1–5. IEEE, Sofia (2018). https://doi.org/10.1109/hitech.2018.8566369
17. Zhen, L., Jiang, X., Lee, L.H., Chew, E.P.: A review on yard management in container terminals. Indust. Eng. Manage. Syst. (2013). https://doi.org/10.7232/iems.2013.12.4.289
18. Liang, Y., Bai, X.: Design of RFID-enabled container yard management system. In: Kacprzyk, J., Huang, G.Q., Mak, K.L., Maropoulos, P.G. (eds.) Proceedings of the 6th CIRP-Sponsored International Conference on Digital Enterprise Technology, AINSC, vol. 66, pp. 1751–1758. Springer, Berlin, (2010). https://doi.org/10.1007/978-3-642-10430-5_131
19. Cekała, T., Telec, Z., Trawiński, B.: Truck loading schedule optimization using genetic algorithm for yard management. In: Nguyen, N.T., Trawiński, B., Kosala, R. (eds.) ACIIDS 2015. LNCS (LNAI), vol. 9011, pp. 536–548. Springer, Cham (2015). https://doi.org/10.1007/978-3-319-15702-3_52
20. Chang, M.-C., Zhao, G., Pandey, A.K., Pulver, A., Tu, P.: Railcar detection, identification and tracking for rail yard management. In: 2020 IEEE International Conference on Image Processing (ICIP), 25.10.2020 - 28.10.2020, pp. 2271–2275. IEEE, Abu Dhabi, United Arab Emirates (2020). https://doi.org/10.1109/ICIP40778.2020.9190763
21. Heiden, B., Alieksieiev, V., Tonino-Heiden, B.: Scalable logistic cell RFID witness model. In: Proceedings of the 5th International Conference on Internet of Things, Big Data and Security, 5/7/2020 - 5/9/2020, pp. 420–427. SCITEPRESS - Science and Technology Publications (2020 - 2020), . Prague, Czech Republic. https://doi.org/10.5220/0009490204200427
22. AnyLogic Homepage. https://www.anylogic.com/. Accessed 1 June 2022

Modelling Thermal Stresses in Laminated Aircraft Elements of a Complex Form with Account of Heat Sources

Natalia Smetankina[1](✉), Alyona Merkulova[1], Dmytro Merkulov[1], Serhii Misiura[1], and Ievgeniia Misiura[2]

[1] A. Pidgornyi Institute for Mechanical Engineering Problems of the National Academy of Sciences of Ukraine, 2/10, Pozharskogo Street, Kharkiv 61046, Ukraine
nsmetankina@ukr.net

[2] Simon Kuznets Kharkiv National University of Economics, 9a Nauky Avenue, Kharkiv 61166, Ukraine

Abstract. A method for calculating thermal stressed state of laminated glazing elements of aircrafts sources is offered. Glazing element is considered as a constant thickness non-closed cylindrical laminated shell when exposed to interlayer film heat sources. The number of layers and their layout and physical properties are arbitrary. Convective heat exchange occurs on the shell surfaces. The temperature influences have been obtained by solving the nonstationary heat conduction problem. Deformation of the shells is considered on the basis of the refined theory of the first-order accounting transverse shear strains in each layer. The problem solution is built in analytical form by the embedding method. According to the method, the complex-shape shell is virtually embedded within an auxiliary enveloping laminated cylindrical non-closed simply supported shell of rectangular planform shape with the same composition of layers. An auxiliary shell is one whose contour shape and boundary conditions yield a simple analytical solution. The thermal stresses in five-layer aircraft elements have been investigated. The results were validated by comparison with test data. The method suggested can be used for designing heating systems and determining temperature stresses in the laminated glazing of different transport vehicles.

Keywords: Laminated shell · Heat source · Thermoelasticity

1 Introduction

A topical problem in modern engineering is reliable determination of temperature fields and stresses in construction elements. A review of models and methods of solving heat conduction and thermoelasticity problems is given in [1–5]. A review of the literature has shown that uniform constructions are the most investigated ones.

If a temperature field changes slowly with time, one can ignore the inertia terms in the equations of motion and the coupling term in the heat conduction equation, and treat the thermoelasticity problem as a quasistatic one [6, 7]. The thermoelasticity problem in the

D. D. Cioboată (Ed.): ICoRSE 2022, LNNS 534, pp. 233–246, 2023.
https://doi.org/10.1007/978-3-031-15944-2_22

quasistatic statement has the greatest practical value. Due to insignificant dynamic effects under common nonstationary heat exchange conditions, one can ignore the coupling of mechanical and thermal energy, and solve the temperature stresses problem in two stages, viz. first solve the heat conduction equation and then determine the field of stresses using the thermal elasticity equation and the temperature distribution found.

For laminated structural elements, the thermoelasticity problems are solved using different hypotheses on the temperature distribution over the laminated pack thickness. Thereat, if the structure has a complex planform shape, the majority of studies use the following numerical methods for analysis: the finite-difference method, and the finite and boundary elements methods [8–10].

Apalak and Demirbas [11] used the finite-difference method and the pseudo singular value method for studying thermal stresses and strains of bi-directional functionally graded clamped plates subjected to constant in-plane heat fluxes along two ceramic edges. The spatial derivatives of thermal and mechanical properties of the material composition were investigated.

Pernica et al. [12] used finite elements method (FEM) to investigate transient temperature fields and longitudinal thermal stresses on the shell and tubes during the start-up process.

Analytical methods are mathematically very involved when describing the geometrical parameters of multilayer bodies with a non-canonical configuration; the conditions of layer conjugation with account of inner heat sources, and with presence of layers with significantly differing properties. Therefore, plates and shells of canonical form are considered most often [2, 3, 7, 9].

Based on Fourier and Laplace transforms, Kushnir et al. [13] obtained an analytic solution of the quasistatic uncoupled problem of thermoelasticity for a functionally graded cylindrical shell of finite length heated by a two-dimensional temperature field. The shell material properties are regarded as analytic functions of the thickness coordinate.

Vattré and Pan [14] investigated thermoelastic stresses in simply supported multilayered anisotropic plates under time-harmonic distributions of temperature. Three-dimensional exact solutions are obtained using the mathematically elegant Stroh formalism and the Fourier series expansions.

Based on the embedding method, this paper suggests an approach of solving thermoelasticity problems for laminated non-canonical shaped shells subject to temperature fields derived by solving the heat conduction problem.

2 Problem Statement

Let us consider a laminated non-closed cylindrical shell with radius R. The shell is composed of I isotropic layers with constant thickness h_i $\left(i = \overline{1, I}\right)$ (Fig. 1). In the Figure, "1" designates the first layer and "2" designates the domain occupied by the film heat sources.

The shell is referenced to the Cartesian system of coordinates linked to the outer surface of the first layer and occupies domain Ω limited by contour Γ on the coordinate surface

$$\Gamma:\ x_\Gamma = x(s),\ y_\Gamma = y(s), \tag{1}$$

where s is current arc length.

Let us denote the upper and bottom shell surfaces as Ω_0 and Ω_I, and the side surface as Ω_Γ. Here, $\Omega = \Omega_0$, $\Omega_\Gamma = \sum_{i=1}^{I} \Omega_\Gamma^i$, $i = \overline{1, I}$. Convective heat exchange occurs on the shell surfaces. The shell is affected by interlayer heat sources that occupy domain Ω_q.

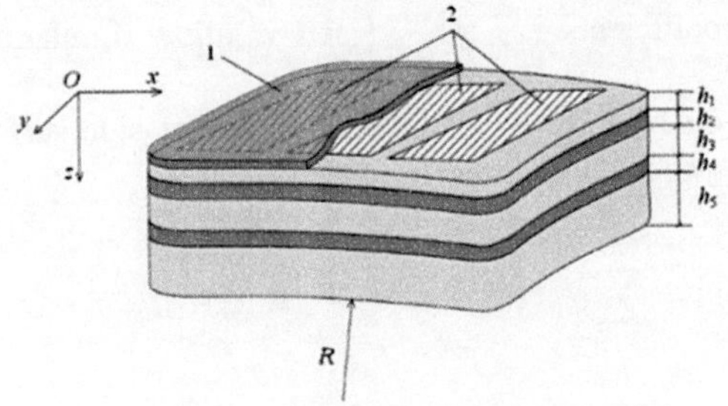

Fig. 1. Geometrical model of the fan blade.

The nonstationary heat conduction equation and the boundary conditions for a laminated shell are derived from the heat balance variational equation:
heat conduction equation

$$\frac{\partial T}{\partial t} = \nu_i \Delta_R T^i + \frac{q^i}{c_i \rho_i},\quad i = \overline{1, I}, \tag{2}$$

boundary conditions on outer surfaces

$$-\chi_1^1 k_1 \frac{\partial T^1}{\partial z} + \chi_2^1 H_1 \left(T^1 - T_t\right) = 0,\ (x, y, z) \in \Omega_0,$$

$$\chi_1^I k_I \frac{\partial T^I}{\partial z} + \chi_2^I H_I \left(T^I - T_b\right) = 0,\ (x, y, z) \in \Omega_I, \tag{3}$$

on layer contact boundary

$$k_i \frac{\partial T^i}{\partial z} - k_{i+1} \frac{\partial T^{i+1}}{\partial z} + q_\Omega^i = 0,\ T^i = T^{i+1},\ z = \delta_i,\ \delta_i = \sum_{j=1}^{i} h_j,\ i = \overline{1, I-1}, \tag{4}$$

on shell side surface

$$k_i \frac{\partial T^i}{\partial \mathbf{n}} + H_\Gamma^i \left(T^i - T_\Gamma^i\right) = 0,\quad (x, y, z) \in \Omega_\Gamma,\ i = \overline{1, I}, \tag{5}$$

where

$$\Delta_R = \frac{\partial^2}{\partial x^2} + \frac{1}{R_z^2}\frac{\partial^2}{\partial y^2} + \frac{1}{RR_z}\frac{\partial}{\partial z} + \frac{\partial^2}{\partial z^2}, \quad R_z = 1 + \frac{z}{R},$$

R is surface Ω_0 curvature radius; $T^i = T^i(x, y, z, \tau)$ is temperature; k_i is thermal conductivity of the i-th layer material; ρ_i is layer material density; c_i is the i-th layer thermal capacity; $\nu_i = k_i/(\rho_i c_i)$ is the i-th layer thermal conductivity; H_Γ^i, H_1 and H_I are convective heat exchange coefficients on the side, upper and bottom surfaces of the plate, respectively; T_Γ^i, T_t and T_b are ambient temperatures on the boundary with the side, top and bottom surfaces; $q^i = q^i(x, y, z, \tau)$ is intensity of the i-th inner heat source; $q_\Omega^i = q_\Omega^i(x, y, \tau)$ is intensity of the i-th film heat source located on the adjacent layers contact boundary; δ_{iI} is Kronecker symbol, t is time; $\mathbf{n} = \mathbf{n}(x, y)$ is external normal to the side surface. Coefficients χ_1^1, χ_2^1, χ_1^I and χ_2^I allow simulating preset boundary conditions.

The temperature distribution in the layers and on the side surface, as well as the density of inner heat sources is given as follows:

$$T^i(x, y, z, t) = \sum_{r=0}^{R} T_r^i(x, y, t) f_r^i(z), \quad (x, y) \in \Omega, \quad z \in [\delta_{i-1}, \delta_i],$$

$$T_\Gamma^i(x, y, z, t) = \sum_{r=0}^{R} T_{\Gamma r}^i(x, y, t) f_r^i(z), \quad (x, y) \in \Gamma,$$

$$q^i(x, y, z, t) = \sum_{r=0}^{R} q_r^i(x, y, t) f_r^i(z), \quad (x, y) \in \Omega_q, \tag{6}$$

where $T_r^i = \int_{\delta_{i-1}}^{\delta_i} T^i f_r^i dz$, $T_{\Gamma r}^i = \int_{\delta_{i-1}}^{\delta_i} T_\Gamma^i f_r^i dz$, $q_r^i = \int_{\delta_{i-1}}^{\delta_i} q^i f_r^i dz$, $j,\ k = \overline{0, r^*}$, $\quad i = \overline{1, I}$; $f_r^i(z)$ is Legendre polynomial with exponent r. Expansions (6) take into account the first four terms of the series $\left(r = \overline{0, 4}\right)$, allowing for a sufficient solution accuracy.

With account of expansion (6), conditions (3)–(5) comprise a system of linear algebraic equations. The system allows expressing coefficients T_2^i and T_3^i via coefficients T_0^i and T_1^i $\left(i = \overline{1, I}\right)$, and write the temperature distribution in shell layers as follows:

$$T^i = \sum_{j=1}^{I} \left[\varphi_{ij}^1 T_0^j + \varphi_{ij}^2 T_1^j\right] + \varphi_i^3, \quad i = \overline{1, I}, \tag{7}$$

where φ_{ij}^1, φ_{ij}^2, φ_i^3 are constants yielded by transformations.

Hence, temperature T^i (7) satisfies the boundary conditions on the outer surfaces Ω_0 and Ω_I of the shell, and on the boundary of contact of adjacent layers.

Heat conduction equation (2) with regard to (7) and the boundary condition on the side surface of the i-th layer (5) yields a system of differential equations solved using the method presented in [15].

3 Research Methodology

The shell behavior is described within the framework of the refined first-order theory. It accounts for lateral shear strain and thickness reduction in each shell layer. The polygonal line hypothesis holds for the pack, and interlayer contact excludes their delamination and mutual slippage. Displacements of a point on the i-th layer are determined as:

$$u^i = u + \sum_{j=1}^{i-1} h_j \psi_x^j + (z - \delta_{i-1})\psi_x^i,\ v^i = v + \sum_{j=1}^{i-1} h_j \psi_y^j + (z - \delta_{i-1})\psi_y^i,$$

$$w^i = w + \sum_{j=1}^{i-1} h_j \psi_z^j + (z - \delta_{i-1})\psi_z^i,\quad i = \overline{1, I}, \tag{8}$$

where $\delta_i = \sum_{j=1}^{i} h_j,\quad \delta_{i-1} \le z \le \delta_i,\quad i = \overline{1,\ I},\ u = u(x, y),\ v = v(x, y),\ w = w(x, y)$ are displacements of a point on the coordinate surface in the direction of coordinate axes; $\psi_x^i = \psi_x^i(x, y),\ \psi_y^i = \psi_y^i(x, y)$ are angles of rotation of a normal element in the i-th layer about X-axis and Y-axis; $\psi_z^i = \psi_z^i(x, y)$ is compression of normal element within the i-th layer.

The shell is exposed to temperature fields and force loads $\mathbf{P} = \{p_j(x, y)\},\ j = \overline{1,\ 3I + 3}$. The temperature distribution in the shell layers is derived by solving heat conduction problem (2)–(5).

Strains are found using Cauchy formulas:

$$\varepsilon_x^i = u_{,x} + \sum_{j=1}^{i-1} h_j \psi_{x,x}^j + (z - \delta_{i-1})\,\psi_{x,x}^i,\ \varepsilon_y^i = \frac{1}{1 + z/R}\left[v_{,y} + \sum_{j=1}^{i-1} h_j \psi_{y,y}^j\right.$$

$$\left. + (z - \delta_{i-1})\psi_{y,y}^i + + \frac{1}{R}\left(w + \sum_{j=1}^{i-1} h_j \psi_z^j + (z - \delta_{i-1})\psi_z^i\right)\right],\ \varepsilon_z^i = \psi_z^i,$$

$$\gamma_{xy}^i = \gamma_{yx}^i = \frac{1}{1 + z/R}\left[u_{,y} + \sum_{j=1}^{i-1} h_j \psi_{x,y}^j + (z - \delta_{i-1})\psi_{x,y}^i\right] + v_{,x} + \sum_{j=1}^{i-1} h_j \psi_{y,x}^j$$

$$+ (z - \delta_{i-1})\psi_{y,x}^i,\ \gamma_{xz}^i = \gamma_{zx}^i = \psi_x^i + w_{,x} + \sum_{j=1}^{i-1} h_j \psi_{z,x}^j + (z - \delta_{i-1})\psi_{z,x}^i,\ \gamma_{yz}^i = \gamma_{zy}^i = \psi_y^i +$$

$$+ \frac{1}{1 + z/R}\left[w_{,y} + \sum_{j=1}^{i-1} h_j \psi_{z,y}^j + (z - \delta_{i-1})\psi_{z,y}^i - \frac{1}{R}\left(v + \sum_{j=1}^{i-1} h_j \psi_y^j + (z - \delta_{i-1})\psi_y^i\right)\right],\quad i = \overline{1, I}.$$

According to the Duhamel-Neumann hypothesis, stresses and strains in the layers obey Hooke's law:

$$\sigma_x^i = \frac{E_i}{1 + \nu_i}\left(\varepsilon_x^i + \frac{\nu_i}{1 - 2\nu_i}\theta^i\right) - \frac{E_i}{1 - \nu_i}\alpha_T^i T^i,\ \sigma_y^i = \frac{E_i}{1 + \nu_i}\left(\varepsilon_y^i + \frac{\nu_i}{1 - 2\nu_i}\theta^i\right) - \frac{E_i}{1 - \nu_i}\alpha_T^i T^i,$$

$$\sigma_z^i = \frac{E_i}{1 + \nu_i}\left(\varepsilon_z^i + \frac{\nu_i}{1 - 2\nu_i}\theta^i\right) - \frac{E_i}{1 - \nu_i}\alpha_T^i T^i,$$

$$\tau_{xy}^{i}=G_i\gamma_{xy}^{i},\ \tau_{xz}^{i}=G_i\gamma_{xz}^{i},\ \tau_{yz}^{i}=G_i\gamma_{yz}^{i},\quad i=\overline{1,I},$$

where $G_i=\frac{E_i}{2(1+\nu_i)}$; $\theta^i=\varepsilon_x^i+\varepsilon_y^i+\varepsilon_z^i$; E_i is Young's modulus of the i-th layer material; ν_i is Poisson's ratio; α_T^i is linear thermal expansion coefficient of the i-th layer material; T^i is temperature distribution obtained by solving the heat conduction problem (2)–(5).

The shell equilibrium equations and the boundary conditions on contour Γ are derived from the Lagrange variational equation [16].

$$\mathbf{\Lambda}\,\mathbf{U}=\mathbf{P}^T-\mathbf{P},\quad (x,y)\in\Omega, \tag{9}$$

$$\mathbf{B}^{\Gamma}\,\mathbf{U}=\mathbf{P}^{\Gamma},\ (x,y)\in\Gamma. \tag{10}$$

In Eqs. (9) and (10), $\mathbf{U}$ is displacement vector:

$$u_1=u,\ u_2=v,\ u_3=w,\ u_{3+i}=\psi_x^i,\ u_{3+I+i}=\psi_y^i,\ u_{3+2I+i}=\psi_z^i,\ i=\overline{1,I};$$

$\mathbf{\Lambda}$ and $\mathbf{B}^{\Gamma}$ are symmetrical matrices of dimension $(3I+3)\times(3I+3)$; $\mathbf{P}$ is force loads vector; and $\mathbf{P}^T$ is thermal loads vector.

The form of the elements of the matrices and vectors depends on the boundary conditions on the shell contour and defines by various values for the coefficients χ_1^1, χ_2^1, χ_1^I and χ_2^I.

According to the embedding method [17], the complex-shape shell (domain Ω) is virtually embedded into an auxiliary laminated cylindrical non-closed simply supported shell of rectangular planform shape with the same composition of layers.

To meet preset boundary conditions (10), the auxiliary shell is exposed to additional compensating loads $\mathbf{Q}^{comp}=\left\{q_j^{comp}(x,y)\right\}, j=\overline{1,\ 3I+3}$ distributed along contour Γ. In equilibrium Eqs. (9), the compensating loads are included as curvilinear distributions $\mathbf{P}^{comp}=\left\{p_j^{comp}(x,y)\right\}$

$$p_j^{comp}(x,y)=\sum_{k=1}^{3I+3}\oint_{\Gamma}L_{jk}\,q_k^{comp}(s)\,\delta(x-x_{\Gamma},y-y_{\Gamma})\,ds,\quad j,k=\overline{1,\ 3I+3}, \tag{11}$$

where $\delta(x-x_{\Gamma},y-y_{\Gamma})$ is the two-dimensional Dirac δ - function.

In (11), the nonzero elements of matrix L_{jk} have the form:

$$L_{11}=L_{22}=L_{3+i\ 3+i}=L_{3+I+i\ 3+I+i}=n_x,$$

$$L_{33}=L_{3+2I+i\ 3+2I+i}=1,\ L_{12}=L_{3+i\ 3+I+i}=-n_y,\ L_{21}=L_{3+I+i\ 3+i}=n_y,$$

where n_x and n_y are direction cosines of the normal to boundary Γ.

Then, with account of (11), the system of Eqs. (9) and (10) is transformed to a system of singular integral-differential equations. The unknown functions are displacement functions $\mathbf{U}$ (8) and compensating load functions $\mathbf{Q}^{\text{comp}}$.

Proceeding from the condition of meeting boundary conditions (10), a system of integral equations is formed on contour Γ to determine the intensities of compensating loads:

$$\mathbf{B}^{\Gamma}\mathbf{U}\left[\mathbf{Q}^{\text{comp}}(x,y)\right] = \mathbf{P}^{\Gamma}, \quad x, y \in \Gamma. \tag{12}$$

Next, displacements and loads are expanded in the rectangular shell domain into trigonometric series for functions that meet simply supported conditions:

$$u_j(x,y) = \sum_{m=1}^{\infty}\sum_{n=1}^{\infty}\Phi_{jmn}B_{jmn}(x,y), \quad p_j(x,y) = \sum_{m=1}^{\infty}\sum_{n=1}^{\infty}p_{jmn}B_{jmn}(x,y),$$

$$p_j^T(x,y) = \sum_{m=1}^{\infty}\sum_{n=1}^{\infty}p_{jmn}^T B_{jmn}(x,y), \quad p_j^{\Gamma}(x,y) = \sum_{m=1}^{\infty}\sum_{n=1}^{\infty}p_{jmn}^{\Gamma} B_{jmn}(x,y),$$

$$p_j^{\text{comp}}(x,y) = \sum_{m=1}^{\infty}\sum_{n=1}^{\infty}p_{jmn}^{\text{comp}} B_{jmn}(x,y), \quad j = \overline{1,\ 3I+3}, \tag{13}$$

$$B_{1mn} = \cos\frac{m\pi x}{A}\sin\frac{n\pi y}{B}, \quad B_{2mn} = \sin\frac{m\pi x}{A}\cos\frac{n\pi y}{B}, \quad B_{3mn} = \sin\frac{m\pi x}{A}\sin\frac{n\pi y}{B},$$

$$B_{3+i\ mn} = B_{1mn},\ B_{3+I+i\ mn} = B_{2mn},\ B_{3+2I+i\ mn} = B_{3mn},\ i = \overline{1,I},\ m = \overline{1,m^*},\ n = \overline{1,n^*},$$

where A is length of auxiliary shell generator; B is length of the shell arc.

Hence, for each pair of values of m and n, system (9) is reduced to a system of linear algebraic equations

$$\mathbf{\Lambda}^{mn}\mathbf{\Phi}_{mn} = \mathbf{P}^{mn} - \mathbf{P}_{mn}^{T} + \mathbf{P}_{mn}^{\text{comp}}, \tag{14}$$

where $\mathbf{\Lambda}^{mn}$ is a symmetrical matrix with dimensions $(3I+3)\times(3I+3)$ whose elements are found by transformations (13)

$$\mathbf{\Phi}_{mn} = \left\{\Phi_{jmn}\right\},\ \mathbf{P}^{mn} = \left\{p_{jmn}\right\},\ \mathbf{P}_{mn}^{T} = \left\{p_{jmn}^{T}\right\},\ \mathbf{P}_{mn}^{\text{comp}} = \left\{p_{jmn}^{\text{comp}}\right\},\ j = \overline{1,\ 3I+3}.$$

The solution of system (14) is written as follows

$$\Phi_{jmn} = \sum_{k=1}^{3I+3}\Delta_{jk}^{mn}\left(p_{jmn} - p_{jmn}^{T} + p_{jmn}^{\text{comp}}\right), \quad j = \overline{1,\ 3I+3}, \tag{15}$$

where Δ_{jk}^{mn} are components of a matrix inverse to matrix $\mathbf{\Lambda}^{mn}$, $\mathbf{\Delta}^{mn} = \mathbf{\Lambda}^{mn^{-1}}$.

Compensating loads $\mathbf{Q}^{\text{comp}}$ are found from the system of integral equations (12). The compensating loads functions are expanded in a series along contour Γ

$$q_j^{\text{comp}}(s) = \sum_{\alpha=1,2}\sum_{\mu=0}^{\infty}q_{j\alpha\mu}\,b_{\alpha\mu}(s), \quad j = \overline{1,\ 3I+3}, \tag{16}$$

where $b_{1\mu} = \sin\left[\mu\gamma(s)\right]$, $b_{2\mu} = \cos\left[\mu\gamma(s)\right]$, $\mu = \overline{0, \mu^*}$, $\gamma(s) = 2\pi \int\limits_0^s d\tilde{s} / \oint\limits_\Gamma d\tilde{s}$, $0 \le \gamma(s) \le 2\pi$.

With account of expansion (16), coefficients (15) are transformed as follows:

$$\Phi_{jmn} = \sum_{k=1}^{3I+3} \Delta_{jk}^{mn} \left[p_{jmn} - p_{jmn}^T + \frac{4}{AB} \sum_{l=1}^{3I+3} \sum_{\alpha=1,2} \sum_{\mu=0}^{\infty} q_{l\alpha\mu}\, \theta_{jl\alpha\mu}^{mn} \right], \quad j = \overline{1,\ 3I+3}, \tag{17}$$

where $\theta_{jl\alpha\mu}^{mn} = \oint\limits_\Gamma L_{jl}(s) B_{jmn}(x_\Gamma, y_\Gamma)\, b_{\alpha\mu}(s)\, ds$.

The boundary functions included in the initial boundary conditions along contour Γ, which form system (12), are also expanded in a series along contour Γ. With account of (17), system (12) takes the form of a system of linear algebraic equations for the coefficients of compensating loads expansion into a series along contour Γ

$$\sum_{j=1}^{3I+3} \sum_{\alpha=1,2} \sum_{\mu=0}^{\infty} H_{ij\alpha\mu\beta\nu} q_{j\alpha\mu} = \Pi_{i\beta\nu},\ i = \overline{1,\ 3I+3},\ \beta = 1,\ 2,\ \nu = \overline{0, \mu^*},\ \nu = \overline{0, \nu^*},$$

where $H_{ij\alpha\mu\beta\nu} = \sum\limits_{m=1}^{\infty} \sum\limits_{n=1}^{\infty} \sum\limits_{l=1}^{3I+3} \overline{\theta}_{il\beta\nu}^{mn} \sum\limits_{k=1}^{3I+3} \Delta_{lk}^{mn} \theta_{kj\alpha\mu}^{mn} + \frac{\chi_i^1 AB}{8} \delta_{ij}, j = \overline{1,\ 3I+3}, \alpha = 1,\ 2,$ $\mu = \overline{0,\ \mu^*}$, $m = \overline{1, m^*}$, $n = \overline{1, n^*}$; δ_{ij} is Kronecker's symbol;

$$\Pi_{i\beta\nu} = -\frac{AB}{4} \sum_{m=1}^{\infty} \sum_{n=1}^{\infty} \sum_{l=1}^{3I+3} \overline{\theta}_{il\beta\nu}^{mn} \eta_{lmn} + \sum_{m=1}^{\infty} \sum_{n=1}^{\infty} p_{imn}^{\Gamma} \overline{\varsigma}_{i\beta\nu}^{mn};$$

$$\overline{\theta}_{il\beta\nu}^{mn} = \frac{1}{\lambda_\nu} \oint\limits_\Gamma \left(\chi_i^1 B_{il}^u + \chi_i^2 B_{il}^\sigma\right) B_{lmn}(x_\Gamma, y_\Gamma)\, b_{\beta\nu}(s) ds; \quad \overline{\zeta}_{i\beta\nu}^{mn} = \frac{1}{\lambda_\nu} \oint\limits_\Gamma B_{imn}(x_\Gamma, y_\Gamma)\, b_{\alpha\mu}(s)\, ds;$$

$$i = \overline{1, 3I+3},\ \eta_{lmn} = \sum_{k=1}^{3I+3} \Delta_{lk}^{mn} \left(p_k^{mn} - p_{kmn}^T\right),\ l = \overline{1, 3I+3}.$$

After the compensating loads have been found by summing series (16), displacements (8), strains and stresses are calculated in the layers of the initial shell.

4 Experimental Setup and Procedure

The technique and instrumentation used for conducting experiments was developed at the A. Pidgornyi Institute for Mechanical Engineering Problems of the National Academy of Sciences of Ukraine. It is used for measuring strains with sufficient accuracy [18].

Figure 2 shows the schematic diagram of the experimental setup. Shell 1 was secured vertically. Small-base (measurement base 1 mm) foil strain gauges 2 were glued in the middle of the outer surface of the first layer. Alternating current 110 V 50 Hz was applied to heating element 3.

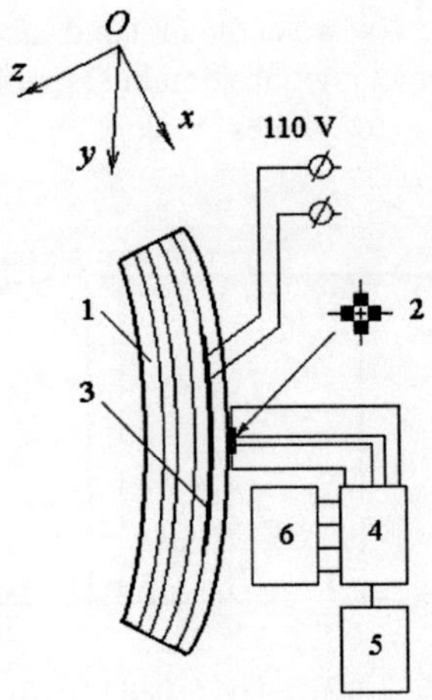

Fig. 2. Schematic diagram of experimental setup.

The temperature was measured with a thermocouple in a point on the outer surface of the shell's first layer. Strains were measured in the same point. Strains caused by shell heating were measured with a strain gauge measurement kit comprising strain gauge station 4, universal voltage tester 5 and calibration device 6. Before starting the experiment, the investigators held the shell at a constant temperature of 20 °C for 24 h. Next, the room temperature and the temperature on the outer surfaces of the first and fifth layers was registered.

The experiment was started when voltage was applied to the film heat source in the shell. Strain and temperature values were recorded every five minutes starting from the instance of applying voltage to the heating element.

5 Results

The numerical results illustrating the effect of the theoretical and experimental approaches were obtained for a simply supported five-layer glazing element affected by an interlayer film heat source. The glazing consists of silicate glass and polymeric material layers. For the simply supported shell, the coefficients in boundary conditions (10) have the form.

$$\chi_1^1 = \chi_2^1 = \chi_3^2 = \chi_{3+i}^1 = \chi_{+I+i}^1 = \chi_{3+2I+i}^1 = 1,$$

$$\chi_1^2 = \chi_2^2 = \chi_3^1 = \chi_{3+i}^2 = \chi_{+I+i}^2 = \chi_{3+2I+i}^2, \quad i=\overline{1,I}.$$

The shell side surface was considered thermally insulated $\partial T^i/\partial \mathbf{n} = 0$. The free convection conditions were satisfied on the outer surfaces of the first and fifth layers. The initial conditions and the ambient temperature during the experiment were such that $T^i = 20$ °C at $t = 0$, $i = \overline{1, I}$, and $T_t = T_b = 20$ °C.

The shell plan shape (1) is described by the Lame curves equation $\Gamma : (x/\alpha)^k + (y/\beta)^k = 1$, where $2\alpha = 0.66$ m, $2\beta = 0.33$ m, $k = 8$. The shell plan shape is shown in

Fig. 3. The heat source is located between the first and second layers, and has the power $q_{\Omega}^{1} = 1340$ W/m^2. It occupies rectangular domain Ω_q with the dimensions $a = 0.56$ m and $b = 0.33$ m. The dashed line designates the domain Ω_q occupied by the heat source.

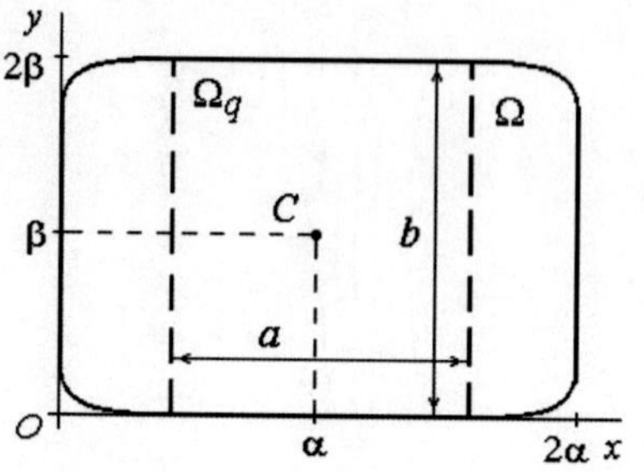

Fig. 3. The five-layer shell plan shape

The shell has the following mechanical, physical and geometric characteristics: $E_i = 68000$ MPa, $\nu_i = 0.22$, $\rho_i = 2520$ kg/m^3, $\alpha_T^i = 9 \times 10^{-6}$ °C^{-1}, $k_i = 1.08$ W/(m·°C), $c_i = 870$ J/(kg·°C) ($i = 1, 3, 5$); $E_i = 220$ MPa, $\nu_i = 0.38$, $\rho_i = 1060$ kg/m^3, $\alpha_T^i = 8.3 \times 10^{-5}$ °C^{-1}, $k_i = 0.22$ W/(m·°C), $c_i = 2344$ J/(kg·°C) ($i = 2, 4$); $h_1 = 0.005$ m, $h_2 = 0.003$ m, $h_3 = 0.015$ m, $h_4 = 0.002$ m, $h_5 = 0.02$ m, $R = 1.34$ m.

First, a comparative analysis of the results of temperature calculations and experimental data in point C with coordinates $x = 0.33$ m, $y = 0.165$ m was performed. The point is located in the center of domain Ω_q (Fig. 3). Figure 4 shows the temperature on the outer surface of the first layer vs. time. The calculation line is the solid one, and the experimental results are shown as points.

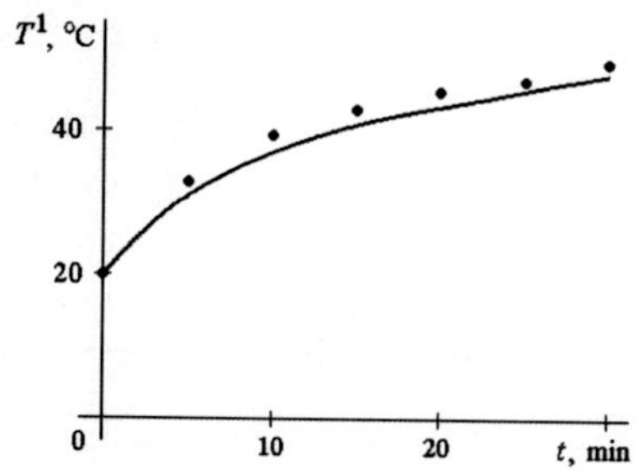

Fig. 4. Temperature T^1 on outer surface of the first layer.

Figure 5 shows strains ε_x (a) and ε_y (b) on the outer surface of the first layer in point C vs. time. The solid lines indicate the calculation results using the method suggested, and the dashed lines show FEM analysis results. The points designate experimental strain values.

Comparison of dependencies shows that strains obtained with different methods are sufficiently close. This confirms the validity of strain calculations based on the method developed.

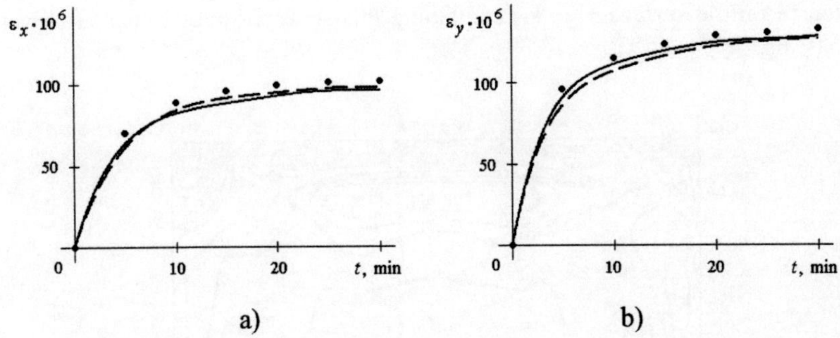

Fig. 5. Computational scheme (a) and finite-element model of the fan blade.

The feasibility of the method suggested was illustrated by the problem of the thermal-stressed state of a seven-layer simply supported shell affected by an interlayer film heat source. This is glazing element of the AN type aircraft. Figure 6 shows the shell planform shape and the heat source. The heat source with the dimensions $a = 0.6$ m, $b = 0.49$ m is located between the first and second layer, and has the power $q_{\Omega}^{1} = 8400$ W/m^2.

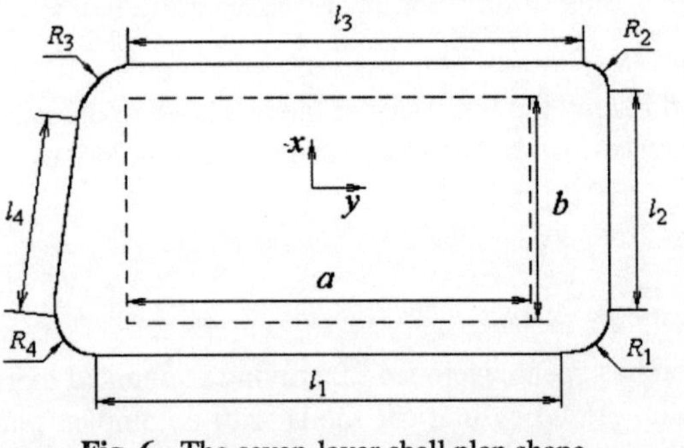

Fig. 6. The seven-layer shell plan shape.

The side surface of the shell was considered thermally insulated. Convective heat exchange occurred on the outer surfaces of the first and seventh layers. The convective heat exchange coefficients, the ambient temperature and the initial temperature were as follows: $H_1 = 150$ W/(m^2·°C), $H_I = 20$ W/(m^2·°C), $T_t = -50$ °C, $T_b = 20$ °C, $T^i = 20$ °C at $t = 0$, $i = \overline{1, I}$.

The geometric and mechanical parameters of the shell are: $R = 1.34$ m, $l_1 = 0.7$ m, $l_2 = 0.54$ m, $l_3 = 0.61$ m, $l_4 = 0.55$ m, $R_1 = 0.05$ m, $R_2 = 0.3$ m, $R_3 = 0.04$ m, $R_1 = 0.045$ m; $h_1 = 0.005$ m, $h_2 = 0.003$ m, $h_3 = h_5 = h_7 = 0.01$ m, $h_4 = h_6 = 0.002$ m; $E_i = 68000$ MPa, $\nu_i = 0.22$, $\rho_i = 2520$ kg/m^3, $\alpha_T^i = 9 \cdot 10^{-6}$ °C^{-1}, $k_i = 1.08$ W/(m·°C), $c_i = 870$ J/(kg·°C) ($i = 1,\ 3, 5, 7$); $E_i = 220$ MPa, $\nu_i = 0.38$, $\rho_i = 1060$ kg/m^3, $\alpha_T^i = 8.3 \cdot 10^{-5}$ °C^{-1}, $k_i = 0.22$ W/(m·°C), $c_i = 2344$ J/(kg·°C) ($i = 2$, 4, 6).

Distributions of the principal stress σ_1 at the time when they reach their highest values are obtained. Figure 7 shows the distribution of the stress in the location of the heat source on the surface $z = \delta_1$, containing this source. Stress values are given in megapascals.

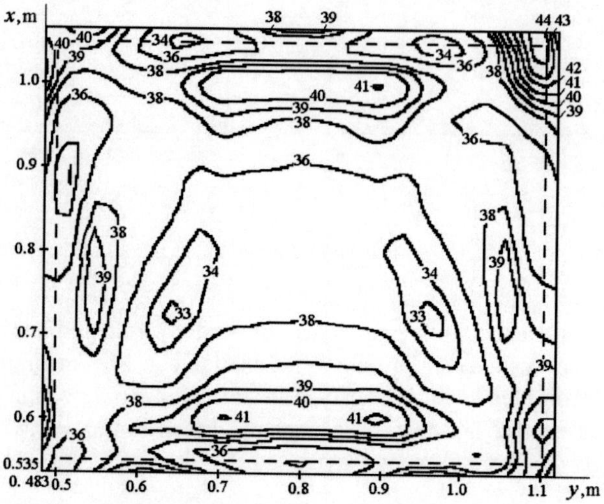

Fig. 7. Stress distribution on the surface with a heat source.

In the corners and along the boundary of the heat source domain there is a concentration of thermal stresses. It was found that the stresses did not exceed their allowable values.

6 Conclusions

An analytical method has been developed for solving uncoupled thermoelasticity problems in laminated non-closed cylindrical shells with a complex planform shape. The shells are affected by inner interlayer film heat sources.

The behavior of shells was described within the framework of the linear refined first-order theory accounting for lateral shear strain and thickness reduction in each shell layer while invoking the polygonal line hypothesis for a pack.

The method feasibility was tested on five- and seven-layer shell. The stresses in shell layers were determined on exposure to temperature fields obtained by solving the nonstationary heat conduction problem. It allows accounting for the effect of temperature fields without involving additional temperature distribution hypotheses.

Temperature and strain calculation results were compared against experimental data. Comparison of strain values with FEM analysis results and experimental data has shown that they are sufficiently close. This confirms the feasibility and effectiveness of the method suggested. The effect of stress concentration in the corners and along the boundary of the heat source domain is established.

The method offered can be used for further analysis of thermal stresses in the laminated glazing of modern aircraft under the influence of film heat sources, which are part of heating systems for their design [19].

References

1. Ostrowski, P., Jędrysiak, J.: Heat conduction in periodic laminates with probabilistic distribution of material properties. Heat Mass Transf. **53**(4), 1425–1437 (2016). https://doi.org/10.1007/s00231-016-1908-0
2. Kushnir, R.M, Zhydyk, U.V.: Temperature stresses in a functionally graded cylindrical shell. Mater. Sci. **54**(5), 666–677 (2019). https://doi.org/10.1007/s11003-019-00231-0
3. Burlayenko, V.N., Sadowski, T., Dimitrova, S.: Three-dimensional free vibration analysis of thermally loaded FGM sandwich plates. Materials **12**(15), 2377 (2019). https://doi.org/10.3390/ma12152377
4. Abdi, R.E., Pinto, R.L., Guérard, G., Capena, C.: Lifetime of optical fibers submitted to thermo-mechanical stresses. In: Cioboată, D.D. (ed.) ICoRSE 2021. LNNS, vol. 305, pp. 23–31. Springer, Cham (2022). https://doi.org/10.1007/978-3-030-83368-8_3
5. Kondratiev, A., Píštěk, V., Purhina, S., Shevtsova, M., Fomina, A., Kučera, P.: Self-heating mould for the composite manufacturing. Polymers **13**(18), 3074 (2021). https://doi.org/10.3390/polym13183074
6. Carslaw, H.S., Jaeger, J.C.: Conduction of Heat in Solids. Clarendon Press, Oxford (1959)
7. Carrera, E., Fazzolari, F.A.: Thermal Stress Analysis of Beams, Plates and Shells: Computational Modeling and Applications. Academic Press, London (2016)
8. Rivera, M.G., Reddy, J.N.: Nonlinear transient and thermal analysis of functionally graded shells using a seven-parameter shell finite element. J. Model. Mech. Mater. **1**(2), 20170003 (2017). https://doi.org/10.1515/jmmm-2017-0003
9. Swaminathan, K., Sangeetha, D.M.: Thermal analysis of FGM plates – a critical review of various modeling techniques and solution methods. Compos. Struct. **160**, 43–60 (2017). https://doi.org/10.1016/j.compstruct.2016.10.047
10. Guo, Y., Jiang, Y., Wang, J., Huang, B.: 3D thermal stresses in composite laminates under steady-state through-thickness thermal conduction. Int. J. Appl. Mech. **12**(6), 2050065 (2020). https://doi.org/10.1142/S1758825120500659
11. Apalak, M.K., Demirbas, M.D.: Thermal stress analysis of in-plane two-directional functionally graded plates subjected to in-plane edge heat fluxes. Proc. IMechE Part L J. Materi. Des. Appl. **232**(8), 693–716 (2018). https://doi.org/10.1007/978-3-030-83368-8_3
12. Pernica, M., Letal, T., Losak, P., Nad, M., Reppich, M., Jegla, Z.: Transient thermal stress calculation of a shell and tube condenser with fixed tubesheet. Chem. Ing. Tech. **93**(10), 1590–1597 (2021). https://doi.org/10.1002/cite.202100036
13. Kushnir, R.M, Zhydyk, U.V., Flyachok, V.M: Thermoelastic analysis of functionally graded cylindrical shells. J. Math. Sci. **254**(1), 46–58 (2021). https://doi.org/10.1007/s10958-021-05287-5
14. Vattré, A., Pan, E.: Thermoelasticity of multilayered plates with imperfect interfaces. Int. J. Engineering Science **158**(103409), 1–30 (2020). https://doi.org/10.1016/j.ijengsci.2020.103409
15. Smetankina, N.V., Postnyi, O.V., Merkulova, A.I., Merkulov, D.O.: Modeling of non-stationary temperature fields in multilayer shells with film heat sources. In: 2020 IEEE KhPI Week on Advanced Technology (KhPIWeek), pp. 242–246. IEEE (2020) https://doi.org/10.1109/KhPIWeek51551.2020.9250139
16. Washizu, K.: Variational Methods in Elasticity and Plasticity. Pergamon Press, Oxford (1982)
17. Smetankina, N., Kravchenko, I., Merculov, V., Ivchenko, D., Malykhina, A.: Modelling of bird strike on an aircraft glazing. In: Nechyporuk, M., Pavlikov, V., Kritskiy, D. (eds.) Integrated Computer Technologies in Mechanical Engineering. AISC, vol. 1113, pp. 289–297. Springer, Cham (2020). https://doi.org/10.1007/978-3-030-37618-5_25

18. Shupikov, A.N., Smetankina, N.V., Svet, Ye.V.: Nonstationary heat conduction in complex-shape laminated plates. Trans. ASME. J. Heat Transfer **129**(3), 335–341 (2007). https://doi.org/10.1115/1.2427073
19. Samieian, M.A., et al.: Temperature effects on laminated glass at high rate. Int. J. Impact Eng. **111**, 177–186 (2018). https://doi.org/10.1016/j.ijimpeng.2017.09.001

Investigation of the SLAM Performance of the Multirotor MAV Developed for the Analysis of Orchards

Şahin Yıldırım(✉) and Burak Ulu

Erciyes University, Engineering Faculty, Mechatronic Engineering Department, Kayseri, Türkiye
sahiny@erciyes.edu.tr

Abstract. Multirotor Micro Aerial Vehicles (MAVs) can operate in confined environments that are complex and insufficient to Global Positioning System (GPS) signals. In addition, studies on the autonomous implementation of the desired tasks and the development of decision-making features against changing environmental conditions offer the opportunity to apply these robotic systems in a wide field. Operating in such environments presents a lot of problems such as positioning and mapping. In this paper, it is proposed to use the Simultaneous Localization and Mapping (SLAM) method in order to minimize the errors of GPS signal data while following the trajectory determined in the fruit garden of a flying robotic system developed for use in agricultural implementations. For this proposed solution, a lidar sensor is placed on the flying robotic system. As a result, the success of the SLAM method on the trajectory tracking performance of this robotic system between tree corridors was evaluated by testing it in the experimental model environment.

Keywords: Robotic · Micro aerial vehicles · SLAM

1 Introduction

In this paper, the results of the applied SLAM method to increase the position accuracy of the flying robotic system designed for fruit detection and analysis in order to increase the production efficiency in orchards are evaluated. The general purpose of the designed robotic system is to collect data about the state of fruits with image processing methods and to evaluate them, and however a flying robotic system is being developed to perform the task. This robotic system will move autonomously through the corridors consisting of fruit trees shown as in Fig. 1. In this paper, reference trajectory tracking of the robotic system by performing position verification with the SLAM method is examined. As a result, the obtained data will be evaluated using the developed convolutional neural network model, and data such as fruit growth and disease symptoms, will be obtained results about the state of the orchard. In one of the studies reviewed in the literature, Lee et al. [1] aimed to significantly reduce the weight of the robot by carrying the data to a remote cloud system instead of applying the object recognition process using image

D. D. Cioboată (Ed.): ICoRSE 2022, LNNS 534, pp. 247–254, 2023.
https://doi.org/10.1007/978-3-031-15944-2_23

processing technology on the unmanned aerial vehicle. Although this method can only be applied to collect data from the garden, it is thought that it may cause problems due to delays that may occur, since it will also be used to determine the behavior of the robot, such as object detection and obstacle avoidance. In addition, keeping the information in the cloud environment should also be evaluated in terms of data security.

In order to carry out this study, the quadrotor flying robotic system, which is designed to perform the task determined in the optimum trajectory, must be able to detect its position in a partially closed environment (e.g. arboraceous region) with no or insufficient GPS signal. There are various studies in the literature [2–4] on this subject, and it is seen that especially real-time kinematics (RTC) method and sensor fusion techniques are used. In Kok [5], one of the studies on SLAM, it is observed that the Extended Kalman Filter (EKF) gives consistent results in the SLAM problem by using the features of the images obtained from the camera on the vehicle to obtain different landmarks, in single and multiple robots equipped with image sensors. Due to object detection with UAVs is a difficult task compared to other object detection methods, object detection performance; image quality depends on factors such as how far from the ground the image is taken [6]. In this paper, in order to increase the accuracy of the satellite positioning system, the results of the position determination performed in a semi-enclosed environment using SLAM were evaluated.

Fig. 1. Apple orchard where work will be done in real environment.

2 Methodology

The study to increase GPS positioning errors using SLAM method of the flying robotic system designed to analyze fruits with image processing method was carried out in the following steps: Step 1: Simulation Environment Development, Step 2: SLAM Method Application in Gazebo Environment, Step 3: Pixhawk Offboard Control.

2.1 Simulation Environment Development

Because of using novel software, the control of the developed robotic system may be risky. Therefore, the system will firstly run in the simulation environment. For this, the hector_quadrotor package has been modified and used on the Gazebo platform.

2.2 SLAM Method İmplementation in Gazebo Environment

SLAM is a method that estimates a model of the system's environment from a sensor data stream and its own motion according to this model [7]. SLAM is frequently used, especially in indoor areas, to map the environment where robotic systems are located and determine their positions. In this study, by considering regular tree corridors in orchards as a semi-enclosed environment a SLAM-based approach is proposed in order to avoid the negative results of errors in GPS data. In this paper, the results of this proposed approach are evaluated by testing in the Gazebo virtual environment, Fig. 2, before the experimental study.

Fig. 2. SLAM implementation on Gazebo 3D environment.

2.3 Pixhawk Offboard Control

In Fig. 3, the communication of the Pixhawk flight control board used in the experimental system with other programs is shown schematically. This communication is made according to the Mavlink protocol, and this protocol is made compatible using Mavros for the ROS platform. As a result, our ROS-based experimental system will be able to communicate with the flight control card, external controller, and other sub-programs using this protocol.

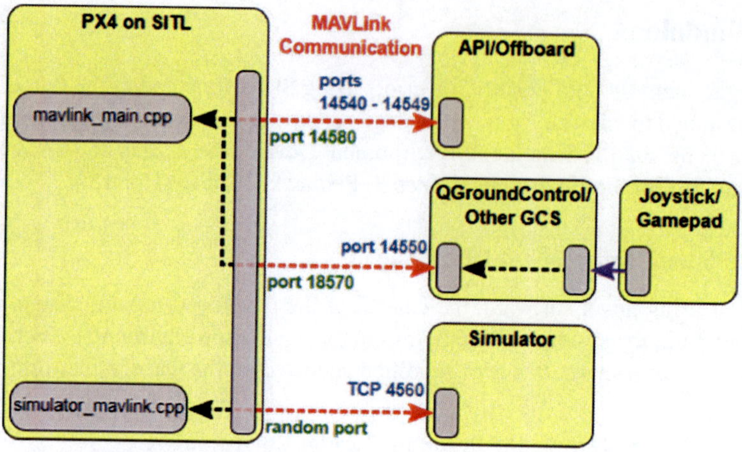

Fig. 3. Communication between robotic sub-systems [8].

3 Experimental System

The flying robotic system used in this experimental study is developed on the ROS platform, and the system architecture of the robot is as in Fig. 4. ROS is a software platform that provides an operating system for robot software developers. The possibility of working in compatible with many robotic systems and sensors and the support of many libraries are made ROS the reason of choice for robot developers.

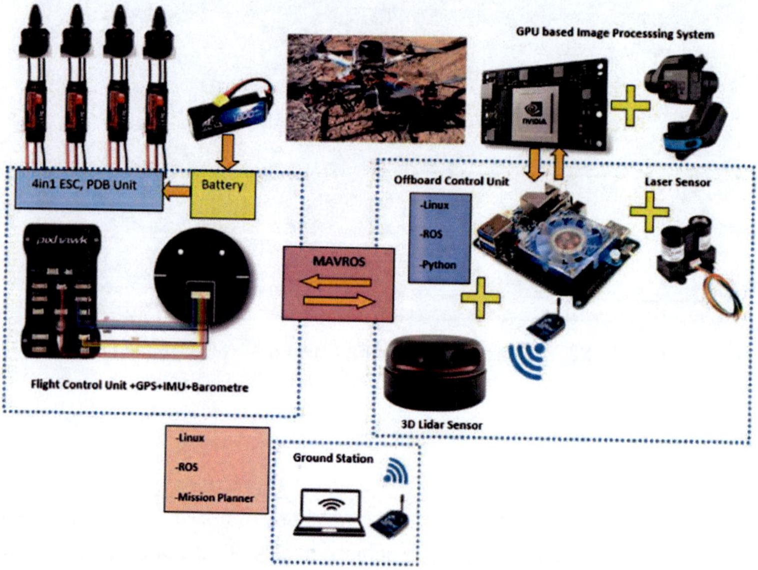

Fig. 4. Communication between robotic sub-systems.

Odroid XU-4, ARM architecture controller is used to provide offboard control on the experimental system, and Pixhawk Cube controller is used to perform flight control of the robot. RP Lidar A2 lidar sensor is preferred as the laser scanner required for SLAM application. In addition, the equipment that will be active in the image processing and artificial intelligence studies shown on the general structure of the system is not used within the scope of this study.

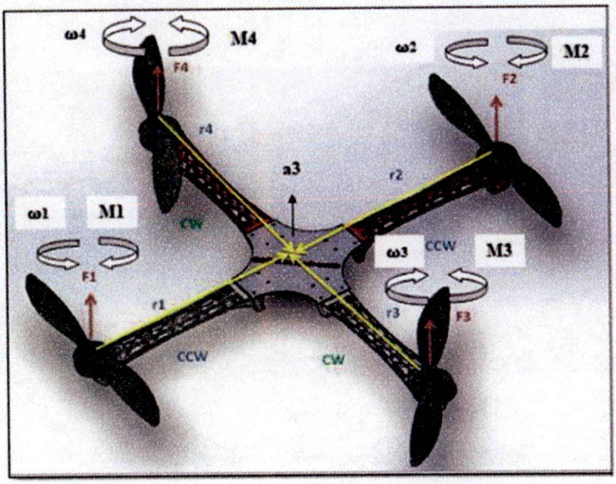

Fig. 5. Physical structure of the quadrotor system.

The basic mathematical equations of the flying robotic system, which physical structure is shown in Fig. 5, are given in (1) and (2):

$$F = F_1 + F_2 + F_3 + F_4 - mga_3 \tag{1}$$

$$M = r_1 \times F_1 + r_2 \times F_2 + r_3 \times F_3 + r_4 \times F_4 + M_1 + M_2 + M_3 + M_4 \tag{2}$$

In this paper, the SLAM performance of the experimental system in the ROS-based Gazebo 3D simulation environment is investigated. For this, the hector_slam ROS package developed by the Hector robot team of Darmstadt Technical University is used in the simulation study [9]. The SLAM problem requires calculating the probability distribution for all k times:

$$P(x_k, m|Z_{0:k}, U_{0:k}, x_0) \tag{3}$$

4 Results

The results obtained as a result of the simulation studies are shown in Fig. 6. It is observed that the SLAM estimation occurs with higher accuracy depending on the motion change of the robot. In addition, the accuracy of the robot's trajectory estimation increases accordingly. It can be said that the simulation results are promising for experimental studies, considering the GPS errors of 1.5–2 m in the outdoor environment between the orchards.

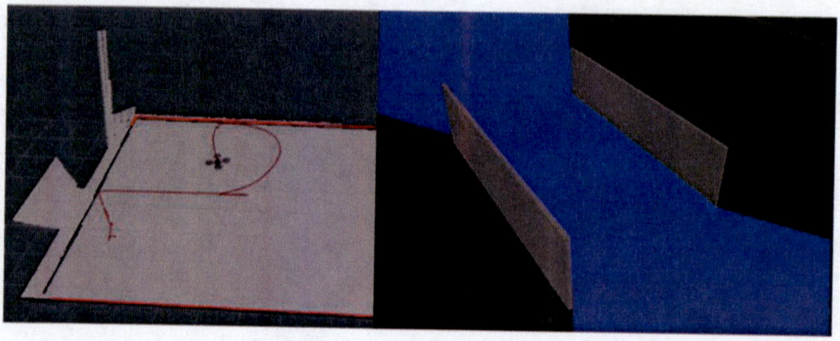

a

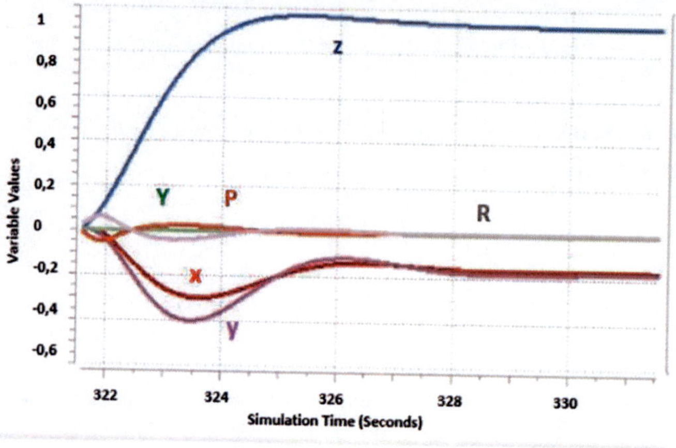

b

Fig. 6. (a) Generating Robot's trajectory using SLAM. (b) Time-dependent change of the robot position on Gazebo Simulation.

In Fig. 6(b), x (red), y (purple), z (blue) are linear position of the aerial robot and Y (green), P (orange), R (grey) are orientation (Yaw, Pitch, Roll) values.

Finally, the robotic system was tested in the apple garden under limited area and conditions. In the first experiment, flight was carried out with GPS signals, Fig. 7(left). The next flight was completed with positioning using the SLAM method, Fig. 7(right).

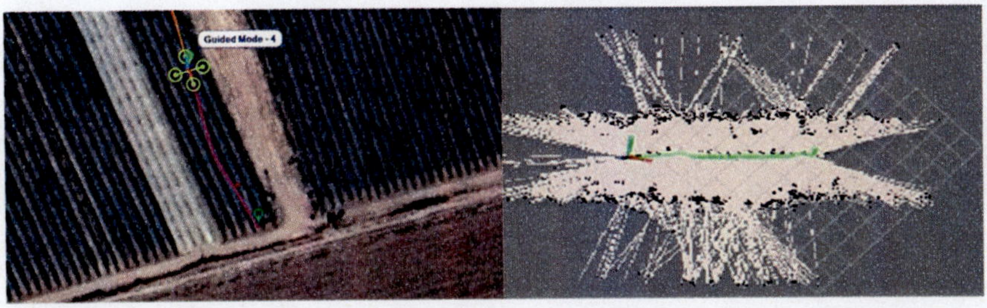

Fig. 7. The generated trajectory using SLAM compare with using GPS signal.

5 Conclusions

In this paper, which is prepared within the scope of the thesis, the test results of the proposed method in the 3D simulation environment are evaluated before the experimental studies to be carried out on the positioning of the robotic system developed to increase agricultural productivity and the creation of the map environment. It is obtained that positive results on the developed robotic system using SLAM method for the location and map detection in the indoor environment and partially closed outdoor area with consisting of trees.

Acknowledgments. We would like to thank Erciyes University Scientific Research Projects Coordination Unit (ERU/BAP).

This work is supported by Erciyes University Scientific Research Projects Coordination Unit (ERU/BAP) under project codes FBA-2021-10882.

References

1. Lee, J., Wang, J., Crandall, D., Sabanovic, S., Fox, G.: Real-time, cloud-based object detection for unmanned aerial vehicles. In: First IEEE International Conference on Robotic Computing 2017, IRC, pp 36–43. IEEE, Taichung (2017)
2. Kiss-Illes, D., Barrado, C., Salami, E.: GPS-SLAM: an augmentation of the ORB-SLAM algorithm. Sensors **19**(22), 1–22 (2019)
3. Matus-Vargas, A., Rodriquez-Gomez, G., Martinez-Carranza, J.: A monocular SLAM-based controller for multirotors with sensor faults under ground effect. Sensors **19**(22), 1–20 (2019)
4. Chan, S.H., Wu, P.T., Fu, L.C.: Robust 2D indoor localization through laser SLAM and visual SLAM fusion. In: International Conference on Systems, Man, and Cybernetics 2018, pp 1263–1268. IEEE, Miyazaki (2018)
5. Kök, M: Simultaneous Localization and Mapping for Unmanned Aerial Vehicles. Bilkent Üniversitesi, Fen Bilimleri Enstitüsü, Ankara (2008)
6. Böyük, M., Duvar, R., Urhan, O.: Deep learning based vehicle detection with images taken from unmanned aerial vehicles. In: Innovations in Intelligent Systems and Applications Conference 2020, ASYU, pp 1–4. IEEE, İstanbul (2020)
7. Concha, A., Loianno G., Kumar, V., Civera J.: Visual-inertial direct SLAM. In: International Conference on Robotics and Automation 2016, ICRA, pp 1331–1338. IEEE, Stockholm (2016)
8. LNCS. https://docs.px4.io/v1.12/en/simulation/. Accessed 20 Mar 2022

9. Meyer, J., Sendobry, A., Kohlbrecher, S., Klingauf, U., von Stryk, O.: Comprehensive simulation of quadrotor UAVs using ROS and Gazebo. In: Noda, I., Ando, N., Brugali, D., Kuffner, J.J. (eds.) Simulation, Modeling, and Programming for Autonomous Robots. LNCS (LNAI), vol. 7628, pp. 400–411. Springer, Heidelberg (2012). https://doi.org/10.1007/978-3-642-34327-8_36

Maintenance of Hydraulic Components on Multifunctional Stands

Liliana Dumitrescu[1](✉), Stefan Sefu[1], Lepadatu Ioan[1], Radu Radoi[1], and Laurențiu Nicolae[2]

[1] INOE 2000 - Subsidiary Hydraulics and Pneumatics Research Institute, 040558 Bucharest, Romania
lilianad.ihp@fluidas.ro
[2] SC PROFLEX Automotive, Reghin, Romania
http://www.ihp.ro, http://www.proflex.ro/

Abstract. Testing in the maintenance process for hydraulic components is usually done on stands dedicated to each type of device. In the event that a company has greater financial possibilities, a multifunctional stand can be made, which allows the evaluation of the performance for several types of devices: pumps and rotary motors, linear motors, distributors and valves, etc. Such a stand is structured around a central hydraulic power generation unit, which works together with dedicated subassemblies (devices). The article presents such a test stand, whose testing capabilities cover the needs of a hydraulic systems maintenance company, but can also be used for teaching purposes in technical faculties where hydrostatic drives are studied.

Keywords: Maintenance · Hydraulics · Testing

1 Introduction

Although in recent years, the Romanian industry has started a modernization process, which also involves the replacement of some old means of production and machinery, there are still many old equipment that are kept in operation, most of the time due to the high price that require their replacement. This situation is also found among hydraulic equipment, whether it is stationary systems (industrial hydraulics) or mobile hydraulics. Many times, hydraulic components are repaired and reintroduced into the productive circuit. The repair of some hydraulic equipment, especially those with a decisive role (pumps, motors) cannot be considered complete without a rigorous test on a specialized stand; unfortunately, many of the companies that offer repair services do not have adequate technical means of testing, and their lack cannot guarantee obtaining the necessary parameters at the level of the equipment or the systems that contain them.

On the other hand, in recent times, the qualified staff, who have adequate knowledge of the use, maintenance and repair of hydraulic equipment, has decreased numerically due to natural causes, leaving the country or due to professional reorientation; this creates a problem for firms that employ specialized hydraulic workers, technicians or engineers.

D. D. Cioboată (Ed.): ICoRSE 2022, LNNS 534, pp. 255–264, 2023.
https://doi.org/10.1007/978-3-031-15944-2_24

These companies want to hire qualified staff or raise the level of specific knowledge of the staff they have. As in any technical field, the deepening of knowledge cannot be done without specific technical means (stands) on which to present the detailed operation of hydraulic equipment.

There are also educational organizations that want to give students a better understanding of theoretical concepts through practical demonstrations. This is why the marketing of industrial equipment such as testing or teaching stands is a direction of development encouraged by the requests of companies in the field of hydraulic maintenance; this is the premise from which the Research Institute for Hydraulics and Pneumatics started, when it offered on the market a range of stands intended for testing hydraulic equipment.

The stands can be dedicated to a family of hydraulic devices or can be used to test several types of devices. At the request of the Polytechnic University of Timișoara, a complex stand was created, which combines the didactic character with the productive character; the stand must be able to be used in 3 main directions of activity:

a) in the practical activity associated with hydrostatic drives courses (laboratories);

b) in the verification process after repair of hydraulic equipment (services for companies);

c) by specialized staff (teachers, engineers) to provide training for employees of some companies.

Taking into account all these requests, the idea of a multifunctional stand was outlined, to be used both for didactic activity with students and to offer services to interested companies. The services mainly consist of testing repaired pumps and hydraulic motors or new linear hydraulic motors produced by the companies. The fields of work are industrial hydraulics, but also mobile hydraulics.

2 Structure and Characteristics of the Equipment

Taking into account the multiple functions that must be fulfilled by the stand and the fact that it was desired to test most of the main hydraulic devices, it was designed in a modular structure [1], which would allow the following devices to be tested in didactic or industrial mode:

- hydraulic pumps of various types, with fixed or variable displacement;
- rotary hydraulic motors;
- linear hydraulic motors (hydraulic cylinders);
- distribution and regulation equipment (classic or proportional distributors, valves, throttles).

The modules of the stand are the following:

A. SBS – the basic structure of the stand;

B. MHL – the module for testing linear hydraulic motors;

C. MHR – module for testing rotary hydraulic machines;

D. BAE – equipped auxiliary tank.

The transmission of commands to the 4 modules, as well as the management of the experimental data generated during the tests, is carried out by a specialized SCAD module - command and data acquisition system.

The modules and their correlation for performing the various tests can be found in the Fig. 1 below.

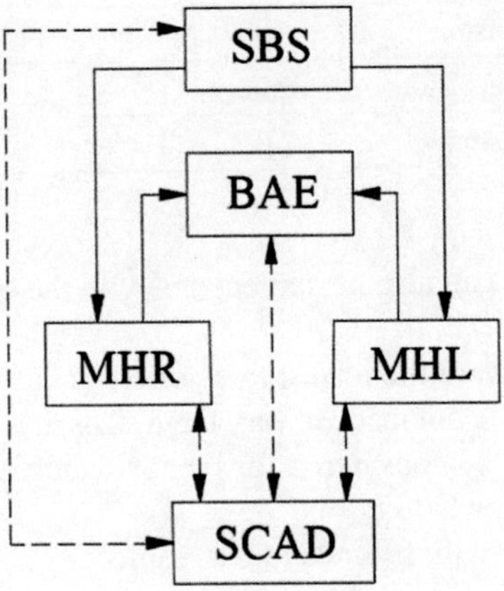

Fig. 1. The structure of the equipment

2.1 Main Features

The most important parameters were established considering the direction of the services provided, as they will only be requested by equipment in operation; for didactic direction, the parameters can be reduced.

Maximum drive power was established starting from a main pump working pressure of 315 bar and a flow rate of 75 l/min which covers most of the common equipment on the market.

The required drive power will be:

$$N = Q \cdot P/(600 \cdot \eta_t)[\text{kW}] \tag{1}$$

where:

Q – pump flow $[l/min]$.
P – outlet pressure $[bar]$.
η_t – the total efficiency, which is calculated with the formula

$$\eta_t = \eta_v \cdot \eta_{mh} \tag{2}$$

where:

η_{v} = volumetric efficiency,
η_{mh} = hydraulical-mechanical efficiency.

Table 1. Efficiency of hydraulic pumps [2]

Pump type	Overall efficiency %
External gear	85
Internal gear	90
Vane	85
Radial piston	90
Bent axis piston	92
Axial piston	91

The total efficiency [2] can also be appreciated with the help of the following Table 1:

As gear pumps, which have the lowest total efficiency according to the table above, are repaired less often and are not made in very large sizes, it is considered that the lowest total efficiency that should be considered for pumps which ensures at least 200 bar, is that of radial piston pumps, where $\eta_t = 0,9$.

In this case, calculating with formula (1), it results:

$$N = 75 \cdot 315/(600 \cdot 0,9) = 23625/540 = 43,75\,[\mathrm{kW}] \tag{3}$$

According to the catalogs of electric motor manufacturers, the closest covering value is 45 kW; therefore, the stand was designed with a 45 kW main pump drive motor.

Due to the modular organization of the stand, most hydraulic devices can be tested using one or two modules, to which the electronic/informatic module is added. With the help of the table below you can determine the possibility of testing and the modules involved (Table 2).

Table 2. Involved modules in devices testing

Tested apparatus	Involved modules
Hydraulic pumps	SBS + BAE + MHR + SCAD
Rotary hydraulic motors	SBS + BAE + MHR + SCAD
Linear hydraulic motors	SBS + BAE + MHL + SCAD
Control and regulation apparatus (distributors, valves, flow regulators)	SBS + SCAD

An auxiliary pump was located on the base module, to provide hydraulic power at low pressure (controls for devices on the hydraulic pumps); the pump can also be used to filter oil from the main tank. This pump has a power of 4 kW.

2.2 Description of the Modules

The SBS Module. It is the basic module of the stand, which generates the hydraulic energy required for the tests. It is based on a variable flow axial piston pump with flow and pressure regulator. The pump capacity variation as well as the maximum working pressure are electrically controlled. The maximum displacement of the pump is 75 cm^3/rot, and the maximum pressure is 320 bar. The auxiliary pump is also part of this module. The hydraulic oil is contained in a tank with a capacity of 320 l.

The auxiliary pump is integrated into a hydraulic circuit that allows the oil to be filtered when the pump is not working as a control pump; in this case the oil passes through a return filter.

On the tank cover there is a T-channel plate that acts as a work table for testing devices such as distributors, valves, etc.

Within this module there are 2 specific devices (transducers) for measuring the flow and pressure supplied by the main pump (Fig. 2).

Fig. 2. The basic structure of the stand

The MHR Module. With the help of this module, rotating hydrostatic machines (pumps and motors) can be tested. Testing a pump or motor involves determining the main parameters (pressure, flow, torque) related to the drive speed [3]. For their determination, the stand was equipped with specific transducers.

The testing process involves driving a hydraulic pump with a hydraulic motor; there are 2 working situations:

a. if it is desired to test a pump, its drive is done with the hydraulic motor that is part of the MHR module, and which receives hydraulic energy from the main pump, from the SBS module. The pump load is achieved hydraulically, with the help of a proportional valve, which is part of the BAE module.
b. if a hydraulic motor is to be tested, it is also powered from the main module and the hydraulic pump in the MHR module acts as a load; the speed and torque of

the hydraulic motor is measured using a transducer integrated in the drive coupling between the pump and the motor.

In conclusion, for the test of a pump or a hydraulic motor, the existing device is replaced in the MHR module with the one to be tested, and the parameters are evaluated with the same transducers, which are located either in this module or in the BAE module (Fig. 3).

Fig. 3. Subassembly for testing rotary hydraulic machines

The MHL Module. This module is used for testing linear hydraulic motors (hydraulic cylinders) [4]. Since in this case too a load is required on the cylinder rod, it is carried out hydraulically, with a load cylinder, which is placed coaxially with the test cylinder; fluid from the load cylinder passes through the proportional valve in the BAE module. Depending on the value to which this valve is adjusted, the resisting force is determined. A force transducer is placed between the rods of the 2 cylinders (Fig. 4).

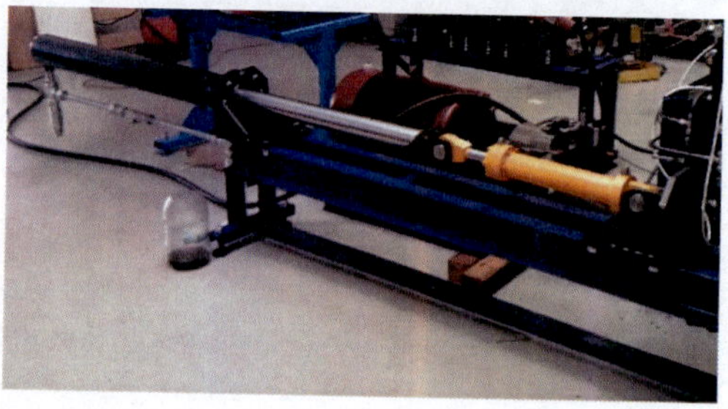

Fig. 4. Subassembly for testing linear hydraulic motors

The BAE Module. This module works in combination with the other testing modules; for this it has 2 important characteristics:

- on it is located the proportional load valve, with electric control;
- is built around an oil tank which is used by the load cylinder (when testing cylinders) and the load pump or test pump if testing a hydraulic pump.

The oil pressure before the proportional valve and the flow through the valve are measured with specific transducers located on this module. The tank that constitutes the basic structure has a capacity of 150*l* and is equipped with filters and an indicator for the level and temperature of the oil inside it. The connection with the other 2 modules is made with the help of hydraulic taps and hoses (Fig. 5).

Fig. 5. Equipped auxiliary tank

The SCAD Module. It has multiple tasks:

- provides the commands for: the electric motors driving the pumps, the distributors that establish the working configuration of the stand for a certain sample, the proportional valve;
- acquires, stores and processes the information received from sensors and transducers (flow, torque, speed, force, temperature).

Commands are transmitted from an electrical panel or, in part, from a computer or mobile device (phone, tablet) [5, 6].

For remote operation and monitoring of the bench (in teaching activity) a computer application has been developed that can run on any computer connected to the LAN. The application has a built-in Web server, which allows access from mobile devices to view parameters and make settings for values of flow and pressure rates (Fig. 6).

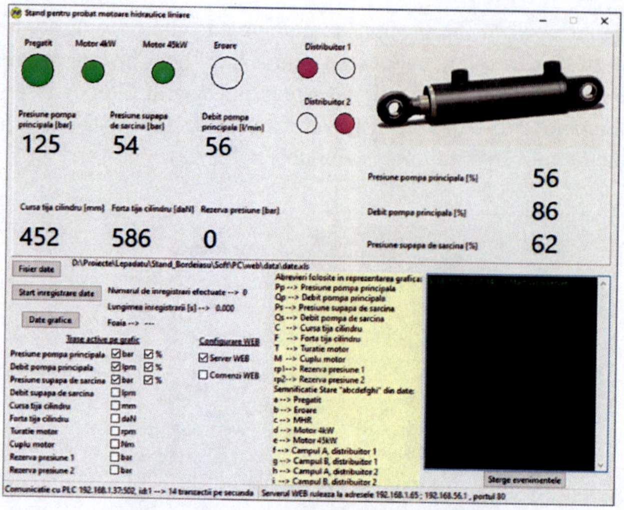

Fig. 6. The main panel of the PC application for control and data acquisition

3 Types of Checks Performed on the Stand for Hydraulic Components

Testing of the hydraulic components on the stand must be carried out in accordance with internal procedure and taking into account (if possible) the instructions from specifications or from technical prescriptions drawn up by the manufacturer.

For MHR, the stand allows the following checks to be performed:

- checking the nominal flow rate;
- checking the flow rate adjustability in idle and load operation mode;
- checking the operation at the set pressure;
- checking the internal and external leakage;
- checking the maximum flow rate (maximum displacement) when idling (no load);
- checking the maximum flow rate (maximum displacement) at maximum pressure;
- checking the torque at the motor/pump shaft.

For MHL, the stand allows the following checks to be performed:

- checking the minimum pressure for uniform and smooth (free shock) displacement of the piston and starting pressure;
- checking the minimum and maximum piston speeds;
- checking the pushing force and the traction force;
- checking the external and internal leakage;
- checking the brake at the end of stroke for cylinders with braking;
- checking the resistance of the cylinder to pressure.

For various hydraulic equipments, the stand allows the following checks to be performed:

- checking the static characteristic of the servovalves;
- checking the internal flow losses at check valves;
- checking the functional diagram of the hydraulic distributors;
- checking the electrohydraulic equipment - determining the frequency response;
- checking the electrohydraulic equipment - determining the response to the step signal;
- checking the electrically piloted valves - determining the adjusting characteristic;
- checking the pressure valves - determining the static characteristic;
- checking the adjustability of the pressure relief valves and determining the characteristic curve;
- checking the differential pressure / flow characteristic at the hydraulic distributors;
- checking the adjustability of the hydraulic throttles;
- checking the flow-stroke characteristic of the hydraulic throttles;
- checking the flow-current characteristic, at zero load for the hydraulic servovalves.

4 Recordings

Both in the educational process and in the testing of equipment for different beneficiaries, data recording is of great importance. The stand allows the recording, storage and processing of experimental data, including graphical presentation. The following parameters can be represented graphically: *Main pump adjusting pressure Pp* (%), *Set flow for main pump Qp* (%), *Set pressure at load valve Ps* (%), *Load cylinder stroke* (mm), *Force from load cylinder transducer* (daN), *Main pump flow rate Q1* (l / min), *Flow rate Q2 of pump from the MHR test subassembly* (l / min), *Torque to the shaft of the MHR test subassembly* (Nm), *Pressure P1 of the main pump (Pp)* (bar), *Pressure P2 of the hydraulic apparatus to be tested - transducer input of 250 bar, Pressure P3 of the hydraulic apparatus to be tested - transducer inlet 400 bar, Pressure P4 (Ps)* (bar) *from the load valve of the test subassembly MHR* and *Speed from pump-motor shaft of the MHR test subassembly* (rpm). Saving data is done by right-clicking in the chart area and then Exporting Data (Fig. 7).

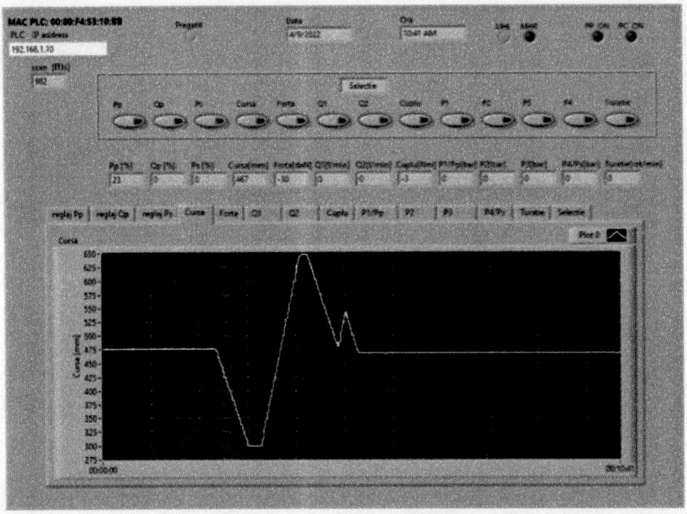

Fig. 7. Example of data acquisitions/records conducted on the stand: the stroke of the tested cylinder

5 Conclusions

The stand responds to growing requests from Romanian society regarding knowledge in the field of hydraulic equipment testing; it was designed for a better understanding of the phenomena associated with hydrostatic drives, but it can also be used by a company offering hydraulic equipment repair services.

Its design in the form of modules, which work correlated, depending on the devices tested, make it more accessible to companies specialized in the maintenance of certain families of devices, or even to manufacturers of hydraulic components (e.g. hydraulic cylinders), for running type tests or batch.

On the other hand, for use in the technical university area, the stand can be brought up to the current trends in remote ordering and tracking of trials.

Regarding the training level of the users, the equipment can be used both for technical personnel with minimal hydraulics qualification and for middle or higher cadres (technicians, engineers), together with a test program adapted to the level of each.

Acknowledgments. This paper has been developed in INOE 2000-IHP, as part of a project co-financed by the European Union through the European Regional Development Fund, under Competitiveness Operational Programme 2014–2020, Priority Axis 1: Research, technological development and innovation (RD&I) to support economic competitiveness and business development, Action 1.2.3 – Partnerships for knowledge transfer, project title: Development of energy efficient technologies in niche applications of the manufacture of on-demand mechanical-hydraulic subassemblies and maintenance of mobile hydraulic equipment, project acronym: MENTEH, SMIS code: 119809, Financial agreement no. 6 /25.06.2018. It has been financed under a project funded by the Ministry of Research, Innovation and Digitalization through Programme 1- Development of the national research & development system, Sub-programme 1.2 - Institutional performance - Projects financing the R&D&I excellence, Financial Agreement no. 18PFE /30.12.2021.

References

1. Dumitrescu, L., Chirita, A.P., Stefan, Ş.: Innovative stand for testing hydraulic pump and motors. In: Proceedings of 2018 International Conference on Hydraulics and Pneumatics - HERVEX, November 7–9, Baile Govora, Romania, pp. 32–37 (2018). ISSN 1454-8003
2. Casey, B.: Hydraulic Pumps and Motors: Considering Efficiency. Machinery Lubrication (3/2011)
3. Vasiliu, N., Vasiliu, D.: Acționări hidraulice și pneumatice, vol. I, Technical Publication, Bucharest (2005)
4. Avram, M.: Acționări hidraulice şi pneumatice. Echipamente şi sisteme clasice şi mecatronice, Univesity Publication, Bucharest (2005)
5. Călinoiu, C.: Sensors and translators (Sensors and transducers), vol. I, Technical Publishing House, Bucharest (2009)
6. Ilie, I., Blejan, M.: Distributed hardware and software architecture for monitoring and control of hydraulic drives. Hidraulica Mag. 4, pp. 46–50, December 2017, ISSN 1453–7303

Novel Features of Special Purpose Induction Electrical Machines Object-Oriented Design

Vladyslav Pliuhin[1], Sergiy Plankovskyy[1], Mykola Zablodskiy[2], Ihor Biletskyi[1], Yevgen Tsegelnyk[1](✉), and Volodymyr Kombarov[1]

[1] O.M. Beketov National University of Urban Economy in Kharkiv, 17 Marshala Bazhanova Street, Kharkiv 61002, Ukraine
y.tsegelnyk@kname.edu.ua

[2] National University of Life and Environmental Sciences of Ukraine, 19 Henerala Rodimtseva Street, Kyiv 03041, Ukraine

Abstract. The paper considers the approach to the design of induction (asynchronous type) electric machines with a solid rotor, based on the method of object-oriented analysis. Such features of the object-oriented programming ideology as hierarchy, inheritance, polymorphism, are transferred to the formation of methods for designing electric machines, invariant to their type. In previous works of the authors, this approach has been partially considered, but this paper reveals the features of induction machines with a solid rotor. The special purpose is the design features of the stator and rotor parts, in particular the rotor, combines the functions of the actuator, heater and mixer of bulk materials. Directly the rotor has the form of a metal tube, on the outer surface of which are welded turns of the auger, and in the middle are two inverted stators with a slot part on the outer cylindrical core surface. A specific feature of induction machines with a solid rotor is the determination of its resistance and reactance parameters.

Keywords: Object-oriented design · Induction motor · Programming · Mathematical model · Solid rotor · Parameters · Methodology

1 Introduction

Nowadays one of the main trends in the development of electrical machines (EM) is the improvement of methods for their designing, which helps to reduce the weight and dimensions of machines, improve their technical and economic indicators [1–4]. The development of new types of electrical machines is carried out by branch research institutes, institutes of the National Academy of Sciences of Ukraine, factories and specialized design bureaus.

Currently, the designer is inextricably linked with a PC, independently solves design problems, or uses third-party software [5, 6]. However, software has come a long way in its development and moved from a procedural to an object-oriented (OO) level a long time ago. As a result, the problem of adapting the existing procedural "book" methodology to the OO software environment constantly arises [7, 8].

D. D. Cioboată (Ed.): ICoRSE 2022, LNNS 534, pp. 265–283, 2023.
https://doi.org/10.1007/978-3-031-15944-2_25

The range of design tasks is so extensive that solving some of them with both standard and applied tools can take a lot of time and be irrational, and sometimes even impossible.

Thus, a progressive design technique must meet the following requirements [9, 10]:

- be as open as possible and include tools for creating a package of user databases;
- focus on market demand.

The transformation of the EM design methodology directly into OO terms [11–13] not only solves these problems, but also allows, simultaneously with the formation of the project structure, to generate a mathematical and software representation, allow to develop modules of various structure to solve narrow-profile and non-standard design problems. This will make it possible to automate complex calculations, select the necessary parameters from databases, exchange data with other projects and, ultimately, automatically receive a package of design documentation with the construction of a 2D or 3D model of unlimited complexity.

The existing design methods [5–8] based on the cascading organization of the calculation stages and the procedural representation of the project do not allow to implement the tasks associated with the following:

- there is no possibility of automated transfer of the entire project or part of it to create a new one, which has both common features with the base project and its own (*for example, replacing a squirrel-cage rotor in an induction motor with a winding rotor involves a complete redesign of the project from start to finish*);
- there is no mechanism for including data and dependencies of other projects into the existing one (for example, including the methodology for calculating reliability indicators in a project where they were not determined or modifying the calculation of reliability when replacing EM components);
- it is impossible to apply modern methods of intellectual optimization, such as the Cartesian product, neural or genetic algorithms [14–16];
- the classification of EM, in addition to being ambiguous, is not related to the organization of design and therefore does not allow, even at the stage of forming the representation of EM, as a kind [9], to synthesize design methods and give evaluation results of the obtained modifications in scientific research;
- the quality of development, the cost of human and material resources, the design time, which are especially relevant in the conditions of market relations and competition, with the traditional approach, are significantly inferior to OO systems;
- there is no synthesis of modern information technologies and design methodology, which would improve the efficiency of modern projects.

In this regard, the role of information technology in electrical engineering has a great importance, and the synthesis of the OO ideology [1–3] of knowledge representation with the methodology for designing and modeling EM will solve the above problems [10].

In most sources studied [5–8], a complex design problem that combines all the stages listed above is never solved. In addition, a unified approach to solving a single design stage has not been formed.

The actual lack of evidence-based approaches that could be used as the basis for the structural classification of EM has led to confusion in electrical engineering regarding the classification of a large number of structural and functional varieties of EM.

Structural-system studies, which have recently been carried out in the direction of creating a generalized theory of physical structures, determining the patterns of the general development of technical systems, cannot be used to solve fundamental and applied problems of electrical engineering, since they are of a general nature and do not take into account the structural features organization and specifics of electromagnetic processes that determine the functioning and behavior of electromechanical systems [9, 10, 17].

The general shortcomings of both traditional information systems and methods for representing the calculation, design and modeling of EM are weak adaptability to changes in the subject area and information needs, the inability to solve poorly formalized tasks that designers constantly deal with:

- analysis of problem situations that arise in computational processes;
- forecasting further use and reliability;
- decision making and choosing the best option in optimization problems.

Moreover, the existing classification and design algorithms for EM are based on outdated "procedural" principles. Modern object-oriented systems and software platforms need to change the principles of classification and structure formation of EM.

The philosophy of representing knowledge about the real world in terms of the interaction of objects provides a convenient environment for solving a large class of problems of designing and modeling complex systems. This allows developers to focus early on choosing the right kinds of objects and their behavior, without getting into the details of implementing calculation formulas and data structures.

Currently, OO programming languages [11, 12, 18, 19] are mainly used as software products and design tools for EM. At the same time, the algorithms described in numerous textbooks and reference books, the sequences for calculating EM ideologically correspond to the outdated and obsolete procedural approach in programming.

The designer takes the extra step of first converting a procedural technique into an object one, and then an object one into a software implementation.

The fundamental idea of object-oriented programming (OOP) is to combine data and actions performed on this data into a single entity, which is called an object.

As an example, in Fig. 1 the structure of the DC motor project is presented. If it is necessary to change the data of an object, then, obviously, this action will also be assigned to the methods of the object. This approach makes it easier to write, edit, and use a project. At the same time, a typical project consists of a set of objects interacting with each other by calling each other's methods.

When approaching the solution of a problem using the OO method, instead of the problem of splitting the task into functions, the question of splitting this task into objects is raised. Thinking in terms of objects is simpler and clearer than thinking in terms of functions, since software objects are like objects in the real world. Here is an example of such objects:

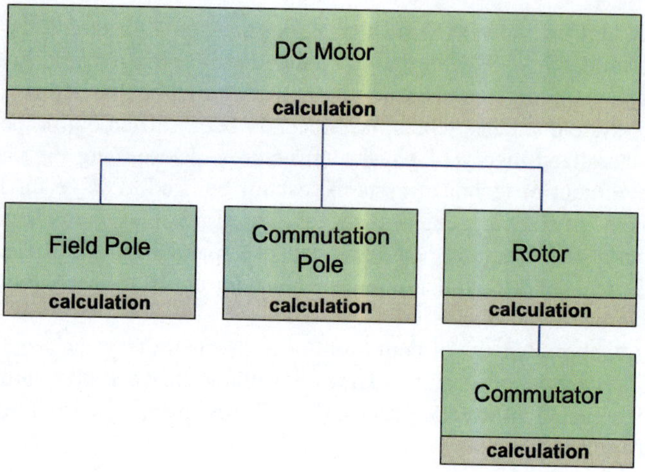

Fig. 1. DC motor project inheritance tree in OOP.

- physical objects: *stator, rotor, winding, electric motor*;
- parametric objects: *complex numbers, arrays, time*;
- graphic objects: *lines, rectangles, circles.*

Ease of reuse is an important advantage of object-oriented design [9, 10, 17]. If we compare a project formed by object-oriented technology (Fig. 2) with a procedural one (Fig. 2), then the modifications will be minimal. Moreover, the changes do not affect the base project at all, and the localization of data and formulas does not change!

Thanks to the inheritance mechanism, the new project will include only new data and formulas, and the rest will be taken from the base one and used as its own. An OO project allows to dynamically change the links between calculation blocks without affecting changes in the project structure. The organization of the object-oriented project is shown in Fig. 3.

The fundamental difference from the procedural scheme is that the phased project life cycle is replaced by a multi-level representation of the problem-solving process, which is obtained by a combination of top-down and bottom-up methods. In a procedural design approach, the calculation process is not iterative, but like the waterfall model, so changing the requirements would mean restarting the entire design process.

On the one hand, the detailing of the upper levels is in progress, on the other hand, the missing components of the upper levels are assembled from the lower levels. Adjacent and other nearby levels may intersect, i.e. have common components.

In the procedural approach, in contrast to the OO one, a consistent representation of the calculation program with the calculation blocks of the higher and lower levels is observed. Any variation in the data, the calculation blocks themselves in the upper levels leads to the inoperability of the entire project, located below the changes made, since the connection in the project with a structural approach is strictly sequential and rigidly connected both by the calculation blocks themselves and by their data. The sequence does

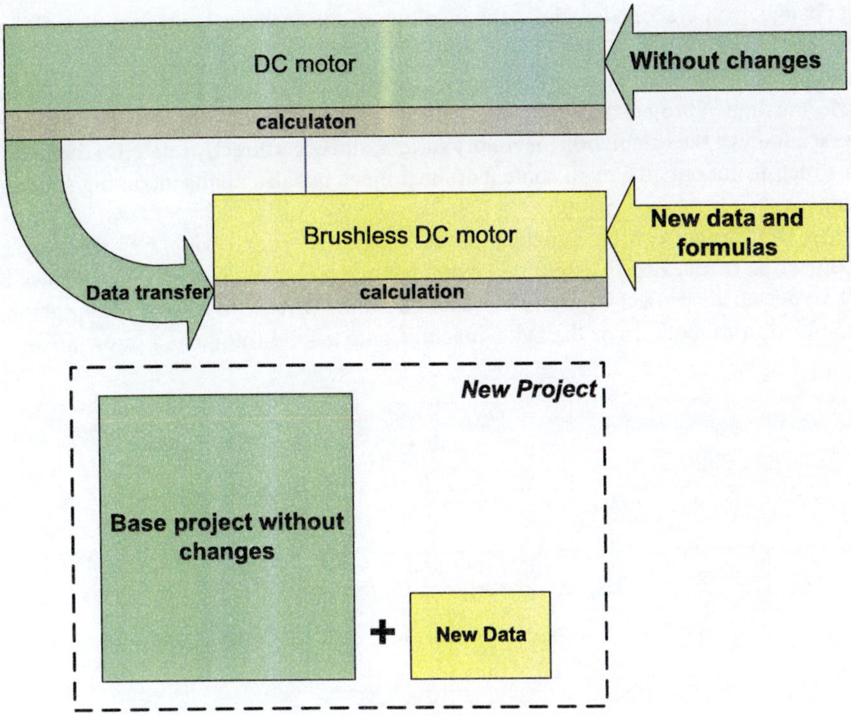

Fig. 2. Modification of the DC motor project into OPP.

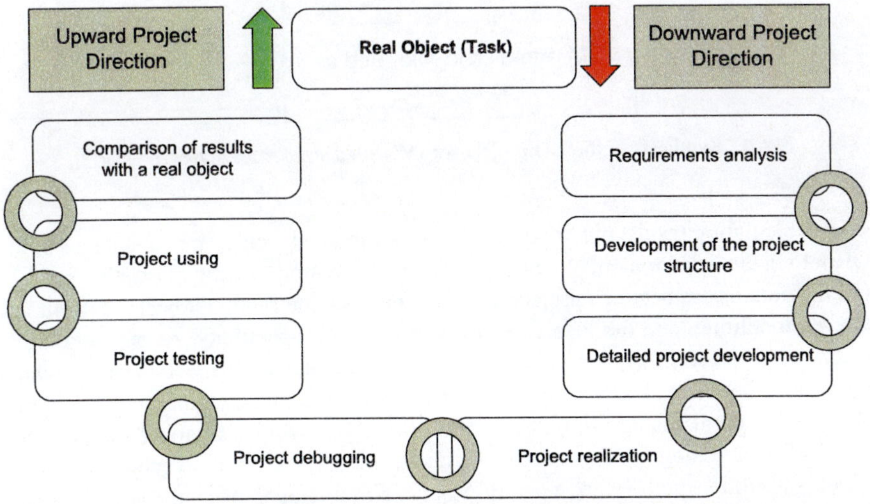

Fig. 3. Implementation of the OO process of creating a project.

not change even branched cycles – the ideology of the structural approach is strikingly different from OOP.

As a result, an increase to a greater or lesser extent, depending on the type of changes made, the time of project development, its debugging, and the quality of work performed. The accuracy of the calculation inevitably suffers, since it is directly related to the logic of constructing not only the main content of the project, but also all the incoming structural components.

An OO project can be imagined as an "*electrical engineering kit*" when a new modification of the EM is assembled from debugged blocks. So, the Fig. 4 shows the OO layout of the project of a screw induction motor (IM) with a solid rotor, which is assembled from modules of the IM, a machine with a solid rotor and a screw mixer.

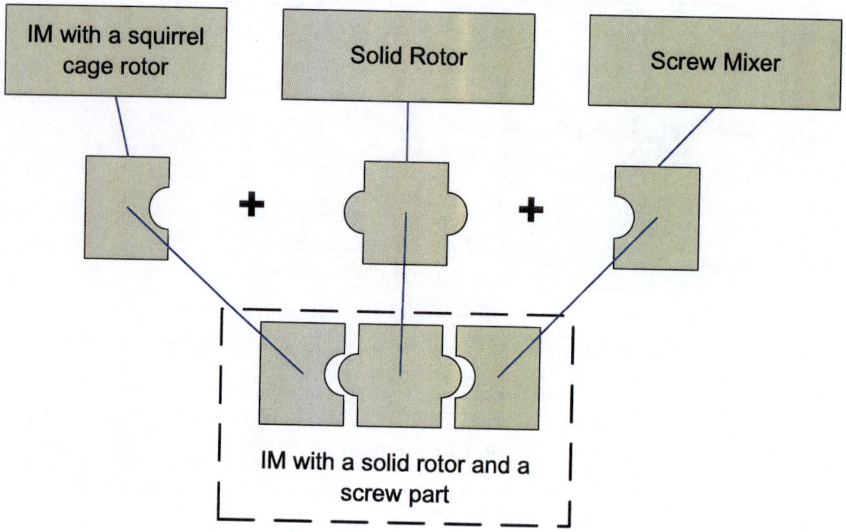

Fig. 4. Synthesis of the OO project for IM with a solid rotor and screw part.

The calculation results obtained by the authors showed significant advantages of the OOP both in the development of new projects and in the modification of existing ones. In OOP, the boundaries between project levels are blurred, their number becomes arbitrary and even indefinite, and the levels themselves lose their specificity. At the same time, the concept of *level diffusion* can be introduced in the design - a special state in which relationships between classes of the object structure led to the fact, that the data of one class cannot be calculated without the data of another, which is part of the common inheritance tree. Using OOP, we create flexible projects written in an economical way. At the same time, there is no separation between the stages of analysis and design, which improves communication between designers, from the beginning to the end of the project.

The creation of new projects in OOP, as well as the modification of existing ones, is based on the experience of previous developments, which leads to a significant reduction in development time, reduces the likelihood of involving errors into a new project.

The capabilities of the OOP allow including modules for assessing the thermal state, analyzing reliability indicators, and giving an economic assessment in the project at the assembly stage, referring to the database on the developed modules of basic projects.

In the proposed method for designing an EM, an OO approach is implemented in organizing data and calculation procedures, OO optimization algorithms are involved, which allow to carry out a complex design of EM, including the design of a base machine, its optimization, calculation of electromagnetic transients, distribution of an electromagnetic field. That's brings the formal representation of the project as close as possible to OO programming languages, which increases design efficiency, reduces time and material costs for design, simplifies project modification, allow to reuse existing developments in new projects, and reduces the time and material costs for developing new projects.

2 Statement of the Design Problem

As an object of design, a Screw Electric Motor (SEM) was chosen, manufactured at factory conditions, and implemented into the production process of mixing and drying coal sludge at several factories of coal-mining industry in Ukraine.

Screw electric motor combines the functions of drying, mixing and transporting bulk materials [20–23]. The design of the SEM is shown in Fig. 5. Photos of the manufactured SEM with the power of the 55-kW driven module and 45-kW of the braking module are shown in Fig. 6.

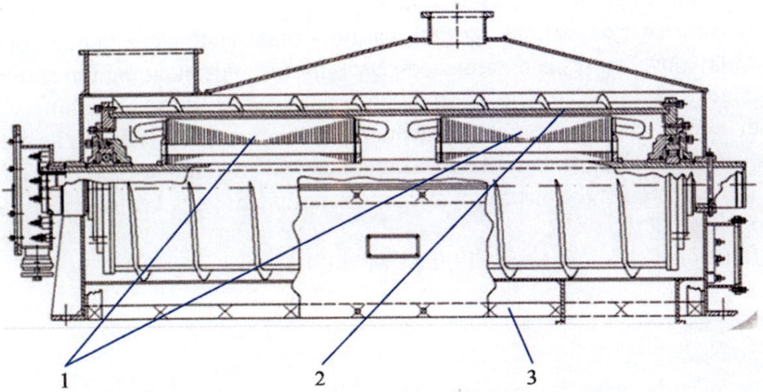

Fig. 5. Structural diagram of a double-stator SEM: 1 – stators; 2 – solid rotor with a screw part; 3 – electric heating system.

SEM (Fig. 5) consists of two modules, operating in the opposition mode. Two stators, mounted on a common hollow shaft, create oppositely directed electromagnetic torques, providing the low rotation speed of the hollow cylinder of the common rotor without the use of a mechanical gearbox. The screw rotor, in addition to the function of moving the working material, simultaneously provides heating of the latter.

The overall efficiency of SEM is very high, which ensures the effective implementation of the energy saving principles [1–3]. The creation of SEM and technologies based

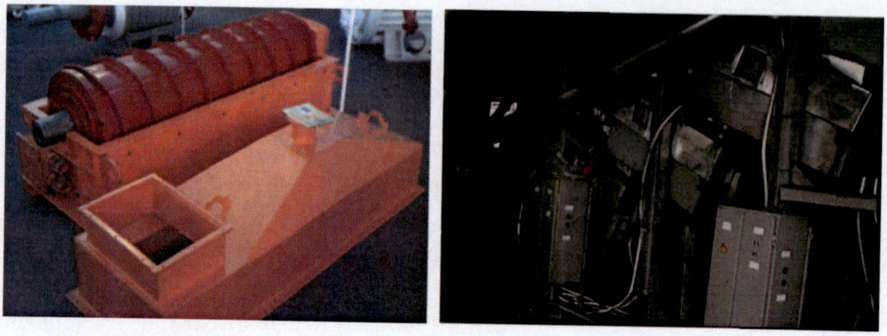

Fig. 6. SEM at the coal-mining exhibition and in the industrial workshop.

on them is grounded on the idea of combining heating, transporting and mixing functions in only one electromechanical device, combining thermal energy and directing the latter to the raw material processing zone [23]. The SEM uses the construction of a hollow solid rotor, which simultaneously performs the functions of an IM rotor, a heater, an actuator and a protective housing. In this case, the surface of the rotor is cooled by the raw material, which is processed. An additional cooling agent in a SEM can be air and low-melting materials with high heat capacity and latent heat of fusion [17, 20].

To obtain a class representation of a SEM, it is necessary to highlight the keywords that are an abstract representation of the EM. The next keywords can be distinguished [10]: *stator; stator internal; rotor; solid rotor; stator slot; winding; internal stator with slots and winding; EM with an external rotor.*

The above keywords are not just a designation of the structural components of the EM and, as it may mistakenly seem, are necessary only for understanding the composition of the designed object. Key words should be considered as links, containing variables and calculation blocks of a separate element of the structure, endowed with its properties and characteristic behavior under known influences. Class variables and computational blocks that characterize the operation and properties of SEM can be determined already at an early stage of the OOP.

The functional characteristics of the SEM will be as follows:

- *parameters*;
- *energy indicators*;
- *coefficients and constants.*

It is advisable to include the selected calculation blocks and variables in the corresponding classes as functions and variables of the class itself, and not as separate classes. The listed properties of the SEM, which determine the design composition, can also be divided into separate classes. However, this would result in multiple nested classes within the respective main classes. For example, for the "*stator*" class, it will be necessary to create nested classes "*stator magnetic system*", "*stator parameters*", "*stator energy indicators*", "*stator coefficients and constants*".

An example of a class diagram of a multilevel hierarchy of a SEM is shown in Fig. 7.

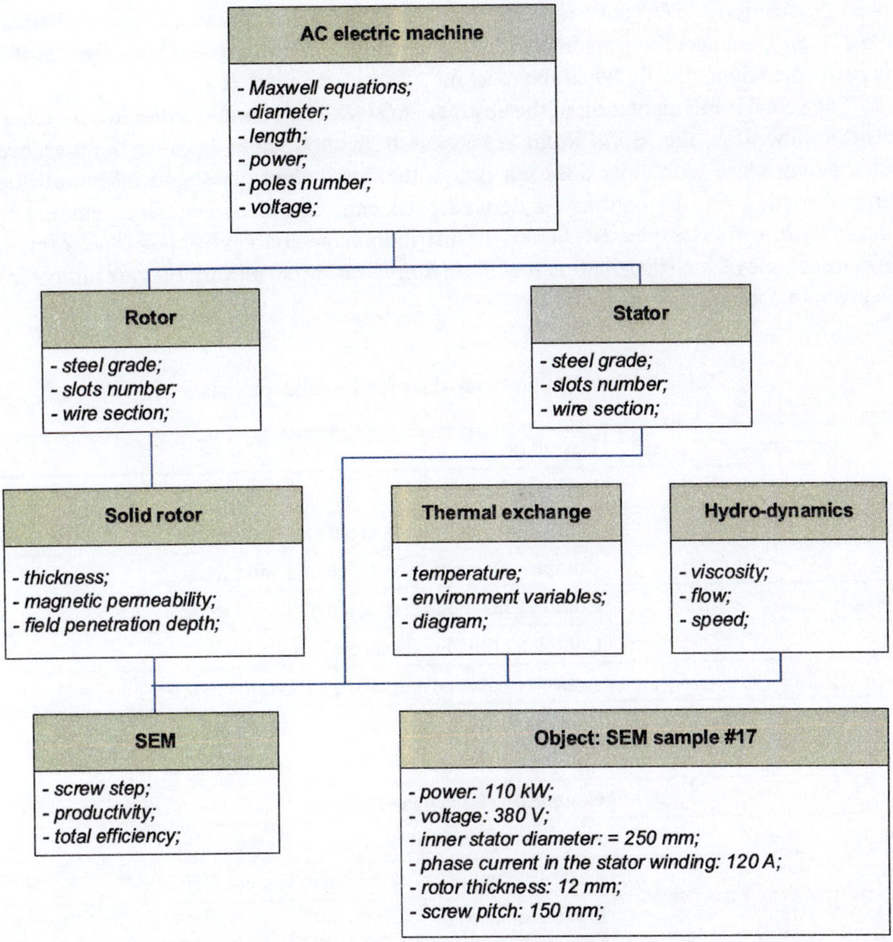

Fig. 7. Hierarchy of a SEM classes.

As a result of inheritance, the class "SEM" will acquire the features of all classes that are higher in the hierarchical tree but taking into account their own individual features.

By deriving classes from base classes, one can effectively use the base class algorithms for own needs. The concept has a parallel in nature: DNA can be seen as the basic material from which any creature can be created. Each organism reuses DNA to reproduce its own species. In design, reuse is done through inheritance.

With regard to electromechanical classes, Maxwell's equations [24–26] can be considered as the basic material, which allow to describe various configurations of electromagnetic fields. Their further development, under given initial and boundary conditions, set restrictions, determines what kind of EM will be created.

On Fig. 8 an example of a class object with inherited parameters of base classes is shown (the figure shows selectively some of them). On Fig. 8 main composition "*Induction motor*", which is a direct descendant of the abstract base class EM, includes two compositions "*STATOR*" and "*ROTOR*", each of which contains a nested abstract class "*Slot*", the specific implementation of which ("*Trapezoidal Slot*" or "*Oval Slot*") is performed dynamically when the machine classes are initialized.

The actual implementation of the abstract "*ROTOR*" class is the "*Solid Rotor*" class. At the same time, the "*Solid Rotor*" class can be without slots, because the presence of a parent class with slots does not oblige the descendant classes to inherit all the characteristics. On the contrary, a derived class can cut off unnecessary features, or introduce its own. For selected classes, the distribution of variables by class is performed. As an example of such distribution for "*Solid Rotor*" class (not all variables are indicated) is given in Table 1.

Table 1. Distribution of variables for the solid rotor class.

Variable name	Description
h_r	Rotor thickness
h_v	Penetration depth of an electromagnetic wave
z_2	Complex electrical resistance of a solid rotor
Z_{2m}	Complex magnetic resistance of a solid rotor
z_{2k}	Complex critical resistance of a solid rotor
r_{2k}	Active critical resistance of a solid rotor
x_{2k}	Reactive critical resistance of a solid rotor
z_{20}	Complex resistance of a solid rotor at idle mode
A_c	Complex Neumann coefficient
a	Active component of the Neumann coefficient
b	Reactive component of the Neumann coefficient
P_{rot}	Losses in the rotor
$s'_{\kappa p}$	Reduced critical slip
μ_2	Magnetic permeability of a solid rotor
ρ_2	Specific electrical resistance of a solid rotor
t_2	Solid rotor temperature

As can be seen from the Table 1, even at the stage of analysis, all calculated variables are distributed into the corresponding classes. A user interface (data exchange with classes) and calculation functions are added to the class.

Thus, a complete OO project of the EM is obtained, expressed by analytical symbols and formulas. The design synthesis of a SEM will differ from an IM with a squirrel cage rotor only by the body of the "*Solid Rotor*" class and its variables and methods, so will only consider this new class added to the hierarchical inheritance tree.

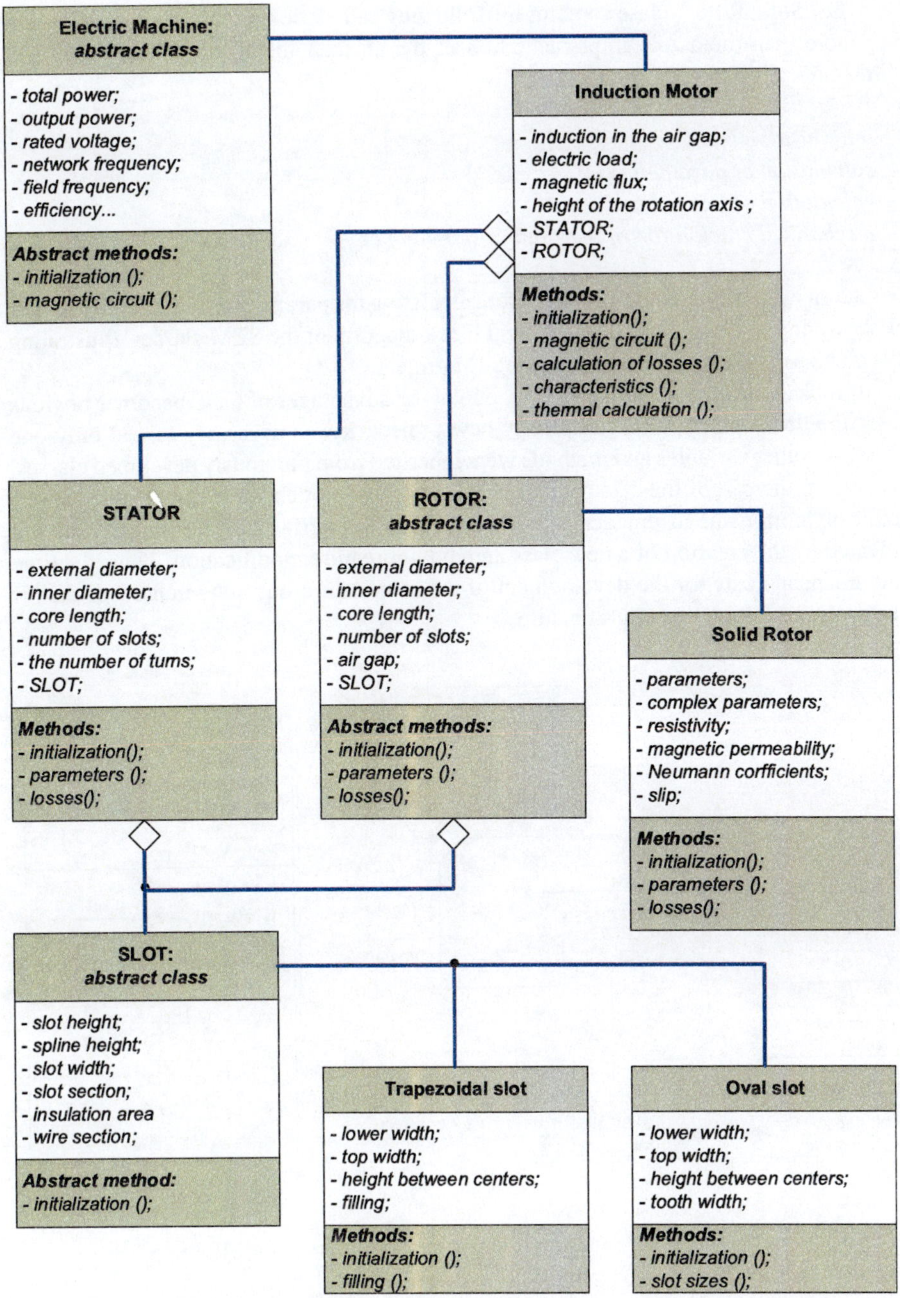

Fig. 8. Hierarchy of inheritance of the electromechanical structure of the SEM.

The "Solid Rotor" class contains the following calculation methods, which are nothing more than the actual implementation of the abstract methods of the parent class "*ROTOR*":

- *initialization ();*
- *calculation of parameters ();*
- *calculation of losses ();*
- *mechanical calculation of the shaft for strength ().*

As an example, the block diagram for calculating the parameters of a solid SEM rotor is shown in Fig. 9. In Fig. 10 is a general block diagram of the SEM classes, illustrating the changes made to an already existing IM project.

If look at the amount of work done, one of the advantages of OOP becomes obvious - for the development of a completely new project, it was necessary to add only one class. All other variables and methods were inherited from previously developed classes. Moreover, the call of the calculated functions has also not changed due to abstractions and polymorphism. In practice, this means that the interface of the project was not affected by the creation of a new class and did not require modification. Thus, the time and financial costs for the development of the project turn out to be much less than if the project was of a procedural nature.

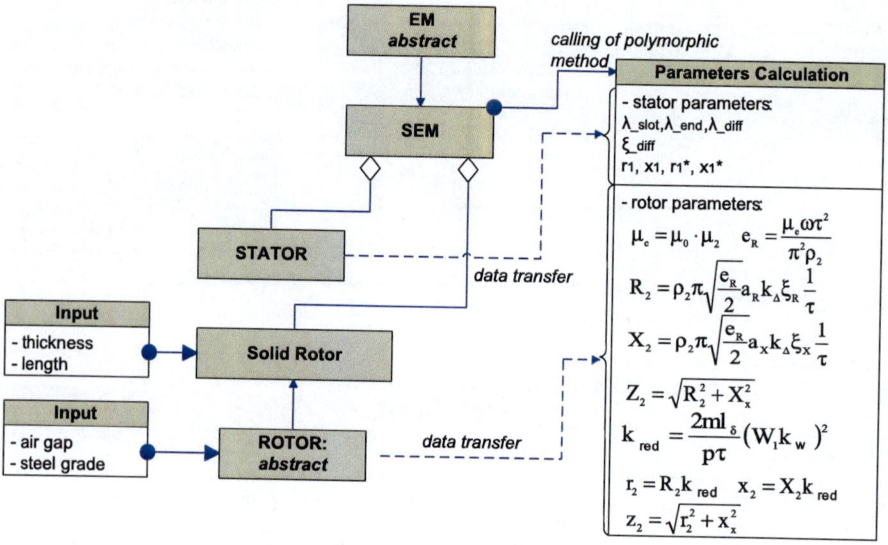

Fig. 9. Block diagram for calculating the parameters of the SEM rotor.

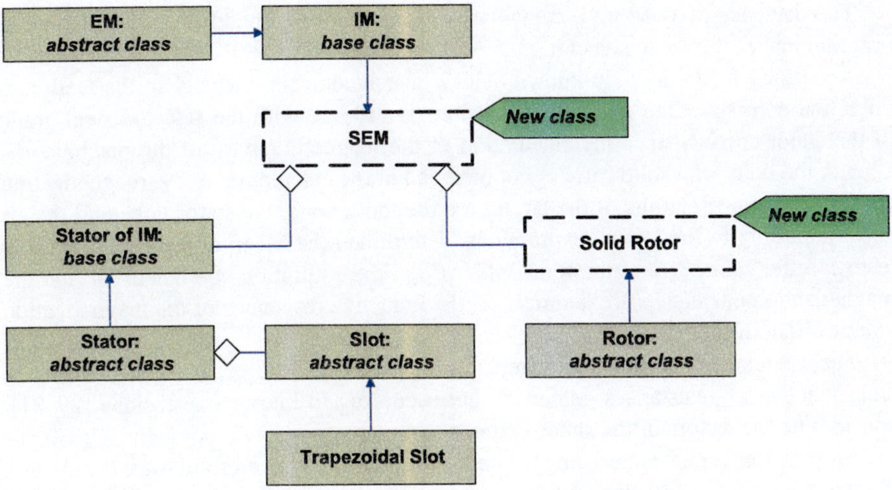

Fig. 10. Simplified block diagram of SEM classes.

3 Calculation Blocks Call Order

After accepting the initial data, objects of the SEM class are formed. First, a base AC machine class is created. It may contain data that common for all AC machines - rated voltage, power, synchronous frequency, number of phases, operating temperature, main frequency. This type is *abstract* and is used to create *descendant* classes.

Since at the initial design stage, apart from setting the initial data, no calculations have yet been made, the objects that are the components of the machine are created in the form of templates with zero data. These data, which are geometric dimensions, materials used, electromagnetic parameters, etc. will be obtained later when calling the calculation methods of the corresponding classes.

The obtained data becomes sufficient to determine the main values, characterizing the SEM stator: the number of turns of the stator winding, the preliminary value of the magnetic flux density in the air gap, the secondary current modulus, the air gap MMF, the air gap magnetic resistance, the MMF of the magnetic circuit of the machine, the total current, the linear load, power factor etc. [27, 28].

The method called from the "*Slot*" class allows calculating the tooth zone, after which the maximum t_{1max} and minimum t_{1min} tooth sizes, as well as the range of the possible number of stator slots $Z_{min} - Z_{max}$, become known.

Further, the control of the calculation is transferred to the hands of the developer - it is possible to select the number of stator slots Z_1, the type of stator slot, the type of winding (single-layer or double-layer). After accepting these data, the calculation and selection of the optimal value of the number of effective conductors U_p is carried out, the number of parallel branches is selected. Specified: number of turns of the stator winding, electric load, main magnetic flux, magnetic flux density in the air gap.

At the next stage, the winding wire is selected, the configuration and filling of the stator slot are calculated.

The database of winding rectangular and round wire embedded in the calculation program makes it easy to select it.

According to the already known values of magnetic flux density in the teeth and yoke, the corresponding magnetic intensities are found, with the selected steel grade of the stator core. Next is the calculation of the magnetic circuit of the machine. As known, the body of a solid rotor is not included in the calculation of the magnetization circuit. The main sections of the circuit are the tooth zone, the stator yoke and the air gap. According to the known expressions, determined the MMF of the air gap F_δ, the teeth F_z, the yoke F_a and the total MMF F_{sum}, the saturation coefficient k_μ and the magnetization current I_μ are determined, the inductive resistance of the magnetization circuit is determined as well.

The calculation of the parameters of the stator winding, namely its active r_1 and inductive x_1 resistances, is carried out according to known expressions [29–31], considering the design of the stator (external or internal).

An important stage concerning the calculation of the reduced solid rotor parameters is carried out taking into account the operating temperature, length, saturation, and, if necessary, curvature of the solid rotor [20, 29–31].

The parameters of the machine for slips other than $s = 1$ are found by recalculation using known expressions, considering the value of the reduced rotor current I'_2.

Based on the found parameters, the values necessary for constructing the operating and starting characteristics of the SEM are determined. Characteristics of IM with solid rotor are calculated considering the effect of steel saturation. This is achieved by introducing the current value of the critical slip s_{cr} into the calculation formulas. The variable in performance and starting performance calculations will be the current $s_i = s_i - 1 + \Delta s$ and the critical slip. The composition of the member methods of the classes allow to carry out a thermal calculation of the SEM, to calculate the starting mode under the influence of variable loads.

As a result of inheritance, the final class "*SEM*" will acquire the characteristics of all classes that are lower in the hierarchical tree but taking into account its own individual features. The next stage of the OOP is to fill the class structure of the SEM with a specific mathematical description:

- the class of hydrodynamic processes introduces functional modules associated with the transport, mixing and swirling of the material in the interblade region of the solid rotor;
- the class of thermal processes, presented in the form of two subclasses, reflects the processes of heat transfer both inside the SEM and on the outer surface of the solid rotor.

The advantage of the class representation is that the base classes of external processes are connected to the SEM only with the help of external independent functional modules. When the operating conditions of the SEM change (for example, in the medium of a viscous low-melting material), the basic structures of hydrodynamics and heat transfer will be completely replaced while maintaining interclass relationships.

The results of calculations of mechanical and starting characteristics are shown in Fig. 11 and 12. The results of calculations [10] are confirmed by comparison with the loss distribution diagram in the SEM, obtained during experimental studies (Table 2).

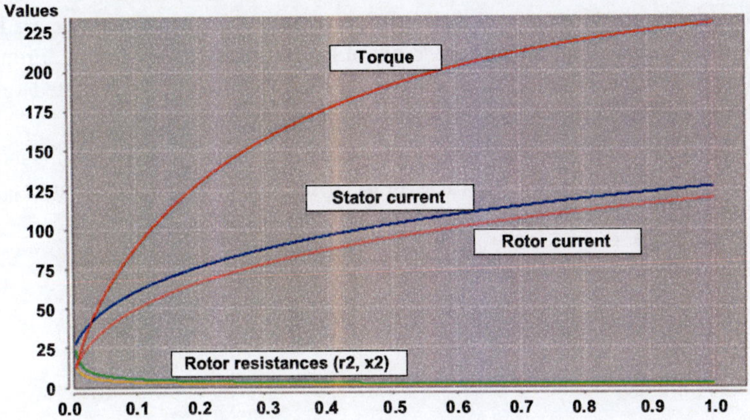

Fig. 11. Estimated starting characteristics of the SEM.

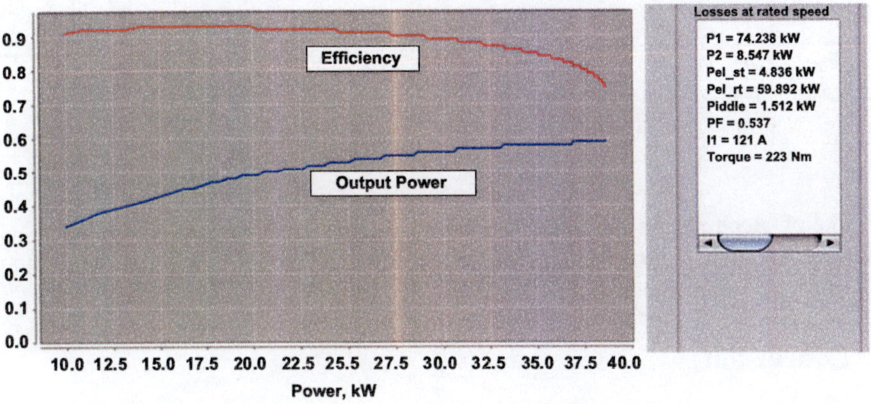

Fig. 12. Calculated mechanical characteristics of SEM.

The results of calculating the characteristics and experimental studies of SEM [10], as mentioned before, are summarized in Table 2. The discrepancy between experimental and calculated values is explained by the fact that the value of losses in the rotor during the experiment was determined indirectly by the temperature difference of the ferromagnetic construction before and after the experiment. As a result, the percentage of losses was redistributed, and, as a result, a smaller value of losses in the rotor compared to the calculation gave a greater percentage weight of losses in the stator winding. From Table 2 it can be seen, that the discrepancy between the measured and calculated values in absolute terms does not exceed 18%.

Table 2. Experimental and calculated values of parameters and characteristics of SEM industrial samples.

Name	Idle mode									Rated mode under the load					
	P_{1drive}, kW	P_{1brake}, kW	I_{1drive}, A	I_{1brake}, A	PF_{drive}	PF_{brake}	M_{drive}, Nm	M_{brake}, Nm	n, rpm	I_{1drive}, A	I_{1brake}, A	PF_{drive}	PF_{brake}	M_{brake}, Nm	n, rpm
Experiment	69.6	52.4	135	100	0.455	0.46	258	215	110	136	104	0.46	0.46	175	88
Calculation	74.2	56.3	121	103	0.537	0.478	223	214	110	124	102	0.533	0.482	203	88
Error, %	6.6	7.4	10.4	3.0	18.0	3.9	13.6	0.47	0	8.8	1.9	15.9	4.8	16.0	0

The obtained results can be used in optimizing the design of the EM, conducting physical modeling of the machine before its practical manufacture, improving the quality of the calculation methods of the EM.

4 Conclusion

The performed analysis of the implementation of existing design methods and mathematical modeling of EM problems allows to highlight their following disadvantages:

- specialization in projects for manufactured products does not make it possible to expand them to new modifications of EM, which are being developed anew;
- the creation of EM projects of special execution requires large resource and time costs;
- the existing design methods do not include mathematical modeling of transients and analysis of the distribution of the electromagnetic field, which does not allow us to assess the complex of problems that arise during the operation of the EM;
- the traditional design of EM during their modernization and repair is accompanied by significant resource and time costs, which is reflected both in the timing of the project and the development of design documentation, and in the cost of products.

The considered concept of an object-oriented approach to design of EM allow not only to solve the problems of traditional design, but also to reorganize the design process itself, turning it into a complex task that takes into account the calculation of machine parameters and characteristics in a block format, automatic generation of engineering design methods and mathematical models for new modifications of EM based on prototype with subsequent optimization and generation of design documentation.

An object-oriented class diagram of the inheritance of SEM with an external solid rotor has been developed, which makes it possible to represent the process of designing a complex electromechanical system in the form of related methods, that combine the calculation of the working environment, the electromagnetic system and the optimization block.

A software implementation of a complex project of a double-stator SEM with a solid rotor has been completed in Java, using the advantages of object-oriented design. It should be noted that developed a software module that performs automated calculation of the machine after the design of the basic prototype is completed. The presence of such a module is indispensable in optimization cycles, where there is a need for repeated recalculation of the EM.

A simulation of a double-stator SEM system with a solid rotor was performed, which is distinguished by consistent and simultaneous simulation of the motor and brake modules. The calculation model was synthesized due to the object-oriented method of inheritance of mathematical models of EM. The obtained results have a high convergence with the data obtained experimentally and can be used in optimizing of SEM design, conducting physical modeling of EM before their manufacture, reducing the design time and improving the technical and economic performance.

Acknowledgment. This research is supported by the Ministry of Education and Science of Ukraine as a part of the scientific research projects No. 0121U109639, 0122U001145.

References

1. Boldea, I.: Electric Machines: Steady State, Transients, and Design with MATLAB. CRC Press, Boca Raton (2009). https://doi.org/10.1201/9781439882979
2. Finley, W.R., Veerkamp, B., Gehring, D., Hanna, P.: Improving motor efficiency levels globally. IEEE Ind. Appl. Mag. **15**(1), 39–49 (2008). https://doi.org/10.1109/MIAS.2008.930894
3. Haataja, J., Härkönen, H.: IEA Hi-motors competition. In: Bertoldi, P., de Almeida, A.T., Falkner, H. (eds.) Energy Efficiency Improvements in Electronic Motors and Drives, pp. 540–545. Springer Berlin Heidelberg, Berlin, Heidelberg (2000). https://doi.org/10.1007/978-3-642-59785-5_52
4. Tsegelnyk, Y., Kombarov, V., Plankovskyy, S., et al.: Investigation of the portal-type machine tool gear-belt gearbox. Int. J. Mech. Appl. Mech. **2022**(11), 295–302 (2022). https://doi.org/10.17683/ijomam/issue11.41
5. Sen, S.K.: Principles of Electrical Machine Design with Computer Programs. Oxford and IBH Publishing, Oxford (2006)
6. Rosu, M., Zhou, P., Lin, D., et al.: Multiphysics Simulation by Design for Electrical Machines, Power Electronics, and Drives. Wiley, Hoboken (2018). https://doi.org/10.1002/9781119103462

7. Upadhyay, K.G.: Design of Electrical Machines. New Age International Ltd., New York (2011)
8. Pyrhönen, J., Jokinen, T., Hrabovcová, V.: Design of Rotating Electrical Machines. Wiley, Chichester (2014). https://doi.org/10.1002/9780470740095
9. Zablodskii, N.N., Pliugin, V.E., Petrenko, A.N.: Using object-oriented design principles in electric machines development. Electr. Eng. Electromech. **2016**(1), 17–20 (2016). https://doi.org/10.20998/2074-272X.2016.1.03
10. Pliuhin, V., Sukhonos, M., Bileckiy, I.: Object oriented mathematical modeling of electrical machines. In: 2020 IEEE 4th International Conference on Intelligent Energy and Power Systems (IEPS), Istanbul, pp. 267–272. IEEE (2020). https://doi.org/10.1109/IEPS51250.2020.9263158
11. Kung, D., Lei, J.: An object-oriented analysis and design environment. In: 2016 IEEE 29th International Conference on Software Engineering Education and Training (CSEET), Dallas, pp. 91–100. IEEE (2016). https://doi.org/10.1109/CSEET.2016.20
12. Bakar, N.S.: The analysis of object-oriented metrics in C++ programs. Lect. Notes Softw. Eng. **4**(1), 48–52 (2016). https://doi.org/10.7763/LNSE.2016.V4.222
13. Aksonov, Y., Kombarov, V., Tsegelnyk, Y., et al.: Visualization and analysis of technological systems experimental operating results. In: 2021 IEEE 16th International Conference on Computer Sciences and Information Technologies (CSIT), Lviv, vol. 2, pp. 141–146. IEEE (2021). https://doi.org/10.1109/CSIT52700.2021.9648592
14. Lei, G., Zhu, J., Guo, Y., et al.: A review of design optimization methods for electrical machines. Energies **10**(12), 1962 (2017). https://doi.org/10.3390/en10121962
15. Kitagawa, W., Inaba, A., Takeshita, T.: Objective function optimization for electrical machine by using multi-objective genetic programming and display method of its results. IEEJ Trans. Electron. Inf. Syst. **139**(7), 796–801 (2019). https://doi.org/10.1541/ieejeiss.139.796
16. Li, Y., Lei, G., Bramerdorfer, G., et al.: Machine learning for design optimization of electromagnetic devices: recent developments and future directions. Appl. Sci. **11**(4), 1627 (2021). https://doi.org/10.3390/app11041627
17. Zablodskiy, M., Pliuhin, V., Chuenko, R.: Simulation of induction machines with common solid rotor. Tech. Electrodynamics **2018**(6), 42–45 (2018). https://doi.org/10.15407/techned2018.06.042
18. Daukaev, K., Rassõlkin, A., Kallaste, A., et al.: A review of electrical machine design processes from the standpoint of software selection. In: 2017 IEEE 58th International Scientific Conference on Power and Electrical Engineering of Riga Technical University (RTUCON), Riga, pp. 1–6. IEEE (2017). https://doi.org/10.1109/RTUCON.2017.8124818
19. Bonneel, P., Le Besnerais, J., Pile, R., Devillers, E.: Pyleecan: an open-source Python object-oriented software for the multiphysic design optimization of electrical machines. In: 2018 XIII International Conference on Electrical Machines (ICEM), Alexandroupoli, pp. 948–954. IEEE (2018). https://doi.org/10.1109/ICELMACH.2018.8506884
20. Nikolaj, Z., Vladyslav, P., Stanislav, F., Jiri, L.: Dynamic simulation of the double-stator induction electromechanical converter with ferromagnetic rotor. In: 4th International Conference on Power Engineering, Energy and Electrical Drives, Istanbul, pp. 1448–1453. IEEE (2013). https://doi.org/10.1109/PowerEng.2013.6635828
21. Pliugin, V., Shilkova, L., Lettl, J., et al.: Analysis of the electromagnetic field of electric machines based on object-oriented design principles. In: Progress in Electromagnetics Research Symposium Proceedings, pp. 2522–2527 (2015). https://ssrn.com/abstract=3201920
22. Zablodskiy, N., Pliugin, V.: 3D magnetic filed distribution in a screw double-stator induction motor. In: 2015 16th International Conference on Computational Problems of Electrical Engineering (CPEE), Lviv, pp. 239–241. IEEE (2015). https://doi.org/10.1109/CPEE.2015.7333386

23. Zablodskiy, M., Gritsyuk, V., Pliuhin, V., Biletskyi, I.: The surface characteristics features of the electromagnetic field of the rotor of a polyfunctional electromechanical converter. In: 2021 International Conference on Electrical, Computer, Communications and Mechatronics Engineering (ICECCME), Mauritius, pp. 1–5. IEEE (2021). https://doi.org/10.1109/ICECCME52200.2021.9590872
24. Schwarz, P., Moeckel, A.: Electric machine design automation with python and ansys maxwell. In: IKMT 2019 – Innovative small Drives and Micro-Motor Systems; 12. ETG/GMM-Symposium, pp. 1–7. VDE, Wuerzburg (2019)
25. Ulu, C., Korman, O., Kömürgöz, G.: Electromagnetic and thermal design/analysis of an induction motor for electric vehicles. Int. J. Mech. Eng. Robot. Res. **8**(2), 239–245 (2019). https://doi.org/10.18178/ijmerr.8.2.239-245
26. Okhrimenko, V., Zbitnieva, M.: Mathematical model of tubular linear induction motor. Math. Model. Eng. Probl. **8**(1), 103–109 (2021). https://doi.org/10.18280/mmep.080113
27. Iegorov, O., Iegorova, O., Kundenko, M., Milenin, A.: Single-phase induction motors winding parameters optimization with maximum efficiency. In: 2020 IEEE Problems of Automated Electrodrive. Theory and Practice (PAEP), Kremenchuk, pp. 1–4. IEEE (2020). https://doi.org/10.1109/PAEP49887.2020.9240878
28. Finkelshtein, V., Iegorov, O., Petrenko, O., Koliada, O.: The analytic-field method for calculating the squirrel-cage induction motor parameters. Naukovyi Visnyk Natsionalnoho Hirnychoho Universytetu **2020**(3), 67–72 (2020). https://doi.org/10.33271/nvngu/2020-3/067
29. Mirzaei, M., Mirsalim, M., Abdollahi, S.E.: Analytical modeling of axial air gap solid rotor induction machines using a quasi-three-dimensional method. IEEE Trans. Magn. **43**(7), 3237–3242 (2007). https://doi.org/10.1109/TMAG.2007.894215
30. Nasiri-Gheidari, Z., Lesani, H.: Optimal design of adjustable air-gap, two-speed, capacitor-run, single-phase axial flux induction motors. IEEE Trans. Energy Convers. **28**(3), 543–552 (2013). https://doi.org/10.1109/TEC.2013.2260826
31. Hong, C., Huang, W., Hu, Z.: Calculation methods of equivalent circuit parameters for a dual stator solid rotor axial flux induction motor. IET Renew. Power Gener. **12**(16), 1977–1983 (2018). https://doi.org/10.1049/iet-rpg.2018.5103

High Speed Actuator for Digital Hydraulics

Nicolae Tanase, Cristinel Ilie(✉), Ionel Chirita, Marius Popa, Lipcinski Daniel, Mihai Gutu, and Romulus Marian Mihai

National Institute for Research and Development in Electrical Engineering ICPE-CA, Bucharest, Romania

cristinel.ilie@icpe-ca.ro

Abstract. This paper presents the construction of an electromagnetic actuator with conical air gap and with a massive plunger, used in digital hydraulics to reduce the power losses and which can be integrated in fast switch on/off valves used in hydraulic circuits. Also, this paper presents the numerical simulations performed on this actuator. The results of the numerical simulations refer to the electromagnetic forces, magnetic field, magnetic energy which characterizes the analyzed electromagnetic actuator. The design of this actuator and the results of the numerical simulations leaded to the manufacturing of an experimental model. Preliminary experimental measurements performed on this model are also presented in this paper.

Keywords: Designing · Manufacturing · Numerical simulation · Electromagnetic actuator · Massive core · Electromagnetic force · Magnetic energy · Digital hydraulics · Experimental testing

1 Introduction

The hydraulic drive systems that are equipped with servo-valves for quick and well-timed movements are indispensable to industrial machines operating in automatic/controlled operation. The very high cost of the servo-valves has led specialists and manufacturers who make hydraulic control and command systems to find alternatives of replacing them. One of the solutions is the digital hydraulics that replaces the servo-drive with fast on/off switch valves, which are running in series or parallel, depending on the system that is operated [1–3].

Most common on/off valves are operated by an electromagnetic actuator. The International Electrotechnical Commission (IEC) define an electromagnet as "a magnet excited by a current" [4–7].

During the contact meetings with experts in hydraulics of the potential beneficiary (IHP - INOE 2000) it turned out that for such applications (digital hydraulics) force of attraction for mobile armature (of the electromagnetic actuator) should fit in range of 50…80 N at approx., 1 mm stroke, and the working rate of the electromagnet mobile subassembly is desirable to reach values of 100…120 Hz (this corresponds to a minimum time value of the period of 8.3…10 ms) [8].

D. D. Cioboată (Ed.): ICoRSE 2022, LNNS 534, pp. 284–297, 2023.
https://doi.org/10.1007/978-3-031-15944-2_26

Regarding the response time, each period comprises a stage of drive time and a stage of recovery time. The drive time is composed of the start time (the current increases without the armature moving) and the movement time (of the armature), and the return time is composed of the time of detachment in which the current decreases but the armature remains still drawn and the effective time of the armature return under the action of the resisting forces (return spring and payload reaction). If for the actuation time there are computational relations with good results, the recovery time is difficult to appreciate depending on a multitude of factors (electromagnetic and mechanical) in close dependence on the active materials used. To be able to advance in the direction of correlating the prediction of theoretical considerations and electromagnetic calculations with experimental behavior, the electromagnetic actuator with massive core and conical plunger for digital hydraulics was manufactured, with conventional ferromagnetic available materials.

The electromagnetic control a digital valve (the most important functional part of a digital valve) is essentially a PWM controlled electromagnetic actuator. This paper presents numerical simulations of such an electromagnetic actuator with massive core and conical plunger, the experimental model that was manufactured and the preliminary tests performed on it [8].

2 The Electromagnetic Actuator with Massive Core and Conical Plunger for Digital Hydraulics

The ferromagnetic circuit can be opened or closed. The armatures are made of ferromagnetic material, most often having the shape of a revolution body (in alternating current, the armature is usually made of sheets). This type of electromagnetic actuator is used when either long stroke of the mobile armature is required (sometimes up to 200 mm) or small strokes but with large load forces, (the forces are higher when the current in the excitation coil is higher).

At this type of actuator, the force is constant throughout the stroke due to the conical shape of the body of the mobile armature [4, 6].

In Fig. 1 the electromagnetic actuator with conical plunger is presented, the 3D model was designed in SolidWorks Premium 2022. In this case the mobile armature stops when hits the fixed armature stopper, when it has made the whole free stroke, the force depending on the height of the stopper.

The design data of the electromagnetic actuator from Fig. 1, taken into consideration, together with potential beneficiaries (IHP - INOE 2000), are:

- The electromagnetic force of the plunger in the range of 50…80 N;
- The plunger stroke of 1 mm.

The electromagnetic actuator consists of excitation windings (coil) (2) and ferromagnetic bodies operated by electromagnetic forces can be called electromagnetic mechanism. The fixed part of the mechanism, made of ferromagnetic material, subjected to the magnetic polarization produced by the magnetic field of the excitation coil is called fixed armature (1) and (3), and the mobile part, mobile armature (plunger) (4).

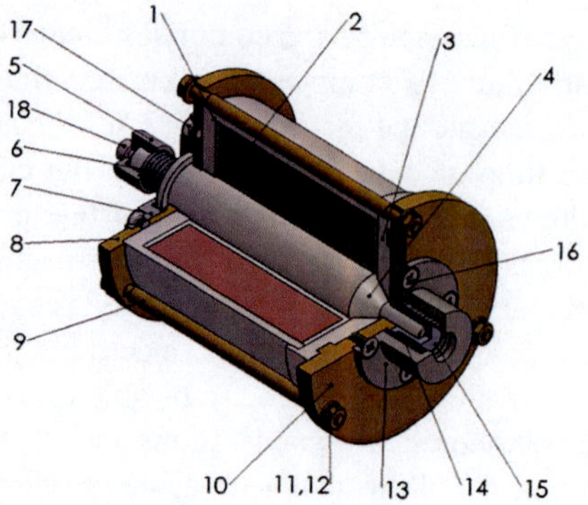

Fig. 1. Electromagnetic actuator 3D design with massive core and conical plunger-ICPE-CA. 1, 3 - fixed armature; 2 - excitation windings (coil); 4 – mobile armature (plunger); 5, 6, 7, 13, 14, 15 – pretension elements (bushings, flanges, springs); 8, 10 – fixing flanges; 9 – fixing rods; 11, 12 – fixing nuts and washers; 16, 17, 18 – fixing screws.

The electromagnetic force corresponding to the maximum distance (maximum air gap) between the mobile armature (4) and fixed armature (1) and (3) it is called the initial attraction force and the force corresponding to the minimum distance between the mobile and the fixed armature is called the final attraction force or the load force. In order to return to the open position of the electromagnetic actuator, the compression spring (6) was introduced. For the simulation of the force developed by the electromagnetic actuator, the compression spring (14) was also introduced.

The electromagnetic actuator stroke is achieved by the mobile armature (4) from the open position to the closed position when the stopper (the back shoulder) of the plunger hits the channel made in the back cover (8).

To guide the plunger, to determine its stroke and to fix the electromagnetic actuator, the covers (8) and (10) are used also three threaded rods (9) that are equidistant disposed and fixed with nuts.

The operation of the electromagnetic actuator is based on the conversion of electromagnetic energy into mechanical energy, the stroke of the mobile armature being limited. The energy consumption does not cease at the end of the stroke of the mobile armature, if it is necessary to maintain an electromagnetic force or an electromagnetic torque, because certain energy losses result from maintaining the existing state (in windings and armature iron) which turns into heat and must be covered by the power supply of the electromagnetic actuator.

Due to the very wide scope of the electromagnetic actuators, characterizing the performances only by the static characteristic, the maximum force developed at the minimum air gap becomes insufficient. For calculating the operating time or frequency and checking the mechanical resistance of the electromagnetic actuators, the closing and opening speeds and times must be known; their determination being done by studying the dynamic regime.

By the dynamic regime of an electromagnetic actuator is understood the behavior in transient mode of the electromagnet when connecting the energy source that supplies the excitation coil. In consideration of the dynamic regime, both electric and magnetic sizes and mechanical sizes are involved.

3 The Mathematical Model and Materials Used for the Electromagnetic Actuator with Massive Core and Conical Plunger

Considering the axial symmetry of the construction of the conical plunger electromagnetic actuator, the simulation can be reduced to a 2D axisymmetric problem, to reduce the computational time and effort. The Maxwell's equations for the magnetic field are [9, 10]:

$$\nabla \times \mathbf{H} = \mathbf{J}, \tag{1}$$

$$\mathrm{curl}\mathbf{E} = -\frac{\partial \mathbf{B}}{\partial t} + \mathrm{curl}(\mathbf{v} \times \mathbf{B}), \tag{2}$$

$$\nabla \times \mathbf{B} = 0, \tag{3}$$

$$\mathbf{B} = \mathbf{B}(\mathbf{H}). \tag{4}$$

where **A** [T*m] - magnetic vector potential; **J** [A/m] – electric current density; **B** [T] - magnetic flux density [T]; **H** [A/m] - magnetic field strength; **E** [V/m] – the electric field strength, and **v** [m/s] the velocity; $\mu_0 = 4\pi 10^{-7}$ H/m - magnetic permeability of free space and μ_r is the relative permeability.

Relying on the symmetry of the system – model and physics - a 2D axisymmetric model may be used, Fig. 2.

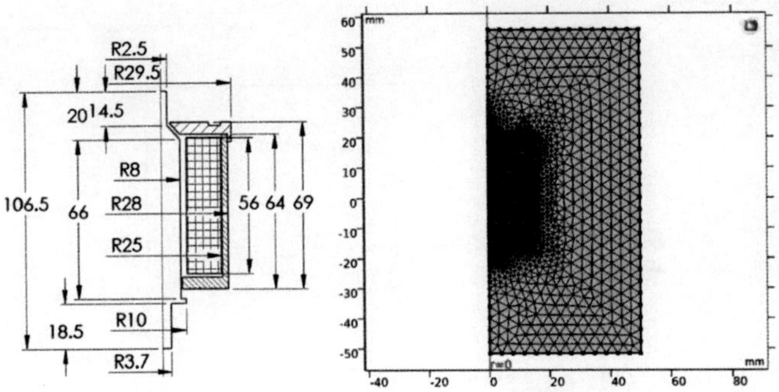

(a) (b)

Fig. 2. The 2D CAD model used for electromagnetic actuator simulation: a) axisymmetric 2D geometry; b) mesh network for the conic plunger electromagnetic actuator (dimensions are in millimeters).

The mathematical model (1) is solved numerically, in finite element technique using Comsol Multiphysics 5.0 [9]. The computational domain is bordered by an air subdomain that implements "infinite" elements. The mesh network comprises 31,320 triangular elements.

The materials chosen for numerical simulation are presented in Fig. 3 (Comsol Multiphysics 5.0 screen capture).

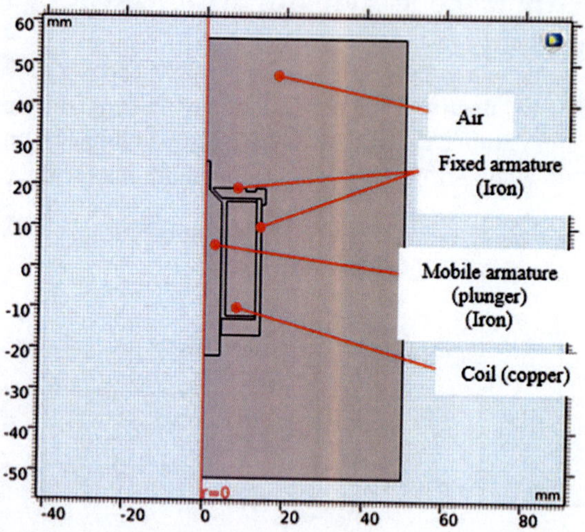

Fig. 3. Materials for computing domains and subdomains of the electromagnetic actuator with conical plunger analyzed.

The magnetic properties of the materials used for the plunger and the iron core of the electromagnetic actuator are very important to obtain a good response time, high electromagnetic force and small dimensions of the device. The material used in this simulation for plunger and fixed armature was soft iron which has the B-H magnetization curve presented in Fig. 4.

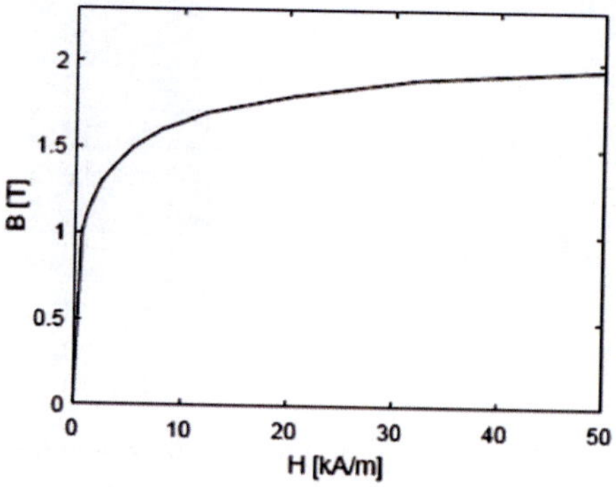

Fig. 4. B-H saturation curves of soft iron material [9, 11].

The simulation is computed using the electromagnetic field problem with the vector potential, A, introduced by Eq. (1) in conjunction with Coulomb's gauge condition, which lead to the following 2D, axial symmetric partial differential equation [9, 10, 12]:

$$\frac{\partial A_\varphi}{\partial t} + (\nabla \times \mathbf{H})_\varphi = \mathbf{J}_\varphi, \tag{5}$$

where the azimuth component is denoted with $(.)_\phi$.

4 Numerical Simulation Results

The following parameters were taken into account in electromagnetic simulation (Fig. 5), so that the electromagnetic actuator develops the necessary characteristics in terms of reaction time and electromagnetic force [6, 12]:

- N = 1365 (no. of turns);
- $I = 2$ A (the maximum excitation current);
- diameter of the winding conductor Ø 0,65 mm.

The plunger (the mobile part) is subjected to the magnetization forces only. We also define the variable F_{mag} for the magnetic force, and use it to compute the Maxwell stress tensor on the plunger. The plunger movement is in Oz direction of motion which is described by Newton's law [12]:

$$\frac{\mathrm{d}u}{\mathrm{d}t} = \frac{F_{mag}}{m} - kp, \tag{6}$$

where u is the velocity of the plunger in Oz direction, F_{mag} is the magnetic force of the mobile plunger in Oz direction, k is the constant of the compression spring (6 from Fig. 1), p is the initial position of the plunger and m is the mass of the mobile plunger (4 from Fig. 1).

The Eqs. (5) and (6) are computed using FEM technique [9]. A third partial differential equation (of Laplace type) that describes the mesh deformation is added to (5) and (6). The last two problems are then solved in the coordinates system provided by the solution of the mesh deformation problem rather than in the general coordinate system, in which the computational domain is constructed. The algebraic system of equations produced by the FEM discretization was solved using the Parallel Sparse Direct Solver (PARDISO) [9, 12].

After the numerical simulations were solved, it was it resulted the magnetic induction for the electromagnet presented in Fig. 5.

Figure 6 shows the displacement of the electromagnetic actuator (the movable plunger 4 from Fig. 1) for a prescribed stroke of 0…1 mm in the Oz vertical direction.

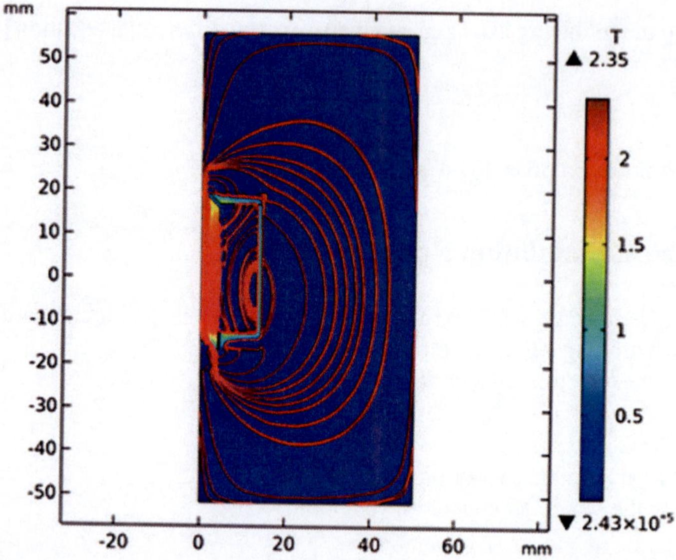

Fig. 5. Field lines and color map for the magnetic induction of the 2D analyzed axisymmetric model, maximum induction |B| = 2.35 T.

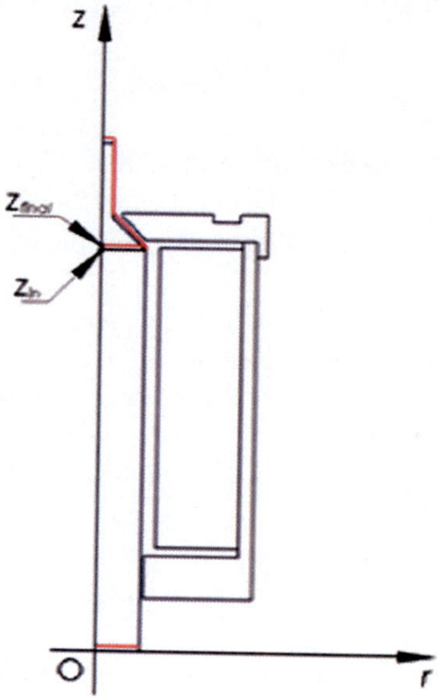

Fig. 6. The limits of the displacement for mobile plunger.

In Fig. 7 is presented the magnetic energy according to supply currents for different positions of the plunger for the actuator analyzed.

Figure 8 shows the magnetic force that appears in the plunger depending on its displacement, the maximum stroke required being 1 mm.

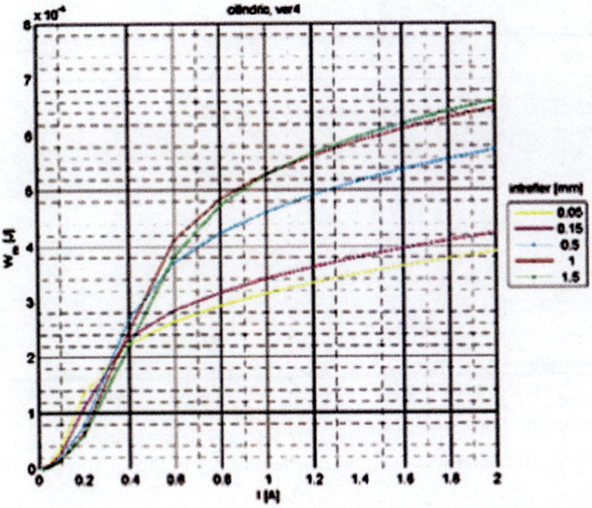

Fig. 7. Magnetic energy vs. current at different positions of the plunger.

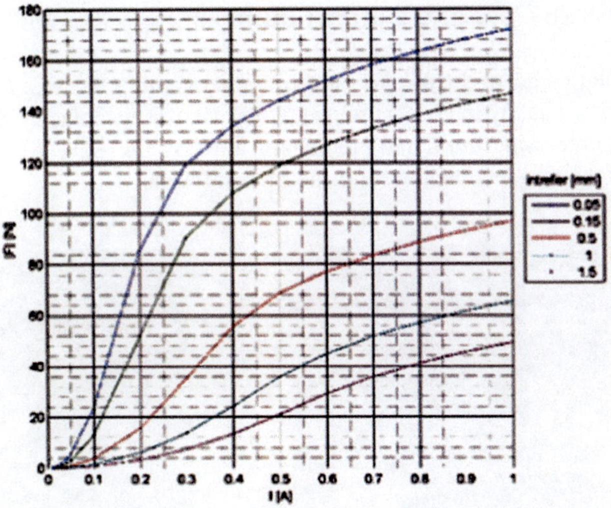

Fig. 8. Magnetic force depending on the current at different positions of the plunger.

In Fig. 9 is shown the axial force in Oz direction of the mobile plunger, the response time for $I = 1$ A.

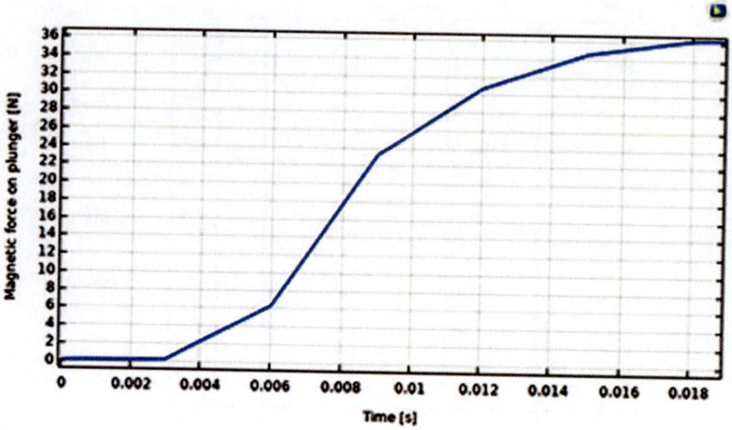

Fig. 9. Magnetic axial force of the mobile plunger in the *Oz* direction in time.

5 Manufacturing and Preliminary Tests Performed on the Experimental Model of the Electromagnetic Actuator with Massive Core and Conical Plunger

The experimental model of the electromagnetic actuator with massive core and conical plunger used for digital hydraulics presented in Fig. 10, was manufactured in the Laboratory of Microprocessing and Rapid Prototyping at ICPE-CA [8]. The plunger and the closing core of the magnetic field were made from soft iron.

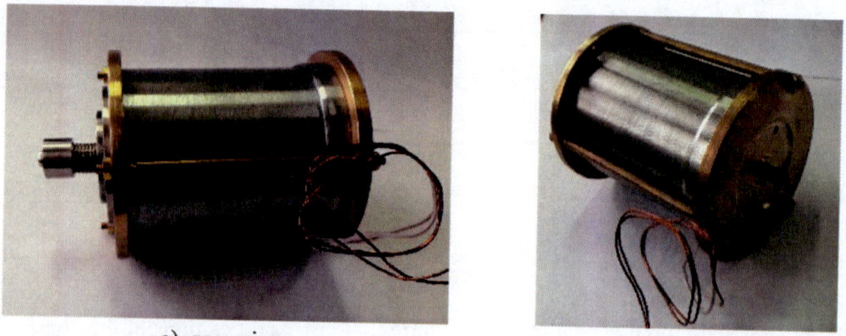

a) rear view b) front view

Fig. 10. Manufacturing of electromagnet with massive core and conical plunger.

The operating principle of the drive system for an electromagnet used for digital hydraulics is based on the use of PWM (Pulse Width Modulator) signals generated by the DRV101 controller.

A rectangular signal with the amplitude $U_{in} = 5$ V and a variable frequency is applied to the input of the PWM circuit: 1–1000 Hz, from an external function generator. This rectangular signal is processed and formatted as a PWM signal and applied to a driver buffer stage for the final output stage. A sufficiently high signal level will be injected onto the solenoid coil via the power amplifier stage to ensure an oscillating motion of the electromagnet. The aim is also to obtain a linear motion on the same frequency [13].

For the functioning of the electromagnet in linear mode, a logic "1" level can be applied via an optocoupler type HCPL2630 to operate the controller on linear mode. The filling factor of the control signal is selectable within the limits: $k = 5\%–90\%$.

In order to ensure the power of the actuator drive, it was necessary to dimension a power amplifier capable of providing a power up to 1 kW on a schematic configuration with IGBT transistors.

The supply of the inductance coil L of the electromagnetic actuator is made in direct voltage, by using a rectified voltage source that can discharge a maximum current of: $I = 10$ A and a voltage of 50 Vdc. The wiring diagram of the PWM drive system for an ultrafast electromagnetic actuator is shown in Fig. 11.

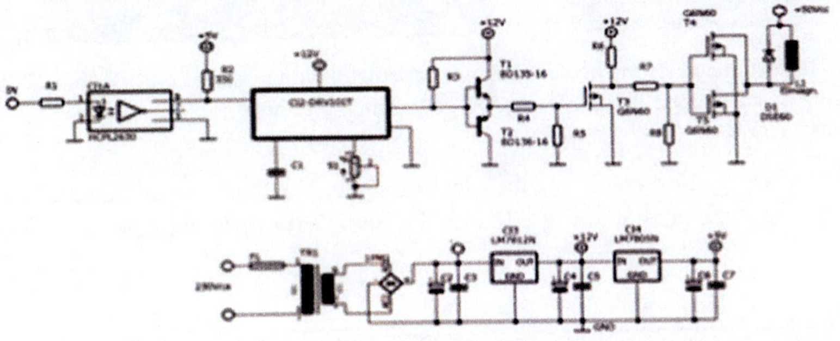

Fig. 11. Wiring diagram of the PWM drive system for the electromagnetic actuator.

From a constructive point of view, the equipment was made in a modular system (Fig. 12), the main components being the following: the module for processing/formatting and signal amplification; the supply stage; the load element.

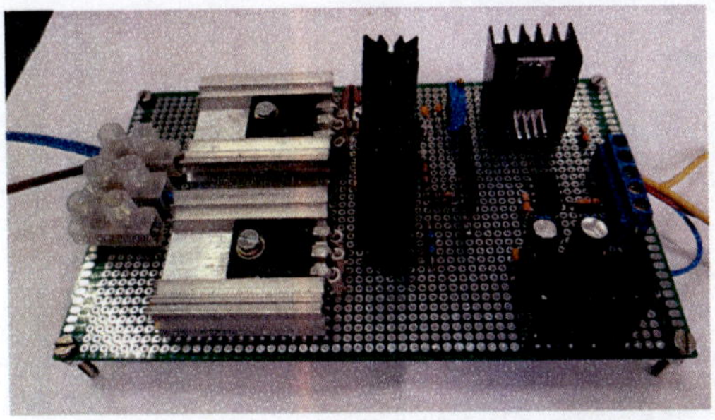

Fig. 12. Electronic drive board for the electromagnetic actuator.

The measurement devices for the preliminary testing of the electromagnetic actuator are presented in Fig. 13. The tests were performed using a laboratory DC power supply type PS5005 (Fig. 13 b), which has an adjustable output voltage 0–50 V, adjustable output current 0–5 A, 1 mV ripple voltage, 5 mA ripple current and an oscilloscope Tektronix TPS2024B [14] (Fig. 13 a) with 4 insulated channels, 200 MHz frequency band, 2 GS/s sampling rate on each channel.

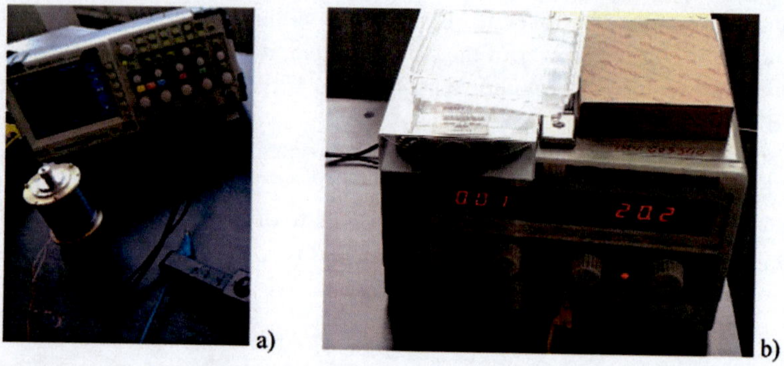

Fig. 13. Experimental testing of electromagnetic actuator with massive core and conical plunger: a) Tektronix oscilloscope TPS2024B; b) DC power supply PS5005.

Preliminary testing results of the electromagnetic actuator with massive core and conical plunger are shown in Figs. 14, 15, 16, and 17 for different currents and supply voltages.

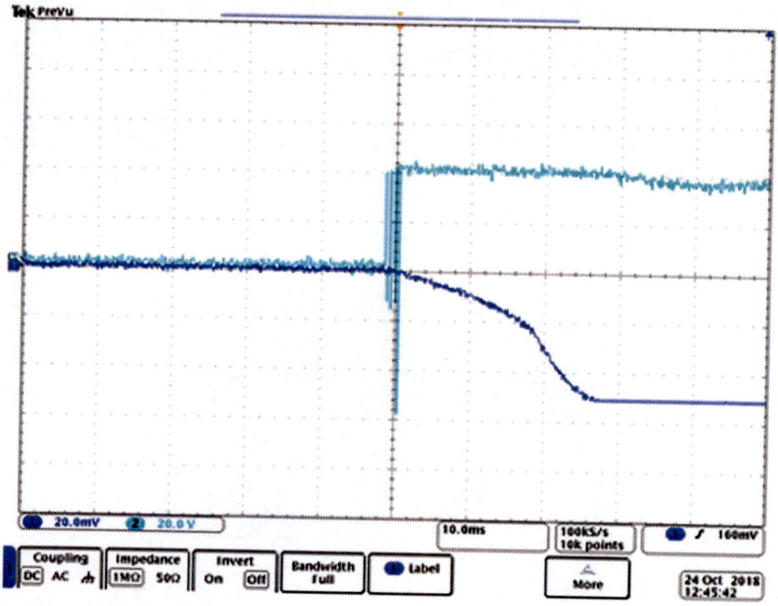

Fig. 14. The response curve of the electromagnetic actuator at $U_{in} = 40$ V drive supply voltage and $I = 5$ A limit of excitation current.

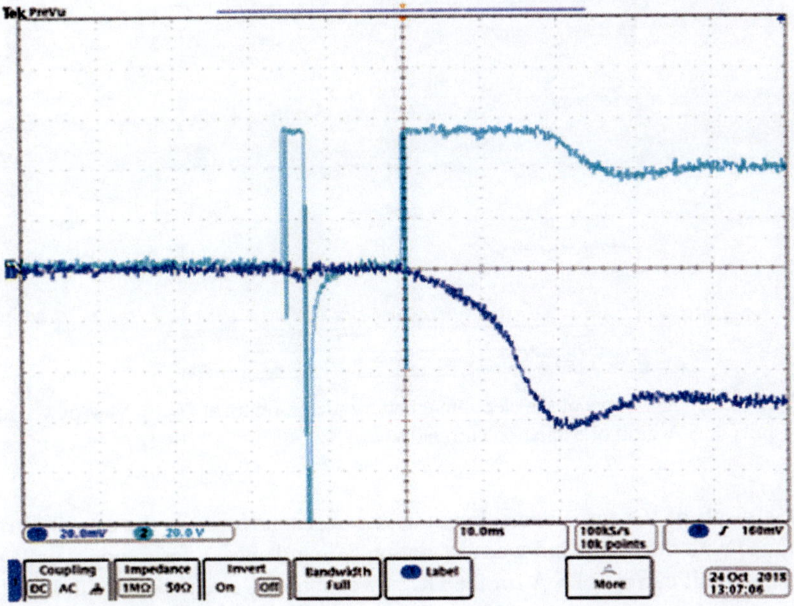

Fig. 15. The response curve of the electromagnetic actuator at $U_{in} = 54$ V drive supply voltage and $I = 5$ A limit of excitation current.

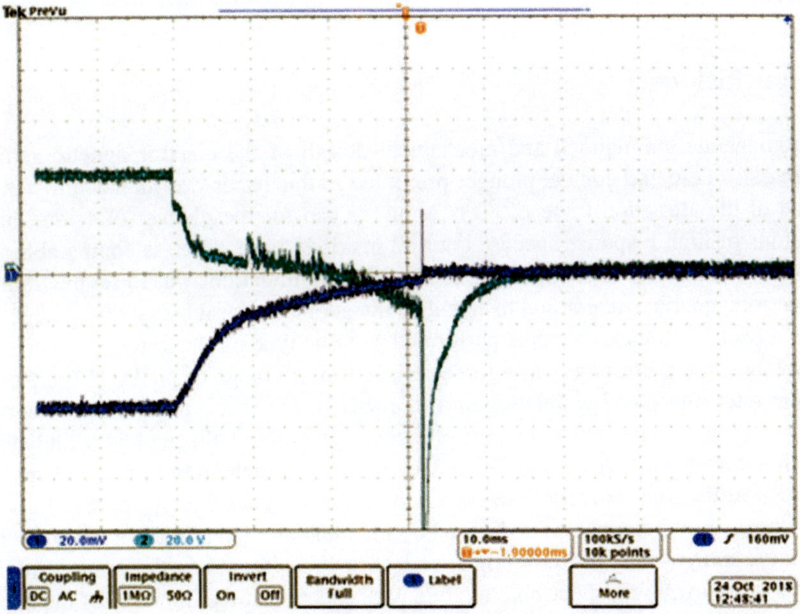

Fig. 16. The response curve of the electromagnetic actuator, return at $U_{in} = 40$ V power supply voltage and $I = 5$ A limit of excitation current.

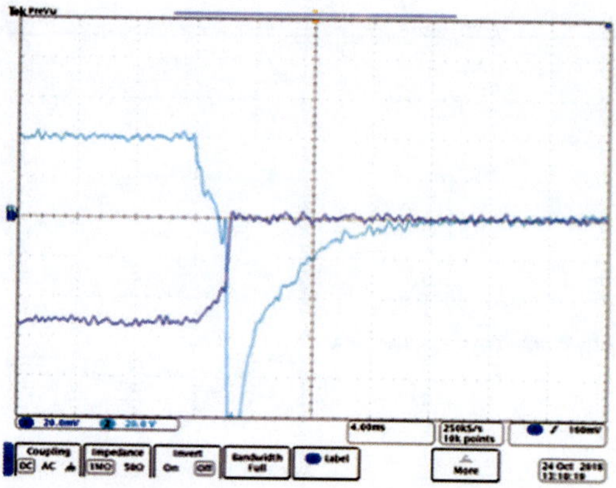

Fig. 17. The response curve of the electromagnetic actuator, return at $U_{in} = 54$ V power supply voltage and $I = 5$ A limit of excitation current.

As a result of the measurements made at a resistant force in the 60–90 N range (performed with the return and load simulation spring devices), for a supply voltage of 54 V and a limit current of 5 A (of the DC power supply), as can be seen in Fig. 15, it has been found for the drive of a 12 ms. Regarding the return time, even if the current is cancelled in approx. 4 ms (Fig. 17), the release of the mobile armature is delayed by the magnetic remanence of the steel used for plunger and yoke (25 ms return time were observed).

6 Conclusions

Electromagnetic simulations and mechanical design of the electromagnetic actuator with massive core and conical plunger presented in this paper was intended to achieve a value of the attraction force of 80 N at an air gap for the plunger of 1 mm, under conditions of high response speed (about 12 ms drive time which is comparable with 18 ms from numerical simulations). The aim of this simulations was to set the limits of the electromagnetic actuator and to establish the geometry for it.

As a result of the experiments performed, for this type of electromagnetic actuator the driving force corresponds to a force resistant in the range of 60–90 N (performed with the return and load simulation spring devices). For this type of electromagnetic actuator with massive core and conical plunger, at an output voltage of 54 V and a limit excitation current of 5 A (of the DC power supply), an actuation time of 12 ms was found. Regarding the recovery time even if the current is cancelled in approx. 4 ms, the release of the mobile armature is delayed by the magnetic remanence of the steel used for plunger and yoke (25 ms return time were observed). Regarding the dynamic regime with the supply from the electronic module with cadence of the impulses of the supply voltage, for this type of electromagnetic actuator could be reached up to the maximum 40 Hz (by increasing the resistant force of the springs).

It was found that the component with the highest value of the duration of a period is determined by the return time of the mobile armature. It is influenced primarily by the existence of a certain magnetic remanence at the level of the yokes and by the speed of decrease of the value of the current through the coil. Regarding these results obtained, is intended to perform the force measurements in the following experiments and compare them with the simulations. It is appreciated that both aspects regarding the response time can be substantially improved by using special materials in further research like ferromagnetic materials (Supermendur) for mobile and fixed armature.

Acknowledgements. The authors acknowledge the support offered through the PN19310301-46N/2019 research grant, also this work was supported by the Romanian Ministry of Education, Research and Digitalization, project number 25PFE/30.12.2021 – Increasing R-D-I capacity for electrical engineering-specific materials and equipment with reference to electromobility and "green" technologies within PNCDI III, Programme 1.

References

1. Tănase, N., et al.: Study of an electromagnet for digital hydraulics - numerical simulation and experimental model testing. In: 2019 11th International Symposium on Advanced Topics in Electrical Engineering (ATEE) Bucharest, Romania, pp. 1–4 (2019).https://doi.org/10.1109/ATEE.2019.8725007
2. Linjama, M., Vilenius, M.: Digital hydraulics, towards perfect valve technology. Digital Hidraulika. **14**, 138–148 (2008)
3. Linjama, M., Laamanen, A., Vilenius, M.: Is it time for digital hydraulics? In: Proceedings of the Eighth Scandinavian International Conference on Fluid Power, 7–9 May 2003, Tampere, Finland, pp. 347–366 (2003)
4. Hortopan, G.: Aparate electrice. Editura didactică și pedagogică, București (1967)
5. Balakrishnan, M., Navaneeth Kumar, N.: Detection of Plunger Movement in DC solenoids. Texas instruments liquid level sensing using hall effect sensors. App. Note Rev. 1(1) (2009)
6. Hortopan, G., Cosmin, G., Huhulescu, M., Panaite, V., Simulescu, D., Tomoioagă, R. : Aparate electrice de joasă tensiune. Editura tehnică, București (1969)
7. *** https://www.iec.ch/
8. Raport NUCLEU - PN18240202–35N/2018
9. *** COMSOL A.B. v. 5.0 (2015)
10. Mocanu, C.I.: Teoria câmpului electromagnetic. Ed. Didactică şi Pedagogică, Bucureşti (1982)
11. Eklund, P., Eriksson, S.: The influence of permanent magnet material properties on generator rotor design. Energies **12**(7), 1314 (2019). https://doi.org/10.3390/en12071314
12. Pîslaru-Dănescu, L., Morega, A.M., Morega, M., Dobre, A.A.: Prototyping a proportionally electromagnetic actuator with wide displacement of its mobile part. In: 2019 11th International Symposium on Advanced Topics in Electrical Engineering (ATEE) Bucharest, Romania, pp. 1–7 (2019).https://doi.org/10.1109/ATEE.2019.8724916
13. Guglielmino, E., Semini, C., Kogler, H., Scheidl, R., Caldwell, D.G.: Power hydraulics, switched mode control of hydraulic actuation. In: The 2010 IEEE/RSJ International Conference on Intelligent Robots and Systems, 18–22 October 2010, Taipei, Taiwan (2010)
14. *** https://ro.farnell.com/tektronix/tps2024b/oscilloscope-4ch-200mhz-2gsps/dp/1877503

Energy Efficient Hydraulic Positioning System for Inverter Driven Pumping Units

Valerian-Emanuel Sârbu[1](✉) and Mihai Avram[2]

[1] Harman International Romania SRL, Bucharest, Romania
vale_sarbu@yahoo.com
[2] Politehnica University of Bucharest, Bucharest, Romania

Abstract. This paper describes an alternative to hydraulic positioning systems that use the resistive method and implicitly proportional equipment. The proposed alternative uses power more efficiently and involves a reduced initial cost. The higher efficiency is derived from the inverter driven pumping unit that can dynamically change the flow rate of the pump by varying its input shaft speed. The usage of low-cost valves, fixed flow pump, simple control algorithm and increased efficiency of the system results in cost reduction for deployment and continuous operation.

Keywords: Energy efficient · Positioning system · Hydraulics · Cost-effective · AMESIM · LabView

1 Introduction

Hydraulic systems exhibit some advantages like being compact, self-lubricating, inherent cooling, reversible and have low inertia due to their low mass to torque ratio. These makes them desirable in rough and space constrained areas like manufacturing.

When the hydraulic actuators employed by the application also need to be positioned with good precision a circuit that uses proportional equipment is typically used. The options of using a pumping unit with variable flow rate, proportional valves or both is the system architect's call. In this paper a different method is being evaluated for improving the cost of the whole positioning system by using the most cost-effective system that employs classic 4/3 hydraulic valves with a fixed flow pump driven by an inverter powered induction motor.

The resistive method for precision speed and position control has already been pursued in previous articles in the department of Mechatronics with results published in 2014 [1]. Further research was done to modify a 1970's pump to be powered by a 230 V inverter that was designed in the department, along with a custom Arduino IDE compatible board that does data acquisition and classic valve control several articles being published respectively [2, 3].

D. D. Cioboată (Ed.): ICoRSE 2022, LNNS 534, pp. 298–304, 2023.
https://doi.org/10.1007/978-3-031-15944-2_27

2 System Overview

The structure of the system on which the algorithm will be tested is presented in Fig. 1. The following components can be distinguished from it:

- Rz – Tank,
- P – Pump,
- S_{sig} – Pressure release valve,
- QT – Flow rate sensor,
- PR – Pressure sensor,
- M3 – Induction motor,
- HM – Hydraulic Motor,
- K1 K2 – relay outputs,
- SE – Signal Encoder.

The system uses a PC running LabView for testing and control purposes, however it may operate independently by connecting the Acquisition Board directly to the Inverter.

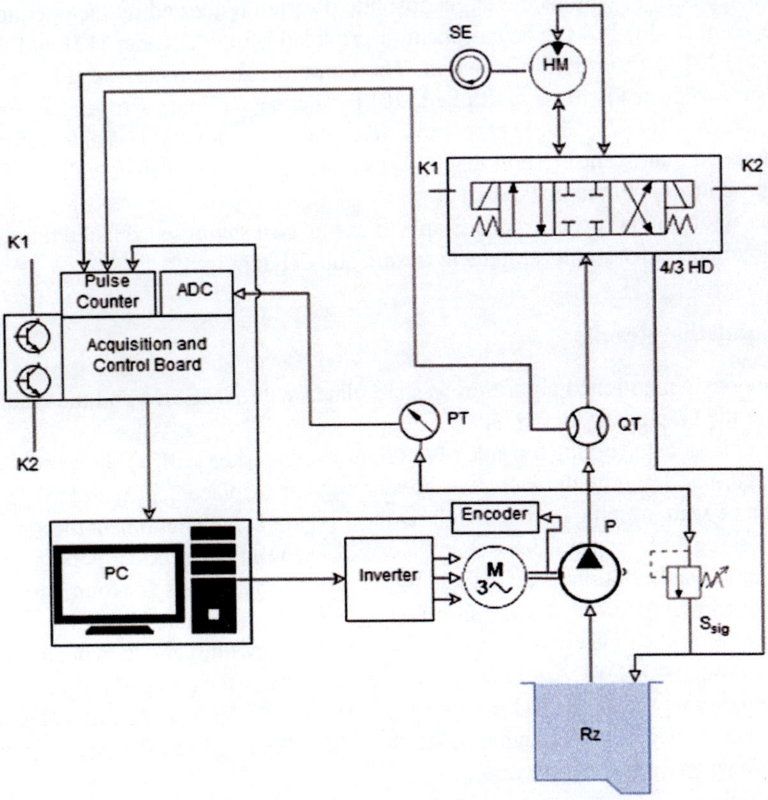

Fig. 1. Complete system setup for position control [2]

The main components required for this low-power low-cost positioning systems are the Inverter, Incremental Encoder on the axle of the Hydraulic Motor and Acquisition board with Pulse counter and Valve control. The rest of the components are standard for simple hydraulic power units.

3 Control Algorithm

3.1 Algorithm Development in AMESIM

The control algorithm was first developed and validated in LMS AMESIM to evaluate the performance of the control loop (see Fig. 2.). The algorithm must control a switch [1] between idle and full speed mode to achieve both low power and reasonable response time. Switching is followed by a rate limiter [2] to protect the induction motor [4] and inverter from stress caused by too fast accelerations. A saturation element [3] is optional but useful as a safety mechanism for limiting induction motor inputs to the range that the pump [5] can operate safely at. Switching logic between high speed and low speed is done by a comparator block [7] that takes the value of the position error [7] and compares it to a predefined constant that triggers the moment of starting deceleration. Reference block [8] outputs the target absolute position requested by the operator, it is further compared [9] to the actual position provided by the encoder [13] and as well subtracted [6] to form the error output. The output of the comparison [9] is either 1 or 0 therefore it needs to be multiplied [10] by a constant value k to match the input requirements of the 4/3 valve [12] solenoid. The hydraulic motor [11] will therefore spin the simulated inertia and friction load [14] as controlled by the valve, with the angular speed proportional to the flow rate.

This algorithm purposely made simple to ensure cost savings also in the future steps of deployment: software development, testing and debugging.

3.2 Simulation Results

The previously mentioned algorithm was simulated with AMESIM and the results are shown in the graphics from Fig. 3.

The chart located on the top side of the figure and marked with **a**) shows two waveforms: setpoint drawn with black respectively angular displacement with red. The setpoint can be seen jumping from 0 to 300° at the 1 [s] mark, at that moment the algorithm starts working and the displacement angle can be seen reaching the set point at around the 5 s mark. On the bottom chart marked **b**) two waveforms can also be found: rotational speed [rev/min] traced with black respectively power [W] with blue. Before the 1 s mark, the system is in Idle mode with rotational speed at 300 [rev/min] and drawing a constant power of around 400 [W]. After the system exits idle mode the pump shaft speed ramps up to the nominal speed of 1500 [rev/min], power can also be seen maxing out at about 1000 W before deceleration begins. After the setpoint is reached the system returns to Idle mode where it draws less current.

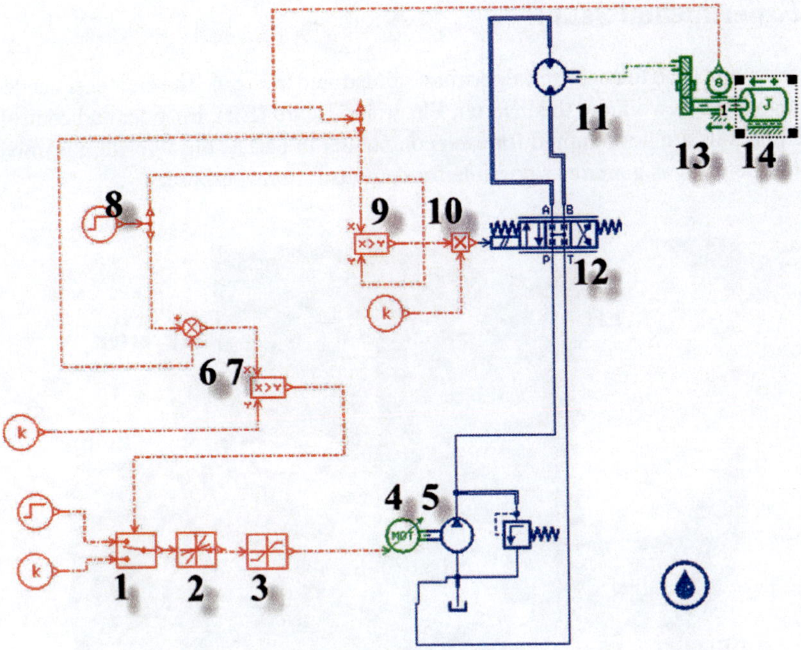

Fig. 2. Simulation of the control algorithm done in Siemens AMESIM

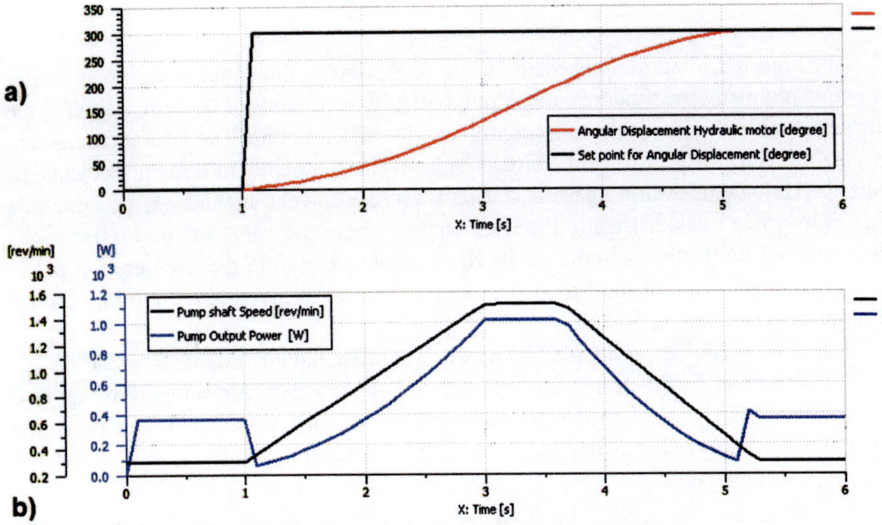

Fig. 3. Results of the simulation with 300[deg] setpoint.

4 Experimental Setup

The test stand used to check the algorithm is illustrated in Fig. 4. The elements needed for this algorithm to work are the Inverter, Electronic Board (EB), Encoder and control PC. Specifications are not required for each component in part as the algorithm is universal and this setup is as generic as possible for demonstrations purpose.

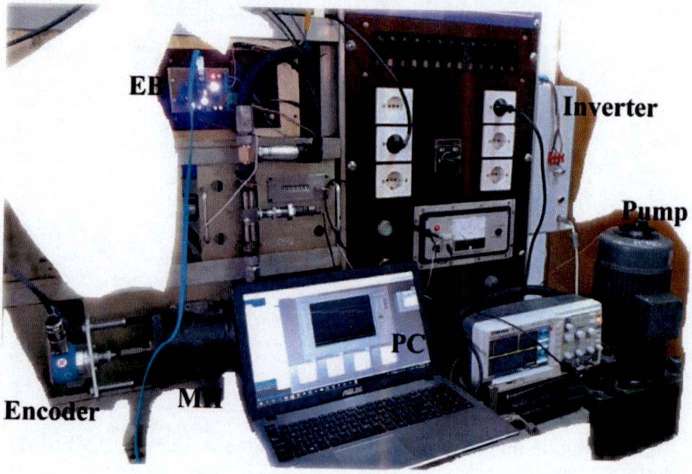

Fig. 4. System setup

The algorithm itself is presented in Fig. 5. Similar to the simulation, a rate limiter is employed using the timer (50 ms) along with the increments or decrements per cycle that are established inside the most central conditional structure. The saturation element is created using two "Min-Max" blocks to limit the frequency in a range between 10 and 50 [Hz]. Deceleration phase is started when the current angular position exceeds the "Ramp neg" value. During the deceleration phase the value of the Frequency is decremented until the minimum of 10 Hz is achieved. Before the deceleration phase is triggered the value of the Frequency is incremented in the acceleration phase until saturation is reached, at 50 Hz.

The whole regulator is enabled by a switch element called "Enable Reg".

The front panel of the program deployed in LabView can be examined in Fig. 6. It contains the basic control needed for the regulator to work: the desired position called Set Point, the buttons that enable the position regulator and hydraulic motor rotation and the negative ramp settings.

Set Point was set to 300 to ease comparison with the simulation. The black waveform shows the progress of the position in time during the positioning cycle. The final position reached was 303 degrees and that implies the existence of an error of about 1%.

The blue waveform represents the frequency output of the inverter as a function of time. During the positioning phase, the algorithm ramps the frequency up until saturation is reached and further maintains the value until the Ramp neg value (250) triggers the start of the deceleration phase.

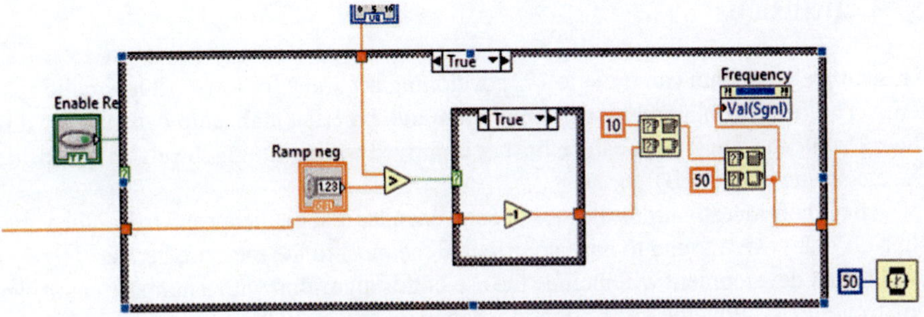

Fig. 5. LabView algorithm

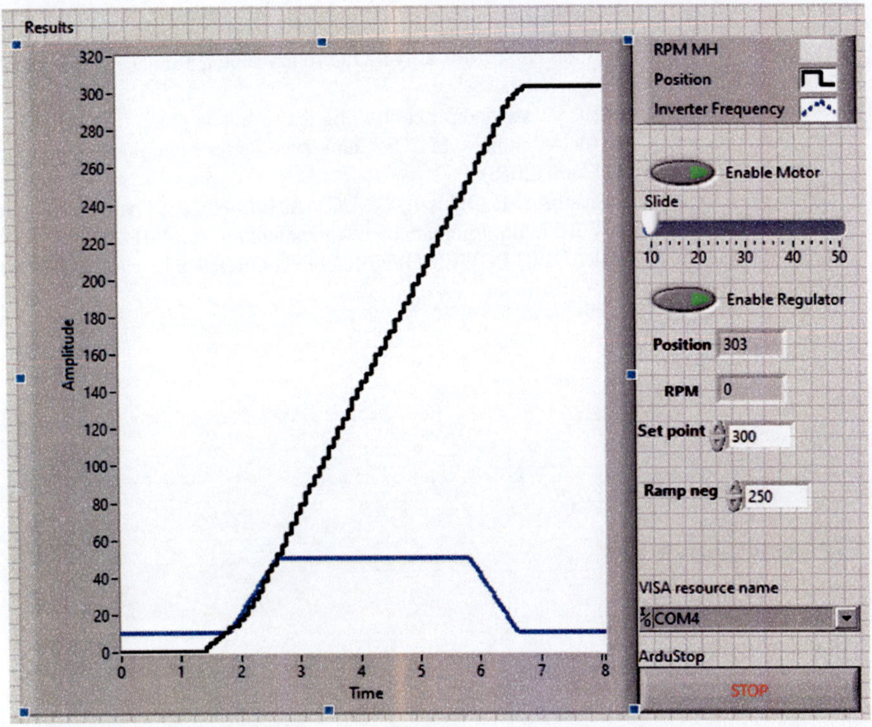

Fig. 6. Results of the LabView algorithm

As soon as the target value set by "Set point" is reached, the Electronic Board (EB) removes power to solenoid of the 4/3 hydraulic valve, essentially cutting off the hydraulic fluid flow going to the hydraulic motor. This functionality is done to mimic the algorithm more closely while also decreasing delays.

5 Conclusions

This simple algorithm can provide 1% positioning accuracy for low-cost hydraulic systems. The results obtained by the simulation and experimental setup can support this theory while the small error can be further improved by fine-tuning the algorithm inside the electronic board (EB).

After short investigations the incremental encoder readout delay and hydraulic valve shut-off delay were found to have contributed the most to the measured error.

Further development will include the use of this algorithm with a numeric controller NC frontend to allow the integration in a complete assembly line.

References

1. Avram, M.: Speed and position control for a hydraulic rotary motor. Roman. REV. Precis. Mech. Opt. Mechatron. **45**, 63 (2014)
2. Avram, M., Spanu, A., Sarbu, V.: Method for controlling the hydraulic pump flow following an imposed frequency law for AC motors. In: IOP Conference Series Materials Science and Engineering, vol. 444, p. 042009 (2018)
3. Avram, M., Sârbu, V.-E., Spânu, A.-R., Bucşan, C.: IJOMAM: Intelligent Hydraulic Power Generating Group, February 2017. http://ijomam.com/wp-content/uploads/2017/02/pag.-157-162_INTELLIGENT-HYDRAULIC-POWER-GENERATING-GROUP.pdf. Accessed 11 2019

A Case-Based Reasoning Based Framework for Developing Computer Aided Remanufacturing Process Planning Systems

Uyi-osa Egbe[1](✉), Chi Hieu Le[1], James Gao[1], Anh My Chu[2], Michael Packianather[3], and Nikolay Zlatov[4]

[1] Faculty of Engineering and Science, University of Greenwich, ME4 4TB Kent, UK
u.egbe@gre.ac.uk
[2] Institute of Simulation Technology, Le Quy Don Technical University, Hanoi, Vietnam
[3] Cardiff School of Engineering, Cardiff University, Cardiff CF24 3AA, UK
[4] Bulgarian Academy of Sciences, 1113 Sofia, Bulgaria

Abstract. There is an emerging need for developing Computer Aided Process Planning (CAPP) systems for rapid retrieval and effective reuse of remanufacturing solutions, especially in the trend of sustainable and smart manufacturing. This paper proposes a framework based on Case-based Reasoning (CBR) for knowledge management to develop CAPP systems for remanufacturing applications, in which the lessons learned during the implementation of CAPP are viewed as a knowledge resource in decision making.

Keywords: Remanufacturing · Sustainable manufacturing · Case-based reasoning · Knowledge management · Computer aided process planning

1 Introduction

Environmental, economic and social benefits have been observed to be the main reasons that organizations engage in remanufacturing, which is defined as a series of processes for returning a used product, component, or part to its original performance with a warranty that is equivalent or better than that of the newly manufactured equivalent (BSI 2009). Restoration is a critical process of remanufacturing involving additive manufacturing processes such as welding, subtractive manufacturing processes (e.g., machining) and surface finishing processes (e.g., polishing and heat treatments).

The success of remanufacturing is to some extent determined by planning the remanufacturing process, called remanufacturing process planning, which determines remanufacturing activities to be carried out on a defective part or sub-system, including operation types, machines, and process parameters, as well as the sequence of operations needed to manufacture a part or sub-system. The outputs of remanufacturing process planning are remanufacturing process plans (RPPs). In practice, remanufacturing process planning is traditionally an ad-hoc process, which is heavily reliant on the skills, knowledge and experiences of process planners and restoration experts.

D. D. Cioboată (Ed.): ICoRSE 2022, LNNS 534, pp. 305–311, 2023.
https://doi.org/10.1007/978-3-031-15944-2_28

Case-based reasoning (CBR) was introduced into the domains of manufacturing and remanufacturing process planning, to promote the use of experiences and know-how from the past planning and restoration activities (Zhou et al. 2014). CBR is a problem-solving methodology, which is used in the field of artificial intelligence (AI), normally with the use of four steps: retrieving similar cases, reusing the retrieved cases, revising the solutions in similar cases, and retaining successful cases for future similar applications. As knowledge-based systems, Computer Aided Process Planning (CAPP) systems, that are developed based on CBR, can process the knowledge resources by retrieving and reusing previous knowledge to finally generate optimal process plans, and also preserve the usefulness of the knowledge resources for solving future similar problems. CBR already possesses at its very core, an in-built mechanism for its maintenance, through its revise and retain steps (Richter and Weber 2013).

It was identified in the preliminary investigations carried out by the authors, especially in automotive industries, that standardised methods for planning the restoration activities in remanufacturing are not available for small or medium scale Enterprises (SMEs), leading to the difficulties when the experienced restoration operators are not available or have left. Most SMEs in remanufacturing industries experience a waste of manufacturing resources, including materials, time and energy, which are required to rework the parts that have already been restored. There is an emerging need for developing CAPP systems to aid the process planning activities in remanufacturing, that can capture the knowledge of experienced restoration operators, to reduce the wastage of manufacturing resources in reworking the restored parts or sub-systems.

In this paper, a framework for effectively developing CAPP systems for remanufacturing based on CBR is proposed (see Fig. 1); with the focus on extending the maintenance mechanism in the CBR-based CAPP systems to generate RPPs. Past knowledges can be learned during the implementation of remanufacturing process planning to generate RPPs, based on CBR, in which previous lessons learned during remanufacturing processes are viewed as a knowledge resource.

2 Literature Review

CBR systems which are developed to generate process plans in manufacturing or remanufacturing need to be maintained. Generally, the maintenance of a CBR system involve the revision of any of its knowledge containers – vocabulary, similarity, adaptation and/or case base (Chebel-Morello et al. 2015). Several methods were developed for the maintenance of these knowledge containers (Mathew and Chakraborti 2017; Ayed et al. 2020); however, the maintenance of the case base knowledge is widely understood to be the most crucial; because it is the knowledge container that is most sensitive to changes in the CBR system. The purpose of case base maintenance (CBM) is to ensure that the efficiency, competence and quality of a CBR system is preserved (Khan et al. 2019). CBM involves the addition of new cases, the revision and updating of existing cases, and the deletion of cases from the case base.

The predominant view amongst researchers in CBR considers CBM policies as either case addition (retention) strategies, case deletion strategies or hybrid strategies. Zhang et al. (2016) proposed a case selection policy that is based on streaming criteria for addition. In this work, a lost function was used to decide on the usefulness of the case at hand. Smiti and Elouedi (2017) proposed a soft case base maintenance method to reduce the size of a case base while preserving as much as possible, the performance and the competence of the CBR system. Khan et al. (2019) proposed a hybrid case base maintenance approach which equally utilizes the benefits of case addition as well as case deletion strategies to maintain the case base in online and offline modes respectively.

One mechanism for implementing a hybrid maintenance mechanism is to continuously carry out case retention up to the point where further case addition no longer improves the CBR system's efficiency or quality. At this point, case deletion is carried out to ensure only important cases remain in the case base. For this mechanism, the case deletion has received more of the attention (Schack 2019), with not enough attention being given to case retention. The few research carried out in case retention – such as Khan et al. (2019) and Zhang et al. (2016) – focused on the suitability of the cases in focus for the CBR base based on selected criteria, but very few of the available literature in this area presented a knowledge management approach for case retention. To the best of our knowledge, there is no available literature that views the implementation of a CBR-generated solution as a resource that can be managed and used to decide which cases gets retained.

3 Functions of the Proposed Framework

As shown in Fig. 1, the Lessons Learned Management System (LLMS) is newly introduced, in which the relationship with its users, data sources and the relevant remanufacturing processes are shown. The LLMS processes data mainly from remanufacturing process functions, based on the use of a four-step process – information extraction, lesson extraction, case reformulation and lesson retention.

3.1 Information Extraction

From the modified CBR-generated RPP which is used to restore a faulty part, the RPP implementation (or the work activity) data, and the data obtained during the testing of a restored part, three kinds of information are extracted, including: the predictability of the restoration process (RPP implementation), the status of restored parts/components after restoration and the occurrence of equipment failure during restoration.

An RPP implementation can be predictable or unpredictable. A Predictable implementation is one where it is observed that the activities carried out during restoration is the same as the instructions contained in the RPP, while for unpredictable implementation, the restoration activities and the instructions are different. The restoration carried out on a defective part can either be successful (if it passes testing) or unsuccessful (if it fails testing).

A combination of the predictability of the restoration process and the restoration status of a restored part/component produces an output of the following four possible information classes: (i) Predictable implementation that led to successful restoration; (ii) Predictable implementation that led to unsuccessful restoration; (ii) Unpredictable implementation that led to unsuccessful restoration; and (iv) Unpredictable implementation that led to successful restoration.

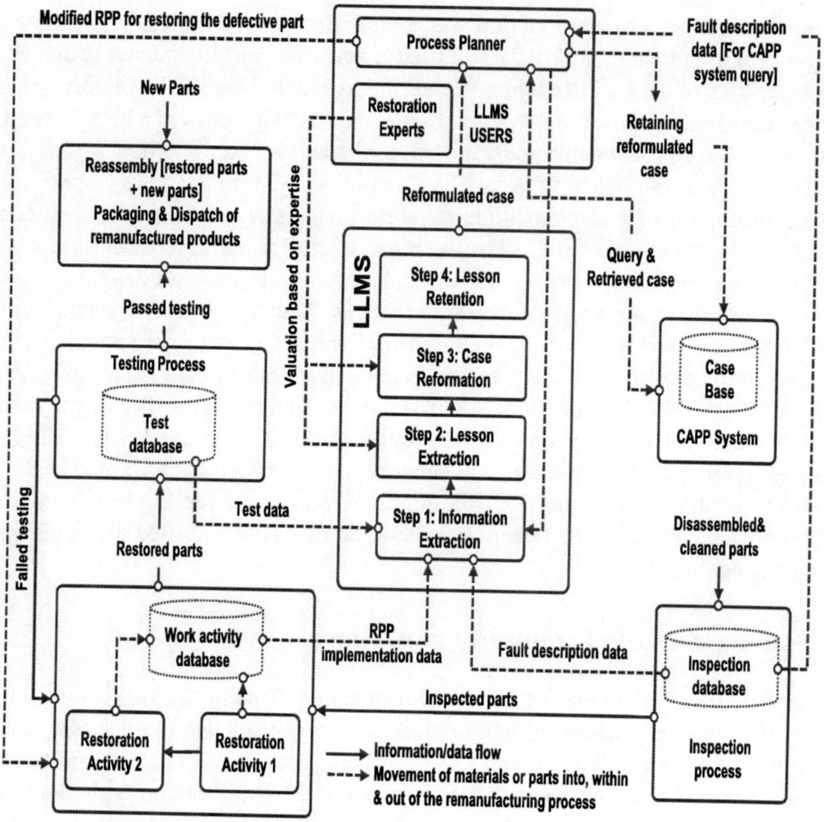

Fig. 1. A proposed framework for development of computer aided process planning systems for remanufacturing based on case-based reasoning (CBR). LLMS stands for Lessons Learned Management System. CAPP stands for Computer Aided Process Planning.

3.2 Extraction of Lessons Learned

The reasons for restoration failures and successes are determined during this step. These reasons are the lessons learned during the implementation of a RPP to restore a defective part. The lessons are the possible cause of restoration failure (if the restoration was unsuccessful), and the possible reasons for restoration success (if the restoration was successful even when unexpected).

The reasons for failure may be RPP failure, operator failure and/or equipment failure while the reasons for success may be the input of tacit knowledge (from experience) by the operator directly responsible for carrying out the restoration process. This tacit knowledge may be in the form of addition and/or substitution of the instructions contained in the RPP.

With the reasons for restoration failures and successes known, corrective actions then need to be prescribed. Table 1 presents the possible reasons for restoration failure and success together with their respective corrective actions.

3.3 Case Reformulation

For the extracted lessons to be truly learned, they must be integrated into the CBR-based CAPP system through a reformulated case. In CBR, a case is a record of a previous experience. It is often divided into a problem aspect and a solution aspect.

Table 1. Corrective actions prescribed to address the reasons for restoration success/failure.

	Lessons	Corrective actions
Restoration success	Addition and/or substitution inputs from operators	Modify the RPP to reflect the inputs from the operators
Restoration failure	RPP failure	Repair the RPP
	Equipment failure	Enrich the RPP with warning of possible occurrence of same failure in the future
	Operator failure	Enrich the RPP with information about the prerequisite training needed to implement the RPP in the future

For CBR-based CAPP systems, a case's problem aspect is the fault description of the defective part to be restored, and its solution aspect is the RPP which was used to restore the fault described in the problem aspect. In the proposed LLMS, the problem aspect of the reformulated case to be reformulated remains the fault description of the defective part, but the solution aspect of the case will now be the new RPP on which the corrective actions (described in Table 1) have been carried out.

3.4 Lesson Retention

This is a straight-forward process of retaining the newly reformulated case into the case base of the CBR-based CAPP system. It is similar to the process of adding a new case to the case base. The decision-making involved in the lesson extraction (Sect. 3.2) and case reformulation (Sect. 3.3) steps require inputs from restoration experts (and process planners) in the form of knowledge validation. Figure 2 presents the procedural flowchart of the proposed LLMS.

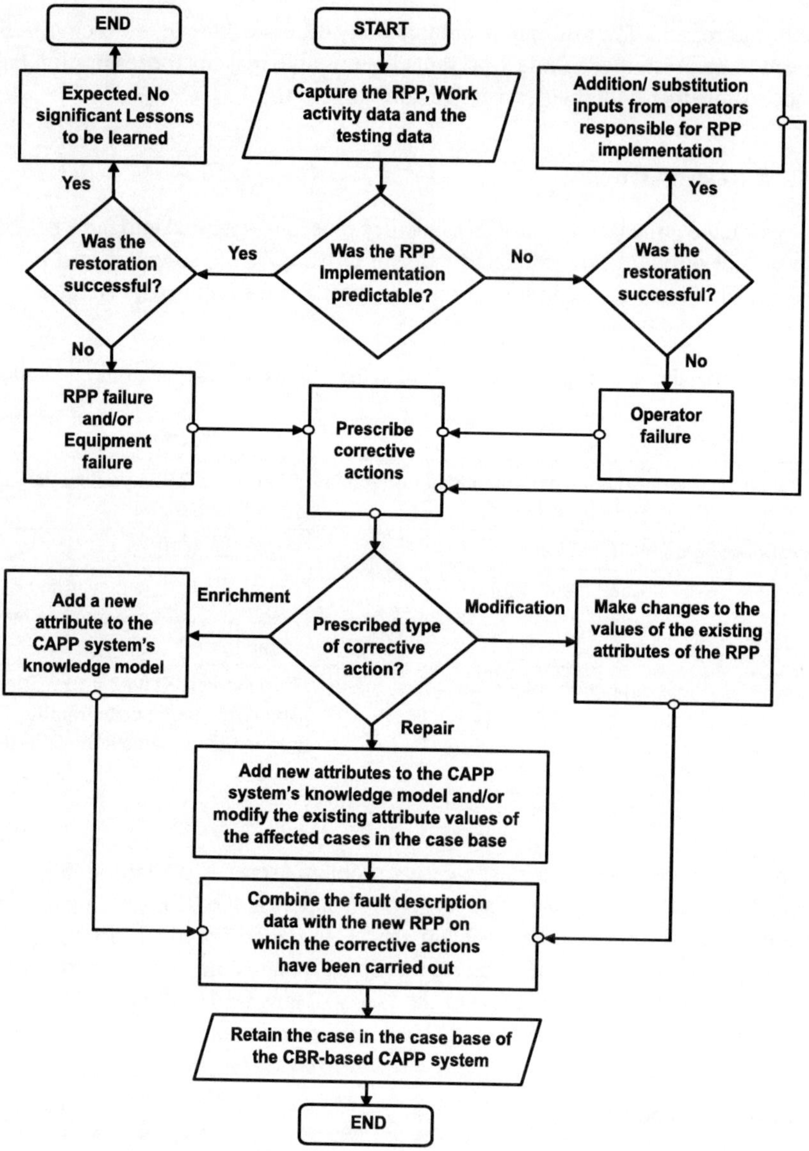

Fig. 2. A procedural flowchart of the proposed lessons learned management system (LLMS).

4 Conclusions

There is a growing need for the development of technical solutions and methods for sustainable design and manufacturing to obtain the sustainable development goals of United Nations (UN 2022). Remanufacturing is one of the sustainable manufacturing technologies, with the focus on reusing materials, parts and sub-systems, and conserving embodied energy in existing products. However, there are technical and practical challenges for SMEs to effectively implement process planning activities in remanufacturing, especially difficulties exists when the experienced restoration operators are not available or have left, leading to the emerging need for developing standardized methods for planning the restoration activities in remanufacturing, especially under the impacts and trends of moving towards sustainable manufacturing and smart manufacturing. In this paper, a framework for knowledge management was proposed, with the focus on developing CBR-based CAPP systems for remanufacturing applications, in which the lessons learned during remanufacturing are viewed as knowledge resources. A knowledge management approach was presented for case retention as a mechanism for the CBM of a CBR-based CAPP system for RPP generation. The LLMS was newly introduced, in which the LLMS operation was based on a four-step process: information extraction, lesson extraction, case reformulation and lesson retention. The framework can potentially be used to effectively develop CAPP systems, to aid the process planning activities in remanufacturing, that can capture the knowledge of experienced restoration operators, to reduce the wastage of manufacturing resources in reworking the restored parts, sub-systems and products.

Acknowledgement. Research was supported by Vingroup Innovation Foundation (VINIF) with the grant ID number VINIF.2019.DA08. It was also supported by a Research Environment Links, with the grant ID number 528085858, under the Newton Fund partnership; the grant was funded by the UK Department for Business, Energy and Industrial Strategy and delivered by the British Council.

References

Ayed, S.B., Elouedi, Z., Lefevre, E.: An evidential integrated method for maintaining case base and vocabulary containers within CBR systems. Inf. Sci. **529**, 214–229 (2020)

BSI - British Standard Institute: Terms and Definitions and BS 8887–220:2010: Design for manufacture, assembly, disassembly and end-of-life processing (MADE). British Standards Publication, UK (2009). ISBN: 978-0-580-68308-4

ChebelMorello, B., Haouchine, M.K., Zerhouni, N.: Case-based maintenance: structuring and incrementing the case base. Knowl.Based Syst. **88**, 165–183 (2015)

Khan, M.J., Hayat, H., Awan, I.: Hybrid case-base maintenance approach for modelling large scale case-based reasoning systems. HCIS **9**, 1–25 (2019)

Mathew, D., Chakraborti, S.: Competence Guided Model for Case Base Maintenance, pp. 4904–4908. AAAI Press, Melbourne (2017)

Richter, M.M., Weber, R.O.: Case-Based Reasoning: A Textbook. Springer, Berlin (2013)

Schack, B.: Case-base maintenance beyond case deletion. Int. Conf. Case Based Reason. **2567**, 191–195 (2019)

Smiti, A., Elouedi, Z.: SCBM: soft case base maintenance method based on competence model. J. Comput. Sci. **25**, 1–25 (2017)

UN: United Nations' Department of Economic and Social Affairs, Sustainable Development (2022). https://sdgs.un.org/goals. Accessed 1 2022

Zhang, Y., Zhang, S., Leake, D.: Case-base maintenance: a steaming approach. Int. Conf. Case Based Reason. **1815**, 222–231 (2016)

Zhou, F., Jiang, Z., Zhang, H., Wang, Y.: A Case-based Reasoning Method for Remanufacturing Process Planning. Discrete Dynamics in Nature and Society, pp. 1–9 (2014)

Configuration of SRR-Metamaterial Based 2 * 1 Array-Type RGW Antenna with Cantilever Beam Switching Technique

Atik Mahabub Fouad[1], Ghazaleh Ramerzani[1], and Ion Stiharu[2](✉)

[1] Maisonneuve Blvd. West, Montreal H3G 1M8, Canada
[2] Concordia University, Montreal, Canada
istih@encs.concordia.ca

Abstract. In this article is presented an MEMS based antenna that generates a radio frequency (RF) signal using switching technology. For the MEMS switching technology, a bi-metallic cantilever beam is used, which will deflect given the thermal properties of the materials. Here, the multi-layer cantilever beam will act like a switch triggered by an electric. Given the thermal properties of the cantilever beam, the beam will be deflected, and the current will flow towards the antenna. The antenna is made of a metallic RGW (Ridge Gap Waveguide). Here, the used method provides low losses for high frequencies to offer low losses at 30 GHz. A single structured RGW antenna can provide 5–6 dB, but using the array technique, the gain can be enhanced and bandwidth, so by using the 2 * 1 array technique, the 3 dB gain has been increased. Later on, a meta-material approach was investigated. Using Split-ring resonator (SRR) meta-material, the gain can be increased by more than 5 dB at specific frequencies. All the analysed antenna structures work for a 5 G range from 27–33 GHz.

Keywords: RF signal antenna · MEMS · Bi-metallic cantilever beam · Ridge gap waveguide · Split-ring resonator · Meta-material

1 Introduction

Microelectromechanical systems (MEMS) have gained importance and are becoming very appealing in demonstrating improved performance in constructing tiny devices and components. MEMS are small devices that combine electrical and mechanical components and are manufactured using batch-processing techniques for integrated circuits (IC) [1, 2]. MEMS switches were first built for microwave applications in the early 1990s [3]. MEMS devices are presently employed in various automotive, biomedical, consumer, aerospace, and military applications. RF MEMS devices such as phase shifters, filters, on-chip antennas, etc., for wideband radios are often employed in wireless communication equipment [4].

RF MEMS switches are excellent for low-power reconfigurable networks and sub-systems because they have low insertion loss and high Q-factor. RF switches can be combined with antennas, filters, and low-loss matching networks [5–7]. Current CMOS

D. D. Cioboată (Ed.): ICoRSE 2022, LNNS 534, pp. 312–328, 2023.
https://doi.org/10.1007/978-3-031-15944-2_29

substrate technologies are unsuitable for integrating antennas; the silicon substrate has a low resistivity – in the range of 10 Ω-cm – which is advantageous for IC design and avoids latch-up currents. However, after the antenna radiates, most of the energy is absorbed/dissipated by the low resistivity substrate rather than radiating to free space [8].

A considerable quantity of surface waves will also penetrate the substrate leading to increased losses. Recent advances in micromachining methods have led to unique, high-performance, high gain antennas for microwave and millimeter-wave applications [9–12]. There are several kinds of micro-machined antennas where MEMS have been used for switching, known as MEMS switching. MEMS switching is used to reconfigure the frequency bands [9, 11]. Another approach is to match the impedance of an antenna to obtain better efficiency. The matching level [10] can also be used in the THz application [12] because of its size. Here, the MEMS switching technique will provide the current flow at the first end of the antenna.

The rapid expansion of wireless communication technologies has fueled demand for more prominent frequency bands and more widespread coverage. Due to these, millimeter-wave frequency bands for short-distance communication should be used, allowing frequency re-use. Hence, mm-wave propagation losses are higher due to the mm-wave oxygen absorption level [13]. High gain antennas must be used to compensate for the undesirable route loss to high frequency [14].

At higher frequencies, microstrip lines have significant dielectric losses, and there is the possibility of undesirable leakage yielding to line bending [15]. The rectangular waveguide is a low-loss guiding device with low attenuation and a high Q-factor. Waves bounce back and forth within the sidewalls while moving in the direction of propagation [16]. The manufacturing difficulty of a rectangular waveguide increases at mm-wave frequencies, affecting the design of the feeding network.

Because both guiding structures support the same dominant mode, a substrate integrated waveguide (SIW) is conceived to complement a rectangular waveguide [17]. However, there are losses with SIW technology in mm-wave bands because it requires a printed circuit board (PCB) material. Kildal presented the Ridge Gap waveguide (RGW) technology, where the essential cut-off is regulated by the PEC-PMC (Perfect Electrical Conductor – Perfect Magnetic Conductor) parallel plate design [18], and there is no wave propagation as long as the air gap spacing between the two PEC- PMC plates is $<\lambda/4$. Based on RGW, several transitions with various feeding processes have been developed [19, 20].

In this work, we demonstrate a new way to excite an antenna with a cantilever beam. Results will demonstrate how a cantilever beam will excite an antenna which will perform, which shows the working principle of the feeding network of the structure. The MEMS technology will be used for RF technology. Rather than that, a metal RGW antenna will be presented. Such a device will create the high needed output for high-frequency wireless communications. In this paper, the use of the array and SRR metamaterial technique for gain enhancement will yield signals above 15dB. SRR metamaterial is a microtechnique used here for the mm-wave application. Such a device could be fabricated by MEMS technology. Surface micromachining and bulk micromachining, both commonly used in MEMS devices fabrication, can be used to fabricate the device using MEMS technology. Design Procedures and Results.

The design of microsystems is carried out with the main objective of being able to create a model and run simulations of its performance. Optimization techniques could also be carried out. The RGW antenna, the cantilever beam, and equipment have been designed and simulated on CST EM Studio Suites. MATLAB software was also used to analyze the deflection of the cantilever beam. The designing steps and the analysis flow is shown in Fig. 1. At first, the unit cell will be designed based on simple matching calculations, and the performance will be checked by its dispersion diagram. In a later stage, this unit cell will be used to create a single structured RGW antenna. Its gain, bandwidth, etc., will indicate the performance; later, a 2 * 1 array will be designed, which acts as a Power Divider using the same unit cell, and the new design will be validated by its bandwidth level. Next, the RGW antenna will be assembled with the power divider, and the results will be evaluated.

Further on, SRR type meta-material will be designed, and its reflection and transmission co-efficient will be reviewed. After that, this meta-material will be installed on top of the 2 * 1 array type RGW antenna; the results will be evaluated. The transition will be added to this substructure at the next stage, such as combining a multi-layer cantilever beam later on, and its bandwidth levels will be assessed. Finally, the multi-layer cantilever beam will be designed by CST, analyzed with MATLAB, and combined with the other antenna structure. The developed systems will be analyzed by MATLAB while combined with the different structure designs at the final stage.

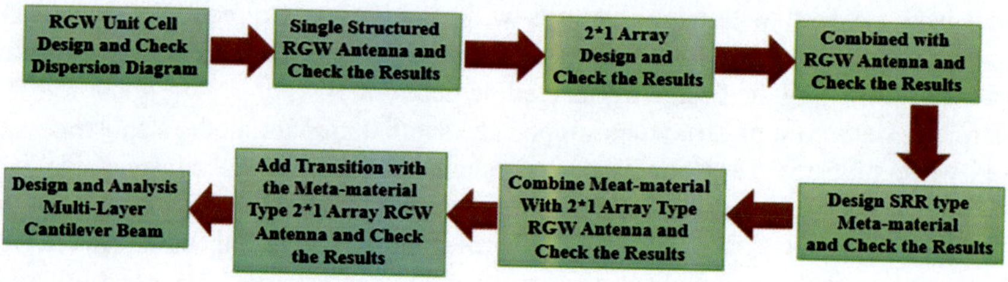

Fig. 1. Designing steps of the multi-layer cantilever beam-based 2 * 1 array RGW antenna with meta-materials.

2 The Unit Cell Design

The unit cell is an essential part of RGW component design. The main characteristic of the RGW unit cell is to act like a stopband structure. It passes through a specific frequency and rejects the others. From Fig. 2(a), the detailed structural design of the unit cell is illustrated. Ground length and Width, Gl = 1.8 mm; Ground height, Gh = 0.5 mm; Pin length and width, Pw = 1.2 mm; Pin height, Ph = 2.5 mm; Airgap, $d_{air} = 0.55$ mm, and Top Layer height, dt = 0.5 mm. The whole structure is made of PEC-type materials like Aluminum, Copper, Steel, etc. Figure 2(b) represents the dispersion diagram of the unit cell. It can be seen that the unit cell is working from 25 GHz–60 GHz and cut off from the other frequencies, and the bandgap is more than 80%. However, only the Q-TEM mode will propagate [21].

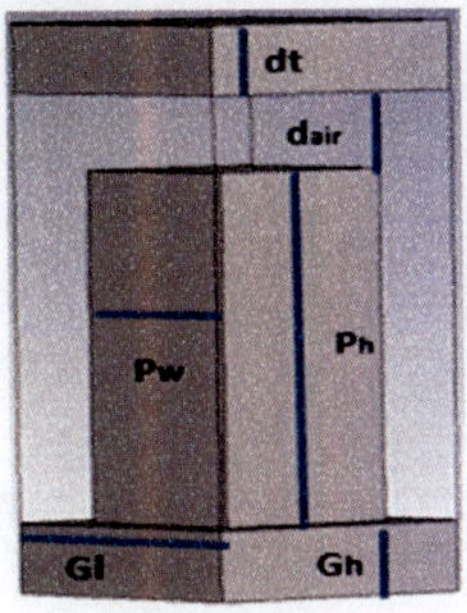

(a)

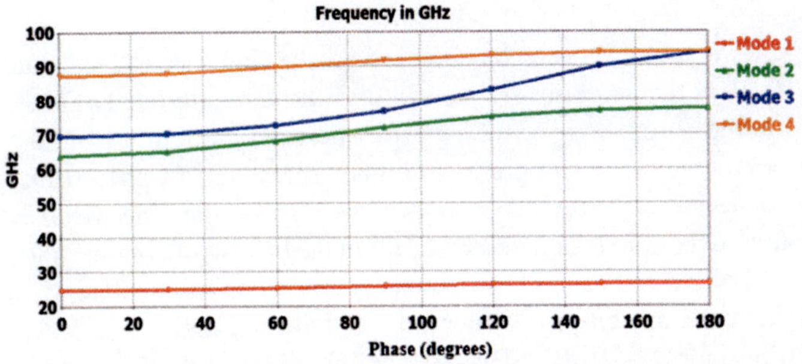

(b)

Fig. 2. (a) Unit-cell structure (b) Dispersion diagram of the unit cell.

3 Single Structured Antenna

After analyzing the unit cell, it is clear that it can be used for the frequency band of 25–60 GHz. The middle ridge has been installed to direct the EM (Electromagnetic) wave. Besides the left and right side of the Unit cell, three rows of unit cells have been installed, which will act as an AMC (Artificial Magnetic Conductor), and the ridge will act like a PEC (Perfect Electric Conductor), so EM wave will propagate through the ridge. The antenna is an EM component where EM waves propagate through space and electric currents, which can contain signals. Here, in this structure, a bow-tie-shaped slot antenna has been chosen because bow-tie-shaped exhibits high gains and compactness [22]. The designing parameters are Ridge width, WR1 = 2.75 mm, Ridge length, LR1 = 10 mm, Bow-tie transition width, WR2 = 0.65 mm, Bow-tie transition length, LR2 = 2 mm, Bow-tie length, LB = 1.3 mm, Bow-tie width, WB = 3.7 mm, Slot length, LW = 2.6 mm, Slot width, SW = 6 mm, point to point pin-pin distance and pin-ridge distance,

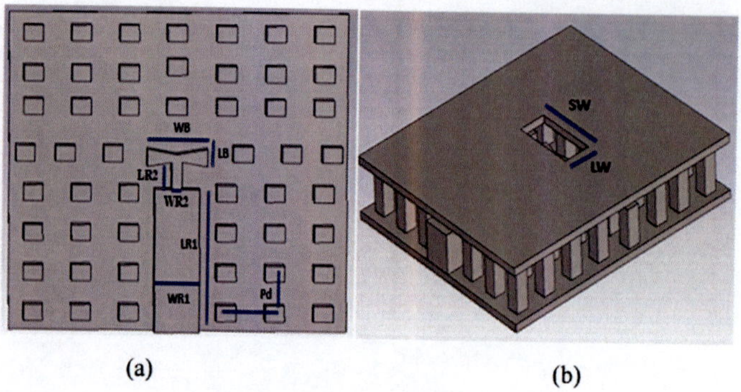

Fig. 3. (a) Top view of the shaped bow-tie-shaped RGW antenna without metal top, (b) Full view of the shaped bow-tie-shaped RGW antenna.

pd = 3 mm, shown in Fig. 3. The slot has been used to propagate the EM wave/signal at the space.

From Fig. 4, the return loss can be observed for the single structured antenna. Return loss below 10dB is the working frequency, and the antenna bandwidth, 27–33 GHz (20%), is observed. From Fig. 5, the gain can be seen, which is from 4.5–6.5 dB at the operating frequency. The slot produces higher gains. The front-to-back ratio should be increased to reduce interference between the signals. For the metal RGW single structured antenna, the gain is higher than 17 dB at the bandwidth region, shown in Fig. 6. From Fig. 7 (a) E-field and H-field, waves can be observed for 30 GHz. Ultimately, this antenna does not offer any cross-polarization at E-field and for H-field except eventually some low shallow below −20 dB and has no side-lobe level (SLL). The cross-polarization and side-lobe level should be down to obtain better performance and signal in the space. The E-field propagation inside the structure is shown in Fig. 7(b), where it can be noticed that the E-field is captured only at the Ridge region, and EM waves are not spreading. This figure illustrates that the E-field is only arrested and charged in the Ridge region, and EM waves do not extend beyond the AMC region. Thus, the unit cell works well for these RGW structures.

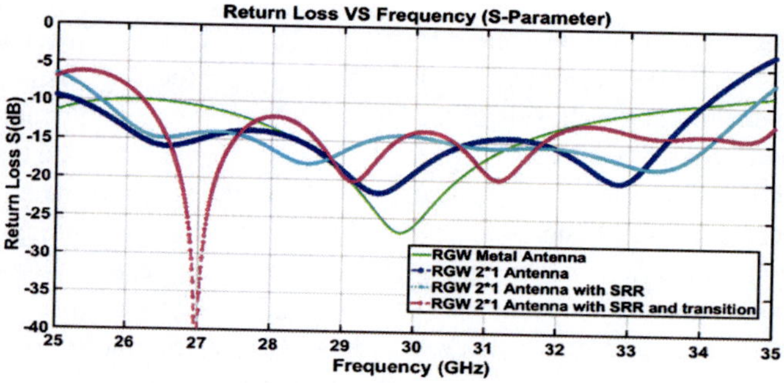

Fig. 4. S-parameters of the different antennas.

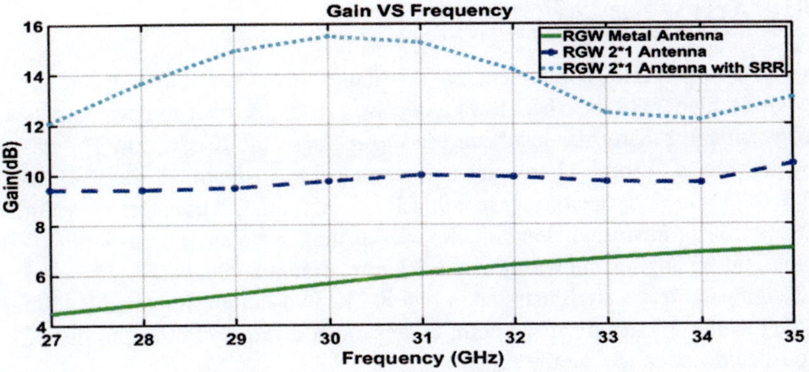

Fig. 5. Gain of the different antennas.

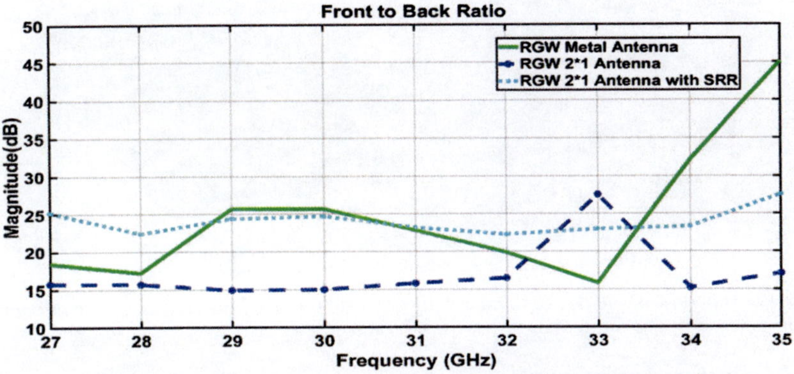

Fig. 6. Front-to-Back Ratio (FBR) of the different antennas.

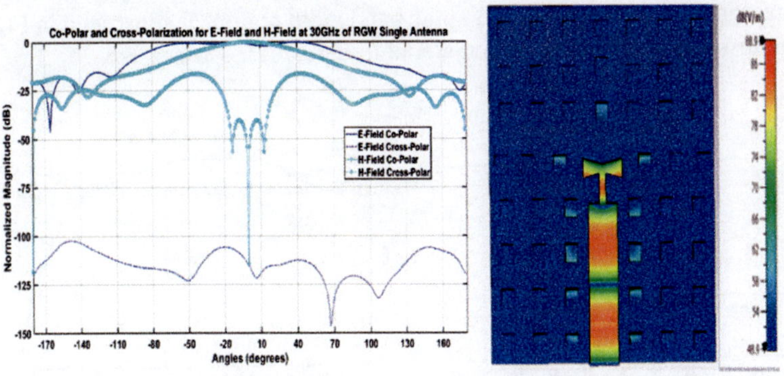

(a) (b)

Fig. 7. (a) E-field and H-field propagation at space (b) E-field propagation inside the structure of the single structured antenna at 30 GHz.

4 The Array/The Power Divider

Power Divider is an EM component that distributes the power uniformly. Here, a 2 * 1 array power divider is used where it was conceived with one input and two outputs. Here, the tri-sectional transformer has been used to increase the bandwidth. The designing parameters are as follows: 1st transformer length, LR1 = 10 mm, 1st transformer width, WR1 = 2.75 mm, 2nd transformer length, LR2 = 3.5 mm, 2nd transformer width, WR2 = 3.6 mm, 3rd transformer length, LR3 = 4.9 mm, 3rd transformer width, WR3 = 4.74 mm, the tri-angular cut width, Cw = 3.6 mm, triangular cut height, Ch = 2.75 mm. The design parameters are illustrated in Fig. 8. The two-output ridge width will remain the same as the 1st transformer width, which impedes 50 Ω. Pin-to-pin distance and pin-to ridge distance will be the same.

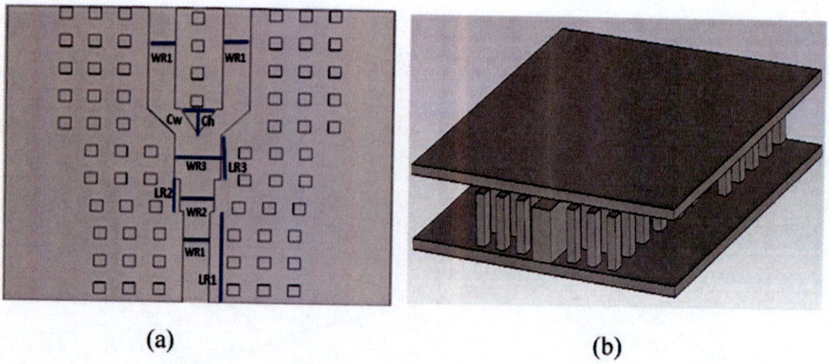

Fig. 8. (a) Top view of the power divider without metal top, (b) Full view of the power divider.

Figure 9 shows the S-parameter of the Power Divider while it is operating from below 25 GHz to 34 GHz from the S1,1 return loss behavior.

For a no-loss 2 * 1-power divider, the outputs insertion loss should be in the 3 dB range, but around −3 dB to −3.3 dB at the bandwidth region, as illustrated in Fig. 9(b). However, losses may happen due to material losses, surface losses, etc.

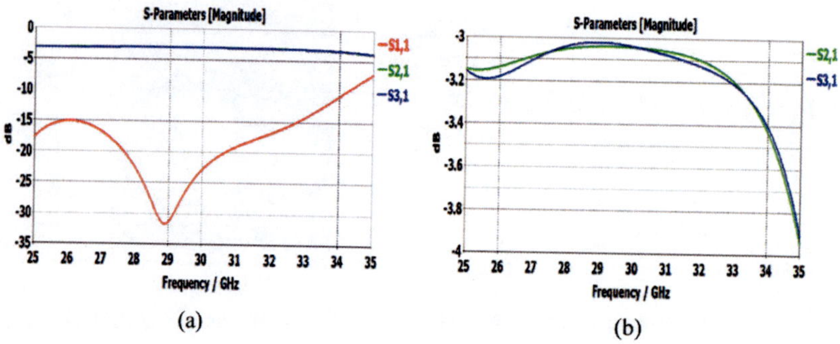

Fig. 9. (a) S-parameter of a 2 * 1-array power divider, (b) Insertion loss of the power divider.

5 The Array-Type RGW Bow-Tie Slot Antenna

After analyzing the behavior of the single structured metal RGW antenna and 2 * 1 power divider, the design of a dual-slot array type RGW Bow-Tie Antenna could be carried out. The designing parameters will be the same as the previous Bow-tie antenna and the 2 * 1-power divider. A double bow-tie slot antenna is used with the spacing between them of 0.75λ point to point, measured at 30 GHz. If it is more than that grating lobe, it can occur at the E-field wave at the space to mismatch the frequency. Here, a single row unit cell was used, which acts as an AMC structure between these two bow-tie slot antennas (Fig. 10).

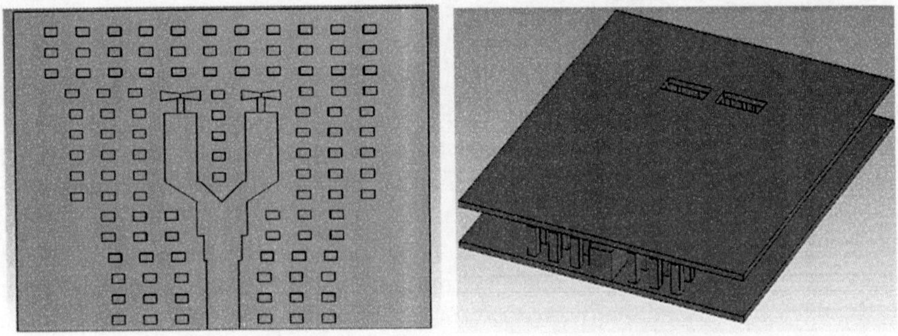

(a) (b)

Fig. 10. (a) Top view of the 2 * 1 array bow-tie shaped RGW antenna without metal top, (b) Full-view of RGW antenna.

From Fig. 4, The bandwidth was increased after using the power divider and the transformer. The antenna is working from 25–34 GHz with an efficiency of about 30%. The primary purpose of this antenna is to increase the gain, which can be seen in Fig. 5.

The growth is almost stable at about 10 dB at the bandwidth region and increases by about 4–5 dB versus the single structure. From Fig. 6, the FBR can be measured as above 15 dB.

Figure 11(a) represents the E-field and the H-field of the 2 * 1 array bow type slot RGW antenna after normalizing. The cross-polarization for the E-field is almost null below −45 dB, and for the H-field, it is a little down −25 dB. Here, E-field does not have any side lobes, but H-field has slightly small side lobes, as well as the H-field has slightly small side lobes below −15 dB. Figure 11(b) shows that the electric field is confined between the ridges and not spreading within the unit cells.

6 The SRR-Metamaterial

The metamaterial can be used for bandwidth enhancement, gain enhancement [24, 25], correlation reduction [26], etc. SRR-type metamaterial has been chosen to increase the gain of this RGW antenna. This metallic SRR type meta-material will help to increase the

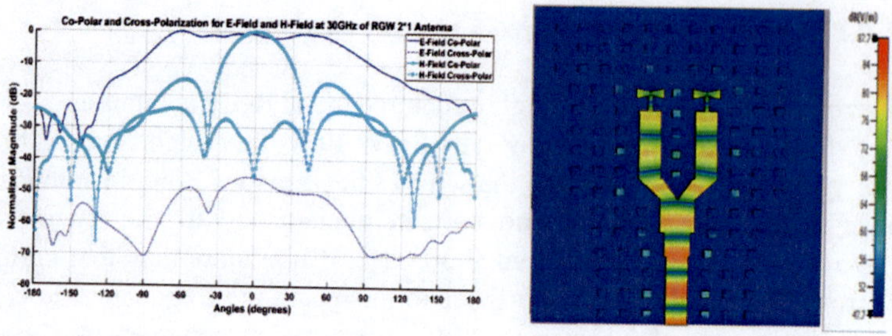

Fig. 11. (a) E-field and H-field propagation in space (b) E-field propagation inside the 2 * 1-array bow-tie-shaped RGW antenna at 30 GHz.

gain [24, 25]. It has been designed in this pattern to only work from 26–32 GHz, representing the bandwidth (BW). Figure 12(b) illustrates the transmission and the reflection coefficients. Here S5,5 represents the reflection co-efficient which is minimum at that BW frequency range, and S6,5 represents the transmission co-efficient which is maximum at that BW frequency range. Therefore, RF signal and radiation will pass through that BW frequency range. This whole SRR metamaterial substrate is made of the laminate Rogers RT-5870, having a relative permittivity of 2.33 and a thickness of 10mils; the structure's backside doesn't have any metals; it's plain. The designing parameter for this SRR-metamaterial [Fig. 12(a)] are: small arm length, LSR2 = 0.3 mm; gap between arm, WSR2 = 0.1 mm; gap from metal to substrate, GSR = 0.35 mm; width of the metal, TSR = 0.08 mm; width of SRR-metamaterial, WSR = 1 mm; length of SRR material, LSR = 0.75 mm. The metal structure is Copper. These metamaterial parameters are narrow, so it would be best to use the MEMS fabrication process to obtain better dimensional and geometric accuracy.

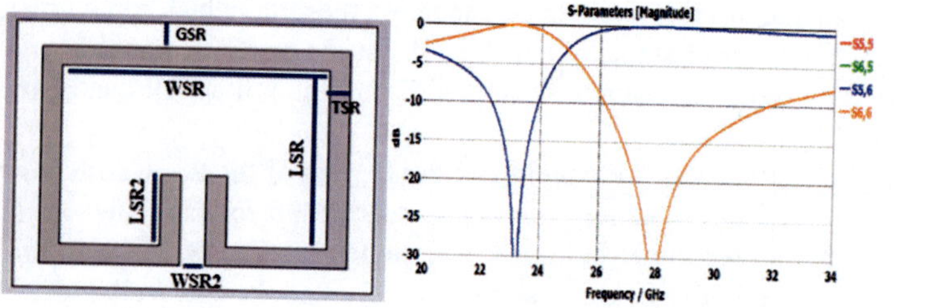

Fig. 12. SRR-type metamaterial (a) Top-view (b) Transmission and reflection co-efficient.

7 SRR-Metamaterial Based RGW Antenna

That designed SRR-metamaterial will be placed on top of the 2 * 1-array type RGW metal antenna. The distance between the top metal and the SRR-metamaterial ground surface should be half full, the wavelength 5 mm. If this distance is impossible to maintain, the potential gain increment would not be possible (Fig. 13).

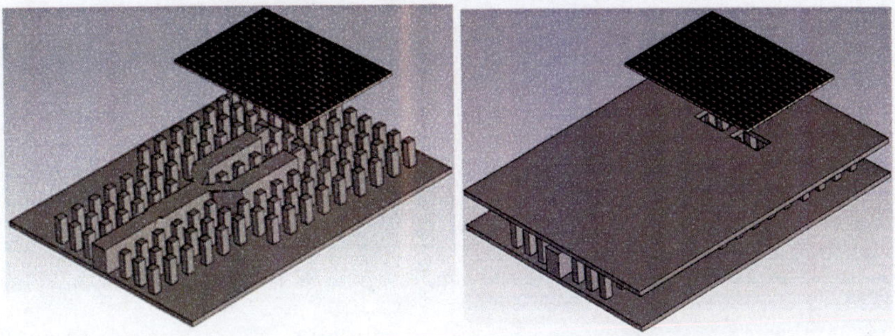

Fig. 13. 2*1-array, type RGW antenna with SRR metamaterials (a) without top (b) with top.

In Fig. 4, the antenna is working from 25.5 GHz–34.5 GHz around 32%. The antenna's primary purpose is to increase the gain, which can be seen in Fig. 5. The gain is above 14 dB, with the maximum at 3 GHz, about 15.5 dB at the SRR-metamaterial operating region from 28 GHz–32 GHz. The increases are about 4–5 dB from the array structure and about 10 dB from the single RGW structure. Due to the SRR metamaterial effect, the gain also increases for the non-operating BW region of the SRR-metamaterial above 12 dB. It mainly happens because of the metal effect on the SRR-metamaterial. Hence, from Fig. 5, it can be observed that the gain of this SRR-based metamaterial-based 2 * 1 array RGW antenna is above 12 dB for the whole bandwidth region and above 14 dB from 28 GHz–32 GHz due to the SRR-metamaterial effects. From Fig. 6, the FBR is significantly improved when using metamaterial, and it is above 22 dB, higher than the one in the single RGW structure.

Figure 11(a) represents the E-field and the H-field of the SRR metamaterial-based 2 * 1- array bow type slot RGW antenna after normalizing. Both E-field and H-field radiation patterns look similar when using SRR-metamaterial. E-field cross-polarization is almost null in the range below −60 dB, and H-field polarization and cross-polarization are very small below −20 dB. The side lobes are reduced after using the SRR metamaterial, nulls for the E-field and −28 dB for the H-field. Figure 11(b) shows that the electric field is confined between the ridges and not spreading because of the unit cells, and the same phenomenon happens to the magnetic field. The wave propagates through the ridge, and using the slot at the top metal, it radiates and reaches the SRR-metamaterial above the antenna (Fig. 14).

8 SRR-Metamaterial Based RGW Antenna with Transition

Transitions are used for connecting the antenna with other Microwave components or with the switch. In this case, the dielectric material is being used as the transition. Rogers-5880 material has been used in this structure, with a relative permittivity of 2.2. The thickness of the air gap and antenna is 0.5 mm. The width of the substrate is 6.5 mm,

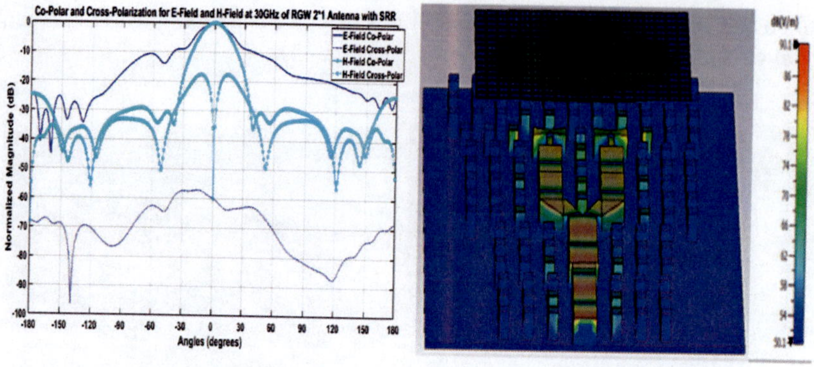

(a) (b)

Fig. 14. (a) E-field and H-field propagation at space (b) E-field propagation inside the structure of the SRR metamaterial stacked 2 * 1-array bow tie-shaped RGW antenna at 30 GHz.

and the length of that is three-unit cells, which is 5.4 mm. Here, two-unit cells will be inside the antenna, and one unit cell will be out of that antenna. Copper can be used for the ground in the outer part. The width of the top metal ridge for the transition is taken as1.5 mm and has been chosen according to the 50 Ω impedance line. For stacking this transition with the previous antenna, the ridge of the substrate will relate to the height of the metal structure, shown in Fig. 15.

The change will be a sandwich between the top metal plate and the ridge antenna structure. The outer part will be used to connect with the switches or other microwave components. It will be associated with the cantilever beam, which is an associated switch.

After using transition, the only part that will be changed is the S-parameter, shown in Fig. 4. Under these conditions, the bandwidth could be improved as it works from 26 GHz-above 35 GHz. This improvement occurs because of the metal ridge, which has been connected with the substrate-based ridge, which compensates for the loss associated with the free space at the transition section. The other parameters like gain, FBR, E-field, H-field, etc., will remain the same as the previous SRR-structure model. This transition can also be used for other antenna structures depending on how much gain is needed.

9 Cantilever Beam Design and Attaching with the Antenna

As a final step, we designed a multi-layer cantilever beam, which is bi-metallic. The fixed part of the beam is made of aluminum (Al), the bottom layer is made of Zinc (Zn), and the top layer is made of silicon (Si).

The cantilever beams will deflect due to thermal energy [27]. Currents flowing through metallic beams produce thermal heat, which will cause deflection. Since one beam is fixed, the thermal energy won't make any effect, while the other beam is free and will deflect. It will act like a mechanical switch triggered by the low current and help trigger the RGW metal antenna.

At this point, Length, L = 10 mm or one wavelength, Width, W = 3 mm, and thickness, t = 0.1 mm. This cantilever beam's deflection is sensitive to its length and its

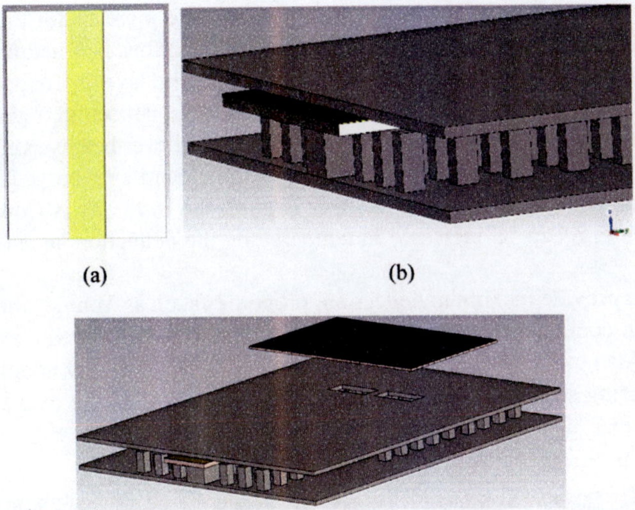

(a) (b) (c)

Fig. 15. SRR metamaterial-based 2 * 1-array type RGW antenna with transition (a) The transition part (b) Stacking the transition part to antenna zoomed view, (c) Full-View with the top.

thickness. As its thickness decreases, the deflection increases, and as its length increases, the deflection also increases.

In this case, Zinc and Silicon have been selected as they have very different thermal expansion coefficients. The bottom layer material zinc has been chosen because its thermal expansion coefficient is the highest among usable materials. The top layer silicon has been selected because its thermal expansion coefficient is the lowest amongst functional materials. Due to its properties, this cantilever beam excites and changes thermal temperature when the current flows from it.

As Zinc has the highest thermal expansion coefficient, it will bend more, causing the cantilever beam to go upwards. If downward bending is needed, Zinc should be used at the top. As the antenna structure transitions, the deflection will occur at the top. While the cantilever beam heats up, it will bend upward. The beam will hit the copper ridge of the transition structure, and an electric field will flow through its height. Next, this field will flow through the metal ridge of the antenna, resulting in RF conversion.

According to this equation, for every 25 °C temperature difference, the deflection occurs around 0.3 mm at the free end. Every degree of temperature difference results in about 0.012 mm of deflection at the top end.

The cantilever can also be multilayered to cause more deflection and be more temperature-sensitive. Adding layers causes more deflection, so it is more sensitive to temperature.

As can be seen in Fig. 16, a comparison between deflection and temperature is shown. The bottom layer of tri-metallic cantilever beams has been selected as Zinc, whereas the top layer is Copper, maintaining the same length, width, and thickness for each metal layer. Aluminum (Al) is used as the middle layer. The deflection has been increased by about 0.35 mm for every 25 °C temperature increase for the tri-metallic cantilever beam structure. For the quad-metallic cantilever beam structure, the bottom layer is Zinc, the second bottom layer is Aluminium, the third layer is Copper, and the top layer is Nickel (Ni). For the quad-metallic cantilever beam, deflection is about 0.45 mm for every 25° temperature increment. Penta-metallic cantilever beams exhibit the same characteristics. The bottom layer of the Penta metallic cantilever beam is silicon; the second bottom layer is steel; the third layer is gold; the top layer is Copper, and the top layer is Zinc. As 25 °C increases in temperature, the deflection increases by about 0.6 mm.

In order to provide maximum deflection, properties such as Young's modulus, thermal expansion coefficients, and Poisson's ratios were selected based on those of the layers used. Also, a greater degree of deflection can be achieved by adding more layers and increasing sensitivity to heat. Figure 17 shows the structure of the multi-layer cantilever beam.

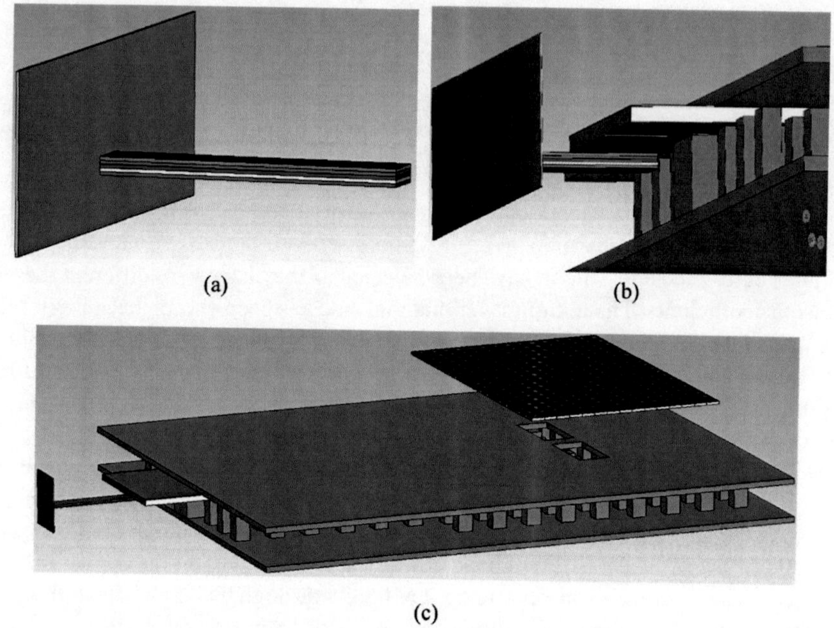

Fig. 16. Deflection of increasing layers for multi-layer cantilever beams at 25° temperature increment.

The calculation and validation of our results were done using MATLAB software. Figure 18 shows the deflections for a bi-metallic structure and a Penta-layer structure. According to the figure, the top end showed the most deflection. It is possible to produce more deflection, which will be more sensitive to temperature, but adding more layers will complicate and make handling the cantilever beam more challenging. Here is the MEMS technology, which is used with antenna RF technology. Mechanical technology is combined with electrical techniques.

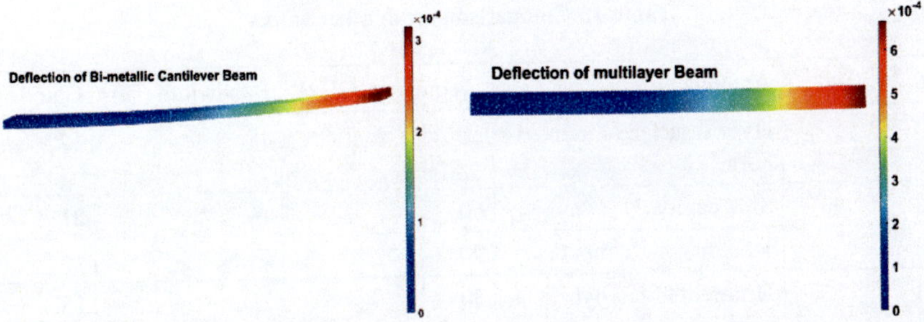

Fig. 17. (a) Multi-layer cantilever beam design, (b) configuration of a cantilever beam with the transition, (c) Full-view.

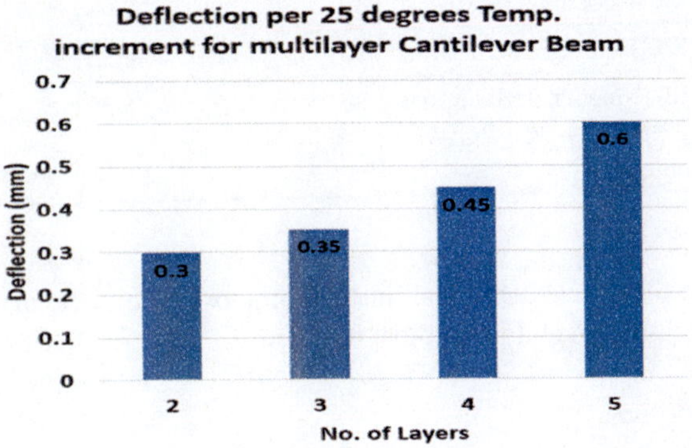

Fig. 18. Multi-layer Cantilever beam deflection at 25° (a) bi-metallic or double layers, (b) Penta-metallic or 5-layers.

10 Results Comparisons & Validation

This work needs to be compared with the other approaches or work done in Table 1.

Different pieces and systems and related works [22, 28, 29, 31, 33] have less bandwidth and gain, associated results [21, 30] have similar increases, but have much less bandwidth, and another proposed related work [32] has more bandwidth but much less gain.

In the reference article [21], the same technique was used for validation, but a lossy dielectric substrate was employed instead of a dielectric structure. Here, a metallic structure was used, which is less lossy than a dielectric structure.

In that article, there are approximately 5–10% or * ± 1 dB differences between the simulated and measured gain sections; there are very few differences in co-polarization and cross-polarization. However, for the S-parameter, there is a more significant deviation between the simulated and the measured structure, which is mostly related to the dielectric material, which is not more than 10–15% or about 2–5 dB difference at some frequencies. But in this article, when fabricated, we are not expecting such differences because here, instead of lossy dielectric material, metal has been used, which has almost

Table 1. Comparisons with other works

Ref.	Antenna type	Frequency (GHz)	Bandwidth (%)	Gain (dB)
[28]	Hybrid surface-mounted horn	30	10	12
[29]	SIW cavity	60	16	21
[30]	1 * 8 magnetic dipole	90	14.5	16.5
[31]	Differential fed hybrid horn	30	25	15
[32]	SIW multilayer horn	90	40	9.5
[33]	2 * 2 magnetic dipole	30	20	10.3
[21]	1 * 4 slot antenna	30	22	15.5
[22]	SIGW integrated horn	30	20.5	11.5
Proposed work	SRR-metamaterial based 2 * 1 array-type RGW antenna	30	>30	15.5

zero losses. Since the article uses metal instead of a lossy dielectric material, there will not be a difference since metal has almost no losses.

List of Abbreviations

MEMS (Micro-electromechanical systems),
RF (Radio Frequency),
IC (integrated circuits),
PCB (Printed Circuit Board),
PEC (Perfect Electric Conductor),
PMC (Perfect Magnetic Conductor),
AMC (Artificial Magnetic Conductor),
RGW (Ridge Gap Waveguide),
EM (electro-magnetics),
BW (Bandwidth),
SLL (Side Lobe Level),
FBR (Front-to-Back Ratio).

11 Conclusions

Finally, by using 2 * 1-array technology and SRR metamaterial at the RGW antenna, 15.5 dB gain has been achieved at 30 GHz, and the bandwidth is about 33% covering from 25.5 GHz–35.5 GHz. The Multi-layer cantilever-switching beam will excite the RGW antenna by using its deflection property, which will carry current and thermal energy and generate an RF wave at the end. Cantilever-switching beams have been used in this article for the power divider and antenna; they can also be integrated with couplers, butler matrix, filters, circulators, etc.

References

1. Bryzek, J., Peterson, K., McCulley, W.: Micro-machines on the march. IEEE Spectr. **31**(5), 20–31 (1994)
2. Brown, E.R.: RF-MEMS switches for reconfigurable integrated circuits. IEEE Trans. MTT. **46**, 1868–1880 (1998)
3. Larson, L.E., Hackett, R.H., Melendes, M.A., Lohr, R.F.: Micromachined microwave actuator (MIMAC) technology-a new tuning approach for integrated microwave circuits. In: IEEE Microwave and Millimeter-Wave Monolithic Circuits Symposium., Boston, MA, USA, pp. 27–30 (1991)
4. Rebeiz, G.M.: RF Mems Theory, Design and Technology. Wiley, New Jersey (2014)
5. Ruby, R.C., et al.: High-Q FBAR filters in a wafer-level chip-scale package. In: IEEE International Solid-State Circuits Conference, San Francisco, CA, USA, pp. 184–458 (2002)
6. Kim, M., Hacker, J.B., Mihailovich, R.E., DeNatale, J.F.: A monolithic MEMS switched dual-path power amplifier. IEEE Microw. Wirel. Compon. Lett. **11**(7), 285–286 (2001)
7. Erdil, E., Topalli, K., Unlu, M., Civi, O.A., Akin, T.: Frequency tunable microstrip patch antenna using RF MEMS technology. IEEE Trans. Antennas Propag. **55**(4), 1193–1196 (2007)
8. Hammad, M.C., Shamim, A.: The Last Barrier, pp.79–91. IEEE Microwave Magazine (2013)
9. Wright, M.D., Baron, W., Miller, J., Tuss, J., Zeppettella, D., Ali, M.: MEMS reconfigurable broadband patch antenna for conformal applications. IEEE Trans. Antennas Propag. **66**(6), 2770–2778 (2018)
10. Yao, S.S., Cheng, Y.J., Zhou, M.M., Wu, Y.F., Fan, Y.: D-band wideband air-filled plate array antenna with multistage impedance matching based on MEMS micromachining technology. IEEE Trans. Antennas Propag. **68**(6), 4502–4511 (2020)
11. Kumar, P.A, Rao, K.S, Sravani, K.G.: Design and simulation of millimeter wave reconfigurable antenna using iterative meandered RF MEMS switch for 5G mobile communications. Microsyst. Technol. **26**(7), 2267–2277 (2020)
12. Boudkhil, A., Chetioui, M., Benabdallah, N., Benahmed, N.: Development and performance enhancement of MEMS helix antenna for THz applications using 3D HFSS-based efficient electromagnetic optimization. TELKOMNIKA (Telecommun. Comput. Electron. Control) **16**(2), 210–216 (2018)
13. MacCartney, G.R., Zhang, J., Nie, S., Rappaport, T.S.: Path loss models for 5G millimeter wave propagation channels in urban microcells. In: IEEE Global Communications Conference, Atlanta, GA, pp. 3948–3953, December 2013
14. Al-Alem, Y, Sifat, S.M, Antar, Y.M.M., Kishk, A.A.: High gain low-cost 20 GHz antenna design based on the utilization of diffracted fields from dielectric edges. In: 19th International Symposium on Antenna Technology Applied Electromagnetics (2021)
15. Pucci, E., Zaman, A.U., Rajo-Iglesias, E., Kildal, P., Kishk, A.: Losses in ridge gap waveguide compared with rectangular waveguides and microstrip transmission lines. In: 4th European Conference on Antennas and Propagation (EuCAP), Barcelona, pp. 1–4, April 2010
16. Iwasaki, T., Kamoda, H., Derham, T., Kuki, T.: A composite right/left-handed rectangular waveguide with tilted corrugations for millimeter-wave frequency scanning antenna. In: 2008 38th European Microwave Conference, pp. 563–566. IEEE (2008)
17. Deslandes, D., Wu, K.: Design consideration and performance analysis of substrate integrated waveguide components. In: Proceedings of 32nd European Microwave Conference, pp. 1–4, September 2002

18. Kildal, P.S., Zaman, A.U., Rajo-Iglesias, E., Alfonso, E., Valero-Nogueira, A.: Design and experimental verification of ridge gap waveguide in bed of nails for parallel-plate mode suppression. IET Microw. Antennas Propag. **5**(3), 262–270 (2011)
19. Shams, S.I., Kishk, A.A.: Wideband coaxial to ridge gap waveguide transition. IEEE Trans. Microw. Theory Techn. **64**(12), 4117–4125 (2016)
20. SharifiSorkherizi, M., Dadgarpour, A., Kishk, A.A.: Planar high-efficiency antenna array using new printed ridge gap waveguide technology. IEEE Trans. Antennas Propag. **65**(7), 3772–3776 (2017)
21. Sifat, S.M., Ali, M.M.M., Shams, S.I., Sebak, A.-R.: High gain bow-tie slot antenna array loaded with grooves based on printed ridge gap waveguide technology. IEEE Access **7**, 36177–36185 (2019)
22. Sifat, S.M., Shams, S.I., Kishk, A.A.: integrated multilayer horn antenna for millimeter-wave application. In: 2020 IEEE International Symposium on Antennas and Propagation and North American Radio Science Meeting, 2020, pp. 1433–1434 (2020). https://doi.org/10.1109/IEEECONF35879.2020.9329449
23. Islam, Md, Islam, M.T., Samsuzzaman, Md., Faruque, M.R.I., Misran, N., Mansor, M.F.: A miniaturized antenna with negative index metamaterial based on modified SRR and CLS unit cell for UWB microwave imaging applications. Materials **8**(2), 392–407 (2015)
24. Kasem, F., Al-Husseini, M., Kabalan, K.Y., El-Hajj, A., Nasser, Y.: A high gain antenna with a single-layer metamaterial superstrate. In: 2013 13th Mediterranean Microwave Symposium (MMS), pp. 1–4. IEEE (2013)
25. Cao, W., Zhang, B., Liu, A., Tongbin, Y., Guo, D., Wei, Y.: Gain enhancement for broadband periodic endfire antenna by using split-ring resonator structures. IEEE Trans. Antennas Propag. **60**(7), 3513–3516 (2012)
26. Zhao, J., Wang, J.: Correlation reduction in antennas with metamaterial based on newly designed SRRs. In: 2010 Asia-Pacific International Symposium on Electromagnetic Compatibility, pp. 981–984. IEEE (2010)
27. Sifat, S.M., Savaj, R., Stiharu, I., Kishk, A.: Wideband bandpass filter design based on RF-mems technology. Int. J. Mechatron. Appl. Mech. **7**, 70–74 (2020)
28. Elboushi, A., Sebak, A.: High-gain hybrid microstrip/conical horn antenna for MMW applications. EEE Antennas Wirel. Propag. Lett. **11**, 129–132 (2012)
29. Razavi, S.A., Kildal, P.-S., Xiang, L., Alós, E.A., Chen, H.: 2 × 2-slot element for 60-GHz planar array antenna realized on two doubled-sided PCBs using SIW cavity and EBG-type soft surface fed by micro-strip ridge gap waveguide. IEEE Trans. Antennas Propag. **62**(9), 4564–4573 (2014)
30. Cao, J., Wang, H., Mou, S., Quan, S., Ye, Z.: W-band high-gain circularly polarized aperture-coupled magneto-electric dipole antenna array with gap waveguide feed network. IEEE Antennas Wirel. Propag. Lett. **16**, 2155–2158 (2017)
31. Alzidani, M., Afifi, I., Asaadi, M., Sebak, A.: Ultra-wideband differential fed hybrid antenna with high-cross polarization discrimination for millimeter wave applications. IEEE Access **8**, 80673–80683 (2020)
32. Ghassemi, N., Wu, K.: Millimeter-wave integrated pyramidal horn antenna made of multilayer printed circuit board (PCB) process. IEEE Trans. Antennas Propag. **60**(9), 4432–4435 (2012)
33. Ali, M.M.M., Sebak, A.: 2-D scanning magneto-electric dipole antenna array fed by RGW Butler matrix. IEEE Trans. Antennas Propag. **66**, 6313–6321 (2018)

Automatic Control of Electrohydraulic Drive for Technological Equipment

Oleksiy Romanchenko(✉), Volodymyr Sokolov, Oleg Krol, Yevhen Baturin, and Oksana Stepanova

Volodymyr Dahl East Ukrainian National University, 59-a Tsentralnyi Pr., Severodonetsk 93400, Ukraine
alexvromanchenko@gmail.com

Abstract. A questions of dynamics characteristics researches and electrohydraulic drives automatic control are considered. The purpose of the work is the electrohydraulic drives' primary dynamic characteristics analysis of technological and productive equipment, and also, the synthesis of automatic control system by drives that takes into account observation noise as well as stochastic perturbation. For these drives, typical mathematical model of dynamic characteristics has been developed, taking into account the specifics of technological equipment of machine-building industries. This typical model makes it possible to identify electrohydraulic drives as an object of automatic control. The system for automatic control of electrohydraulic drives with throttle control has been developed and presented that takes into account observation noise as well as stochastic perturbation. The Kalman-Bucy filter synthesis was carried out in the MATLAB application package environment. When synthesizing the optimal linear controller, the dynamic programming method was used. Study of the quality of regulation of the proposed automatic control system was carried out. For this, analysis of transient processes occurring in the automatic drive control system was performed. Assessment of the influence of the main parameters on the indicators of the regulation quality was carried out, on the basis of which recommendations were formulated for choosing rational values of the transfer coefficients. The results of the work can be used to improve the technical and operational characteristics of automatic electrohydraulic drives of machine-building industries equipment, as well as to expand the functional characteristics of the technological and productive equipment, and the safety level in production.

Keywords: Dynamic characteristics · Observation noise · Stochastic perturbation · Automatic control system · Structural scheme · Transient process

1 Introduction

One of the most effective ways that ensure rapid growth in labor productivity is the mechanization and automation of production. Mechanization significantly increases labor productivity, frees a person from performing difficult and time-consuming operations, allows

D. D. Cioboată (Ed.): ICoRSE 2022, LNNS 534, pp. 329–337, 2023.
https://doi.org/10.1007/978-3-031-15944-2_30

more rational use of raw materials, materials, energy, helps reduce the cost of production, and increase its quality. With automation, the functions of management and control over the production process, which were previously performed by workers-operators, are transferred to devices and automatic devices [1–4].

Until recently, the main direction of automation in mechanical engineering was the automation of technological processes of mechanical processing: the creation of turning, grinding, milling automatic and semi-automatic machines, assembly machines and automatic lines from assembly machines, which allows creating automated departments and workshops, as well as significantly reducing the number of production workers, directly engaged in the maintenance of machines [5–8].

In recent years, work on the automation of control and assembly processes has been widely deployed. Automation of these processes allows, first of all, to improve the quality of manufactured products, as well as to eliminate the situation when more workers are employed in the assembly and control of products than in their manufacture. The introduction of automation in procurement shops: foundry, blacksmithing, etc., has great prospects. Automation increasingly covers auxiliary workshops (tool, repair, etc.), whose products are individual and serial, even with mass production. The development of equipment with automatic electrohydraulic drives (AEHD) and machines tools with automatic control systems (ACS) and software control allowed to solve the problem of automated production of complex products [9–12].

Automatic regulation maintains the constancy of the mode of operation of machines and devices, stabilizes their operation or changes this mode according to a predetermined regulation algorithm. When automating production, control functions are also transferred to automatic devices. Machines monitor the position of parts, their dimensions, tool condition, processing parameters, etc. [13–16].

In this regard, the question of studying the dynamic characteristics of AEHD for technological and productive equipment of machine-building industries and the development of ACS for them is topical.

2 Literature Review

The investigation, modeling and studying the working processes characteristics in technological and productive equipment of machine-building industries is connected with the certain difficulties for describing dynamic characteristics [17, 18]. For AEHD there are there are fluctuations of the devices some parts, pressures and flow rates due to the fluid compressibility, the influence of working fluid environment on regulation elements etc. This leads to the complex nonstationary hydromechanics processes, that must to consider in designing and development of the technological and productive equipment with AEHD [19–22].

The problems of technological and productive equipment calculation and design with AEHD, the ACS synthesis and their research for technological equipment are presented in number of works. It is important to note that the known approaches to modeling and studying workflows use incomplete mathematical models of dynamic characteristics, which do not take into account and do not consider observation noise, as well as stochastic perturbations [23–26].

The technological and productive equipment of machine-building industries operation, in particular, machining equipment, occurs under influences on system, laws of which are stochastic. ACS developed and designed on the basis of such methods will satisfy the requirements not for one deterministic impact, but for the entire set of impacts, that are given by statistical characteristics [27–30].

The purpose of the work is the primary dynamic characteristics analysis of the AEHD of throttle regulation for technological and productive equipment of machine-building industries, and synthesis of the ACS for drives that takes into account observation noise as well as stochastic perturbation.

3 Research Methodology

The settlement scheme for the AEHD of throttle regulation includes the main elements (Fig. 1): hydraulic cylinder (HC), electrohydraulic amplifier (EHA), that contains hydraulic amplifier (HA) and electromechanical transducer (EMT), electronic block (EB), feedback gauge (FB). There is spool hydraulic amplifier in the EHA output cascade, which is also shown in Fig. 1.

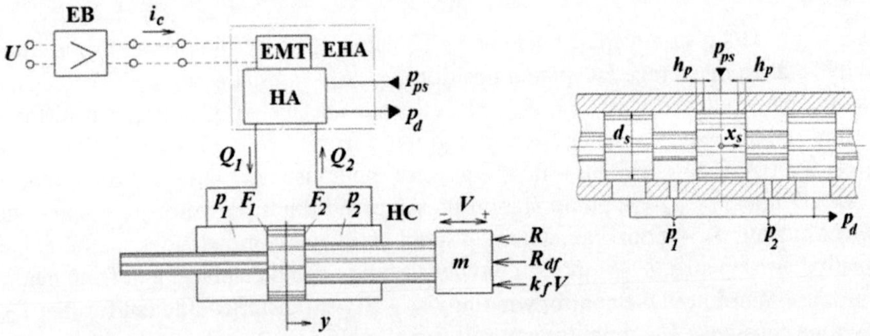

Fig. 1. The settlement scheme for AEHD and output cascade of the EHA.

According to the settlement scheme for AEHD and EHA output cascade the mathematical model has the form:

$$m\frac{dV}{dt} = p_1F_1 - p_2F_2 - cy - k_fV - R - R_{df}\,signV; \tag{1}$$

$$\frac{dy}{dt} = V, \quad -H/2 \le y \le H/2; \tag{2}$$

$$\frac{W_p + F_1(H/2 + y)}{E_f}\frac{dp_1}{dt} = Q_1 - F_1V; \tag{3}$$

$$\frac{W_d + F_2(H/2 - y)}{E_f}\frac{dp_2}{dt} = -Q_1 + F_1V; \tag{4}$$

$$T_{2a}^2\frac{d^2x_s}{dt^2}+T_{1a}\frac{dx_s}{dt}+x_s=k_{xi}i_c; \quad (5)$$

$$L_c\frac{di_c}{dt}+R_e i_c=U; \quad (6)$$

$$Q_1=\begin{cases}\mu_s\pi d_s k_n\left(x_s-h_p\right)\sqrt{\frac{2}{\rho}\left|p_{ps}-p_1\right|}\,sign\left(p_{ps}-p_1\right), x_s>h_p;\\ 0, \left|x_s\right|\le h_p;\\ \mu_s\pi d_s k_n\left(x_s+h_p\right)\sqrt{\frac{2}{\rho}\left|p_1-p_d\right|}\,sign\left(p_1-p_d\right), x_s<-h_p;\end{cases} \quad (7)$$

$$Q_2=\begin{cases}\mu_s\pi d_s k_n\left(x_s-h_p\right)\sqrt{\frac{2}{\rho}\left|p_2-p_d\right|}\,sign\left(p_2-p_d\right), x_s>h_p;\\ 0, \left|x_s\right|\le h_p;\\ \mu_s\pi d_s k_n\left(x_s+h_p\right)\sqrt{\frac{2}{\rho}\left|p_{ps}-p_2\right|}\,sign\left(p_{ps}-p_2\right), x_s<-h_p;\end{cases} \quad (8)$$

where y, V – HC piston displacement and HC piston velocity (the piston displacement usually is taken in average HC piston position); p_1, p_2 – pressure in the HC cavities; m – moving parts reduced mass; F_1, F_2 – HC effective areas; c – positional load stiffness; k_f – force coefficient of viscous friction; R_{df} – dry friction force; R – load; H – stroke; Q_1, Q_2 – EHA flow rates; E_f – fluid elasticity modulus; W_p, W_d – "dead" volumes for AEHD lines; p_{ps}, p_d – pump station pressure and drain pressure; μ_s – spool slot flow coefficient; d_s – spool diameter; k_n – spool perimeter completeness coefficient; h_p – positive overlap size; ρ – working fluid density; x_s – spool displacement from neutral position; i_c – current in the control winding; k_{xi} – EHA transfer coefficient for the; T_{2a}, T_{1a} – time constants; U – input (or control) signal (voltage); R_e – electrical circuit active resistance; L_c – control winding inductance.

For these drives, typical mathematical model of dynamic characteristics has been developed, taking into account the specifics of technological and productive equipment of machine-building industries. The typical mathematical model structural scheme for AEHD is shown in Fig. 2, marked here: u - input signal; y – output variable; k_0 – transfer coefficient for AEHD control unit; T_0 – time constant for AEHD control unit; k – transfer coefficient for AEHD; T_1, T_2 – time constants for AEHD; $V_o(t)$ – stochastic perturbation action on the control object.

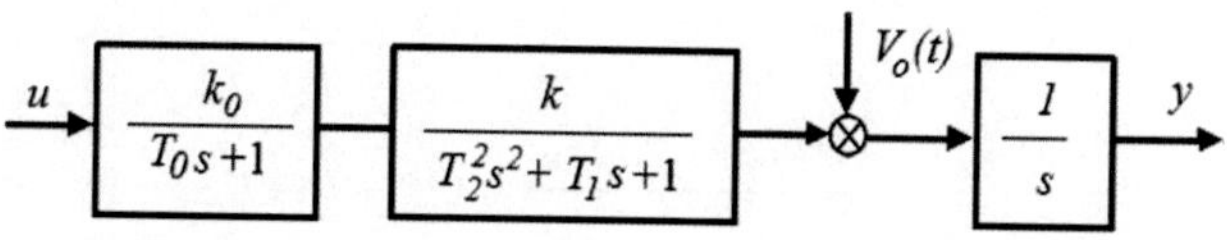

Fig. 2. The typical mathematical model structural scheme for AEHD as the automatic control object.

The control object equations in the state space are represented in the matrix form

$$\dot{x} = Ax + Bu + V_0(t); \tag{9}$$

$$y = Cx + V_n(t), \tag{10}$$

where x – state variables vector; A – control object parameters matrix; B – input parameters matrix; C – output parameters matrix; $V_o(t)$ – color noise; $V_n(t)$ – white observation noise;

$$A = \begin{bmatrix} 0 & 1 & 0 & 0 \\ 0 & 0 & 1 & 0 \\ 0 & 0 & 0 & 1 \\ 0 & -1/T_2^2 T_0 & -(T_1 + T_0)/T_2^2 T_0 & -(T_2^2 + T_0 T_1)/T_2^2 T_0 \end{bmatrix};$$

$$B = \begin{bmatrix} 0 \\ 0 \\ 0 \\ K_0 K / T_0 T_2^2 \end{bmatrix};$$

$$C = \begin{bmatrix} 1\,0\,0\,0 \end{bmatrix}.$$

The former filter generates color noise for control object with transfer function

$$W_f(s) = \frac{b_0^* s + b_1^*}{a_0^* s^2 + a_1^* s + a_2^*}, \tag{11}$$

the input is a stochastic signal $V(t)$ which has the form of white noise with a spectral density $S_V(\omega) = L_V$.

Control optimality criterion

$$J = \int_0^{t_f} (\tilde{Q}x^2(t) + \tilde{R}u^2(t))dt \rightarrow \min, \tag{12}$$

where $\tilde{Q}$ – matrix that characterize control quality; $\tilde{R}$ – parameter that characterize control action limitation value.

The ACS synthesis for AEHD technological and productive equipment of machine-building industries is divided into two tasks: the optimal observer synthesis task and the deterministic optimal controller synthesis task.

To synthesize the Kalman-Bucy filter the MATLAB application package [29, 30] is applied. In accordance with the task, for synthesizing the optimal linear controller the dynamic programming method is used.

For optimal control AEHD obtained

$$u_c = -\left(K_1'' x_1 + K_2'' x_2 + K_3'' x_3 + K_4'' x_4\right), \tag{13}$$

where K_1'', K_2'', K_3'', K_4'' – feedback transfer coefficient of feedback; x_1, x_2, x_3, x_4 – phase variables.

The ACS structural scheme for AEHD is shown in Fig. 3, on scheme K_1', K_2', K_3', K_4' – Kalman-Bucy filter coefficients.

4 Results

For the stochastic ACS for the AEHD of throttle regulation using and without the Kalman-Bucy filter in the stochastic perturbation presence as the white noise with the spectral density $S_v(\omega) = 1$ and $W_f(s) = 0.025$, transient processes are shown in Fig. 4. The investigations have shown that the Kalman-Bucy filter carries out the optimal filtering function in possible parameters of perturbation, supplies required quality control for the AEHD of technological and productive equipment of machine-building industries.

The ACS research was carried out, in particular, the influence of the parameters for control object and controller on transient process quality. In Fig. 5 shows the transient processes in the ACS at different values of the coefficient k_{QU} ($m^3/(s \cdot V)$) and coefficient K_2'' of the controller at transfer coefficient $K_1'' = 108$ (V/m).

When conducting research, the AEHD of throttle regulation for special technological equipment was considered. For this AEHD were accepted next the time constants values of: $T = 0{,}01$ s; $T_1 = 1{,}3 \cdot 10^{-3}$ s; $T_2 = 2{,}62 \cdot 10^{-3}$ s.

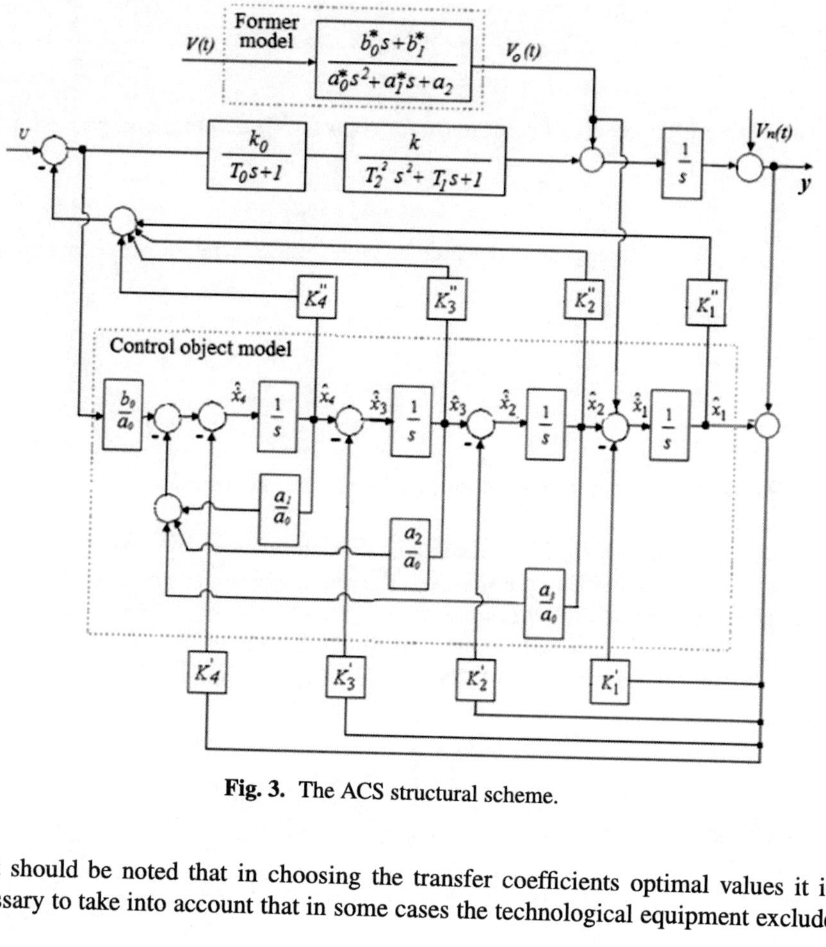

Fig. 3. The ACS structural scheme.

It should be noted that in choosing the transfer coefficients optimal values it is necessary to take into account that in some cases the technological equipment exclude the re-regulation of the working body displacement (or sign change of its velocity). Obviously, that for the researches results presented in Fig. 5, in this case the transfer coefficients recommended values will not match to the best response time of equipment.

The results of performed researches can be used to improve the technical and operational characteristics of AEHD of machine-building industries equipment, as well as to expand the functional characteristics of the technological and productive equipment, and the safety level in production.

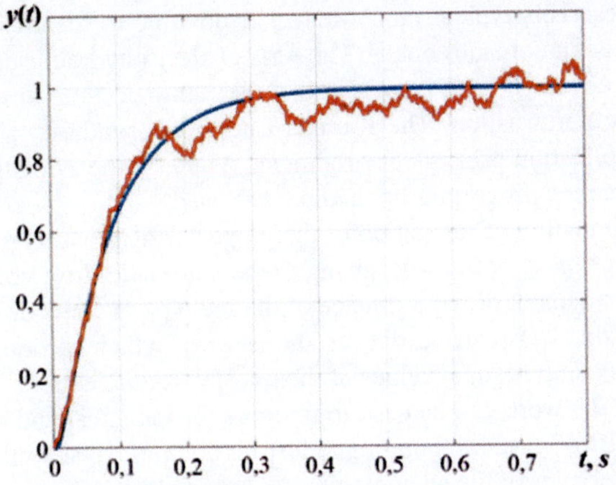

Fig. 4. The transient processes in the ACS using and without using the Kalman-Bucy filter.

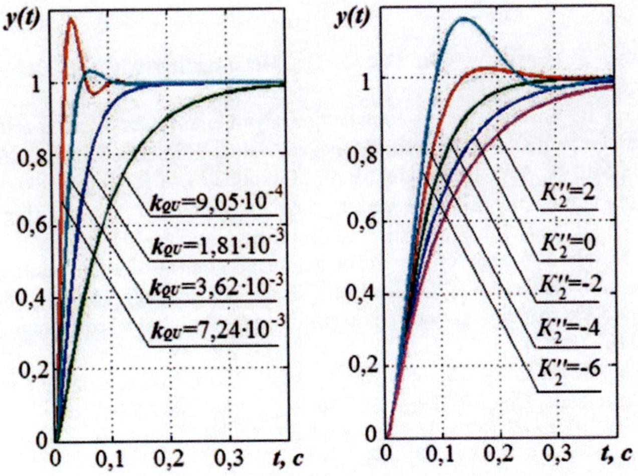

Fig. 5. The transient processes in system.

5 Conclusions

Thus, a questions of dynamics characteristics researches as well as electrohydraulic drives automatic control are considered. The electrohydraulic drives' primary significant dynamic characteristics analysis of technological and productive equipment with throttle regulation carried out. Also, the purpose of the work is and synthesis of ACS by drives that takes into account observation noise as well as stochastic perturbation.

For these drives, a typical mathematical model of dynamic characteristics has been developed, taking into account the specifics of technological equipment of machine-building industries. This typical model makes it possible to identify electrohydraulic drives as an object of automatic control. The ASC of electrohydraulic drives with throttle control has been developed and presented that takes into account observation noise as well as stochastic perturbation. The Kalman-Bucy filter synthesis was carried out in the MATLAB application package environment. When synthesizing the optimal linear controller, the dynamic programming method was used.

A study of the quality of regulation of the proposed ACS was carried out. For this, an analysis of transient processes occurring in the automatic drive control system was performed. An assessment of the influence of the main parameters on the indicators of the regulation quality was carried out, on the basis of which recommendations were formulated for choosing rational values of the transfer coefficients.

The results of the work can be used to improve the technical and operational characteristics of AEHD of machine-building industries equipment, as well as to expand the functional characteristics of the technological and productive equipment, and the safety level in production.

References

1. Abrahamova, T., Bushuyev, V., Gilova, L.: Metal-cutting Machine Tools. Machinery Engineering, Moscow (2012)
2. Kotliar, A., et al.: Ensuring the economic efficiency of enterprises by multi-criteria selection of the optimal manufacturing process. Manage. Product. Eng. Rev. **11**(1), 52–61 (2020)
3. Kundrák, J., Mitsyk, A.V., Fedorovich, V.A., Markopoulos, A.P., Grabchenko, A.I.: Simulation of the circulating motion of the working medium and metal removal during multi-energy processing under the action of vibration and centrifugal forces. Machines **9**(6), 1–22 (2021)
4. Sokolova, Y.: The synthesis of system of automatic control of equipment for machining materials with hydraulic drive. Eastern-Eur. J. Enterprise Technol. **2**(2(68)), 56–60 (2014)
5. Romanchenko, O.: Principles of design of specialized technological equipment. Diagnostyka **23**(1), 2022109 (2022)
6. Guan, C., Pan, S.: Adaptive sliding mode control of electro-hydraulic system with nonlinear unknown parameters. Control. Eng. Pract. **16**(11), 1275–1284 (2008)
7. Sveshnikov, V.: Hydrodrives in modern mechanical engineering. Hydraulic Pneumatic **28**, 10–16 (2007)
8. Krol, O., Sokolov, V.: 3D modelling of angular spindle's head for machining centre. J. Phys. Conf. Ser. **1278**, 012002 (2019)
9. Kundrák, J., Mitsyk, A.V., Fedorovich, V.A., Markopoulos, A.P., Grabchenko, A.I.: Modeling the Energy Action of vibration and centrifugal forces on the Working Medium and Parts in a Vibration Machine Oscillating Reservoir with an Impeller. Manuf. Technol. **21**(3), 364–372 (2021)

10. Krol, O., Sokolov, V.: Research of modified gear drive for multioperational machine with increased load capacity. Diagnostyka **21**(3), 87–93 (2020)
11. Burennikov, Y., Kozlov, L., Pyliavets, V., Piontkevych, O.: Mechatronic hydraulic drive with regulator, based on artificial neural network. IOP Conf. Ser. Mater. Sci. Eng. **209**, 012071 (2017)
12. Fomin, O., Lovska, A.: Establishing patterns in determining the dynamics and strength of a covered freight car, which exhausted its resource. Eastern-Eur. J. Enterp. Technol. **6**(7(108)), 21–29 (2020)
13. Sokolov, V., Krol, O.: Installations criterion of deceleration device in volumetric hydraulic drive. Procedia Eng. **206**, 936–943 (2017)
14. Lovska, A., Fomin, O.: A new fastener to ensure the reliability of a passenger car body on a train ferry. Acta Polytech. **60**(6), 478–485 (2020)
15. Shevchenko, S., Mukhovaty, A., Krol, O.: Gear transmission with conic axoid on parallel axes. In: Radionov, A.A., Kravchenko, O.A., Guzeev, V.I., Rozhdestvenskiy, Y.V. (eds.) ICIE 2019. LNME, pp. 1–10. Springer, Cham (2020). https://doi.org/10.1007/978-3-030-22041-9_1
16. Tsankov, P.: Modeling of vertical spindle head for machining center. J. Phys. Conf. Ser. **1553**, 012012 (2020)
17. Golubovsky, V., Konovalov, V., Doncova, M.: Modelling the force action of a liquid on the shutter of a measuring transducer. E3S Web Conf. **157**, 02007 (2020)
18. Sokolov, V.: Transfer functions for shearing stress in nonstationary fluid friction. In: Radionov, A.A., Kravchenko, O.A., Guzeev, V.I., Rozhdestvenskiy, Y.V. (eds.) ICIE 2019. LNME, pp. 707–715. Springer, Cham (2020). https://doi.org/10.1007/978-3-030-22041-9_76
19. Popov, D.: Mechanics of Hydro- and Pneumodrives. MSTU, Moscow (2001)
20. Golubovsky, V., Konovalov, V., Doncova, M.: Modeling static characteristics of angular velocity measuring transducer of the "nozzle-damper" type. J. Phys. Conf. Ser. **1614**(01), 012084 (2020)
21. Rydberg, K.-E.: Hydraulic Servo Systems: Dynamic Properties and Control. Linköping University Electronic Press, Linköping (2016)
22. Rogovyi, A., Korohodskyi, V., Khovanskyi, S., Hrechka, I., Medvediev, Y.: Optimal design of vortex chamber pump. J. Phys. Conf. Ser. **1741**, 012018 (2021)
23. Sokolov, V., Porkuian, O., Krol, O., Baturin, Y.: Design calculation of electrohydraulic servo drive for technological equipment. In: Ivanov, V., Trojanowska, J., Pavlenko, I., Zajac, J., Peraković, D. (eds.) DSMIE 2020. LNME, pp. 75–84. Springer, Cham (2020). https://doi.org/10.1007/978-3-030-50794-7_8
24. Rasskazova, Y.: Automation of control processes of technological equipment with rotary hydraulic drive. Eastern-Eur. J. Enterp. Technol. **2**(2(80)), 44–50 (2016)
25. Sveshnikov, V.: Hydrodrives of tools. Machinery Engineering, Moscow (2008)
26. Sokolov, V., Porkuian, O., Krol, O., Stepanova, O.: Design calculation of automatic rotary motion electrohydraulic drive for technological equipment. In: Ivanov, V., Trojanowska, J., Pavlenko, I., Zajac, J., Peraković, D. (eds.) DSMIE 2021. LNME, pp. 133–142. Springer, Cham (2021). https://doi.org/10.1007/978-3-030-77719-7_14
27. Pupkov, K., Egupov, N.: Methods of Classical and Modern Theory for Automation Control, vol. 3. Synthesis of Controllers for Automation Control Systems. MSTU, Moscow (2004)
28. Lurie, B., Enright, P.: Classical Automation Control Methods. BHV-Petersburg, Saint-Petersburg (2004)
29. Nuruzzaman, M.: Modeling and Simulating in Simulink for Engineers and Scientists. AuthorHouse, Bloomington (2004)
30. Tewari, A.: Modern Control Design with MATLAB and Simulink. Wiley, Weinheim (2002)

Influence of the Deposition Method on the Hardness and Elastic Modulus of Biocompatible Thin Layers Deposited on Metallic Substrates

Liliana-Laura Badita[1](✉), Aurel Zapciu[1], Catalin Vitelaru[2], Anca Constantina Parau[2], Lidia Ruxandra Constantin[2], Arcadie Sobetkii[3], and Iulian Sorin Munteanu[1]

[1] National Institute of Research and Development in Mechatronics and Measurement Technique INCDMTM, Pantelimon Str. 6-8, 021361 Bucharest, Romania
badita_l@yahoo.com

[2] National Institute for Research and Development in Optoelectronics INOE 2000, Atomistilor Str. 409, 077125 Magurele, Romania

[3] SC MGM Star Construct SRL, Pancota Str. 7, 022773 Bucharest, Romania

Abstract. The quality evaluation of the micro and nanostructured thin layers for biomedical applications, deposited on metallic substrates, by physical methods was the main objective of the research presented in this article. For this reason, micro and nanostructured TiN, TiO_2 and CrN thin layers were deposited on CoCr and M30NW steel substrates by the Direct Current (DC) sputtering, High Power Impulse Magnetron Sputtering (HiPIMS) and cathodic arc methods. Nanoindentation tests were performed in order to establish the hardness and the elastic modulus of the TiN, TiO_2 and CrN thin layers. It was observed that the highest values of the hardness and elastic modulus of TiO_2 and CrN layers are obtained when the DC sputtering method is used for deposition. Cathodic arc method is the one that helps to obtain TiN layers with superior properties on both types of substrates and HiP sputtering can be used to obtain TiO_2 and CrN layers with superior elastic modulus. Thus, the main conclusion of this study was that DC sputtering is the best method for deposition of biocompatible thin layers on metallic substrates. In the future, this method can be used for improving the quality and increasing the use period of biomedical components by thin layers deposition.

Keywords: Thin layers · Hardness · Elastic modulus

1 Introduction

The issue of developing materials with superior characteristics, such as high mechanical strength is an important topic of current research in the field of materials engineering for biomedical applications.

Mechanical properties are an important criterion to consider in the case of these applications, being a factor that can negatively influence the stability of a biomedical

D. D. Cioboată (Ed.): ICoRSE 2022, LNNS 534, pp. 338–357, 2023.
https://doi.org/10.1007/978-3-031-15944-2_31

component, for example an implant. An implant must have sufficient mechanical strength to withstand the stresses that arise over time. Implants are subjected to both static and dynamic stresses, depending on the patient's activity. In the case of orthopedic implants, the degree of stress depends on the position of the body while walking, and reaches the maximum point in the hip and knee joint, four and three times the weight of the body, respectively.

For this reason, the usual materials used in this field must have certain properties, such as excellent resistance to mechanical degradation or corrosion resistance. Materials used for prostheses manufacturing must also have a biocompatible chemical composition to avoid adverse reactions of the human body, and excellent resistance to degradation in the human body to minimize the generation of particles that may adversely affect the osseointegrability of the implant.

In the case of metallic materials used in implants, their structure and properties, like the elastic modulus, which must be similar to that of the organ whose function is replaced, corrosion and wear resistance, long term fatigue resistance, biocompatibility or bioadhesion, must be taken into account depending on the use.

Pure metals, stainless steels, cobalt or titanium alloys and ultra-high molecular weight polyethylene are currently used. Titanium, iron, tantalum, chromium, cobalt, molybdenum, niobium are the main metals used to create alloys for in body-tolerant implants. Stainless steels are steels that contain 12% chromium and less than 0.1% carbon. Chromium has the role of providing resistance to oxidation and corrosion for steels, due to the coating with a passive layer of chromium oxide in aggressive environments: water, air, industrial atmosphere, salts, acids, etc. If other elements, such as titanium, nickel, aluminum, silicon, boron, manganese, copper, niobium, silicon, etc., are added in the composition of stainless steels, other steels are obtained with increasing resistance to high stresses, high temperatures, high pressures, corrosion and aggressive environments. Cobalt and chromium-based alloys have a high corrosion resistance and mechanical strength, which is why they are used for dental, hip and knee prostheses. In order to obtain mechanical strength and corrosion resistance, the percentage of chromium is between 10 and 30%. Also, in order to obtain the superior parameters for corrosion and resistance, iron, molybdenum, and nickel are added to these alloys.

Over time, the use of these materials, depending on the tissue (for example, bone tissue, in the case of prostheses) can lead to their wear. For example, the movement of the body or of the replaced segment causes wear of the material, and ultimately the destruction of a prosthesis. The major disadvantages of implants are also determined by the nature, the way of synthesis and the different physical and chemical properties of the materials used, in relation to living tissues, such as the ability to change its structure and properties according to the stresses it supports (mechanical loading for bone tissue) and self-repair capacity. Most failures occur due to improper properties or incorrect choice of materials used.

In order to eliminate this problem, in time different methods of materials improvement have been tried [1–4]. It was observed that surfaces treatments with metals or other materials can improve corrosion resistance, adhesion and many other properties [5, 6].

Deposition of micro and nanostructured thin layers from materials with directed properties, through intelligent mechatronic processes and technologies constitutes a solution for optimizing the constructive characteristics and increasing the operational duration for biomedical components (e.g. implants) [7, 8].

Thin layers offer the alternative possibility of reducing wear, wear particles production and ions release in metal-to-metal couplings. Factors that could contribute to this reduction are: increased component durability, a different surface chemistry to reduce the adhesive friction, and coated components that remain undamaged. For example, in the case of hip prostheses, sliding surfaces have to be usually hard in order to keep wear to minimum values. Wear decreases sharply with increasing resistance to cracking of surfaces. The need to have resistant prostheses, with anti-corrosion composition and improved mechanical properties have led to the application of thin layers of materials with superior properties on the prostheses surface.

Studies and progresses have been made by using hard biocompatible alloys and materials. There are experiments performed with amorphous carbon [9], even if it has lower stability [10], with nitride-based compounds, such as TiN, ZrN, CrN, and TiAlN [11], with calcium-phosphate [12] or with diamond-like carbon (DLC), graphite-like carbon (GLC) and tantalum (Ta) [13]. The mechanical properties of the CrN layer have so far been studied on various types of substrates – for example mirrorpolished M2 high speed steel substrates – on which it was deposited by arc bond sputtering (ABS) or high-power impulse magnetron sputtering (HIPIMS) techniques [14].

Multi-layer depositions have also been used and studied, such as hydroxyapatite/Ti-alloy [15], zirconium oxynitride/zirconium oxide [16], Mo/Ni, TiN/Cu and ZrN/W [17], Ti/TiC multilayer [18].

Some studies have shown that active antimicrobial coatings of inorganic metal oxides, such as ZnO [19], TiO_2 [20], TaO [21], etc. are very promising in terms of biomedical applications, as these oxides contain essential mineral elements. In addition, these have special functional properties that, combined with antimicrobial ones, lead to a bacteriostatically functionalized surface, and prevent bacterial colonization and proliferation.

Based on previous researches, this article presents the results obtained from the study of micro and nanostructured thin layers made of biocompatible materials, realized to determine their ability to be used in order to increase the durability of prosthetic components.

2 Experiments

Nanoindentation tests were performed to determine the hardness and elastic modulus of the deposited thin layers. These are micro and nanostructured thin layers of TiN, TiO_2, CrN, deposited by three different methods – cathodic arc, Direct Current (DC) sputtering and High Power Impulse Magnetron Sputtering (HiPIMS). These layers were deposited on CoCr and M30NW steel substrates with 3 different thicknesses – 0.5 μm, 1 μm, and 1.5 μm.

2.1 Materials

CrN and TiN were chosen because previous studies have shown that these materials increase wear resistance. TiO_2 – antimicrobial material – was studied in this project to compare its mechanical properties with those of TiN. In this way, it can be determined which of the two materials – oxide or nitride – has a higher hardness and could be used, in the future, for the real biomedical components.

The two types of substrates – CoCr and M30NW steel – are biocompatible materials, generally used for production of surgical implants and devices for osteosynthesis [22, 23]. It is known that these have corrosion resistance combined with good mechanical properties.

CoCr substrate disks, 8 mm thick and 10 mm in diameter, were made by 3D printing technology, using a laser metal sintering prototyping equipment – Shining 3D EP M250 3D Printer (Fig. 1) – from the Research Center for Intelligent Mechatronic Systems used for Securing Objectives and Intervention – CERMISO of INCDMTM.

Fig. 1. Shining 3D EP M250 3D printer.

Other disks, made of biocompatible M30NW type steel used as substrates, were made by machining from bars using the MECATOME T202 micro-cutting machine (Fig. 2a). The pieces of substrate obtained after cutting have been polished using the MECATECH SPI polishing machine (Fig. 2b) to ensure a uniform surface on which the layers are deposited.

(a) (b)

Fig. 2. (a) Cutting samples using the MECATOME T202 micro-cutting machine. (b) MECATECH SPI polishing machine.

2.2 Thin Layers Deposition

Micro and nanostructured layers were obtained by various physical vapor deposition (PVD) techniques, respectively cathodic arc (CA), direct current (DC) sputtering, High Power Impulse Magnetron Sputtering (HiPIMS) on the substrates selected.

Cathodic arc (thermal evaporation with electric arc) is a PVD technique in which an electric arc is used to vaporize material from a cathodic target, in vacuum (≈1 Pa). The vaporized material then condenses on a substrate, forming a thin layer [24, 25].

DC Sputtering is a deposition technique where a target material to be used as the coating is bombarded with ionized gas molecules. These hit the target and the atoms are sputtered off into the plasma. The vaporized atoms condense as a thin film on the substrate to be coated [26, 27].

HiPIMS is a method for the physical deposition of thin layers vapors that is based on magnetron sputtering deposition. In magnetron sputtering, increased plasma densities are created near the target, which increase the spraying rate. HIPIMS has a high degree of ionization of the pulverized metal and a high rate of molecular gas dissociation, which lead to high density of the deposited films [24, 28].

The deposition of thin layers by cathodic arc, DC sputtering and HiPIMS was performed at SC MGM Star Construct SRL using a DC and HiPIMS equipment Leybiold Z-550-S (Fig. 3) and a Cathodic-Arc UVN-MGM equipment (Fig. 4). The deposition processes by the three methods had the technological parameters presented in Table 1.

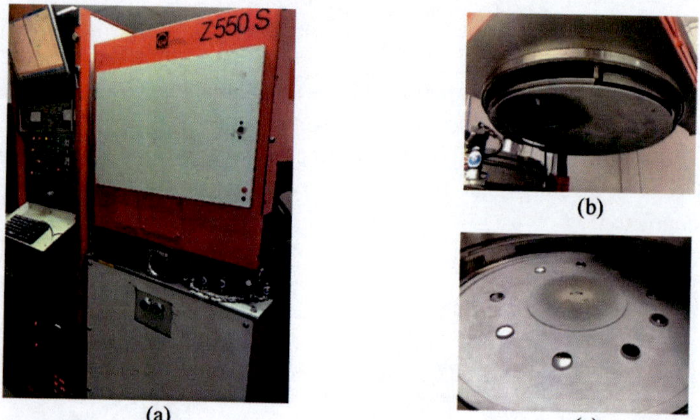

(a) (b) (c)

Fig. 3. (a) DC and HiPIMS equipment Leybiold Z-550-S. (b) Enclosure with 3 cathodes with a diameter of 150 mm - PK-150 Leybold. (c) Samples mounting plate.

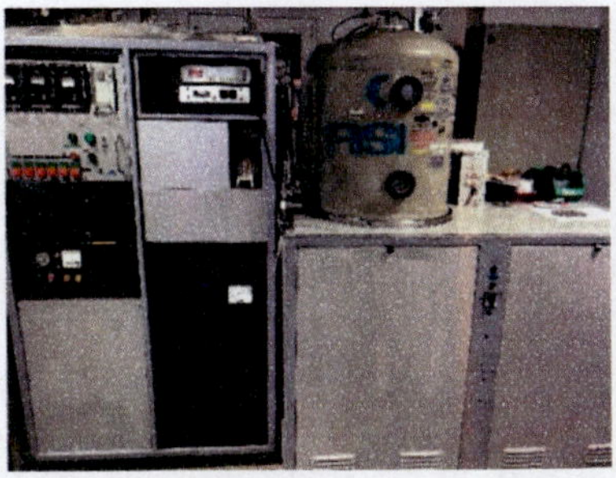

(a) (b)

Fig. 4. (a) Cathodic-Arc UVN-MGM equipment with centrally positioned linear Ti cathode. (b) Samples rotation system centered with respect to the cathode.

Table 1. Technological parameters of the deposition processes for the three methods used.

No.	Deposited material	Layer thickness μm	HiPIMS time, min	DC sputtering time, min	Cathodic Arc time, min	Process gas
1	TiO_2	0.5	5	5	5	O_2
2	TiO_2	1	10	10	10	O_2
3	TiO_2	1.5	15	15	15	O_2
4	TiN	0.5	5	5	5	N_2
5	TiN	1	10	10	10	N_2
6	TiN	1.5	15	15	15	N_2
7	CrN	0.5	5	5		N_2
8	CrN	1	10	10		N_2
9	CrN	1.5	15	15		N_2
Power, W_t			1000	1000	1000	
BIAS, V			100	100		
Current, A					40	
Voltage, V					24–26	
Pressure, mbar			$10^{(-3)}$	$10^{(-3)}$		
Pressure, Pa					$4 \times 10^{(-1)}$	

Thus, layers of CrN, TiN and TiO_2 with thicknesses of 0.5 μm, 1 μm and 1.5 μm were deposited on CoCr substrate and biocompatible M30NW type steel substrate. The TiN and TiO_2 layers were deposited by the 3 types of methods studied, while the CrN layers were deposited by DC sputtering and HiPIMS.

2.3 Nanoindentation Tests

The physico-mechanical, structural, and micro-nanotopographic characterization of the deposited micro and nanostructured layers allow to highlight their properties, as well as to observe the defects that may appear on these surfaces. Following a comparative analysis of the experimental results, information regarding the degree of influence of each deposition method can be obtained.

From the multitude of characterization possibilities, nanoindentation tests were performed to determine the hardness and elastic modulus of micro and nanostructured thin layers for biomedical applications, deposited on metallic substrates, by the physical procedures described above. The nanoindentation tests were performed at the ReCAST Laboratory within the National Institute for Research and Development in Optoelectronics INOE 2000 using a Hysitron TI Premier nanoindenter – a system that can be applied to polymers, metals, nanostructures, alloys and thin films (Fig. 5).

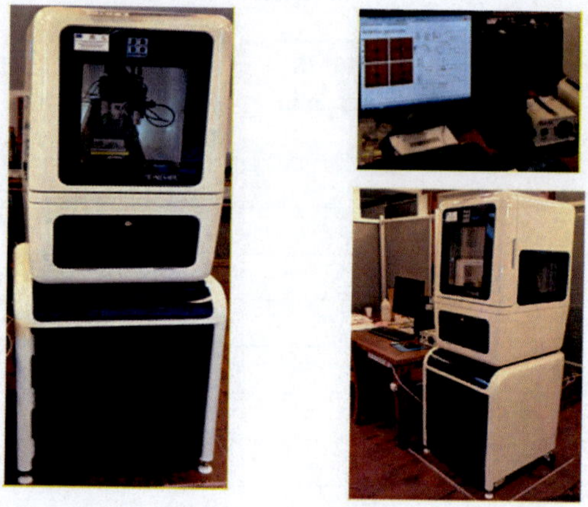

Fig. 5. Nanoindentor hysitron TI premier.

In the case of tests for determining the hardness (H) and the elastic modulus (E_r), for recording the force-displacement curves and extracting the parameters of interest, a Hysitron TI Premier nanoindenter, equipped with a Berkovich peak 142.3°, was used. The nanoindentation tests were performed using a maximum compressive force of 0.7 mN (except for the CrN layer – force of 2 mN).

3 Results and Discussions

3.1 Thin Layers

Initially, the surfaces of the uncovered substrates were characterized in order to be able to make a comparative analysis between the properties of these materials and those of the micro and nanostructured layers that were deposited and can be applied in the biomedical field.

The micro and nanostructured thin layers were characterized at the macro level, respectively by an optical description (Fig. 6).

0.5 μm TiO_2 layer deposited on CoCr substrate by cathodic arc	1 μm TiN layer deposited on CoCr substrate by cathodic arc	1 μm TiO_2 layer deposited on M30NW steel substrate by DC sputtering	1,5 μm TiN layer deposited on M30NW steel substrate by DC sputtering

1 μm CrN layer deposited on CoCr substrate by DC sputtering	1,5 μm TiO_2 layer deposited on M30NW steel substrate by HiPIMS	1,5 μm TiN layer deposited on CoCr substrate by HiPIMS	0.5 μm CrN layer deposited on M30NW steel substrate by HiPIMS

Fig. 6. Thin layers deposited on CoCr and M30NW substrates by cathodic arc, DC sputtering and HiPIMS.

For a more detailed characterization, the deposited micro and nanostructured thin layers were also analyzed at microscopic scale using atomic force microscopy. The surface of interaction between the deposited layers and substrates was observed by the characterizations made. It has been seen that there was a good connection between the layer and the substrate, the thin layers materials completely penetrating the irregularities of the substrate material. The conclusion reached at the end of the analysis by atomic force microscopy is that the thin layers were uniformly deposited on the substrates used.

3.2 Nanoindentation Tests

Optical images of the surfaces, 10 × 10 microns SPM images before the nanoindentation test and 2 × 2 microns SPM images with indentation trace were obtained at the end of the tests.

The analysis results of the images obtained from the optical and SPM characterization of the surfaces showed that the surface of the samples had a high roughness. The SPM characterization of the samples coated with layers with a thickness of 1.5 μm demonstrated this and highlighted the fact that the surface modifications varied depending on the deposition method used for each type of layer.

The nanoindentation graphs obtained are presented in the Tables 2, 3, 4, and Tables 5, 6, 7 for the CoCr substrate and M30NW type steel substrate, respectively.

Table 2. Nanoindentation graphs of the TN layers deposited on CoCr substrate

<table>
<tr><td colspan="3">CoCr_TiN_nanoindentation curves</td></tr>
<tr><td colspan="3">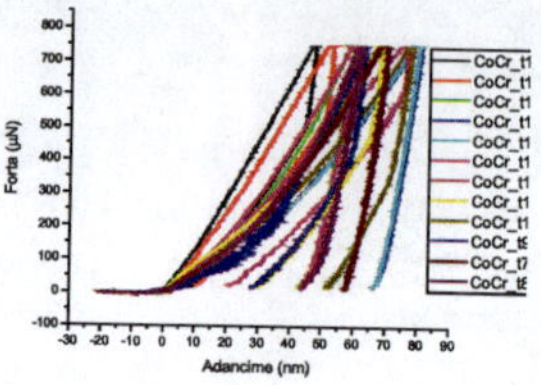
</td></tr>
<tr><td colspan="3">CoCr</td></tr>
<tr><td>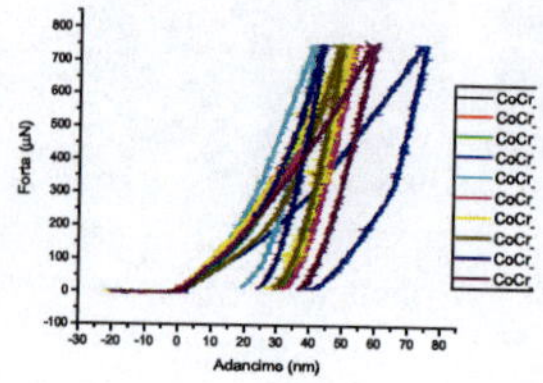
</td><td>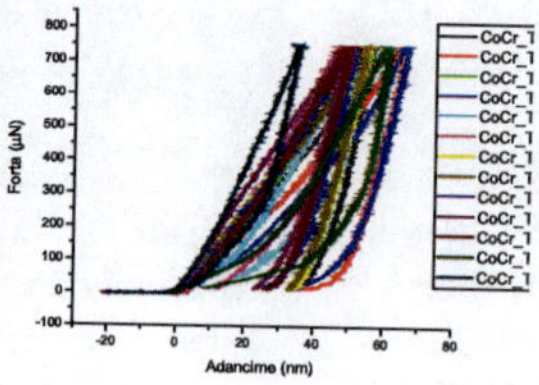
</td><td>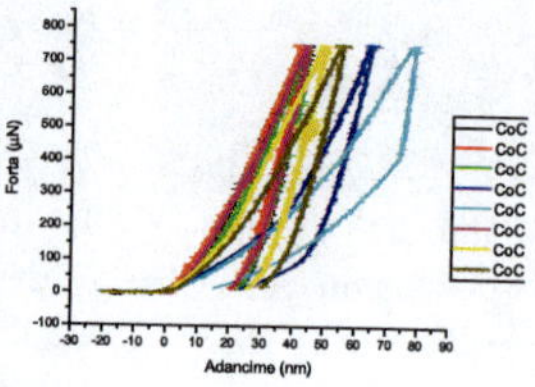
</td></tr>
<tr><td>DC 0.5</td><td>DC 1</td><td>DC 1.5</td></tr>
<tr><td>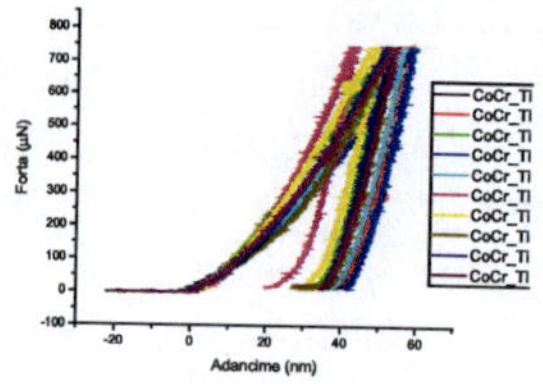
</td><td>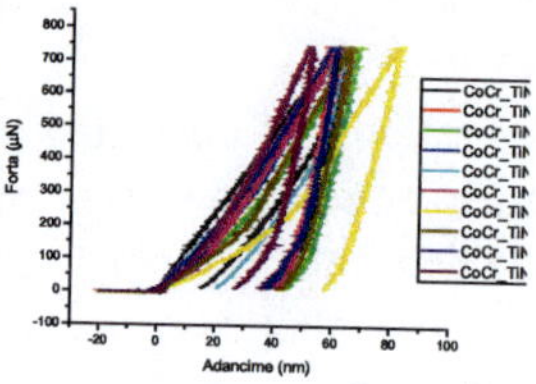
</td><td>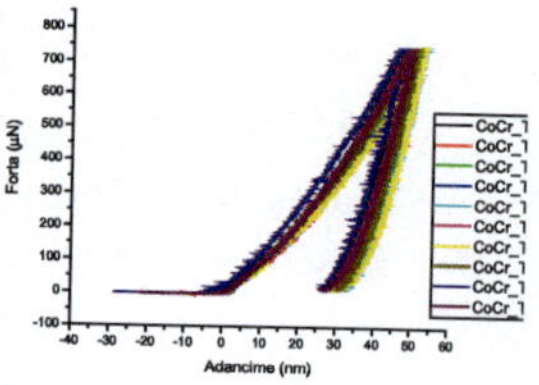
</td></tr>
<tr><td>HiPIMS 0.5</td><td>HiPIMS 1</td><td>HiPIMS 1.5</td></tr>
<tr><td>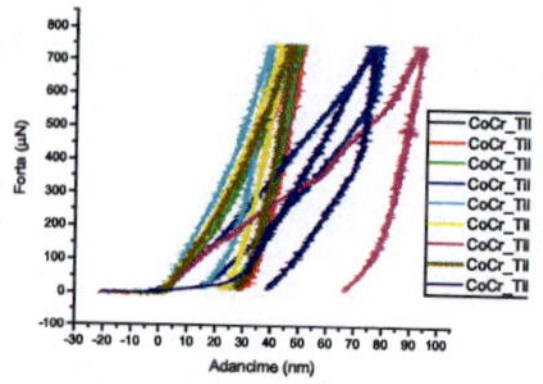
</td><td>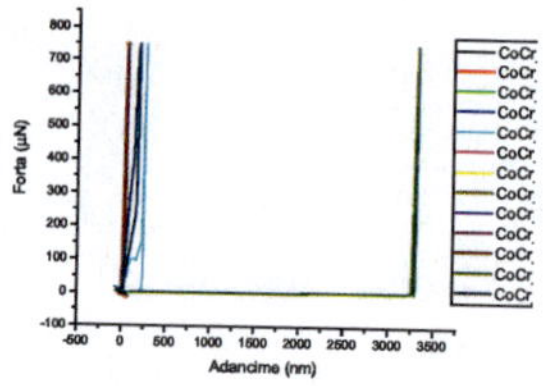
</td><td>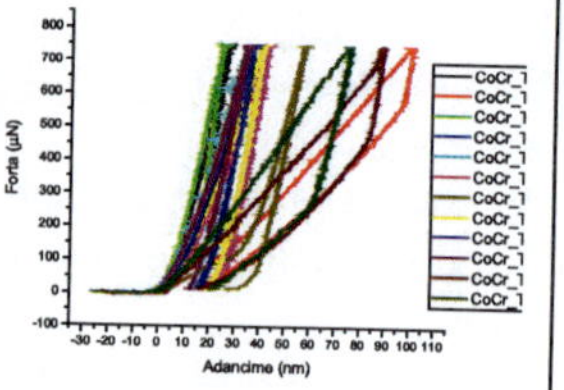
</td></tr>
<tr><td>CA 0.5</td><td>CA 1</td><td>CA 1.5</td></tr>
</table>

Table 3. Nanoindentation graphs of the CrN layers deposited on CoCr substrate

CoCr_CrN_nanoindentation curves		
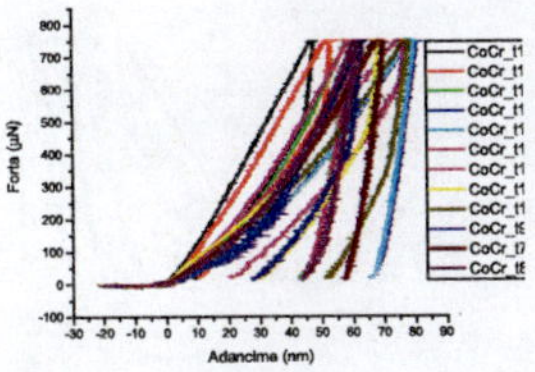		
CoCr		
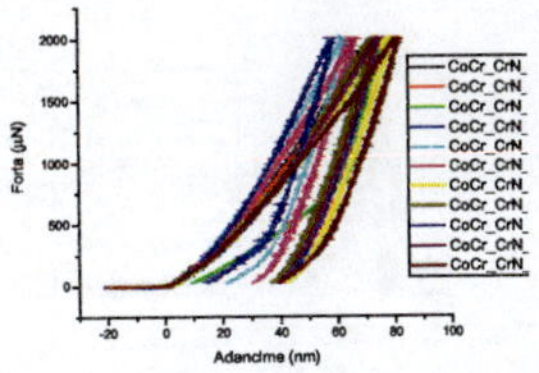	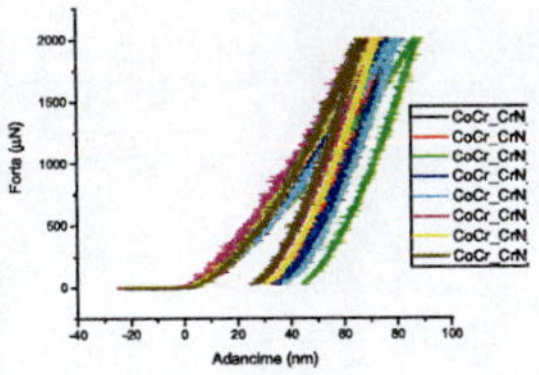	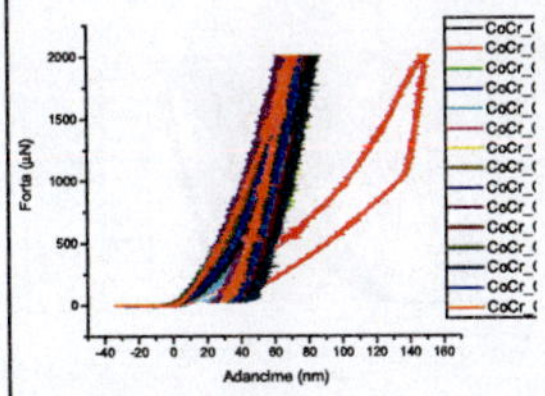
DC 0.5	DC 1	DC 1.5
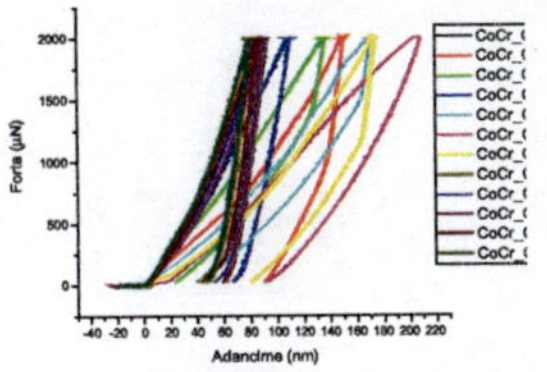	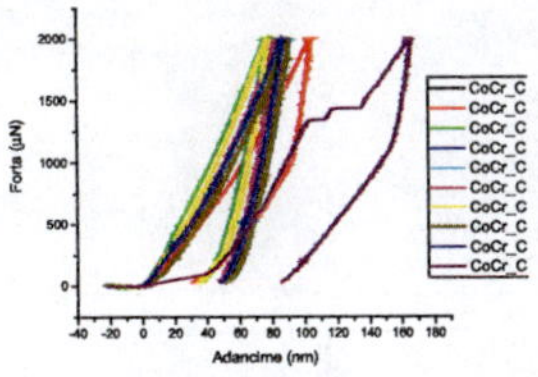	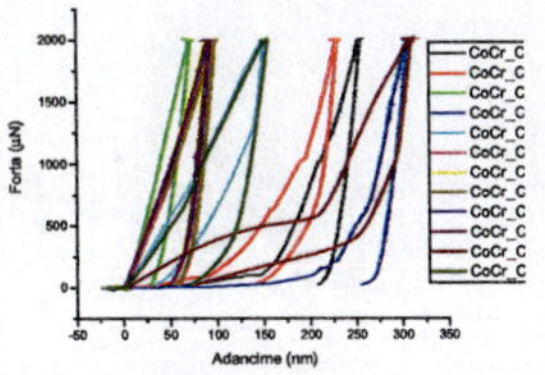
HiPIMS 0.5	HiPIMS 1	HiPIMS 1.5

From the obtained force-displacement curves, the ones with smaller errors (presented grouped in the graph) were selected and fitted to calculate the average values of the hardness H and the elastic modulus E_r.

The experimental results were compared and information were obtained regarding the influence degree of each deposition method on the properties of the deposited layers, respectively hardness and elasticity.

Hardness H of the Deposited Layers – Influence of the Deposition Method. The use of certain deposition methods of thin layers with different thicknesses on the CoCr, and M30NW type steel, substrates has led to different values of layers hardness, which has an important influence on their wear resistance (Fig. 7, 8 and 9).

The hardness of all 0.5 μm thick layers deposited on CoCr substrate and steel substrate had the highest values when the cathodic arc deposition method was used. The use of DC sputtering and HiPIMS methods has led to a decrease in the hardness of all types of analyzed materials (Fig. 7).

Table 4. Nanoindentation graphs of the TiO_2 layers deposited on CoCr substrate

CoCr_TiO_2_nanoindentation curves		
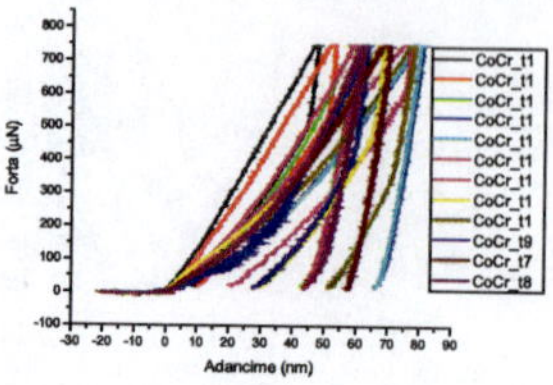		
CoCr		
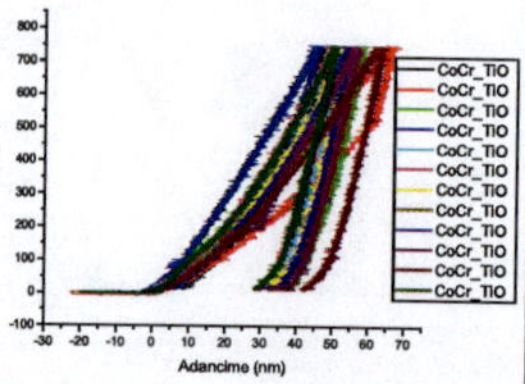	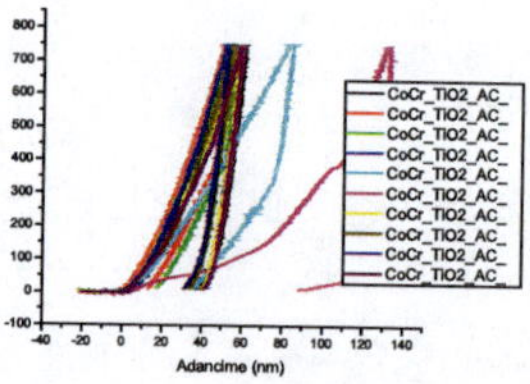	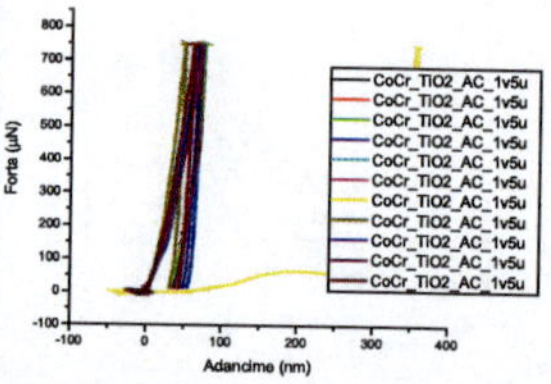
DC 0.5	DC 1	DC 1.5
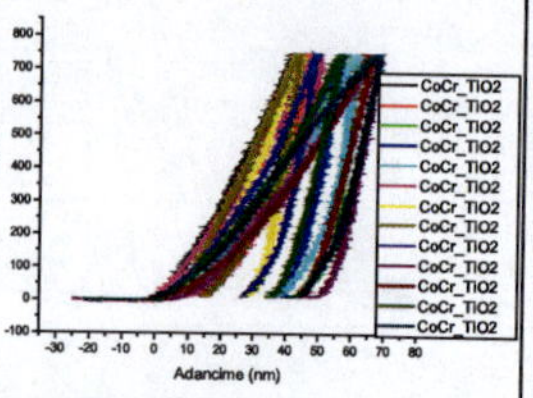	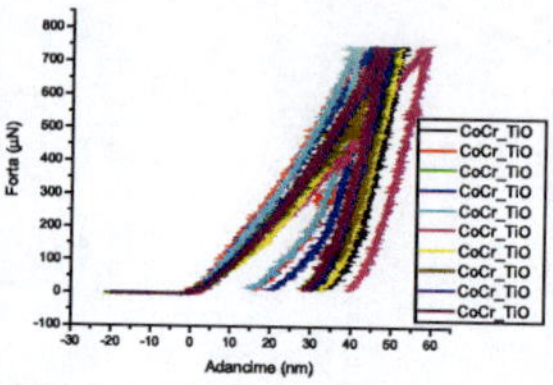	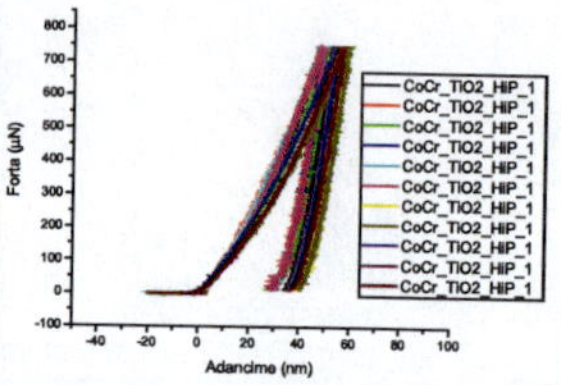
HiPIMS 0.5	HiPIMS 1	HiPIMS 1.5
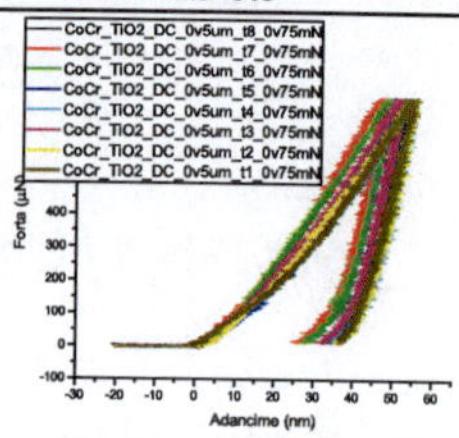	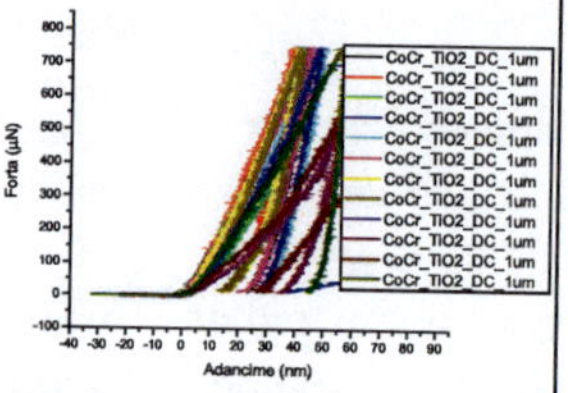	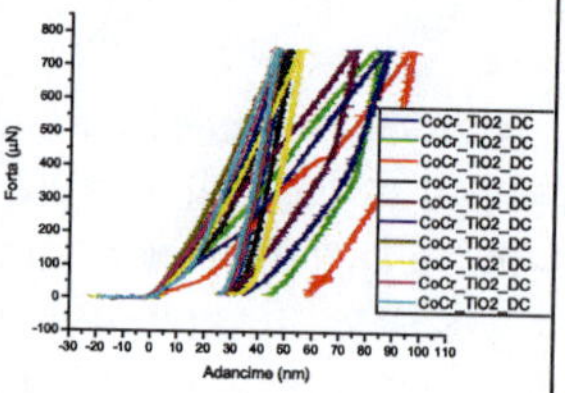
CA 0.5	CA 1	CA 1.5

The hardness of the 1 μm thick layers deposited on the CoCr substrate (Fig. 8a) had the highest values when the DC sputtering method was used. The use of HiPIMS and cathodic arc methods has led to a decrease in the hardness of the TiO_2 and CrN layers. The TiN layer had the highest hardness when it was deposited by cathodic arc. The hardness of the 1 μm thick layers deposited on the steel substrate (Fig. 8b) had the highest values when the DC sputtering method was used. The use of cathodic arc and HiPIMS methods has reduced the hardness of all types of analyzed materials.

Table 5. Nanoindentation graphs of the CrN layers deposited on M30NW steel substrate

Steel_CrN_nanoindentation curves		
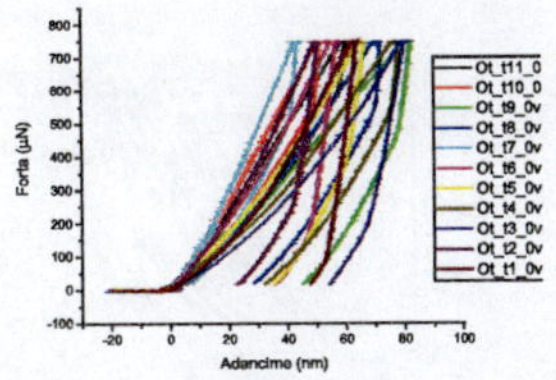		
Steel		
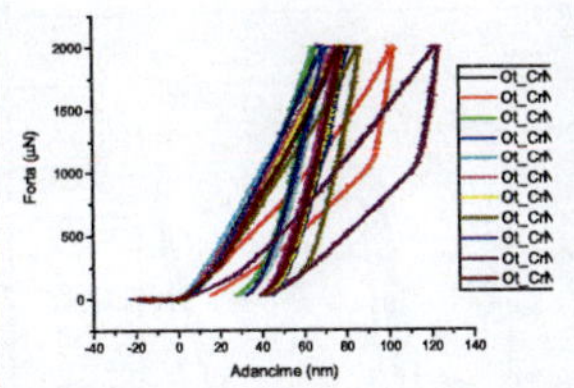	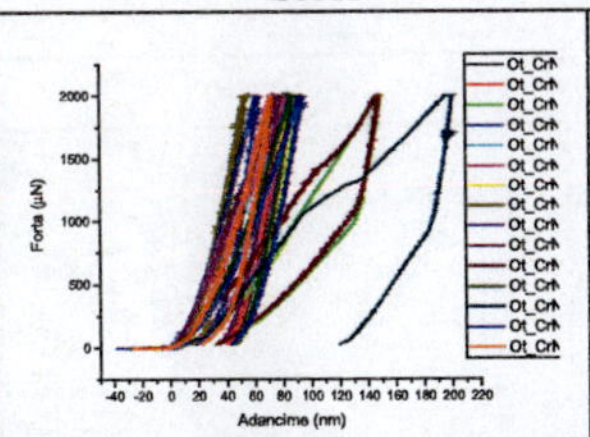	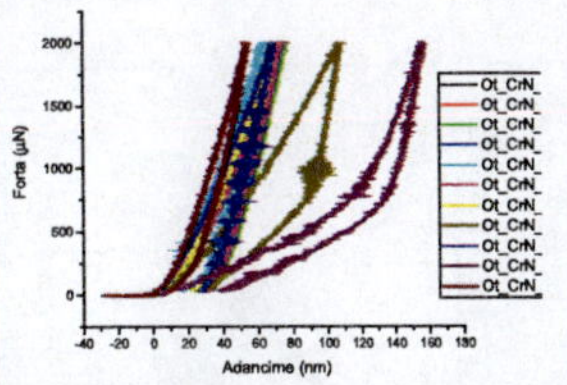
DC 0.5	DC 1	DC 1.5
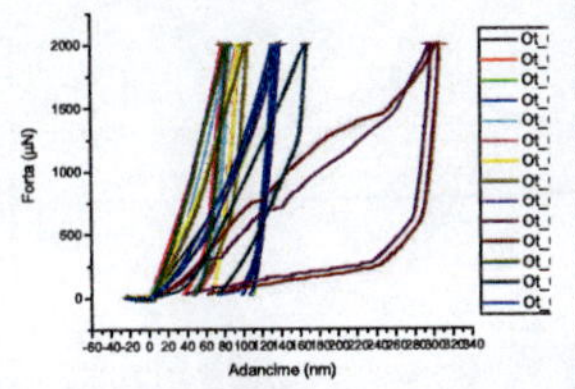	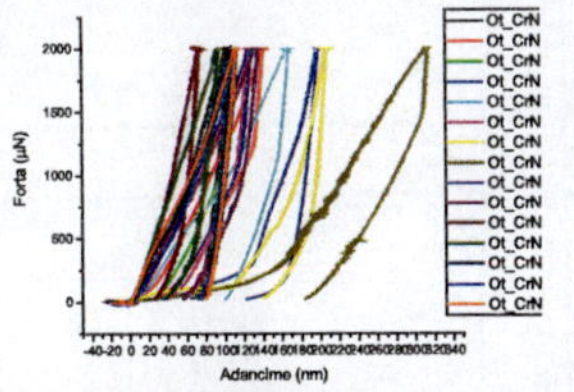	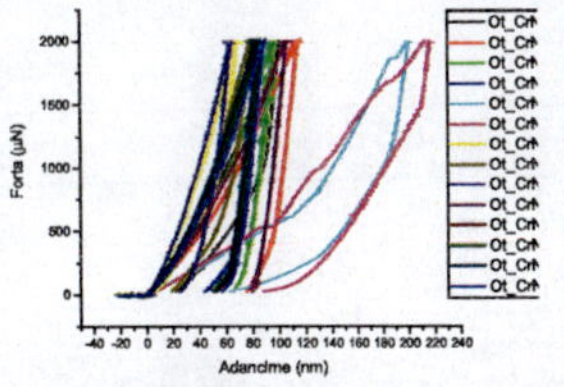
HiPIMS 0.5	HiPIMS 1	HiPIMS 1.5

The hardness of the 1.5 μm thick layers deposited on the CoCr substrate (Fig. 9a) and of the layers deposited on the steel substrate (Fig. 9b) had the highest values when the DC sputtering method was used. The use of HiPIMS and cathodic arc methods has, in both cases, reduced the hardness of the TiO_2 and CrN layers. The TiN layer had the highest hardness when it was deposited by cathodic arc.

Elastic Modulus of the Deposited Layers E_r – Influence of the Deposition Method. The use of certain deposition methods of thin layers with different thicknesses on the CoCr, and M30NW type steel, substrates has led to different values of the elastic modulus, which has an important influence on their wear resistance (Fig. 10, 11 and 12).

Table 6. Nanoindentation graphs of the TiN layers deposited on M30NW steel substrate

Steel_TiN_nanoindentation curves		
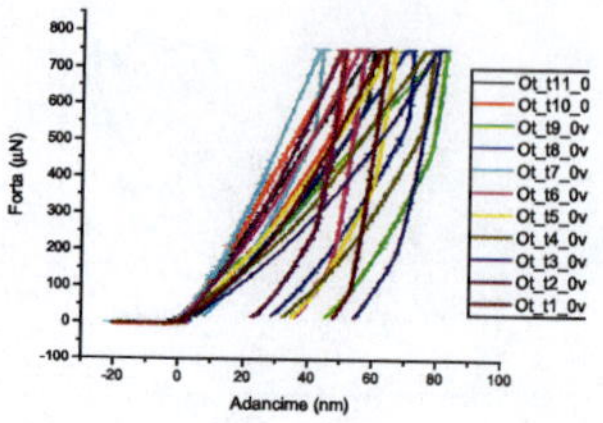		
Steel		
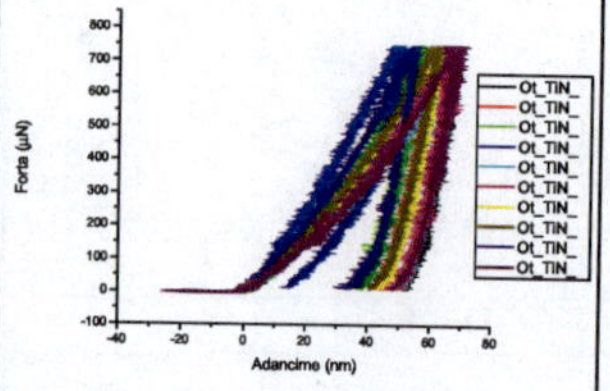	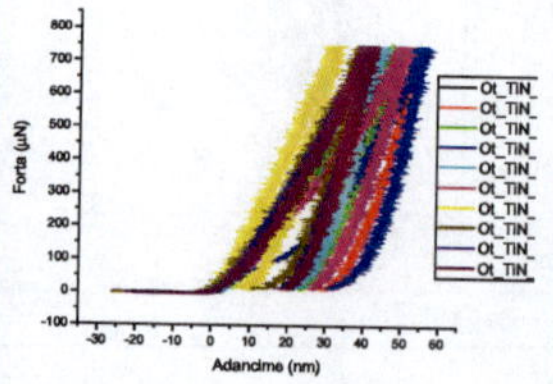	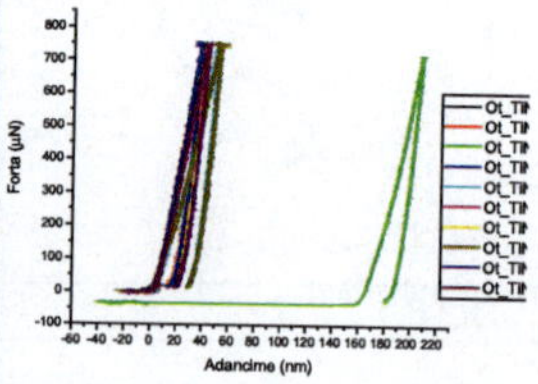
DC 0.5	DC 1	DC 1.5
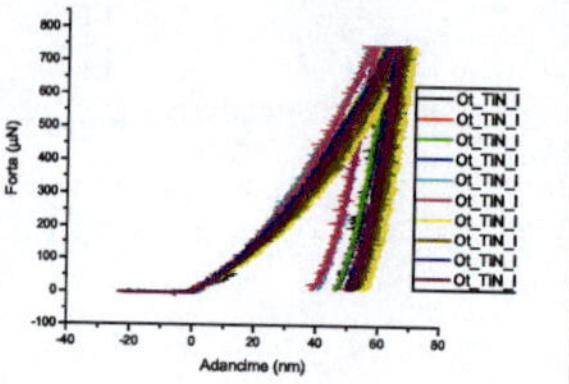	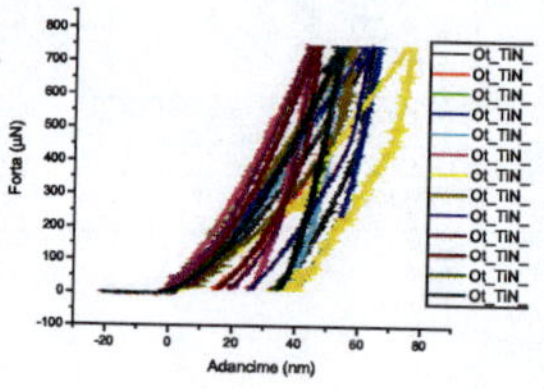	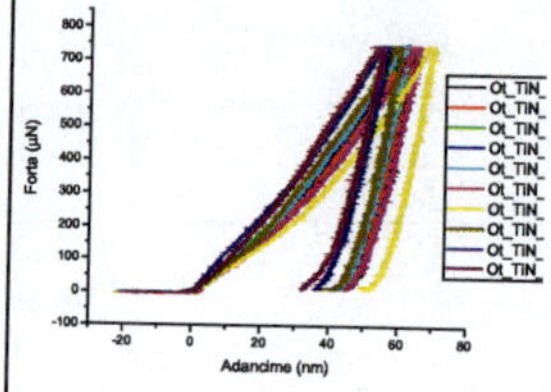
HiPIMS 0.5	HiPIMS 1	HiPIMS 1.5
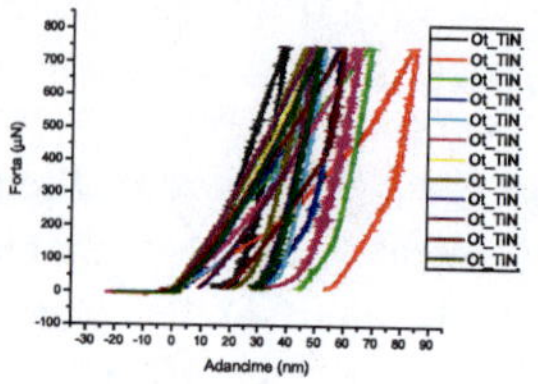	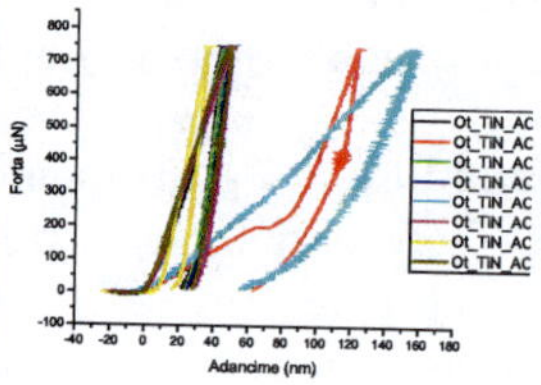	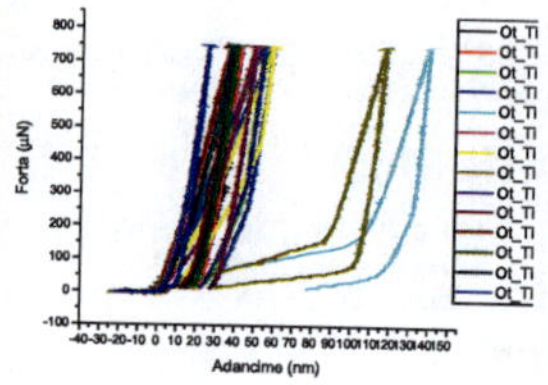
CA 0.5	CA 1	CA 1.5

Table 7. Nanoindentation graphs of the TiO_2 layers deposited on M30NW steel substrate

<table>
<tr><td colspan="3">Steel_TiO2 nanoindentation curves</td></tr>
<tr><td colspan="3">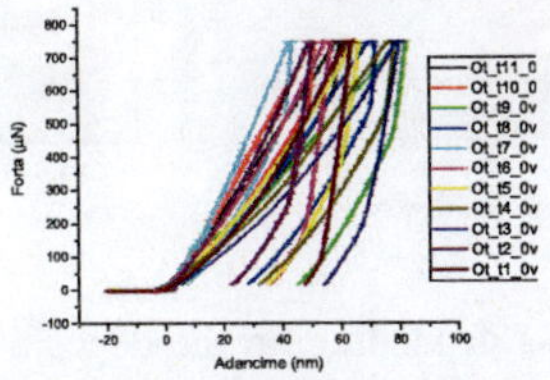
</td></tr>
<tr><td colspan="3">Steel</td></tr>
<tr><td>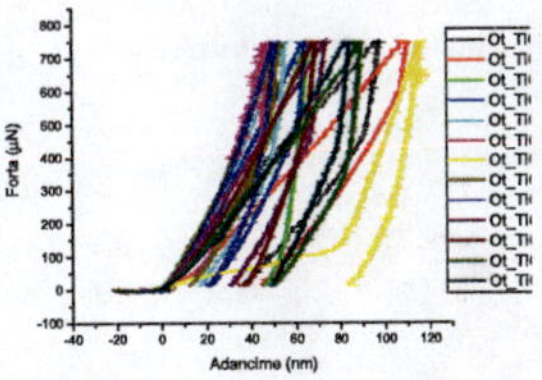
</td><td>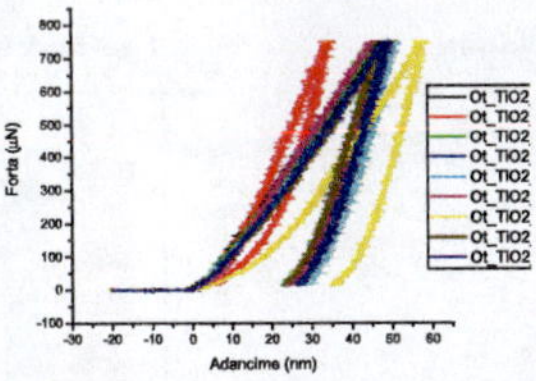
</td><td>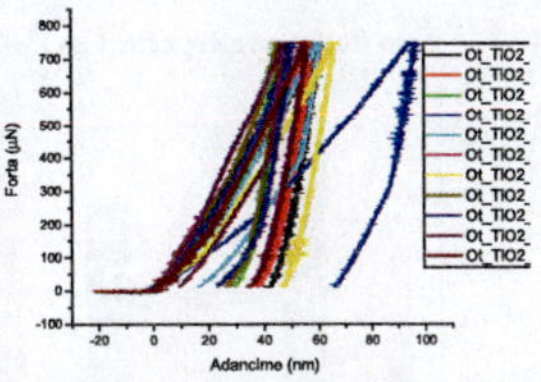
</td></tr>
<tr><td>DC 0.5</td><td>DC 1</td><td>DC 1.5</td></tr>
<tr><td>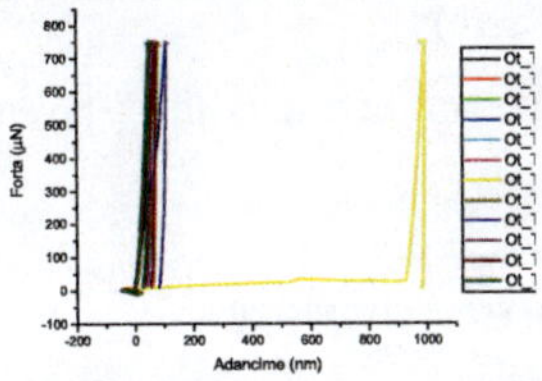
</td><td>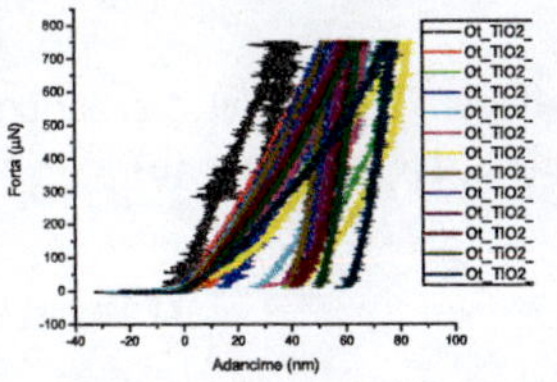
</td><td>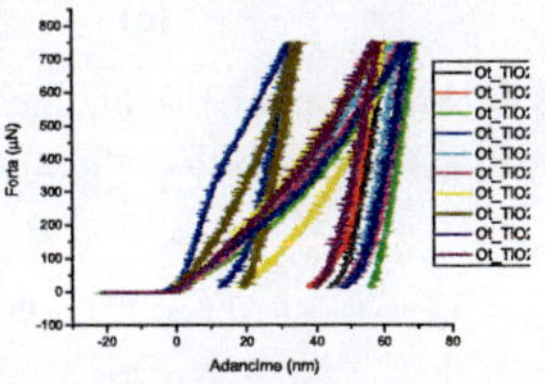
</td></tr>
<tr><td>HiPIMS 0.5</td><td>HiPIMS 1</td><td>HiPIMS 1.5</td></tr>
<tr><td>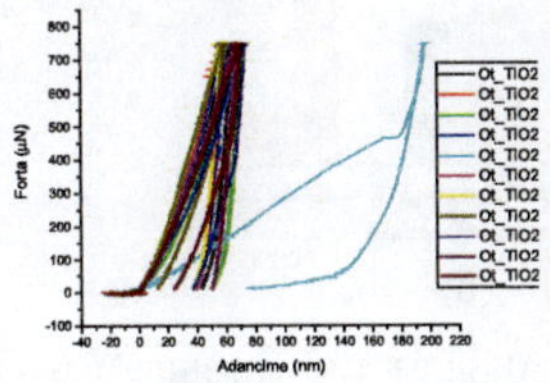
</td><td>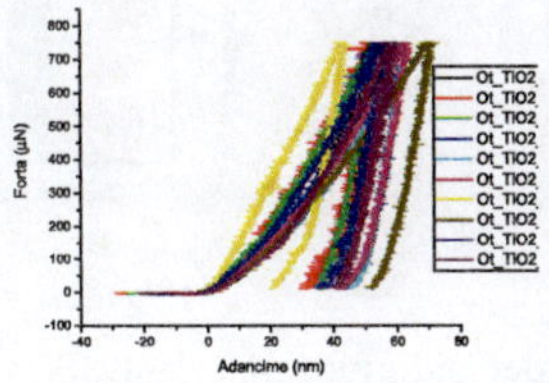
</td><td>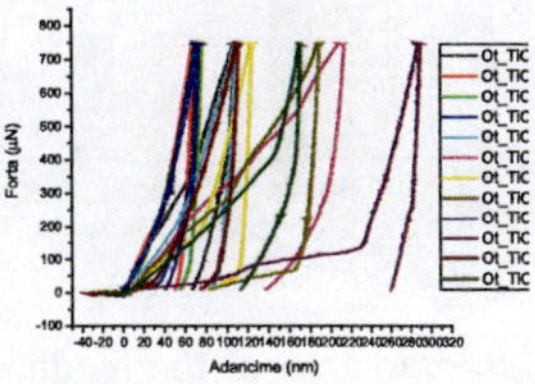
</td></tr>
<tr><td>CA 0.5</td><td>CA 1</td><td>CA 1.5</td></tr>
</table>

The elastic modulus of the 0.5 μm thick layers deposited on the CoCr substrate (Fig. 10a) had the highest values when the DC sputtering method was used. The use of cathodic arc and HiPIMS methods has led to a decrease in the elastic modulus of the TiO_2 and CrN layers. The TiN layer has a maximum value of the elastic modulus when the cathodic arc deposition method was used. The elastic modulus of the 0.5 μm thick layers deposited on the steel substrate (Fig. 10b) had the highest values when the cathodic arc deposition method was used. The use of HiPIMS and DC sputtering methods has led to a decrease in the elastic modulus of the TiN and TiO_2 layers. The CrN layer has a maximum value of the elastic modulus when the DC sputtering method was used.

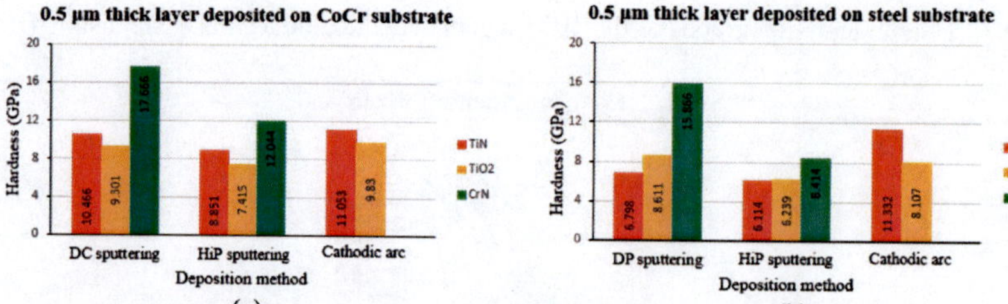

(a) (b)

Fig. 7. Variation of the hardness depending on the deposition method of 0.5 μm thick layers deposited on CoCr substrate (a) and on steel substrate (b).

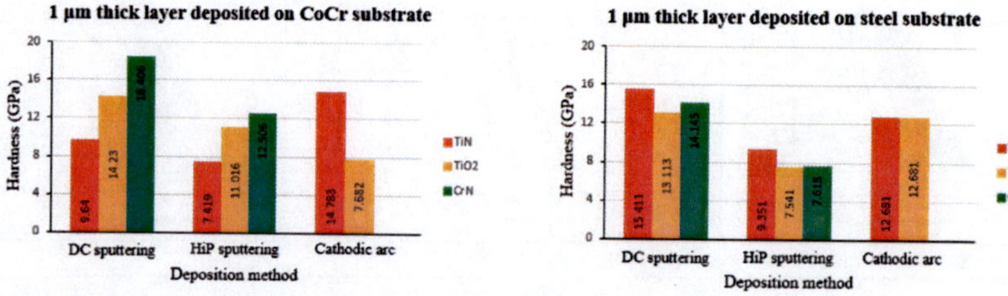

(a) (b)

Fig. 8. Variation of the hardness depending on the deposition method of 1 μm thick layers deposited on CoCr substrate (a) and on steel substrate (b).

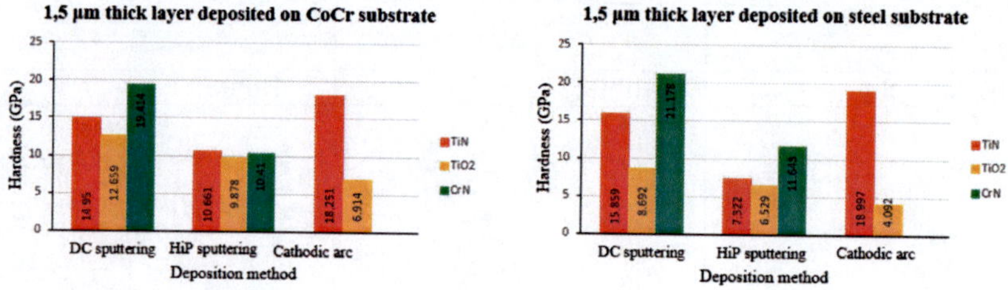

(a) (b)

Fig. 9. Variation of the hardness depending on the deposition method of 1,5 μm thick layers deposited on CoCr substrate (a) and on steel substrate (b).

The elastic modulus of the 1 μm thick layers deposited both on the CoCr substrate (Fig. 11a) and on the steel substrate (Fig. 11b) had maximum variable values depending on the deposited material. Thus, the elastic modulus of the TiN layer deposited on CoCr is maximum when it was deposited by the cathodic arc method. The TiO_2 layer deposited on CoCr by HiPIMS has a maximum value of elastic modulus, and the CrN layer deposited on CoCr by DC sputtering has this high value. In the case of layers deposited on the steel substrate, the elastic modulus of the TiN layer is maximum when it has been deposited by the DC sputtering method. The TiO_2 layer deposited by HiPIMS has a maximum value of the elastic modulus, and the CrN layer deposited by DC sputtering has this high value.

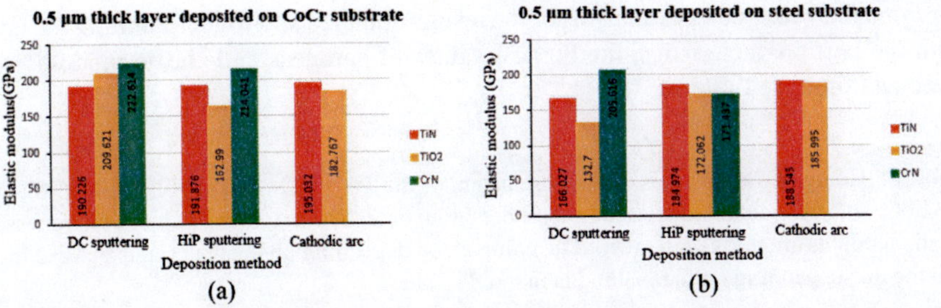

Fig. 10. Variation of the elastic modulus depending on the deposition method of 0.5 µm thick layers deposited on CoCr substrate (a) and on steel substrate (b).

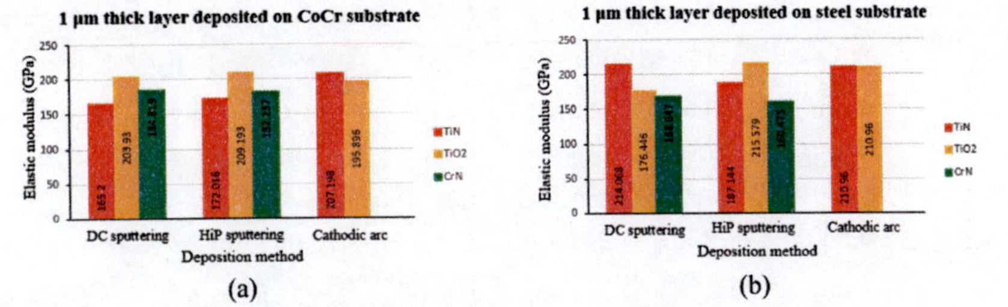

Fig. 11. Variation of the elastic modulus depending on to the deposition method of 1 µm thick layers deposited on CoCr substrate (a) and on steel substrate (b).

The elastic modulus of the 1.5 µm thick layers deposited on the CoCr substrate (Fig. 12a) had the highest values when the DC sputtering method was used. The use of HiPIMS and cathodic arc methods has led to a decrease of the elastic modulus of the TiO_2 and CrN layers. The TiN layer had the highest elastic modulus when it was deposited by the cathodic arc method. The elastic modulus of the 1.5 µm thick layers deposited on the steel substrate (Fig. 12b) had variable maximum values depending on the deposited material. Thus, the elastic modulus of the TiN layer is maximum when it was deposited by the cathodic arc method. The TiO_2 and CrN layers deposited by HiPIMS have the maximum value of the elastic modulus.

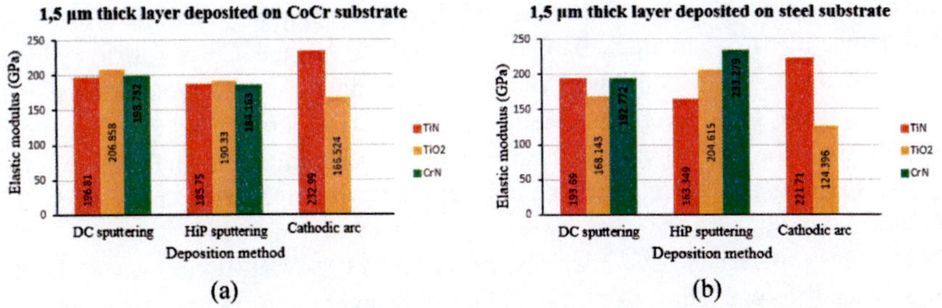

Fig. 12. Variation of the elastic modulus depending on to the deposition method of 1,5 µm thick layers deposited on CoCr substrate (a) and on steel substrate (b).

For a final comparative analysis of the results obtained and for determining the layers with the best characteristics, the highest values of hardness and elastic modulus were selected (Table 8, Table 9).

Table 8. Comparative analysis of the parameters of the TiN, TiO_2 and CrN layers deposited on the CoCr substrate obtained after the nanoindentation tests (The maximum values of hardness and elastic modulus are marked in green. The comparison depending on the material is marked in red, and the one depending on the method is marked in blue).

Material	Layer thickness	Deposition method					
		DC sputtering		HiPIMS		Cathodic arc	
		H	E_r	H	E_r	H	E_r
TiN	0.5 μm	10.466	190.226	8.851	191.876	11.053	195.032
	1 μm	9.64	165.2	7.419	172.016	14.783	207.198
	1.5 μm	14.95	196.81	10.661	185.75	**18.251**	**232.99**/met
TiO_2	0.5 μm	9.301	209.621	7.415	162.99	9.83	182.767
	1 μm	**14.23**	203.93	11.016	**209.193**	7.682	195.896
	1.5 μm	12.659	**206.858** /met	9.878	190.33	6.914	166.524
CrN	0.5 μm	**17.666**	**222.614** /met	12.044	214.041	-	-
	1 μm	**18.406**	184.819	12.506	182.237	-	-
	1.5 μm	**19.414** /met	198.732	10.41	184.163	-	-

CoCr SUBSTRATE → Hardness = 4.496; Elastic modulus = 144.337

Generally, the hardness and elastic modulus of the CoCr substrate and of the M30NW steel substrate increase when TiN, TiO_2 and CrN layers studied are deposited on it.

Taking into account the results obtained from the nanoindentation tests, it was observed that the 1.5 μm thick TiN layer deposited on the CoCr substrate by the cathodic arc method has the highest value of the hardness and elastic modulus. 1 μm and 1.5 μm thick TiO_2 layers deposited on the CoCr substrate by the DC sputtering method presents the highest values of hardness and elastic modulus. In the case of CrN, using DC sputtering, 1.5 μm and 0.5 μm layers have the highest hardness, and elastic modulus.

When these materials have been deposited on M30NW steel substrates, the final situation was different. Thus, the 1.5 μm thick TiN layer deposited by the cathodic arc method has an increased hardness and elastic modulus. 1 μm thick TiO_2 layers deposited by DC sputtering and HiPIMS had the highest values of hardness, and elastic modulus, respectively. 1.5 μm and 0.5 μm CrN layers deposited by DC sputtering and HiPIMS had the highest values of hardness, and elastic modulus, respectively.

It can be observed that cathodic arc method is the one that helps to obtain TiN layers with superior properties on both types of substrates. DC sputtering contributes to the increase of hardness and elastic modulus of TiO_2 and CrN layers deposited on CoCr and M30NW substrates. HiPIMS can be used to obtain TiO_2 and CrN layers with superior elastic modulus.

Table 9. Comparative analysis of the parameters of the TiN, TiO_2 and CrN layers deposited on the M30NW steel substrate obtained after the nanoindentation tests (The maximum values of hardness and elastic modulus are marked in green. The comparison depending on the material is marked in red, and the one depending on the method is marked in blue).

Material	Layer thickness	Deposition method					
		DC sputtering		HiPIMS		Cathodic arc	
		H	E_r	H	E_r	H	E_r
TiN	0.5 μm	6.798	166.027	6.114	184.974	11.332	188.545
	1 μm	15.411	214.068	9.351	187.144	12.681	210.96
	1.5 μm	15.895	193.69	7.322	163.349	18.997 /met	221.71/ met
TiO_2	0.5 μm	8.611	132.7	6.239	172.062	8.107	185.995
	1 μm	13.113	176.446	7.541	215.579 /met	12.681	210.96
	1.5 μm	8.692	168.143	6.529	204.615	4.092	124.396
CrN	0.5 μm	15.866	205.616	8.414	171.437	-	-
	1 μm	14.145	168.647	7.618	160.475	-	-
	1.5 μm	21.178/ met	192.772	11.643	233.279 /met	-	-

M30NW Steel SUBSTRATE → Hardness = 5.914; Elastic modulus = 97.618

Taking into account the values of all parameters obtained analyzed, the main conclusion of this analysis is that the layers deposited by DC sputtering have superior mechanical properties, this being considered the best method for depositing the layers. The following are the cathodic arc and HiPIMS deposition methods, the first one not being able to be used for deposition of CrN layers.

4 Conclusions

The aim of the researches presented in this article was to choose the optimal variants of substrate material – deposited material couple, which would satisfy the requirements of different biomedical applications in terms of the durability and sustainability of a normal patients' life.

For this reason, nanoindentation tests were performed in order to establish the hardness and the elastic modulus of the micro and nanostructured thin layers. The physico-mechanical characterization of the realized samples provided information related to the quality of the deposited material and the influence of the deposition technique.

Thus, it was shown that DC sputtering contributes to the increase of hardness and elastic modulus of TiO_2 and CrN layers deposited on CoCr and M30NW substrates. Cathodic arc method is the one that helps to obtain TiN layers with superior properties on both types of substrates. HiPIMS can be used to obtain TiO_2 and CrN layers with superior elastic modulus.

The main conclusion of the study is that the layers deposited by DC sputtering have superior mechanical properties, this being considered the best method for layers deposition.

The information about the materials, obtained from these complex research activities, can be used later, so as to help improve the quality of some products needed by society, respectively biomedical components (for example, hip prostheses).

Acknowledgements. The authors would like to present their sincere gratitude to the SC MGM Star Construct SRL for PVD coatings and gratefully acknowledge to the employees of Research Center for Advanced Surface Processing and Analysis by Vacuum Technologies from National Institute of Research and Development for Optoelectronics - INOE 2000, who have done the nanoindentation tests. The authors would like to thank Dr. Eng. Dorin Angelescu for his collaboration in the realization of CoCr substrate disks in the Research Center for Intelligent Mechatronic Systems used for Securing Objectives and Intervention - CERMISO of INCDMTM.

References

1. Wang, L., Wang, C., Wu, S., Fan, Y., Li, X.: Influence of the mechanical properties of biomaterials on degradability, cell behaviors and signaling pathways: current progress and challenges. Biomater. Sci. **8**, 2714–2733 (2020)
2. Goharian, A., Abdullah, M.R.: Bioinert metals (stainless steel, titanium, cobalt chromium). In: Trauma Plating Systems, pp. 115–142, Elsevier (2017)
3. Li, H., Zheng, Y.: Recent advances in bulk metallic glasses for biomedical applications. Acta Biomater. **36**, 1–20 (2016)
4. Hussain, M.A., Maqbool, A., Khalid, F.A., Bakhsh, N.: Mechanical properties of CNT reinforced hybrid functionally graded materials for bioimplants. Trans. Nonferrous Metals Soc. China **24**(1), 90–98 (2014)
5. Auciello, O., Renou, S., Kang, K., Tasat, D., Olmedo, D.: A biocompatible Ultrananocrystalline Diamond (UNCD) coating for a new generation of dental implants. Nanomaterials **12**(5), 782–796 (2022)
6. Rahmati, B., Sarhan, A.D., Basirun, W.J., Abas, W.A.B.W.: Ceramic tantalum oxide thin film coating to enhance the corrosion and wear characteristics of Ti-6Al-4V alloy. J. Alloy. Compd. **676**, 369–376 (2016)
7. Catledge, S.A., et al.: Nanostructured ceramics for biomedical implants. J. Nanosci. Nanotechnol. **2**(3–4), 293–312 (2002)
8. Strąkowska, P., Beutner, R., Gnyba, M., Zielinski, A., Scharnweber, D.: Electrochemically assisted deposition of hydroxyapatite on Ti6Al4V substrates covered by CVD diamond films - coating characterization and first cell biological results. Mater. Sci. Eng., C **59**, 624–635 (2016)
9. Österle, W., Klaffke, D., Griepentrog, M., Gross, U., Kranz, I., Knabe, C.: Potential of wear resistant coatings on Ti–6Al–4V for artificial hip joint bearing surfaces. Wear **264**(7–8), 505–517 (2008)

10. Zeng, H., Lacefield, W.R.: The study of surface transformation of pulsed laser deposited hydroxyapatite coatings. J. Biomed. Mater. Res. **50**(2), 239–247 (2000)
11. Subramanian, B., Muraleedharan, C.V., Ananthakumar, R., Jayachandran, M.: A comparative study of titanium nitride (TiN), titanium oxy nitride (TiON) and titanium aluminum nitride (TiAlN), as surface coatings for bio implants. Surf. Coat. Technol. **205**(21–22), 5014–5020 (2011)
12. Mahmoodian, R., Rafieerad, A.R., Ashra, M.R., Bushroa, A.R.: Surface characterization and corrosion behavior of calcium phosphate-base composite layer on titanium and its alloys via plasma electrolytic oxidation: a review paper. Mater. Sci. Eng., C **57**, 397–413 (2015)
13. Ching, H.A., Choudhury, D., Nine, M.J., Abu Osman, N.A.: Effects of surface coating on reducing friction and wear of orthopaedic implants. Sci. Technol. Adv. Mater. **15**(1), 014402 (2014)
14. Ehiasarian, A.P., Hovsepian, P.Eh., Hultman, L., Helmersson, U.: Comparison of microstructure and mechanical properties of chromium nitride-based coatings deposited by high power impulse magnetron sputtering and by the combined steered cathodic arc/unbalanced magnetron technique. Thin Solid Films **457**(2), 270–277 (2004)
15. Sygnatowicz, M., Tiwari, A.: Controlled synthesis of hydroxyapatite-based coatings for biomedical application. Mater. Sci. Eng., C **29**(3), 1071–1076 (2009)
16. Cubillos, G.I., Bethencourt, M., Olaya, J.J., Alfonso, J.E., Marco, J.F.: The influence of deposition temperature on microstructure and corrosion resistance of ZrOxNy/ZrO2 coatings deposited using RF sputtering. Appl. Surf. Sci. **309**, 181–187 (2014)
17. Abadias, G., Michel, A., Tromas, C., Jaouen, C., Dub, S.N.: Stress, interfacial effects and mechanical properties of nanoscale multilayered coatings. Surf. Coat. Technol. **202**(4–7), 844–853 (2007)
18. Shanaghi, A., Chu, P.K., Rouh Aghdam, A.S., Xu, R.: Structure and corrosion resistance of Ti/TiC coatings fabricated by plasma immersion ion implantation and deposition on nickel–titanium. Surf. Coat. Technol. **229**, 151–155 (2013)
19. Sollazzo, V., et al.: Zirconium oxide coating improves implant osseointegration in vivo. Dental Mater. **24**(3), 357–361 (2008)
20. Giavaresi, G., et al.: In vitro biocompatibility of titanium oxide for prosthetic devices nanostructured by low pressure metal-organic chemical vapor deposition. Int. J. Artif. Organs **26**(8), 774–780 (2003)
21. Rahmati, B., Sarhan, A.A.D., Zalnezhad, E., Kamiab, Z.: Development of tantalum oxide (Ta-O) thin film coating on biomedical Ti-6Al-4V alloy to enhance mechanical properties and biocompatibility. Ceram. Int. **42**(1), 466–480 (2016)
22. Hunt, J.A., Callaghan, J.T., Sutcliffe, C.J., Morgan, R.H., Halford, B., Black, R.A.: The design and production of Co-Cr alloy implants with controlled surface topography by CAD-CAM method and their effects on osseointegration. Biomaterials **26**(29), 5890–5897 (2005)
23. Guezmil, M., Bensalah, W., Mezlini, S.: Effect of bio-lubrication on the tribological behavior of UHMWPE against M30NW stainless steel. Tribol. Int. **94**, 550–559 (2016)
24. Anders, A.: A review comparing cathodic arcs and high-power impulse magnetron sputtering (HiPIMS). Surf. Coat. Technol. **257**, 308–325 (2014)
25. Brown, I.G.: Cathodic arc deposition of films. Annu. Rev. Mater. Sci. **28**, 243–269 (1998)
26. Wasa, K., Kanno, I., Kotera, H.: Handbook of sputter deposition technology, 2nd edn. Elsevier Inc. (2012)
27. Gobbi, A.L., Nascente, P.A.P.: D.C. Sputtering. In: Wang, Q.J., Chung, Y.W. (eds.) Encyclopedia of Tribology, pp. 699–706, Springer Science and Business Media, New York (2013)
28. Ehiasarian, A.: High-power impulse magnetron sputtering and its applications. Pure Appl. Chem. **82**(6), 1247–1258 (2010)

Forward Dynamics of the Five-Bar Parallel Mechanism in Presence of Singularities

Maurizio Ruggiu, Elio Frau, and Pierluigi Rea(✉)

University of Cagliari, Cagliari, Italy
pierluigi.rea@unica.it

Abstract. Simulation of motion of dynamic systems is addressing the interest of the robotic and mechatronic communities with the goal to have a full digitalization of the physical system (digital twin). The final goal consists of exchanging data from/to the system and to use the model as predictor and even corrector of the system motion. The first step of the application described is to implement a dynamic model able to simulate the behaviour of the system under several conditions. This paper deals with the solution of the forward dynamics problem of a planar five-bar parallel mechanism in presence of singular configurations. In the paper the index-3 augmented Lagrangian formulation with velocity and acceleration projections is implemented to overcome the kinematic combined and the constraint singularities.

Keywords: Multibody system · Singularities · ODE

1 Introduction

The two degrees of freedom five-bar planar parallel manipulator addressed in this paper, was extensively studied and finds applications for pick and place operations. A kinematic analysis was first developed by Liu et al. [1] who investigated the workspace, singularities of the mechanism. They solved the inverse and forward position problems determining the working and assembly modes of the mechanism. In the paper the maximal inscribed workspace (MIW) was detected as the maximum workspace with no singularities. Further, they conducted a deep investigation on the nature of the singularities of the mechanisms. They found the stationary (open chain) singularities, a.k.a. inverse, the uncertainty (closed) singularities, a.k.a. direct and the combined singularities. Campos et al. [2] built DexTAR, a double SCARA robot (a.k.a. five-bar parallel mechanism) with all four links with equal lengths. With this geometry the number of singularities increases such that the authors had to implement a strategy for avoiding them by switching working modes. DexTAR has a workspace at least one third larger than that of a similarly sized five-bar parallel robot with the ability of crossing the open chain singularities. Another mechatronic system dedicated to control a five-bar parallel mechanism was then proposed by Giberti et al. [3]. They built a

D. D. Ciobotă (Ed.): ICoRSE 2022, LNNS 534, pp. 358–366, 2023.
https://doi.org/10.1007/978-3-031-15944-2_32

laboratory prototype with coaxial motors, i.e., the ground link with zero length. This geometry was selected as outcome of a kinematic synthesis based on the optimization of the condition number of the kinematic Jacobian.

Finally, it is worth mentioning the great attention addressed to this mechanism by the research groups in the field of the biomechanics. For example, recently, Yamine et al. [4] proposed the mechanism as an alternative design to the planar rehabilitative robots as the MIT-Manus. The mechanism seems to be an acceptable compromise in terms of workspace symmetry with respect to the sagittal plane, relatively large workspace, portability and affordability. Although the five-bar planar parallel manipulator has been extensively studied and has numerous applications, there is a lack of work that addresses the solution of the forward dynamics problem, that can be useful as simulation tool, but also as a step of a design process, not based only on kinematic parameters, and it can even become mandatory for applications where the mechanism is driven by the operator as reported in [5] for the rehabilitation mirror therapy, i.e., rehabilitative machines, haptic devices costituited by spatial mechanisms as reported in [6]. Further it may represent a first step needed to build a digital twin of the system to finally predict the behaviour and eventually the faults. In other words, solution of the forward dynamic problem may allow building a cyber-physical sytem.

However, simulation of the motion of a mechanism may become challenging for the presence of singular configurations [7]. Indeed, the goal of this paper is to solve the forward dynamics problem for the five-bar mechanism by implementing a numerical algorithm able to cross the singular configurations. Further, in the paper, the singular configurations of the mechanism are clearly classified distinguishing between kinematic and constraint singularities.

The paper is organized as follows. First a review of the numerical algorithm implemented is presented, then a decription of the mechanism with its geometry and with some details on the nature of the singularities is given. Finally, the results from the simulations are shown and the conclusions are drawn.

2 Index-3 Augmented Lagrangian Formulation

In this section a brief description of the method used to model the dynamics of the mechanism is presented. The method was developed by Bayo [8–10] and extensively applied in [11–13].

Equation of motion of a mechanical system described by a set of n Cartesian coordinates $\mathbf{q}(t)$ subjected to m holonomic kinematic constraints $\boldsymbol{\Phi}(\mathbf{q}(t))$ can be expressed as:

$$\mathbf{M}\ddot{\mathbf{q}} + \boldsymbol{\Phi}_q^T \boldsymbol{\lambda} = \mathbf{f}. \tag{1}$$

In Eq. (1), $-\boldsymbol{\Phi}_q^T \boldsymbol{\lambda}$ represents the constraints forces with $\boldsymbol{\Phi}_q = \partial \boldsymbol{\Phi} / \partial \mathbf{q}$ the Jacobian of the constraints and $\boldsymbol{\lambda}$ vector of the unknown Lagrangian multipliers. $\mathbf{M}$ is the mass matrix that becomes constant according to the coordinates used, $\mathbf{f}$ is the vector of forces applied to the system.

Equation (1) can be treated according to numerous formulations in order to convert the differential-algeabric into the standard ordinary-differential equations (ODE). In this paper we follow the penalty formulation combined with the Newmark integration scheme to obtain the index-3 Lagrangian augmented formulation. In the penalty formulation the constraint is modeled as a spring-damper unit such that

$$\boldsymbol{\lambda} = \alpha(\ddot{\boldsymbol{\Phi}} + 2\mu\omega\dot{\boldsymbol{\Phi}} + \omega^2\boldsymbol{\Phi}), \tag{2}$$

where α is the penalty factor, μ and ω are the Baumgarte stabilization parameters. By substituting $\ddot{\boldsymbol{\Phi}} = \dot{\boldsymbol{\Phi}}_q\dot{\mathbf{q}} + \boldsymbol{\Phi}_q\ddot{\mathbf{q}}$, $\dot{\boldsymbol{\Phi}} = \boldsymbol{\Phi}_q\dot{\mathbf{q}}$ into Eq. (2), we obtain the ODE:

$$(\mathbf{M} + \boldsymbol{\Phi}_q^T\alpha\boldsymbol{\Phi}_q)\ddot{\mathbf{q}} = \mathbf{f} - \boldsymbol{\Phi}_q^T\alpha(\dot{\boldsymbol{\Phi}}_q\dot{\mathbf{q}} + 2\mu\omega\dot{\boldsymbol{\Phi}} + \omega^2\boldsymbol{\Phi}) \tag{3}$$

Equation (3) can be written by following the Lagrangian multiplier method leading to the index-1 iterative algorithm:

$$\begin{aligned}(\mathbf{M} + \boldsymbol{\Phi}_q^T\alpha\boldsymbol{\Phi}_q)\ddot{\mathbf{q}} + \boldsymbol{\Phi}_q^T\boldsymbol{\lambda}^* &= \mathbf{f} - \boldsymbol{\Phi}_q^T\alpha(\dot{\boldsymbol{\Phi}}_q\dot{\mathbf{q}} + 2\mu\omega\dot{\boldsymbol{\Phi}} + \omega^2\boldsymbol{\Phi})\\ \boldsymbol{\lambda}^*_{i+1} &= \boldsymbol{\lambda}^*_i + \alpha(\ddot{\boldsymbol{\Phi}} + 2\mu\omega\dot{\boldsymbol{\Phi}} + \omega^2\boldsymbol{\Phi})\end{aligned} \tag{4}$$

The algorithm in Eq. (4) is then combined with the implicit single-step trapezoidal rule:

$$\dot{\mathbf{q}}_{k+1} = \frac{2}{h}\mathbf{q}_{k+1} + \hat{\dot{\mathbf{q}}}_k \text{ with } \hat{\dot{\mathbf{q}}}_k = -(\frac{2}{h}\mathbf{q}_k + \dot{\mathbf{q}}_k),$$

$$\ddot{\mathbf{q}}_{k+1} = \frac{2}{h^2}\mathbf{q}_{k+1} + \hat{\ddot{\mathbf{q}}}_k \text{ with } \hat{\ddot{\mathbf{q}}}_k = -(\frac{2}{h^2}\mathbf{q}_k + \frac{4}{h}\dot{\mathbf{q}}_k + \ddot{\mathbf{q}}_k).$$

and with the velocity and acceleration projections such that dynamic equilibrium can be established at time step $k+1$ as:

$$\mathbf{M}\mathbf{q}_{k+1} + \frac{h^2}{4}\boldsymbol{\Phi}^T_{q_{k+1}}(\alpha\boldsymbol{\Phi}_{k+1} + \boldsymbol{\lambda}^*_{k+1}) - \frac{h^2}{4}\mathbf{f}_{k+1} + \frac{h^2}{4}\mathbf{M}\hat{\ddot{\mathbf{q}}}_k = \mathbf{0} \tag{5}$$

The system of nonlinear equations in Eq. (5) is then solved via Newton-Raphson iterative approach (i is the iteration counter)

$$\left[\frac{\partial\mathbf{f}(\mathbf{q})}{\partial\mathbf{q}}\right]_i \Delta\mathbf{q}_{i+1} = [\mathbf{f}(\mathbf{q})]_i,$$

with

$$[\mathbf{f}(\mathbf{q})] = \frac{h^2}{4}(\mathbf{M}\ddot{\mathbf{q}} + \boldsymbol{\Phi}_q^T\alpha\boldsymbol{\Phi}_q + \boldsymbol{\Phi}_q^T\boldsymbol{\lambda}^* - \mathbf{f})$$

and the approximated tangent matrix

$$\left[\frac{\partial\mathbf{f}(\mathbf{q})}{\partial\mathbf{q}}\right] = \mathbf{M} + \frac{h^2}{4}(\boldsymbol{\Phi}_q^T\alpha\boldsymbol{\Phi}_q).$$

What is relevant in the formulation presented is that the leading matrix in Eq. (3) is always invertible although the constraint Jacobian $\boldsymbol{\Phi}_q$ becomes rank deficient whenever the mechanism crosses a constraint singular configuration.

3 Description of the Mechanism

The planar five-bar parallel mechanism is shown in Fig. 1. The mechanism is commonly used for positioning point Q on the plane. The model considered has the moving links with L=210 mm with uniform mass $m = 0.017$ Kg and moment of inertia of $I = 6.23 \cdot 10^{-5}$ Kgm^2. In order to implement the algorithm presented, a reference system was attached to each link, thus the mechanism motion was parametrized by a vector

$$\mathbf{q} = \left(r_{x_1}, r_{y_1}, \theta_1, r_{x_2}, r_{y_2}, \theta_2, \ldots, r_{x_{n_b}}, r_{y_{n_b}}, \theta_{n_b}\right)^{\mathrm{T}},$$

where $(r_{x_i}, r_{y_i}, \theta_i)$, $i = 1, \ldots, n_b$, identifies the position of the reference system with origin in the center of mass of the ith body and $n_b = 5$.

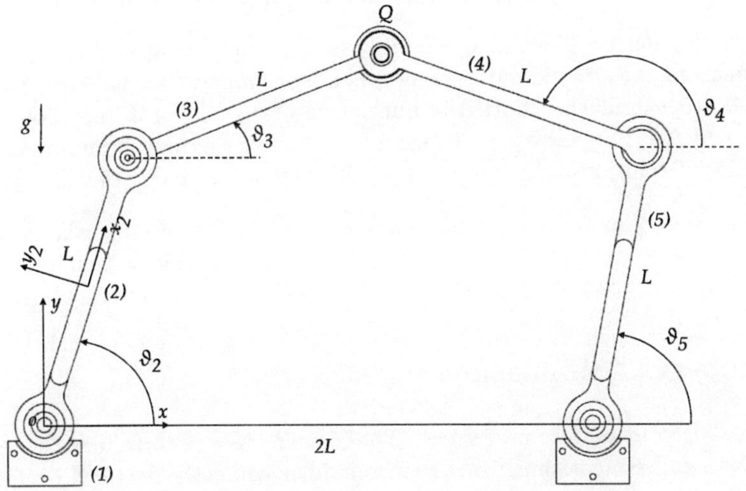

Fig. 1. Planar five-bar parallel mechanism.

3.1 Constraint and Kinematic Singular Configurations

The ability of the algorithm used to simulate the mechanism motion is to overcome the constraint singularities. Constraint singularities occur whenever $\mathbf{\Phi}_q$ loses full rank. That means that the constraint equations are no longer linear independent, one or more constraints are needless. This fact has impact on the mobility of the mechanism that, in a singular configuration, may change its dofs and may switch the working mode. On the other hand, the mechanism motion presents kinematic singularities, as well. Usually, kinematic and constraint singular configurations are addressed by the kinematic and dynamic communities independently. In this case we treat them all together to try to clarify their connection in the case of the five-bar mechanism.

It is well known that for a parallel mechanism the velocity mapping is expressed by

$$\mathbf{J}_x\dot{\mathbf{x}} = \mathbf{J}_q\dot{\mathbf{q}}.$$

$\mathbf{x} = (x, y)^{\mathrm{T}}$ is the vector of the point Q coordinates, $\mathbf{J}_x$, $\mathbf{J}_q$ are the Jacobian matrices given as:

$$\mathbf{J}_x = \begin{bmatrix} L(yc_{\theta_2} - xs_{\theta_2}) & 0 \\ 0 & L(yc_{\theta_4} - xs_{\theta_4} + 2Ls_{\theta_4}) \end{bmatrix},$$

$$\mathbf{J}_q = \begin{bmatrix} x - Lc_{\theta_2} & y - Ls_{\theta_2} \\ x - Lc_{\theta_4} - 2L & y - Ls_{\theta_4} \end{bmatrix} \tag{6}$$

and

$$x = r_{x3} + \frac{L}{2}c_{\theta_3}, \quad y = r_{y3} + \frac{L}{2}s_{\theta_3}.$$

The mechanism can attain three kinematic singularities during its motion:

- Direct singularity: the null space of $\mathbf{J}_x$ is not empty, that is $det(\mathbf{J}_x) = 0$;
- Inverse singularity: the null space of $\mathbf{J}_q$ is not empty, that is $det(\mathbf{J}_q) = 0$;
- Combined singularity: both the null space of $\mathbf{J}_x$ and of $\mathbf{J}_q$ are not emplty, that is $det(\mathbf{J}_x) = det(\mathbf{J}_q) = 0$. In the combined singular configurations the number of solutions either of the forward or inverse kinematics change.

For the geometry under study, combined singular configurations take on importance as it was found from the simulations that they coincide with the constraint singularities, hereafter they will be called combined-constraint singularities.

4 Numerical Simulations

Two simulation tests were performed. The former was run to validate the implementation of the algorithm i3-ALP in our own code, the latter to stress our code when the mechanism passes through combined-constraint singularities. In the i3-ALP algorithm we chose $\alpha = 10^7$ and a fixed time integration step of $h = 0.001$ s.

4.1 Code Validation

In this case an external torque of 0.25 Nm was applied to link (2) with initial configuration given by $\theta_2 = \theta_5 = \pi/2$ and $\dot{\mathbf{q}} = \mathbf{0}$. The simulation time was set to $t_{span} = 2.5$ s. The results from the i3-ALP algorithm were compared with those obtained by a code written via a commercial software, namely Simscape, with Runge-Kutta order 4 as integration method and same fixed time integration step.

Figures 2, 3 show $x(t)$, $y(t)$ obtained from the simulations. It is evident from the plots that our code produces results almost coincident with those from the commercial software.

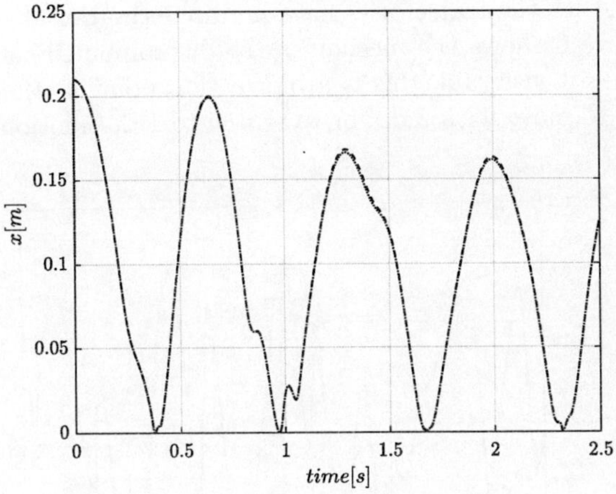

Fig. 2. $x(t)$: −− RK4 commercial, ··· i3-ALP.

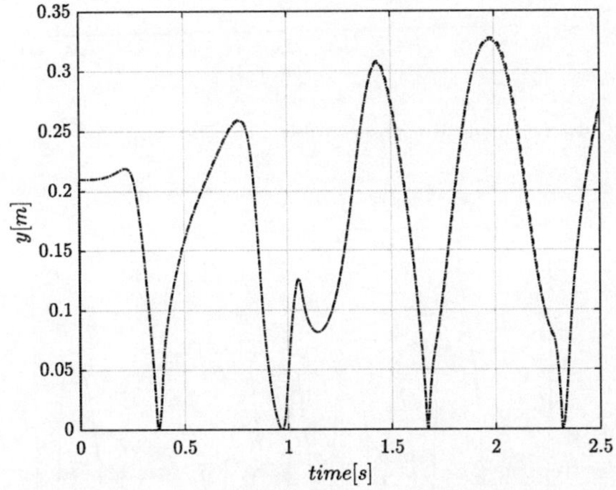

Fig. 3. $y(t)$: −− RK4 commercial, ··· i3-ALP.

4.2 Tests on Singularities

In this case the mechanism was subjected only to gravity leaving it to fall freely. This motion can be considered very challenging from a numerical point of view as the mechanism may attain any kind of singular configurations. The types of singularities met was checked calculating the singular values of the Jacobians, $\mathbf{J}_x$, $\mathbf{J}_q$ and $\mathbf{\Phi}_q$. The simulation parameters were held identical to those in the validation test except for the simulation time that was extended to t_{span}=50 s. The combined-constraint singularities occured, for example, at t=41.84 s without stopping abrutely the computation. Figures 4, 5 show $x(t)$ and $y(t)$, respectively.

The plots show that the trajectories have no discontinuties at the singular configuration. Figure 6 shows the mechanism at the combined-constraint singular configuration. As it was said, this is a bifuraction point as the mechanism can either continue to move with 2 dof or to switch to 1 dof motion.

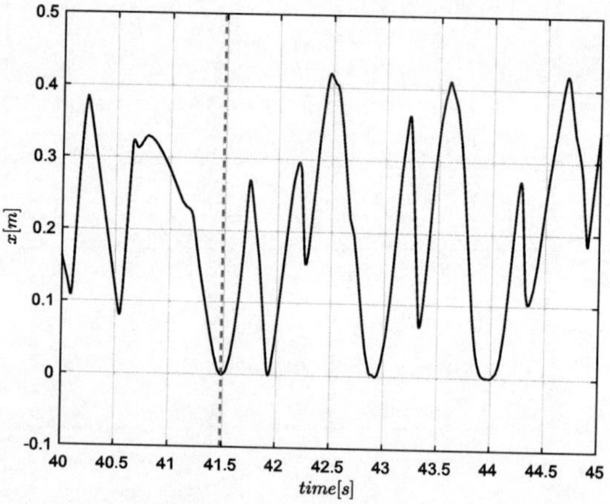

Fig. 4. x(t) crossing the combined-constraint singularity.

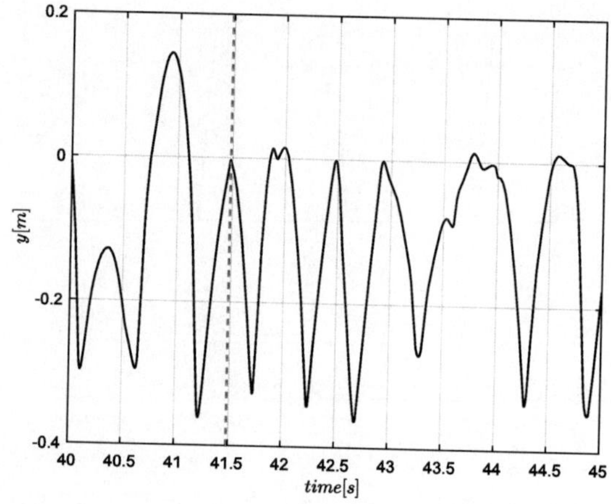

Fig. 5. x(t) crossing the combined-constraint singularity.

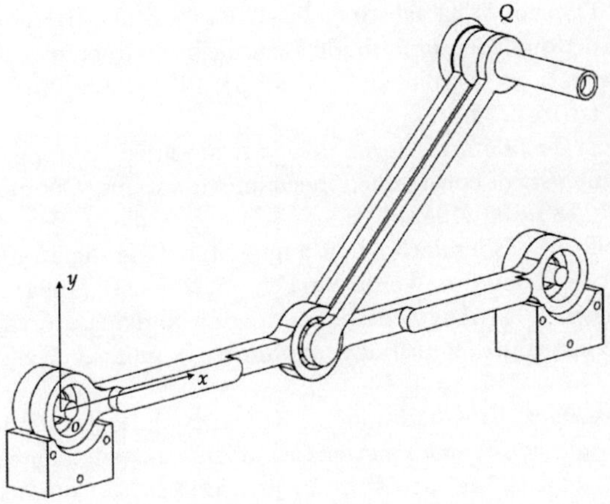

Fig. 6. Costrained-combined singularity configuration.

5 Conclusions

The paper presents the solution of the forward dynamics problem for a planar five-bar mechanism in the presence of singular configurations. The algorithm i3-ALP was used successfully allowing us to simulate the mechanism motion when crossing the constraint singularities. For the geometry studied, it was found that the constraint singular points coincide with the kinematic combined singularities. Future work will be dedicate to arrange a mechatronic system with a 3D printed prototype. The final goal will be dedicated to develop a digital twin of the printed mechanism to have a cyber sytem working as predictor/corrector of the prototype motion.

References

1. Liu, X.-J., Wang, J., Pritschow, G.: Kinematics, singularity and workspace of planar 5R symmetrical parallel mechanisms. Mech. Mach. Theory **41**, 145–169 (2006)
2. Campos, L., Bourbonnais, F., Bonev, I.A., Bigras, P.: Development of a five-bar parallel robot with large workspace. In: Proceedings of IDETC/CIE 2010, Montreal (2010)
3. Giberti, H., Cinquemani, S., Ambrosetti, S.: 5R 2dof parallel kinematic manipulator-l multisiciplinary test case in mechatronics. Mechatronics **23**, 949–959 (2013)
4. Yamine, J., Prini, A., Lavit Nicora, M., Dinon, T., Giberti, H., Malosio, M.: A planar parallel device for neurorehabilitation. Robotics **9**(104), (2020)
5. Ruggiu, M., Rea, P.: Development of a Mechatronic system for the Mirror Therapy. Actuat. **11**(1), (2022)
6. Figliolini G., Rea P., Angeles J.: The synthesis of the axodes of RCCC linkages. J. Mech. Robot. **8**(2), (2016)

7. González, F., Dopico, D., Pastorino, R., Cuadrado, J.: Behaviour of augmented Lagrangian and Hamiltonian methods for multibody dynamics in the proximity of singular configurations. Nonlinear Dyn. **85**(3), 1491–1508 (2016). https://doi.org/10.1007/s11071-016-2774-5
8. Bayo, E., García de Jalón, L., Serna, M.A.: A modified Lagrangian formulation for the dynamic analysis of constrained mechanical systems. Comput. Methods Appl. Mech. Eng. **71**, 183–195 (1988)
9. Bayo, E., Avello, A.: Singularity free augmented Lagrangian algorithms for constraint multibody dynamics. Nonlinear Dyn. **5**, 209–231 (1994)
10. Bayo, E., Ledesma, R.: Augmented Lagrangian and mass orthogonal projection methods for constrained multibody dynamics. Nonlinear Dyn. **9**(1–2), 113–130 (1996)
11. Dopico, D., González, F., Cuadrado, J., Kövecses, J.: Determination of holonomic and nonholonomic constraint reactions in an index-3 augmented Lagrangian formulation with velocity and acceleration projections. J. Comput. Nonlinear Dyn. **9**(4), 041006 (2014)
12. González, F., Dopico, D., Pastorino, R., Cuadrado, J.: Benchmarking of augmented Lagrangian and Hamiltonian formulations for multibody system dynamics. In: Proceedings of ECCOMAS Thematic Conference on Multibody Dynamics, Barcelona (2015)
13. Ruggiu, M., González, F.: Assessment of variable step-size integration of multibody systems. In: Proceedings of 10th ECCOMAS Thematic Conference on Multibody Dynamics, Budapest (2021)

Check for updates

Industrial Networks Protocols PROFIBUS and RS485 – A Description of the Most Common Problems

Vítor da Cunha[1], Vítor Carvalho[1,2], José Machado[3](✉), and Filomena Soares[2]

[1] 2Ai, School of Technology, IPCA, Barcelos, Portugal
vcarvalho@ipca.pt
[2] Algoritmi Research Centre, University of Minho, Guimarães, Portugal
[3] Metrics Research Centre, University of Minho, Guimarães, Portugal
jmachado@dem.uminho.pt

Abstract. Fieldbus standards are industrial networks that establish hardware interfaces and communication protocols. Central automation structures were used to undertake automation prior to these standards. Among others, it was required direct wiring to the stations of central automation which increased errors and faults. Networked Control Systems are one category of distributed control systems where controllers, sensors, and actuators are connected by communication networks. This organization type is advantageous because of its distribution and its independence from other computational systems. Therefore, it is fundamental to describe the main problems in the most used protocols for industrial networks.

Keywords: Industrial networks problems · PROFIBUS · RS485

1 Introduction

This paper presents a study of the main problems in the most used industrial networks protocols implemented in distributed control systems and in the physical medium, namely, PROFIBUS and RS485 protocols.

Industrial networks allow the transmission of data in a fast and reliable way. They are applied in distributed systems, that is, processes that work in a synchronous, decentralized way and carried out by two or more computational systems. This type of organization is advantageous because of its distribution and its independence from other computational systems [1].

Operation of distributed control systems (DCS) is highly dependent of the performance of the communication system. So, in cases where the network operates in faulty environments and the communications have real-time constrains its analysis turns out to be a relevant point [2].

Integrating control, communication, and computing into distinctive information processes and machine/factory operations is a major tendency in recent commercial and

D. D. Cioboată (Ed.): ICoRSE 2022, LNNS 534, pp. 367–374, 2023.
https://doi.org/10.1007/978-3-031-15944-2_33

industrial systems. These DCS are named Networked Control Systems (NCSs) and integrate controllers, actuators, and sensors connected by communication networks. Modification of point-to-point to common-bus approach communication architecture generates diverse ways in the closed-loop system dynamics of time-delay uncertainty. This results from the communication processing, the computational time required for physical signal coding and the time sharing of the communication medium. In a control application the time delays can cause system instability and reduce the system's performance [3].

Following the background presented, the purpose of this paper is to find and list the main problems of the industrial networks protocols PROFIBUS and RS485.

This paper is organized into 4 sections. Section 2 presents the Theoretical Concepts, Sect. 3 presents the Networks Problems and Sect. 4, emanates the Final Remarks.

2 Theoretical Concepts

Control systems performance is not only measured by the programmable logic controllers (PLCs) but considers as well their location in the physical environment. This requires a high-performance communication system besides monitoring, operation and plant visualization [4].

2.1 Distributed Automation Systems

Process automation and manufacturing are increasing their use of distributed automation systems allowing the split of complex control tasks into smaller and less complex subtasks. Although, certain requirements must be fulfilled for communication between distributed automation systems. These systems present the next advantages:

- Simultaneous and independent start-up of specific regions of a system/plant;
- Small and more objective programs;
- Parallel processing, which allows:
  - Quick reaction times;
  - Lower load on the single processing units.
- Systemwide configuration;
- Superior system/plant disposal.

A comprehensive and powerful communication system is crucial to a distributed plant structure. Cost-effective, reliable and real-time solutions are provide by current fieldbus technologies to industrial automation systems [5].

From the point of view of protocol, generality, throughput, frame size and network dimension fieldbus technologies have reduced the complexity of local area networks. Moreover, in order to achieve hard real-time communication with low-level devices, such as actuators, intelligent sensors and PLCs, other mechanisms and services are considered [6].

2.2 PROFIBUS

PROcess FIeldBUS (PROFIBUS) is a fieldbus open standard with purposes in building automation, processing, and manufacturing. It uses the EN 50170 standard (the Euronorm for field communications) assuring transparency and vendor independence operation. With this protocol it is possible to establish a connection between devices of different manufactures without any special interface adjustment [7].

This protocol is supported by the international standards IEC 61158 and IEC 61784 and is built in 3 layers of the reference model ISO/OSI [8]. Moreover, being a multi-master system allows parallel operation of visualizing or engineering systems and several automations at a bus, supporting Multicast and Broadcast communication [4].

There are 3 PROFIBUS family compatible versions: PROFIBUS Fieldbus Message Specification, Decentralized Periphery and Process Automation.

Fieldbus Message Specification (FMS) is a high-level extensive and complex communication tasks general-purpose solution providing high flexibility and services for multiple applications.

PROFIBUS Decentralized Periphery (DP) was developed as a low-cost and high speed interface being proper for device level data exchange between distributed I/O and automation control systems.

Process Automation (PA) allows actuators and sensors to be linked to a common bus as well as it allows supplying data and power in the bus following the international standard IEC 1158-2 [7].

The physical installation of PROFIBUS communication considers:

- twisted pair (shielded) cables (150 Ω impedance);
- Fibre optic.

Communication can combine different types of networks, besides from being used individually [4]. Master and slave devices are distinguished in the PROFIBUS protocol. The bus data communication can be determined by the master device. No external request is necessary when sending a message by the master device (when possessing the bus token – access rights). In the PROFIBUS standard a master device is considered an active device.

PROFIBUS DP Master devices can be of the following classes:

- Class 1 Master (DPM1) – a central controller trades messages with DP slaves with a determined cycle;
- class 2 Master (DPM2) - considered for commissioning in the configuration, monitoring or operation of the DP system.

Slaves (peripherical devices) do not include bus access rights (passive operation).

The PROFIBUS DP-V1 establishes non cyclic message servicing between master and slave devices, allowing trading of alarm info, diagnostics and parameter settings in common exchange of I/O data. PROFIBUS DP devices can be supported by PROFIBUS DP-V1 [7].

2.3 EIA Standard RS-485

The EIA standard RS485 is a balanced data transmission mandatory in IEC 61158-2/EN 61158-2 for twisted-pair cables data transmission (Fig. 1) [4].

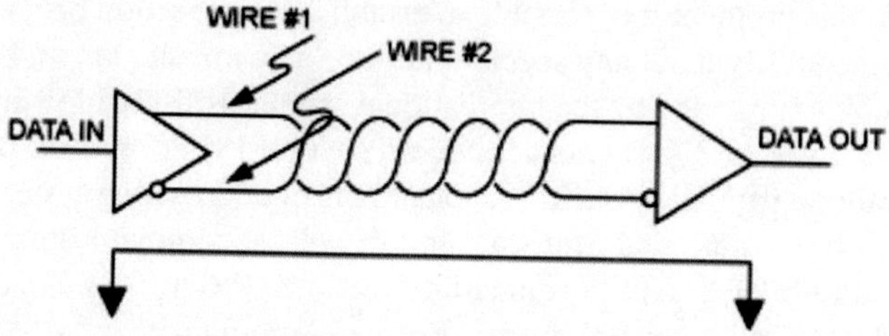

Fig. 1. RS-485 considers two cables to transmit data (besides the ground cable). Grounding helps to avoid problems of communication due to a difference in potential from one location to another [9].

Adding or removing devices is allowed by the BUS structure of the RS-485 protocol. Future extensions do not influence nodes that are in operation [7]. A twisted and shielded pair cable is used as medium being the bus cable (segment) terminated with the characteristic impedance at both ends [4]. The speed of transmission can vary between 9.6 kbit/s and 12 Mbit/s (Table 1). A single speed should be considered for all the bus devices when the system is commissioned. The maximum length of the cable is dependent on the speed of transmission which can be increased by repeaters (no more than 3 in series should be used) [7].

Table 1. RS485 cable length dependent on transmission speed.

Baud rate (kbit/s)	Distance/Segment (m)
9.6; 19.2; 45.45; 93.75	1200
187.5	1000
500	400
1500	200
3000; 6000; 12000	100

3 Industrial Networks Protocols Most Common Problems

This section describes the most common problems in the industrial networks protocols PROFIBUS and RS485.

3.1 Configurations and Parameter Sets

The connection of distributed field devices or drives requiring very high response times uses Decentralise Peripheral (DP). A PROFIBUS DP system includes permanently a Master and Slave relation being possible to have multiple master's on a network. The DP-Master is able to control remote IOs [10].

Clear error messages are provided by the PROFIBUS Master DTM if failing parameters downloading the to the PROFIBUS Master Unit. This can be a result of:

- Slave parameter sets errors or inconsistencies (checked before download);
- DTM Master unable to establish communication with CS1/CJ1W-PRM21 PROFIBUS Master Unit;
- Download process communication interruption.

Common faults parameter sets in slave can be:

- Number of slave devices (>= 1);
- Number of I/O modules/slave (>= 1);
- Maximum I/O data size (<= 7168 words);
- Maximum I/O modules (<= 4000);
- I/O Areas configured overlap;
- Allocated PLC memory areas overlap;
- EM banks availability (dependant of the PLC/CPU type).

Faults which are detected by the PLC and displayed can be:

- When establishing communication;
- During download;
- Recovery (following download failing) [7].

A maximum of 32 devices can be connected in a PROFIBUS segment. If more devices are required up to 126 stations, a repeater must be included in the network to "refresh" the communication signal [10].

3.2 RS485

A RS485 physical medium can be a part of a PROFIBUS connecting devices with a fibre optic cable or daisy chain with line, tree, or ring network configurations. The speed bus is up to 12 Mbps; the cable length of the PROFIBUS segment is determined by the transmission speed (a transmission of 1.5 kbps allows a 200 m maximum length - a repeater is necessary to cover other 200 m recursively) [10].

Typical fieldbus faults (>90%) are associated to problems in the physical layer including power supplies, terminations and wiring [11]. The physical characteristics include the: topology, module size, distances and number of nodes, which are known requirements provided by the vendors [12].

RS485 is a type of network broadly used. Though being simple, its installation and characteristic depends on the nodes and equipment being connected. In order to avoid trivial mistakes there are presented certain steps to debug and plan a RS485 network:

- If possible, the network should be created gradually. Start with a slave and add the others one by one, checking whether the operation of the network is still OK. In the event of any problem, debugging is much easier;
- All slaves must be configured with the same speed (baud rate) and the same parity. In contrast, each slave must be configured with a unique address. The duplication of addresses can cause problems in communication with that address, because both devices try to respond to commands;
- No routing cables for communication with power cables;
- When connecting the cables to the devices, the wires must not be reversed. Connection inversion at a single point can prevent operation of the entire network;
- In long networks where the devices are distributed at points distant from each other, consider using the third wire grounding all devices properly. This helps to avoid problems of communication due to a difference in potential from one location to another;
- In long networks where the devices are distributed at points distant from each other, it is recommended using the third wire grounding all devices properly; This helps to avoid problems of communication due to a difference in potential from one location to another (Fig. 1);
- If not sure that a certain device is communicating properly try changing its physical position in relation to another one that is working properly, to see if the problem is the equipment itself (typically configuration) or is in the physical position (typically, cabling problem or noise (see Fig. 2 and Fig. 3);
- Often, a little slower network can render more efficient communication because with low error rates it will require fewer retransmissions. Likewise, make sure that the distances involved will allow the desired speed. Long cables and very high speeds usually do not get along well;
- When troubleshooting the LEDs signals can give valuable information on what is working and what is not.

It is important to refer that many transmission errors occur due to cabling problems. The physical treatment of the cable highly influences its durability as well as its performance is reduced if under continuous movement or if being bent back and forth. This may cause system lowering of its lifetime and performance [12–14].

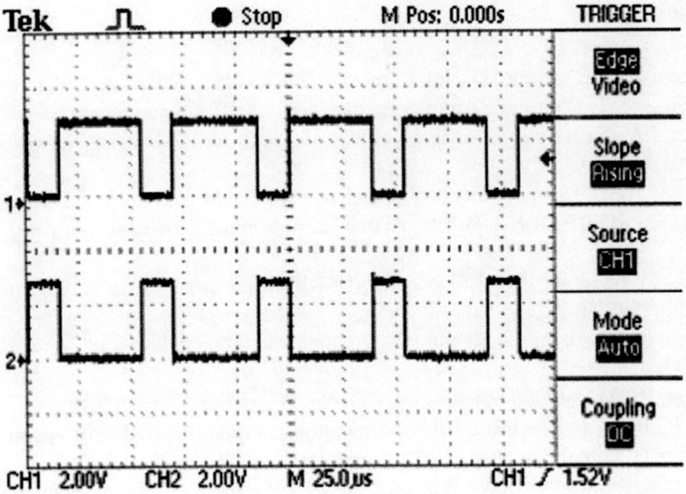

Fig. 2. RS-485 - a balanced system ideally opposites the two wires signals [9].

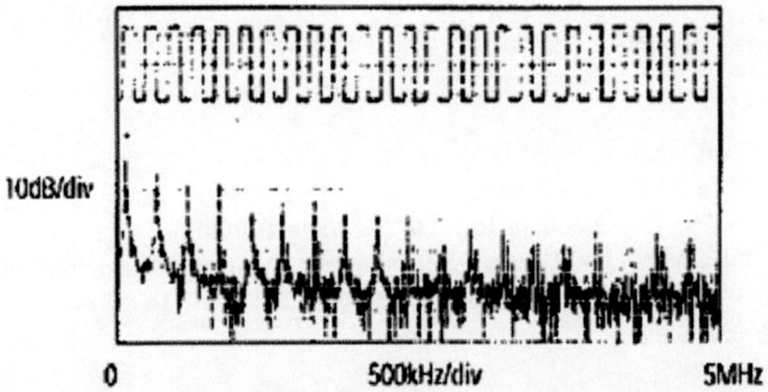

Fig. 3. RS-485 – EMI is radiate from the fast edges high-frequency components coupled with long wires [9].

4 Final Remarks

The paper presents an overview on the PROFIBUS protocol and the most common errors and faults in the PROFIBUS and RS485 industrial networks. These problems are usually associated to configuration and parameter sets as well as related to the physical medium. In the physical medium beside the human factor and the time factor (examples: poorly executed junctions on the connectors; loosen up junctions due to vibrations), there are disturbances that can occur when operating in faulty environments such as industrial ones. A considerable number of fieldbus problems (>90%) are associated to faults in the physical layer, such as power supplies, terminations and wiring,

Further developments will consider the description of case studies associated to errors and faults in the PROFIBUS and RS485 industrial networks.

References

1. De Engenharia, E.: Nuno Daniel Carneiro de Carvalho WALC-AI : Laboratório para Aprendizagem do Controlo e Automação Industrial (2010) (in portuguese)
2. Carvalho, J.A.: Experimental Analysis of Outage Times for PROFIBUS Networks, pp. 421–426 (2005)
3. Lian, F.: Analysis, Design, Modeling, and Control of Networked Control Systems by (2001)
4. SIEMENS: PROFIBUS Network Manual. System Manual - Siemens, no. April, pp. 1–350 (2009)
5. Tovar, E.: Engineering PROFIBUS networks with heterogeneous transmission media, vol. 30, pp. 17–32 (2006)
6. Verissimo, P.: How hard is hard real-time communication on field-buses? Network, pp. 112–121 (1997)
7. Operation, P.C.: Operation Manual, vol. Cat. No. W, no. 608, p. 137 (2014)
8. Drahoš, P., Bélai, I.: The PROFIBUS protocol observation. In: IFAC Proceedings Volumes (IFAC-PapersOnline), vol. 9, no. PART 1, pp. 258–263 (2012)
9. Parts, R.: Guidelines for Proper Wiring of an RS-485 (TIA/EIA-485-A) Network, vol. 485, pp. 1–11 (2018)
10. Updates, I.: Installation Rules For PROFIBUS, Industrial Automation Asia, no. ENQUIRY NO. 7101 (2012)
11. Rezabek, J.: Troubleshooting FOUNDATION Fieldbus networks, InTech Magazine, no. September/October (2012)
12. Alholt, G., Wang, X.: Examining Fieldbus Quality A comparative study of fieldbus attributes (2013)
13. Kjellsson, J., Vallestad, A.E., Steigmann, R., Dzung, D.: Integration of a wireless I/O interface for PROFIBUS and PROFINET for factory automation. IEEE Trans. Industr. Electron. **56**(10), 4279–4287 (2009). https://doi.org/10.1109/TIE.2009.2017098
14. Yaqoo, I., et al.: Internet of Things architecture: recent advances, taxonomy, requirements, and open challenges. IEEE Wireless Commun. **24** (6 2017), 10–16 (2017)

Step, Servo and Hub Motor Based Hybrid PCB Processing and Prototyping Device Design and Analysis

Atakan Yerli[1](✉), Fuad Aliew[1], José Mendes Machado[2], Eurico Augusto Rodrigues Saebra[2], and António Alberto Caetano Monteiro[2]

[1] Gebze Technical University, 41400 Kocaeli, Turkey
a.yerli2019@gtu.edu.tr
[2] Universidade do Minho, 4800-042 Guimaraes, Portugal

Abstract. Aim of this research is to design the central processing unit, which is indispensable today, with 3 different motor types and to control the ideal motor in ideal torque and rpm units in the market, and to complete the process with the same precision with the minimum unit required in the energy and time domain. The stepper motors used in the hobby and light industry sector in the market offer a 2% step-miss sensitivity and have minimum energy consumption, while the servo motors have low energy efficiency against 1/10000 step-miss sensitivity and the high rpm speeds of magnetic hall sensor hub servo systems and the sensitivity that is the same as servo motors. The mission of this research is to establish a system that can respond to the increasing prototyping demands in the domestic market due to the exchange rate by making comparison analyzes of the principles in the energy and time domain. In this research one of another purpose is designing hybrid machine with 3D Printing abilities and CNC PCB prototyping abilities under the same chassis. In this research the parts that used had been chosen to form available market products in Turkey. In this research another software explained detailly with advantages and disadvantages in any perspective.

Keywords: CNC · Hub motor · Servo motor · Step motor · Klipper · Hybrid systems · FDM

1 State of Art 3D Printers and CNC

In manufacturing process there is different types of machines that works by a definition of an occupation. CNC, 3D Printer and the Lasers are the main names of the processes that required. CNC principle is eliminating the material from main frame of the material and obtaining the final desired structure. subsequently, it can be called as a Top to Bottom process.

3D Printer's works with the soft material or materials. With those raw materials, it obtains the final desired structure or structures by add-in method. Subsequently, it can be called as a Bottom to Top process.

D. D. Cioboată (Ed.): ICoRSE 2022, LNNS 534, pp. 375–387, 2023.
https://doi.org/10.1007/978-3-031-15944-2_34

Laser's principle lay on cutting the piece from desired piece in 2d dimension mostly. But with the increase software and hardware lasers can cut 3d pieces in desired lengths and depths mostly on all materials. In this research objectives are observing different motor types and comparing them with respect of criterias. Mounting and designing hybrid machine will be elucidated with details.

1.1 3D Printer Technologies

In this section the 3D printer basis and the chase designs will be explained. There is FDM, SLA and SLS technologies (Fig. 1).

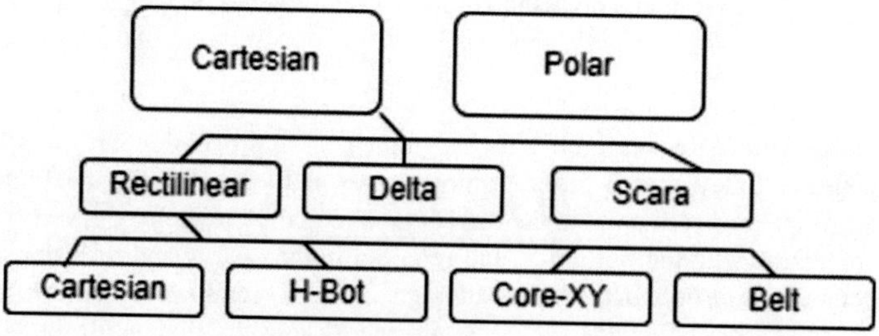

Fig. 1. FDM technology sub-divisions

Rectilinear 3D Printers. Rectilinear 3D Printer Models Have 4 Sub-division that Uses X, Y and Z Axis-Based g-codes. They Are Cartesians, H-bot, Core X-Y and Belt.

Cartesian 3D Printers

This rectilinear cartesian model lies on the 3 axes as a X.Y and Z. In this model Z and the X axes are the axes of the extruder head and Y axis is the printing bed's axis. In the mostly of these systems are used 4 electrical motors in order of: extruder, z axis, y axis, x axis.

H-Bot

This type of 3d printer lies on the co-operations of the 2-electric motor on 1 continuous belt for moving extruder on X-Y axes together. This system offers high accuracy with the advantageous demand all metal parts except belt system (Fig. 2).

Core X-Y

This type of 3d printer lies on the co-operations of the 2-electric motor on 1 continuous virtual belt, 2 un-continued real belt for moving extruder on X-Y axes together or separately. In 45-degree operations on x-y axis on opposite directions only 1 electrical motor can work separately. For another movements 2 motor of X-Y axes should be work together for working properly (Fig. 3).

Belt Type

In this type 3d printer model is the one which engineered for the continuous manufacturing scenarios and long object that can be able to fit in 2 other dimensions as like as

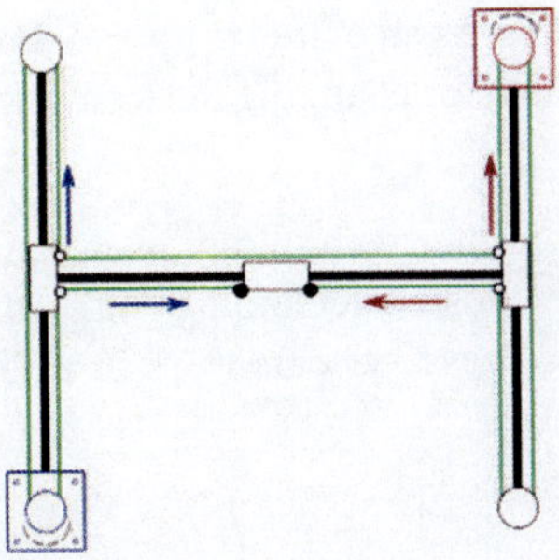

Fig. 2. H-Bot 3D printers working principle

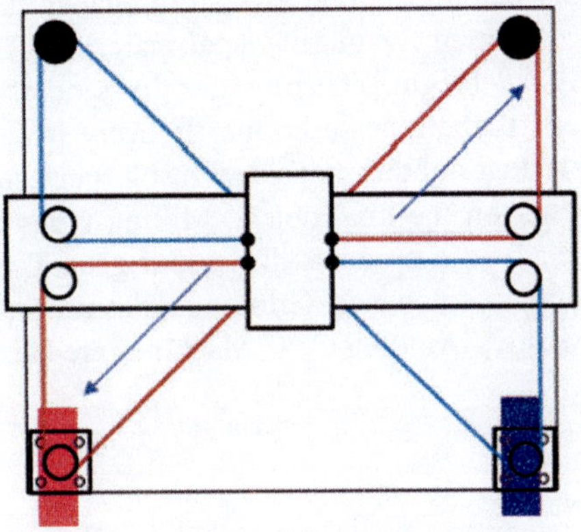

Fig. 3. Core XY 3d printer principle

industrial pieces. In this model x axis works same but z axis work with the 45° to the Y axis to make it continues all Y line.

Delta Type

In this type of printer, the G-code use Z supported three arms system instead of cartesian planes. Arms support extruder and they are attached to the three vertical posts arranged in a triangle. Arms con move only on Z axis subsequently vertically to the up and down independently each other subsequently it can create X and Y axes, it makes an atmosphere to extruder move in all directions.

Scara Type

Selective Compliance Assembly Robotic Arm (SCARA) 3D unique type printer use cartesian plane and takes its power from controlled arms. These arms move in X and Y axes. The bed moves on Z axis. These types of printers do not support direct drive extrusions.

Polar Type

In this 3d printer model the location is not determining with X, Y and Z axis. There is only angle, and the length dimensions are in use. In this type of 3d printer there is 3 another motors that moves z axis as an extruder up and down, rotating plate and another motor that connected to this plate for changing the location of the plate respect to coding.

SLA & MJF Type
Stereolithography Apparatus printer technology uses resin as a raw material to obtain desired final object. Multi Jet Fusion technology uses the powder and the inkjet for obtaining desired final object. These systems work in additive methods.

1.2 CNC Technologies

There are 5 types of CNC machines. Drilling CNC Machines, Router CNC Machines, Lathe CNC Machines, Milling CNC Machines, Grinding CNC Machines. Drilling systems mostly obey for only drilling on one axis. Tool change times is minimum then others. Router Types mostly used for out-of-metal materials. The system bed size is much greater than their Z axis height. This provides this kind of machines a flexibility of working on large layers. Lathe types used mostly every material with the required tools for curving. The advantages of this machine it results perfectly accurate as an outer diameter and inner diameter on the final object. Milling types consist of 3 to 6 axis configurations to produce advance products like a spur gear. These types of machines are easy to use and have mid maintenance. Grinding types uses tool that rotating wheel to cut material surface. Mostly, Grinding CNC Machines are used for typical accuracy required final works.

1.3 System Build

The system analysis of the 3D Printer geometries and the working principles are different from each other. With their slicer programs may vary with the system selection of the inserted 3d Printer specifications. 3D Printer specifications had to be advantageous determined before choosing. The first criterion of the selection base should be technology basement of the 3D printer. There is 3 another technology basement; FDM, SLA, MJF. FDM printer's layers should be to layer by layer for project selection criterion and all these technologies offer that future. Secondly, the criterion of the system bottom to top layering future should be supported, FDM and MJF technologies contain it. The third criterion is the using same bed criterion, only FDM technology-based system designs can use it.

FDM technology obtains sub-division geometrical supports. The first criterion is cartesian coordinate planes-based firmware should be supported. Cartesian and Polar designs both allow that criterion but the polar system working principle requires a fully rotated bed in a circle shape. Cartesian system supports stable bed geometric de-sign conditions. The second criterion is an easy attachable or fully attached extrusion head geometrical design. This design criterion is only available for the rectilinear model's subdivisions; Cartesian, H-Bot, and Core XY. The third criterion is stable X-Y-Z carrier availability. In this criterion H-bot, Cartesian, and Core-XY all passes the criterion but there are geometrical problems that occur. In H-BOT and Core XY systems collapse in both 2 another possibility.

The first possibility is staying the bed unmovable for x or y or z axes for heavy material, in this possibility the x and y and z axes are all carriers for the printer head and CNC drilling/engraving head. In this situation, the CNC spindle will be rotated more than 3.000 rpm to 20.000 rpm and the vibration of the spindle cannot be handled by the carriers in this complexity. In the second possibility, the movable bed system works only on the z-axis and this gives x and y complexity and may work with the spindle vibration but here the z-axis cannot be stable under the load and force of the spindle. The only geometrical design model that consists of all the criteria is the Cartesian 3D printer model.

The perspective of the CNC machine's first criterion is the top to bottom layering future consistent. The machine types that matched are drilling, router, and milling CNC machine types. The second criterion is based on the basic 3 axes of usability. The CNC Machines that match that future are router and milling CNC Machines.

The combination of CNC and 3D Printer elements makes a hybrid system that can work in one geometrical design. The only matched one with the criterion mentioned on the 3D Printing side is the Cartesian 3D Printer base model. On the CNC Machine side with the mentioned criteria the ones that match with criterion are Router CNC Machine and Milling CNC Machine types. The hybrid system matches with Milling CNC Machine type cause of the systematic geometrical relatives.

1.4 Control Boards

Technological developments offer new features with the complex control systems in the different package with another feature. In the light of this taking advantage of the control theory principles as a package with our specific personal requirements to obtain sustainable projects is the main aim. The purpose of the research is not choosing the perfect and the best system, the purpose is understanding the choosing standards with the technological developments offered to obtain the sustainable system parts for the system requirements and configuring them. Order of the criteria must be handled. The system must be handled with its properties, which are licensing degrees, extruder counts, fan numbers, heater counts, end stop counts and features, temperature Sense, ethernet port support, internal or external Wi-Fi support, SPI [Serial Peripheral Interface] support, I2C support, rs485port support or with Arduino supported port, CPU model bit rate and MHz of it, supported stepper drivers. It can be added to other I/O's such as a microSD card slot, micro-USB, USB, and built-in add-on pins.

2 Mechanical and Control Systems Research

In this research manufacturing processes of the CNC will be determined and discussed with 3 motor configurations, and they are Stepper Motor, Servo Motor, and BLDC Hub Motor.

2.1 Stepper Motors

Stepper motors are in an availably usable position on the high market shares in 3D Printing Industry. 3D printing mostly does not require a closed-loop feedback feature.

There is 2 other main reason for that. Initially, it is the price differences between the servo motor and stepper motors. Stepper motors are highly cheaper than servo motors. A second big reason for the 3d printers is the maximum load of the Z-axis may be changeable, in mostly Z carrier is usable for only head toolsets not heating bed and final object material. This makes the high-pressure ratios of the Z-axis really decrease. This low-pressure advantage on all axes step motors makes themselves available without mostly no step-missing and degeneration on the last object with high accuracy of more than 200 steps and 1,80 accuracies in 360-degree, full cycle with open-loop, not close loop feedback.

Stepper motors work with UART and Step-Dir interface not with a bus interface because of the latency. This gives swift actionability with justified protocols. Market availability is wide and considerably cheaper with IC support.

2.2 Servo Motors

Servo Motors share the same specifications as steppe motors in theory. The difference of this motor type, it supports AC and DC energizing methods with close loop feedback. This motor type provides feedback signals with the support of a control circuit. The Servo motor has the same physical standards as the stepper motor and provides high efficiency and higher torques with higher operating angular speeds in precision. The main basis of this phenomenon is the control circuit design of servos. It provides the required amperage to the coils to achieve desired location change in desired workloads without a step missing. Accuracy might be in the same demand with the stepper motors in most usage criteria but the projection of the servo's load based current principle. This projection determines high efficiency with error and success feedback.

Servo motors work with UART and Step-Dir interface with serial protocols (RS-485). Control circuits of servo motors have wide market availability with expensive price tags in comparison with stepper motors.

2.3 BLDC Hub Motors

HUB motors are electric motors (Direct current motors). HUB motors, which are brushless DC motors, have one rotor and one stator element, like all other DC motors. The design process this type of dc motor change with slot to pole ratio and application area [1].

The main objective of this motor type is a significant torque increase in comparison with same-sized AC and DC motors [2]. The hub motor indicated in Fig. 4 describes the main physical entities as a demonstration. The driving methods of BLDC Hub motors are Field Oriented Control (FOC) and Electronic Speed Controller (ESC).

ESC methodology does not support closed-loop feedback operation and this decreases accuracy. FOC supports close-loop features by hall sensor BJT circuits [3]. This feature defines exact location and speed data to the main control board, this feature gives augmented efficiency, torque, and accuracy with extreme speeds such as 100 kV [4]. This feature gives future perspective to the new generation CNC and sapphire laser-based heating element embedded in 3D printers for commercial products.

Fig. 4. 24 slot 16 pole 2 whole coil designed 2 winded layer with 4 strand 1500-W BLDC HUB motor

Hub motor has a narrow market share and a narrow open-source database on the development side. Products on the market do not match the requirements of the designed system.

2.4 Control Board Selection Criteria

In the light of the specifications of the system, software must be provided with wide g-code database standards and it must be designed for multi-head signaling optimization. Control board has chosen with 32-bit base micro-controller unit and PWM signal output featured with fuse and electrical feedback protected for pins. 32-bit microcontroller supported with BCM2710A1 ARM-Cortex A53 based Raspberry Pi Zero 2W. This configuration provides extended features.

2.5 Screen and Bl Touch Connection

Screen and Bl touch connection optimized for configured system mainboards. Usability and cost reference ratio criteria matched with 12864lcd screen. This screen offers enough information to provide a range and instant system intervention for CNC & Printer farm projects. BL touch software will be configured on the system and it will provide industrial sustainable production on Z-axis operation accuracy.

3 Chassis Design

This part of the research elucidates system analysis and choice of the parts used on the project. The PCB engraving project has been named a CNC project and determined the chosen criteria of the allowable CNC model for PCB engraving and drilling with the meantime 3D printers. After deciding on the model, the chassis material should be chosen. Materials that can be used for chassis are laser engraved full metal body, aluminum sigma profiles, or concrete plasters.

Concrete plaster is designed for rough CNC machining projects. Concrete plaster norms a full weight of the profile with the strength of the plaster after dried fully. To apply concrete plaster the CAD software and calculations are not enough. This type of Gypsum Casting process requires some qualifications and an industrial development budget. This type of CNC chassis design type is not a regular and minded feature for research projects cause mold manufacturing and casting require high budgets and qualification experiences to establish this type.

Laser engraved full metal chassis is an advantageous alternative model for using CNC chassis. The advantages of the laser engraved metal parts are high precisions, easy assembling process, and extensive lifetime. Disadvantages of the full metal chassis are Extreme heavyweight, less strength on long pieces, and shock damping losses. The advantage of the easy assembling option may not be beneficial for thicker parts than 4 mm with respect to being bumped disadvantage (Fig. 5).

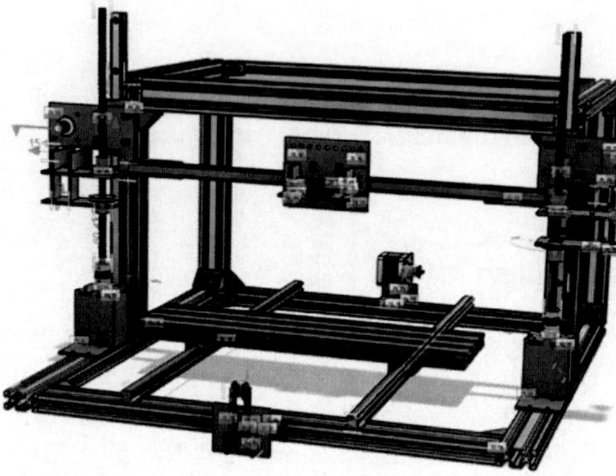

Fig. 5. Last render of the hybrid machine design

The aluminum sigma profile is an advantageous alternative for designing chassis. The advantages of the aluminum sigma profile low bumping ratios until 1-m length for every size, high strength, low-weight squared and rounded edges with the geometric design, easy the design and easy to be assembling, and high market availability. Disadvantages of this modeling are high prices compared to the laser engraved metal; soft aluminum metal may have dysfunctionality after teething operation re-assemble operations.

The chassis will be designed on the aluminum sigma profiles and laser engraved 3 mm metal sheets. In this project 20 * 40, 8-channel aluminum sigma profiles will be used in the chassis and 20 * 20 4 channel sigma profiles will be used on the x axis for attaching linear ray and mounting hot-end or spindle for occupation titles. 20 * 20 choosing criteria is the minimum weight load for the core XZ design. The system design computer-aided design (CAD) program will be Fusion 360.

The chassis design must be compatible with the PCB plaques that are available on the market. In the tur-key market standards of PCB plaque change from 50 mm * 50 mm to 400 mm * 400 mm. The fixing barriers of the CNC must support those cards subsequently it must be a minimum of 500 mm to 500 mm on the x and y axes. Z-axis is depending on the 3d printer standards minimum and other axes must obtain minimum standards of CNC. In the market average 3d printer dimension is 300 mm * 300 mm vs 250 mm. Subsequently, the chassis must be able to provide a minimum of 250 mm highness on Z-axis.

3.1 Z Axis Design

Z-axis designed with respect to raising and lowering torque calculations [5]. The given elements of the system are Ʈ = Torque; Φ = Angle of friction; F = load on the screw; Λ = lead angle; L = lead; dm = mean diameter; μ = coefficient of friction (common values are found in the friction table. Brass as a nut material and machine oiled bronze as steel).

$$Ʈraise=\frac{F*dm}{2}*(\frac{l+pi*\mu*dm}{pi*dm-\mu*l}) = 2.74 N/cm \tag{1}$$

$$Ʈlower=\frac{F*dm}{2}*(\frac{pi*\mu*dm-l}{pi*dm+\mu*l}) = 0.50 N/cm \tag{2}$$

Weight has been determined as 10 kg for load; pitch diameter as a 2 mm; thread density 10 threads per centimeter square; coefficient of friction as an 0.11. The results are only for moving the head 2.74 for raising and 0,50 for lowering the level. Also, a Drilling force must be added for requirements. But as the main issue, the pushing force is not the only parameter here. The system's feed speed of the main spindle, feed per revolution, main axis spindle speed, cutting speed, and drill diameter are crucial for pushing net force. For calculating the pushing force these should be known and calculated.

The results are in; Vf(mm/min) = feed speed of the main spindle = 40000 mm/min, fr(mm/rev) = feed per revolution = 0,2 mm, n(min-1) = main axis spindle speed = 20000 mm/rev. vc(mm/min) = main axis spindle speed = 125.6 mm/min, DC (mm) = drill diameter = 2 mm, Kc = K Factor = 3.33. Ff = Feed Force = 2.9 N/cm

$$vc = (pi \ * \ DC \ * \ N) \div 1000 \tag{3}$$

$$vf = fr \ * \ n \tag{4}$$

$$\mathrm{Pc} = \mathrm{DC}/4 * \mathrm{fr} * \mathrm{vc} * \mathrm{kc}\ 33{,}000 * \eta(\mathrm{Hp}) \tag{5}$$

$$\mathrm{Ff} = 0.7 * \mathrm{DC}/2 * \mathrm{fr} * \mathrm{kc}\ (\mathrm{lbs}) \tag{6}$$

These calculations consider Feeding Force and Lowering Force multiplication is 3,4 N/cm. This calculation elucidates for drilling PCB for a 0.2 mm layer is not require high torque for 20.000 rotation/min spindle speed. Nema 11 and a more powerful one will be enough for Z-axis. Dynamic torque is 80% percent of holding torque's in maximum. Also, this characteristic N/cm may be changeable with voltage to the step motor driver. In the light of this information, Nema 17 must be an advantageous option for Z-axis. For vibration effect 2 Z-axis supporter and shaft might be advantageous choose criteria. The

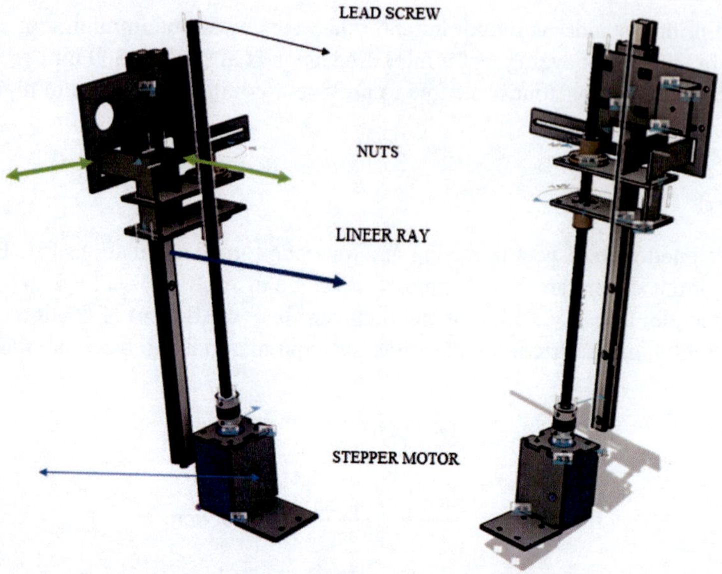

Fig. 6. Z axis with the left and right holder designs.

chosen NEMA 17's specifications are 103 ons/inch or 72.734 N/cm for holding torque, 42 cm × 42 cm * 75 cm with 1.8° (Fig. 6).

Z-axis supported with 2 sides and linear rail motion actuators. The motor gives direct drive with shaft system with 300 mm 4 wing lead screw 300 mm Linear rail was preferred due to the low-price advantage. With the motor highness the size of the Z-Axis is 380 mm and because of that linear ray is placed in the end point of the Lead Screw.

3.2 X and Z-axis Design

X axis designed under low-weight criterias respect to carriers and motor designs (Fig. 7).

3.3 Y Axis Belt System Design

X-axis and Y-axis will work with GT2 belt system gear ratios 1:1 (20tooth to 20 tooth). Radial speed to mechanical speed calculations and arrangement will be done on the software part of this project and it will be integrated in software part (Fig. 8).

3.4 Spindle System Design

Spindle and hot-end mounting system will be with the changeable head design the design works with screw system.8 point screw and 2-point mounting process eliminate the change conflicts with the high accuracy. The spindle system spindle attaches with 4 screws and a 2-point mounting surface that is coated with cohesive glue (Fig. 9).

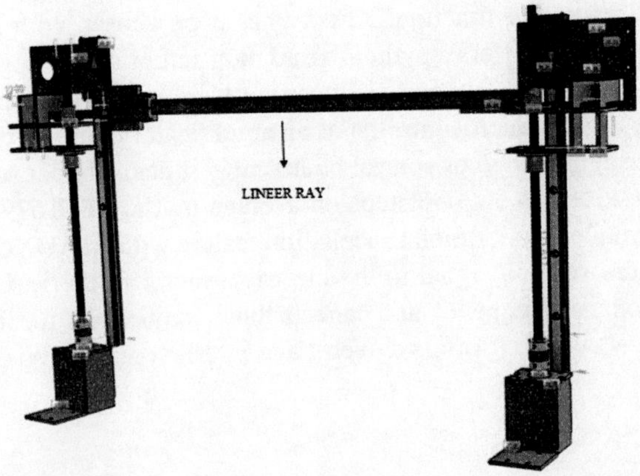

Fig. 7. The design of combination of X and Z axes with the system of linear ray.

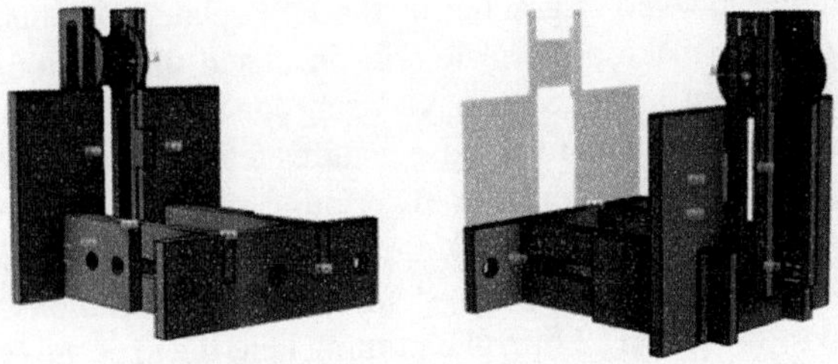

Fig. 8. Y axis belt system design

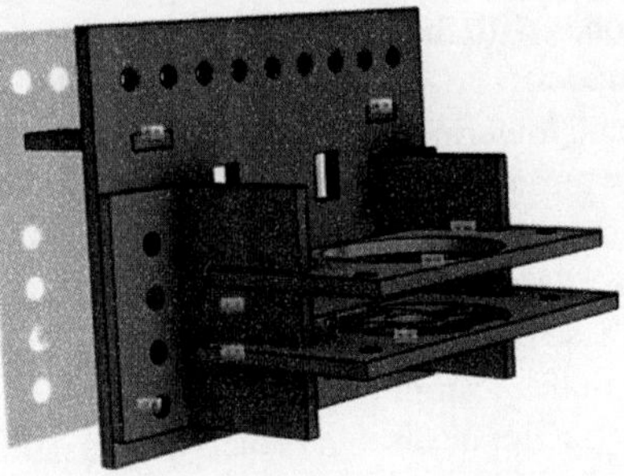

Fig. 9. Spindle and Hot end mounting system

4 Software Configuration

System designed and analyzed with Marlin 1, Marlin 2, and Marlin 2 with Octoprint System, and Klipper system. Marlin 1 system has a bottleneck with the CNC and motor driving partitions. Marlin 2 works fluently with the de-signed way. Octoprint add-on upgrade offers new features such as Wi-Fi-based control and community support with add-on formatted g-code and stl viewer & designer features. Klipper software has a unique workload operating function. Klipper has been chosen with respect to design criteria. Klipper software offers a python-based plug-in library for upgrading the developed systems with software support. Klipper software improves stepper drivers' step multiplication by using a microcontroller as an amplificatory. The software uses its own processing power in the range of control board capabilities. Klipper software increases step counts from 25600 to 500000 steps on average on the LPC1679 microcontroller. Klipper offers a frequency vibration cancellation feature with an ADXL345 acceleration sensor. This feature sends a signal to the electrical motor in the variation of all steps and finds the resonance frequency and cancels those frequencies for that motor, which improves system accuracy. It offers a Web-Page interface with a Wi-Fi connection on IP.

5 Conclusions

This project has defined as a PCB manufacturing machine with two options. These functions are spindle-powered engraving of the PCB plate and obtaining the circuit board, another option is making the plate with insulated filament powered by the 3d printer option of the system, in the second layer using conductive filament gives the power to design multi-layered circuit boards. Those features feature other using scenarios such as 3D printing and machining option after 3D printing engineering geometrical objects.

The machine has obtained linear rays on axes; in X and Z 2 linear rays and in Y 1 linear ray. In the power transmission system, 2 different variations are used. Those variations are lead screw and GT2 fiberglass timing belt, the axes the GT2 belt used are X and Y, on Z-axis 2 lead screw had used.

Motor selection criteria had been chosen with the range of ideal torque with the ideal speed with respect to minimum financial impact. Respect to those criteria Nema series stepper motors are ideal ones with the feature of a no-step missing guarantee for a range of torque of the Nema model. The mathematical calculation had done with respect to raising torque with feed and lowering torque summation. The sum of feed and lowering torque is 3.55 N/cm, this torque must be minimum in the range of %5–10 percent of the chosen motor with 4 tooth lead screw. Nema 17 long shaft series offer 70–75 N/cm holding torque for one motor. In dynamic movement, the power transmission may handle extreme conditions. For these cases, there are 2 other Z-axis for extra accuracy with the extra power linearized with the 15 mm linear rays.

Software obtained as a Klipper on this project for sustaining extra features of the open-source developer community of the Klipper software. Klipper software might be used on future projects to develop BLDC Hub Motor designs with Linux/Python container supported software for CNC applications that includes Field Oriented Control software development on Python with self-integration on Klipper software.

As a conclusion system has been designed with these parameters with the assistance of open-source market available software and product for competitive scenarios on a low financial budget with counted features.

References

1. Chlebosz, W., Ombach, G., ve Junak, J.: Comparison of permanent magnet brushless motor with outer and inner rotor used in e-bike. In: 2010 XIX International Conference on Electrical Machines (ICEM), pp. 1–5 (2010)
2. Chen, J., Li, J., Qu, R., Ge, M.: Magnet-frozen-permeability FEA and DC-biased measurement for machine inductance: application on a variable-flux PM machine. IEEE Trans. Indust. Electron. **65**, 4599–4607 (2017)
3. Demagnetization curve of a permanent magnet. https://www.coolmagnetman.com/magfund07. Accessed 21 Dec 2021
4. Hendershot J.R., ve Miller, T.J.E.: Design of Brushless Permanent Magnet Motors, Magna Physics Publishing and Oxford University Press, USA (1994)
5. Richard G.B., Nisbett, J.K.: Shigley's Mechanical Engineering Design. 9th edn. The McGraw-Hill Comp (2012)

Author Index

D. D. Cioboată (Ed.): ICoRSE 2022, LNNS 534, pp. 389–390, 2023.
https://doi.org/10.1007/978-3-031-15944-2